2018 International SoC Design Conference (ISOCC 2018)

Daegu, South Korea
12 – 15 November 2018

IEEE Catalog Number: CFP1869E-POD
ISBN: 978-1-5386-7961-6

**Copyright © 2018 by the Institute of Electrical and Electronics Engineers, Inc.
All Rights Reserved**

Copyright and Reprint Permissions: Abstracting is permitted with credit to the source. Libraries are permitted to photocopy beyond the limit of U.S. copyright law for private use of patrons those articles in this volume that carry a code at the bottom of the first page, provided the per-copy fee indicated in the code is paid through Copyright Clearance Center, 222 Rosewood Drive, Danvers, MA 01923.

For other copying, reprint or republication permission, write to IEEE Copyrights Manager, IEEE Service Center, 445 Hoes Lane, Piscataway, NJ 08854. All rights reserved.

*** This is a print representation of what appears in the IEEE Digital Library. Some format issues inherent in the e-media version may also appear in this print version.

IEEE Catalog Number: CFP1869E-POD
ISBN (Print-On-Demand):   978-1-5386-7961-6
ISBN (Online):   978-1-5386-7960-9
ISSN: 2163-9612

Additional Copies of This Publication Are Available From:

Curran Associates, Inc
57 Morehouse Lane
Red Hook, NY 12571 USA
Phone: (845) 758-0400
Fax: (845) 758-2633
E-mail: curran@proceedings.com
Web: www.proceedings.com

TABLE OF CONTENTS

12-BIT 20M-S/S SAR ADC USING C-R DAC AND CAPACITOR CALIBRATION 1
Eunji Youn ; Young-Chan Jang

CURRENT FEEDBACK-BASED HIGH LOAD CURRENT LOW DROP-OUT VOLTAGE
REGULATOR IN 65NM CMOS TECHNOLOGY 3
Melody Teoh ; Sotoudeh Hamedi-Hagh

DYNAMIC VOLTAGE DROP INDUCED PATH DELAY ANALYSIS FOR STV AND NTV
CIRCUITS DURING AT-SPEED SCAN TEST 7
Hyunggoy Oh ; Heetae Kim ; Sangjun Lee ; Sungho Kang

A MULTI-FAULT DYNAMIC COMPACTION TECHNIQUE FOR TEST PATTERN COUNT
REDUCTION 9
Bo-Yi Li ; Jiun-Lang Huang

A TEST METHODOLOGY FOR NEURAL COMPUTING UNIT 11
Minho Cheong ; Ingeol Lee ; Sungho Kang

AN EFFECTIVE APPROACH FOR BUILDING LOW-POWER GENERAL ACTIVITY-DRIVEN
CLOCK TREES 13
Chen-Hsien Lin ; Shih-Hsu Huang ; Wei-Kai Cheng

FAST STEADY-STATE THERMAL ANALYSIS 15
Lih-Yih Chiou ; Chun-Hao Chang ; Liang-Ying Lu ; Wei-Hsuan Yang ; Yeong-Jar Chang ; Juin-Ming Lu

A RULE-OF-THUMB CONDITION TO AVOID LARGE HRS CURRENT IN RERAM CROSSBAR
ARRAY DESIGN 17
Yelim Youn ; Kwangmin Kim ; Byungsub Kim

SPIN ORBIT TORQUE-RAM WRITE ENERGY REDUCTION WITH SELF-VERIFICATION
SCHEME 19
Taehwan Kim ; Jongsun Park

2-D FAILURE BITMAP COMPRESSION USING LINE FAULT MARKING METHOD 21
Keewon Cho ; Young-Woo Lee ; Sungyoul Seo ; Sungho Kang

DIAGNOSIS OF RESISTIVE NONVOLATILE-8T SRAMS 23
Yu-Ting Li ; Jin-Fu Li ; Chun-Lung Hsu ; Chi-Tien Sun

RELIABILITY OPTIMIZATION OF RERAM ARCHITECTURE USING HETEROGENEOUS
ERROR CORRECTING CODE SCHEME 25
Kwangjin Lee ; Tae Hee Han

A DIGITAL ΣΔ MODULATED CLASS-S TRANSMITTER WITH TWO-STEP UP-CONVERSION
AND FILTER-LESS FRONT END 27
Pan Xue ; Haijun Shao ; Dan Fang ; Gan Guo ; Wei Che ; Zhiliang Hong

A NOVEL HIGH PRECISION SDR-BASED POSITIONING SYSTEM 34
G. Piccinni ; G. Avitabile ; G. Coviello ; C. Talarico

IMPLEMENTATION OF SYSTOLIC CO-PROCESSOR FOR DEEP NEURAL NETWORK
INFERENCE BASED ON SOC 36
Erwin Setiawan ; Trio Adiono

APPROXIMATE ADDER GENERATION FOR IMAGE PROCESSING USING
CONVOLUTIONAL NEURAL NETWORK 38
Ryuta Ishida ; Toshinori Sato ; Tomoaki Ukezono

DESIGN OF A GENERIC SECURITY INTERFACE FOR RISC-V PROCESSORS AND ITS
APPLICATIONS 40
Hyunyoung Oh ; Junmo Park ; Myonghoon Yang ; Dongil Hwang ; Yunheung Paek

CONVOLUTIONAL NEURAL NETWORK ACCELERATOR WITH RECONFIGURABLE
DATAFLOW 42
Myungwoo Oh ; Chaeeun Lee ; Sanghun Lee ; Youngho Seo ; Sunwoo Kim ; Jooho Wang ; Chester Sungchung

NOVEL LOW POWER FINFET SRAM CELL DESIGN WITH BETTER READ AND
WRITABILTY FOR CACHE MEMORY 44
Shreyash Patel ; Youngbae Kim ; Ken Choi

LOW COST HARDWARE IMPLEMENTATION OF LEA-128 ENCRYPTION USING BIT-
SERIAL TECHNIQUE 46
Byungjun Choi ; Bohun Kim ; Jongsun Park

A COST-EFFECTIVE HIGH ACCURACY AUTO-TRIMMING SYSTEM WITHOUT TESTER CONSTRAINT FOR LOW-END EMBEDDED FLASH MEMORY 48

Junichi Suzuki ; Junichi Yamashita ; Masami Hanyu ; Masamichi Ido ; Tatsuya Saito ; Yasuhiro Nakashima ; Masanori Hayashikoshi ; Yukiyoshi Kiyota

INTEGRATION OF RETENTION-AWARE REFRESH AND BISR TECHNIQUES FOR DRAM REFRESH POWER REDUCTION 50

Wei-Kai Cheng ; Jian-Kai Chen ; Shih-Hsu Huang

STUDY ON INTEL CPU-FPGA ARCHITECTURE: SECURITY PERSPECTIVE 52

Jinyong Lee

A 13.56 MHZ ACTIVE RECTIFIER WITH SELF-SWITCHING COMPARATOR FOR WIRELESS POWER TRANSFER SYSTEMS 54

Yingfei Xiang ; Yu Wang ; C.-J. Richard Shi

1.6-PPM/ °C REFERENCE VOLTAGE GENERATOR WITH PSRR OF -93DB BASED ON THRESHOLD VOLTAGE DIFFERENCE OF LVT AND SVT DEVICES 60

Behnam Samadpoor Rikan ; Hamed Abbasizadeh ; Reza E. Rad ; Arash Hejazi ; Kang Yoon Lee

A 4.86 μW/CHANNEL FULLY DIFFERENTIAL MULTI-CHANNEL NEURAL RECORDING SYSTEM 68

Taeju Lee ; Ji-Hyoung Cha ; Su-Hyun Han ; Seong-Jin Kim ; Minkyu Je

TAPERED-RATIO COMPRESSION FOR RESIDUAL NETWORK 72

Sungbum Kang ; Joonsang Yu ; Kiyoung Choi

SEGMENTATION-BASED DISPARITY REFINEMENT 74

Gyujin Bae ; Young Hwan Kim

HUMAN VISUAL ATTENTION ANALYSIS-BASED IMAGE SEGMENTATION USING COLOR HISTOGRAM 76

Ho Sub Lee ; Young Hwan Kim

A REAL-TIME TRACKING ALGORITHM FOR HUMAN FOLLOWING MOBILE ROBOT 78

Tsung-Han Tsai ; Chia-Hsiang Yao

DRIVABLE AREA DETECTION METHOD CAPABLE OF DISTINGUISHING VEGETATION AREA ON COUNTRY ROAD 80

Sangjae Lee ; Byungin Moon

ON-CHIP MEMORY OPTIMIZATION OF HIGH EFFICIENCY ACCELERATOR FOR DEEP CONVOLUTIONAL NEURAL NETWORKS 82

Tzu-Yi Lai ; Kuan-Hung Chen

PERFORMANCE METRICS OF INEXACT MULTIPLIERS BASED ON APPROXIMATE 5:2 COMPRESSORS 84

Lavanya Maddisetti ; Jvr Ravindra

EXPLOITING CONFIGURABILITY FOR CORRECT SIGN CALCULATION IN AN APPROXIMATE ADDER 86

Toshinori Sato ; Tomoaki Ukezono

A FOLDED LOCKING SCHEME FOR THE LONG-RANGE DELAY BLOCK IN A WIDE-RANGE DLL 90

Yu-Chi Wei ; Shi-Yu Huang

TEMPERATURE INDEPENDENT SUBTHRESHOLD CIRCUITS DESIGN 92

Morteza Nabavi ; Maitham Shams ; Mohammad Sawan

OPTIMIZING THE PERFORMANCE OF A LOW POWER – AREA EFFICIENT OTA DESIGN THAT IS BASED ON HYBRID CURRENT SHUNTING TECHNIQUE 95

Imtinan B. Attili ; Soliman A. Mahmoud

A LOW-POWER LOW-NOISE OPEN-LOOP CONFIGURED SIGNAL FOLDING NEURAL RECORDING AMPLIFIER 99

Gauri Punekar ; Venkateswarlu Gonuguntla ; Palagani Yellappa ; Jun Rim Choi ; Ramesh Vaddi

NANO-AMPERE CURRENT SENSING TECHNIQUE FOR OLED MOBILE DISPLAYS 101

Minhyun Jin ; Hyejin Im ; Minkyu Song ; Soo Youn Kim

ELLIPTIC OTA-C LOW-PASS FILTERS FOR ANALOG FRONT-END OF BIOSIGNAL DETECTION SYSTEM 103

Maha S. Diab ; Soliman Mahmoud

KA-BAND RF FRONT-END WITH 5DB NF AND 16DB CONVERSION GAIN IN 45NM CMOS TECHNOLOGY 105

Hyunki Jung ; Dzuhri Radityo Utomo ; Saebyeok Shin ; Seok-Kyun Han ; Sang-Gug Lee ; Jusung Kim

A 22.8-TO-32.4 GHZ INJECTION-LOCKED FREQUENCY TRIPLER WITH SOURCE DEGENERATION 107

Saebyeok Shin ; Dzuhri Radityo Utomo ; Hyunki Jung ; Seok-Kyun Han ; Sang-Gug Lee ; Jusung Kim

DESIGN OF A LOW-POWER COMPLEX BASEBAND FILTER WITH TUNABLE GAIN AND BANDWIDTH IN 65NM CMOS 109

Jaegyeong Choi ; Jungah Kim ; Yongho Lee ; Seungsoo Kim ; Jongsik Kim ; Hyunchol Shin

A 24-28GHZ RECONFIGURABLE CMOS POWER AMPLIFIER IN 22NM FD-SOI FOR INTELLIGENT SOC APPLICATIONS 111

Jill C. Mayeda ; Donald Y. C. Lie ; Jerry Lopez

PHY LAYER DESIGN OF OFDM-VLC SYSTEM BASED ON SOC USING REUSE METHODOLOGY 115

Erwin Setiawan ; Trio Adiono ; Syifaul Fuada

MODEL-BASED PARALLELIZER FOR EMBEDDED CONTROL SYSTEMS ON SINGLE-ISA HETEROGENEOUS MULTICORE PROCESSORS 117

Zhaoqian Zhong ; Masato Edahiro

EFFICIENT IMPLEMENTATION OF MULTIPLE INTERLEAVERS IN IDMA FOR 5G 119

Byeong Yong Kong ; In-Cheol Park

HYBRID DECODING FOR POLAR CODES 121

Soyeon Choi ; Hoyoung Yoo

THE HARDWARE ACCELERATION OF SC DECODER FOR POLAR CODE TOWARDS HLS OPTIMIZATION 123

Yujie Huang ; Yujie Cai ; Minge Jing ; Jun Han ; Yibo Fan ; Xiaoyang Zeng

FPGA-BASED OPTICAL CHARACTER RECOGNITION FOR HANDWRITTEN MATHEMATICAL EXPRESSIONS 125

Bobbi Winema Yogatama ; Jhonson Lee ; Suksmandhira Harimurti ; Trio Adiono

TRANSFER LEARNING-BASED VEHICLE CLASSIFICATION 127

So Yeon Jo ; Namhyun Ahn ; Yunsoo Lee ; Suk-Ju Kang

FIXED-POINT QUANTIZATION OF 3D CONVOLUTIONAL NEURAL NETWORKS FOR ENERGY-EFFICIENT ACTION RECOGNITION 129

Hyunhoon Lee ; Younghoon Byun ; Seokha Hwang ; Sunggu Lee ; Youngjoo Lee

IMPLEMENTATION OF 3D HAND GESTURE RECOGNITION SYSTEM USING FPGA 131

Tsung-Han Tsai ; Yuan-Chen Ho ; Yih-Ru Tsai

MULTI-MODE LSTM NETWORK FOR ENERGY-EFFICIENT SPEECH RECOGNITION 133

Junseo Jo ; Seokha Hwang ; Sunggu Lee ; Youngjoo Lee

A WIDEBAND DIFFERENTIAL VCO BASED ON MULTIPLE-PATH LOOP ARCHITECTURE 135

Tomotaka Tanaka ; Fumiya Naito ; Makoto Nakamura ; Daisuke Ito ; Keiji Kishine

DIGITAL PHY DESIGN METHODOLOGIES FOR HIGH-SPEED AND LOW-POWER MEMORY INTERFACE 140

Kwanyeob Chae ; Billy Koo ; Jihun Oh ; Sanghune Park ; Jongshin Shin ; Jaehong Park

A TWO-STEP TIME-TO-DIGITAL CONVERTER USING RING OSCILLATOR TIME AMPLIFIER 143

Min Kim ; Kyung-Sub Son ; Namhoon Kim ; Chang Hang Rho ; Jin-Ku Kang

LOW POWER - HIGH SPEED MAGNITUDE COMPARATOR CIRCUIT USING 12 CNFETS 145

Jitendra Kumar Saini ; Avireni Srinivasulu ; Renu Kumawat

ESTIMATION OF LEAKAGE DISTRIBUTION UTILIZING GAUSSIAN MIXTURE MODEL 149

Hyunjeong Kwon ; Young Hwan Kim ; Seokhyeong Kang

SYSTEM LEVEL POWER REDUCTION FOR YOLO2 SUB-MODULES FOR OBJECT DETECTION OF FUTURE AUTONOMOUS VEHICLES 151

Youngbae Kim ; Qiang Tong ; Ken Choi ; Eunchong Lee ; Sung-Joon Jang ; Byeong-Ho Choi

A SIMPLE CIRCUIT MODEL FOR PWM-1-CONTROLLED DC-DC CONVERTER AND ITS ANALYSIS 158

Kojiro Tamura ; Yuki Kametaka ; Takuji Kousaka ; Hirokazu Ohtagaki ; Hiroyuki Asahara

ANALYSIS AND DESIGN OF PHASE-CONTROLLED CLASS-D ZVS INVERTER 160

Tatsuki Ohsato ; Yuta Yamada ; Xiuqin Wei ; Hiroo Sekiya

DESIGN OF TWO TEMPLATE CELLULAR NEURAL NETWORKS FOR COLOR IMAGE PROCESSING 162

Yasuteru Hosokawa ; Yoko Uwate ; Yoshifumi Nishio

SYNCHRONIZATION PHENOMENA OF COUPLED CHAOTIC CIRCUITS NETWORK WITH COUPLING STRENGTH DEPENDING ON NUMBER OF DEGREE 164

Kyohei Fujii ; Shuhei Hashimoto ; Yoko Uwate ; Yoshifumi Nishio

DESIGN OF CONVOLUTIONAL NEURAL NETWORK FOR CLASSIFYING DEPTH PREDICTION IMAGES FROM OVERHEAD 166

Shu Sumimoto ; Yuichi Miyata ; Ryuta Yoshimura ; Yoko Uwate ; Yoshifumi Nishio

ANALYSIS OF CHAOTIC CIRCUIT NETWORKS WITH ONE-WAY COUPLING 168

Akari Oura ; Kyohei Fujii ; Yoko Uwate ; Yoshifumi Nishio

AREA-DELAY PRODUCT EFFICIENT DESIGN FOR CONVOLUTIONAL NEURAL NETWORK CIRCUITS USING LOGARITHMIC NUMBER SYSTEMS 170

Tso-Bing Juang ; Cong-Yi Lin ; Guan-Zhong Lin

NEUROMORPHIC PROPERTIES OF MEMRISTOR TOWARDS ARTIFICIAL INTELLIGENCE 172

Chun Zhao ; Zong Jie Shen ; Guang You Zhou ; Ce Zhou Zhao ; Li Yang ; Ka Lok Man ; Eng Gee Lim

MEMRISTOR-BASED NEUROMORPHIC IMPLEMENTATIONS FOR ARTIFICIAL NEURAL NETWORKS 174

Chun Zhao ; Guang You Zhou ; Ce Zhou Zhao ; Li Yang ; Ka Lok Man ; Eng Gee Lim

AUTOMATIC SHADING DETECTION SYSTEM FOR PHOTOVOLTAIC STRINGS 176

Jieming Ma ; Ziqiang Bi ; Ka Lok Man ; Yong Yue ; Jeremy S. Smith

A MULTIBAND RECTENNA FOR SELF-SUSTAINABLE DEVICES 178

Zhao Wang ; Heng Zhang ; Zhenzhen Jiang ; Mark Leach ; Jingchen Wang ; Kalok Man ; Eng Gee Lim

OSCILLATION QUENCHING IN COUPLED VAN DER POL OSCILLATORS WITH DIFFERENT FREQUENCIES 180

Yoko Uwate ; Yoshifumi Nishio

A DESIGN OF RECTIFIER FOR 13.56MHZ WIRELESS POWER TRANSFER RECEIVER WITH ALL DIGITAL DELAY-LOCKED LOOP 194

Joonho Park ; Yong Moon

LEAKAGE CONTROL SYSTEM USING DATA ESTIMATION OF RESISTIVE MEMORY 196

Junyoung Kweon ; Juntae Choi ; Yun-Heup Song ; Tony Tae-Hyoung Kim

STEREO VISION-BASED COLLISION AVOIDANCE FOR UNMANNED SYSTEMS 204

Sungick Kong ; Sang-Seol Lee ; Sung-Joon Jang

NIGHT-TIME VEHICLE DETECTION BASED ON BRAKE/TAIL LIGHT COLOR 206

Thathupara Subramanyan Kavya ; Erdenetuya Tsogtbaatar ; Young-Min Jang ; Sang-Bock Cho

IMPLEMENTATION OF MULTI-CHANNEL FM REPEATER USING DIGITAL SIGNAL PROCESSING ALGORITHM IN FPGA 208

Mangi Han ; Ji Min Song ; Hoon Gee Yang ; Youngmin Kim

EFFICIENT FOUR-WAY ROW-SPLITTING LAYERED QC-LDPC DECODER ARCHITECTURE 210

Tram Thi Bao Nguyen ; Hanho Lee

RECONFIGURABLE MULTI-INPUT ADDER DESIGN FOR DEEP NEURAL NETWORK ACCELERATORS 212

Hossein Moradian ; Sujeong Jo ; Kiyoung Choi

A METHOD OF PREVENT LOSS OF INFORMATION IN ILL-POSED PROBLEM BASED APPLICATION USING ATMOSPHERIC SCATTERING MODEL 214

Geun-Jun Kim ; Bongsoon Kang

4-BIT DATA ARRANGEMENT ALGORITHM FOR CAN COMPRESSION 216

Yeon-Jin Kim ; Ho-Yun Lee ; Jin-Gyun Chung

GENERALIZED ADAPTIVE VARIABLE BIT TRUNCATION METHOD FOR APPROXIMATE STOCHASTIC COMPUTING 218

Keerthana Pamidimukkala ; Kyung Ki Kim ; Yong-Bin Kim ; Minsu Choi

THE ANALYSIS OF CNN STRUCTURE FOR IMAGE DENOISING 220

Jae Hyeon Park ; Jeong Hyeon Kim ; Sung In Cho

OPTIMIZED IMAGE CROP-BASED VIDEO RETARGETING 224

Jeong Hyeon Kim ; Jae Hyeon Park ; Sung In Cho

HIGH-THROUGHPUT HW-SW IMPLEMENTATION FOR MV-HEVC DECODER 226

Wei Liu ; Wei Li ; Park Sang Un ; Yong Beom Cho

ACCURATE STOCHASTIC COMPUTING USING A WIRE EXCHANGING UNIPOLAR MULTIPLIER 229

Houghun Joe ; Manhee Cho ; Youngmin Kim

TRUE RANDOM NUMBER GENERATOR USING BIO-RELATED SIGNALS IN WEARABLE DEVICES 231

Hoyoung Yu ; Youngmin Kim

LOW-POWER NULL CONVENTION LOGIC MULTIPLIER DESIGN BASED ON GATE DIFFUSION INPUT TECHNIQUE 233

Prashanthi Metku ; Kyung Ki Kim ; Yong-Bin Kim ; Minsu Choi

A RADIATION HARDENED SRAM WITH SELF-REFRESH AND COMPACT ERROR CORRECTION 235

Sultan M. Siddiqui ; Ruchi Sharma ; Van Loi Le ; Taegeun Yoo ; Ik-Joon Chang ; Tony Tae-Hyoung Kim

EXPERIMENTAL VERIFICATION OF A SIMPLE, INTUITIVE, AND ACCURATE CLOSED-FORM TRANSFER FUNCTION MODEL FOR DIVERSE HIGH-SPEED INTERCONNECTS 239

Kyunghyun Lim ; Minsoo Choi ; Myat-Thu-Linn Aung ; Kyunghwan Kim ; Ji-Seong Kim ; Rock-Hyun Baek ; Ho-Jin Song ; Tony Tae-Hyoung Kim ; Byungsub Kim

A WEARABLE ELECTROCARDIOGRAM MONITORING SYSTEM ROBUST TO MOTION ARTIFACTS 241
Taeryoung Seol ; Sehwan Lee ; Junghyup Lee

INFRARED AND VISIBLE IMAGE FUSION USING MULTI-SCALE DECOMPOSITION AND VISUAL SALIENCY MAP 243
Yunfan Chen ; Han Xie ; Donghoon Yeo ; Hyunchul Shin

WEATHER CLASSIFICATION USING CONVOLUTIONAL NEURAL NETWORKS 245
Jehong An ; Yunfan Chen ; Hyunchul Shin

KOREAN TRAFFIC SIGN DETECTION USING DEEP LEARNING 247
Prateek Manocha ; Ayush Kumar ; Jameel Ahmed Khan ; Hyunchul Shin

DESIGN OF ROAD SURFACE LIGHTING SYSTEM FOR REAR LAMP USING AUTOMOTIVE ULTRASONIC SENSOR 249
Donghee Han ; Hyo Bin Choi ; Yong Sin Kim

DESIGN OF 3D INDUCTORS FOR IOT SECURITY 251
Bruce Kim ; Sang-Bock Cho

LOW POWER NEAR-SENSOR COARSE TO FINE XOR BASED MEMRISTIVE EDGE DETECTION 253
Kamilya Smagulova ; Aidana Irmanova ; Alex Pappachen James

A KA-BAND LOW NOISE AMPLIFIER IN 0.15µM GAAS E-MODE PHEMT TECHNOLOGY 255
Jeongsoo Park ; Jinhyun Kim ; Jeong-Geun Kim

A 28-GHZ 28.5-DBM POWER AMPLIFIER USING 0.15-µM INGAAS E-MODE PHEMT TECHNOLOGY 257
Hui Dong Lee ; Sunwoo Kong ; Bonghyuk Park ; Kwang Chun Lee ; Jeong-Soo Park ; Jeong-Geun Kim

A CMOS RECTIFIER WITH 72.3% RF-TO-DC CONVERSION EFFICIENCY EMPLOYING TUNABLE IMPEDANCE MATCHING NETWORK FOR AMBIENT RF ENERGY HARVESTING 259
Donggu Lee ; Taejong Kim ; Sinyoung Kim ; Kanghyeon Byun ; Kuduck Kwon

DESIGN AND ANALYSIS OF DIGITAL PID CONTROLLER IN MCU AND FPGA 261
Kyungnam Lee ; Youngmin Kim

OPTIMAL MODEL ANALYSIS FOR DENOISING MONTE CALRO RENDERING NOISE 263
Min-Cheol Kim ; Kwang-Yeob Lee

A SOFTWARE-BASED SCAN CHAIN DIAGNOSIS FOR DOUBLE FAULTS IN A SCAN CHAIN 265
Hyeonchan Lim ; Seokjun Jang ; Sungho Kang

LOW POWER SCAN CHAIN ARCHITECTURE BASED ON CIRCUIT TOPOLOGY 267
Heetae Kim ; Hyunggoy Oh ; Sangjun Lee ; Sungho Kang

LIFETIME IMPROVEMENT METHOD USING THRESHOLD-BASED PARTIAL DATA COMPRESSION IN NOC 269
Ju Sung Kim ; Jeong Beom Hong ; Ju Yeon Kang ; Tae Hee Han

ENERGY EFFICIENT ANALOG SYNAPSE/NEURON CIRCUIT FOR BINARIZED NEURAL NETWORKS 271
Jaehyun Kim ; Chaeun Lee ; Kiyoung Choi

METHOD OF RTL DEBUGGING WHEN USING HLS FOR HW DESIGN: DIFFERENT SIMULATION RESULT OF VERILOG & VHDL 273
Sang Un Park ; Tae Pyeong Kim ; Mee Zee Lee ; Yong Beom Cho

Author Index

International SoC Design Conference 2018

PROCEEDINGS

International SoC Design Conference 2018
(ISOCC 2018)

PROCEEDINGS OF TECHNICAL PAPERS

November 12-15, 2018
Hotel Inter-Burgo Daegu, Daegu, Korea

CONFERENCE INFORMATION

"Welcome to ISOCC 2018"

On behalf of the Organizing Committee, it is my great pleasure to welcome you to the 15th International SoC Conference (ISOCC 2018). ISOCC has established a long tradition as an annual conference providing the world's premier SoC design forum for leading researchers from academia and industries. This year at ISOCC 2018, many excellent articles are presented in the field of semiconductor circuits and systems such as new advanced concept and developments in technology of analog/digital circuits or systems, theory, simulation, modeling, advanced experimental results and experience of SoC with SW, and an emerging technology for the future.

ISOCC 2018 is held from November 12th to 15th, 2018 at the Hotel Inter-Burgo Daegu, Daegu, Korea. The city of Daegu is surrounded by UNESCO World Heritage Sites (Gyeongju Seokkuram and Bulguksa, Gyeongju Historical Areas, Haeinsa Temple Tripitaka Koreana, and the historic villages of Hahoe and Andong). We are truly glad to host such a prestigious event in the central hub of world heritage sites of Korea.

The organizing committee is gearing up for an exciting and informative conference program including plenary lectures, symposia, workshops, tutorials on a variety of topics, poster presentations and various social programs for over 1,000 participants from around the world.

Researchers, engineers, and students, from industry, universities and government agencies are invited to present their latest work and to discuss research and applications for system on chip designs. All participants will be sure to have a meaningful experience with industry peoples and scholars from around the world.

All members of the ISOCC'2018 organizing committee are happy to meet you in Daegu, Korea.

Kwang-Hyun Baek
General Chair
ISOCC 2018

International SoC Design Conference 2018

CONFERENCE INFORMATION

Message from TPC Chair

On behalf of the Technical Program Committee, it is my great pleasure to welcome you to the 15th International SoC Design Conference in Daegu, South Korea. ISOCC continues our tradition of showcasing the recent innovations and advancements in the System on Chip (SoC) area. The conference theme is "Intelligent SoC driving the Fourth Industrial Revolution". Recent IoT for the Fourth Industrial Revolution trend heavily depends on small things with various sensing capabilities, a certain level of intelligence, seamless wireless connectivity, ultra-low power consumption and sometimes batteryless operation.

Such small intelligent things for the Fourth Industrial Revolution cannot be realized without a semiconductor system on a chip or SoC. Intelligent SoC senses and aggregates wanted information from outside, processes them to a desired way, and create values for humans. It can be applied to human wellness, robot industry, bio/medical health, and automotive, automated factory, and so much more. In this conference, we try to open a forum to discuss on key issues driving the Fourth Industrial Revolution and hurdles to realize the intelligent SoCs and see what the future SoC will look like.

The ISOCC 2018 technical program consists of 141 outstanding papers over 21 technical sessions. Among them, 106 regular papers and 11 invited papers are presented at 15 oral sessions and 1 poster session, and 24 special session papers are presented at 5 special sessions. The 106 excellent papers are selected out of 158 regular submissions from 22 countries, which reflects the accept rate of 67.09 % and continuing internationalization of this conference.

The conference features 3 keynote speeches by outstanding leading innovators of LG Electronics, Synopsys and Virginia Tech to share their visions and prospects. It also features 3 industrial talks, 4 educational tutorials, industrial demos, and chip design contest demos (CDC) to promote the technical updates in SoC research and development. I would like to thank the keynote speakers, industrial talk speakers and tutorial speakers for contributing to these important parts of the program.

I would also like to express our sincerest thanks the TPC Co-Chairs, Meng-Fan(Marvin) Chang and Chip Hong Chang, the TPC Vice-Chairs, Kang-Yoon Lee and Jongsun Park, TPC Track-Chairs, Special Session Co-Chairs, Poster Session&Tutorial Co-Chairs, CDC Chairs, and reviewers for their dedication. Without their help, this conference would not be possible.

I hope you will enjoy the technical program. I also hope that all the attendees will have great time in Daegu.

Kyung Ki Kim
Technical Program Committee Chair
ISOCC 2018

CONFERENCE INFORMATION

Invitation to ISOCC 2019

As General Chair of ISOCC 2019, let me extend my early welcome to ISOCC 2019. It is my pleasure to invite all of you to the next year's conference. The sixteenth event of the International SoC Conference (ISOCC) will be held in Jeju island from October 6 through October 8 in 2019. I hope that not only all of you but including your family, colleagues, and friends could come and travel again to Jeju for this exciting ISOCC 2019.

Since its inception, ISOCC has been continuing to showcase the most recent innovations and trends in the semiconductor system-on-a-chip area with active participations from worldwide researchers in academia, industry, and institutes. We believe this tradition will be continued and furthermore developed and advanced to a higher quality and larger scale in the next year's event. The organizing and technical committee of ISOCC 2019 is gearing up for an exciting and advanced program including plenary speeches, invited talks, tutorials, technical paper presentations, and various social programs.

Jeju island has been the most favorite venue in ISOCC history for its beautiful volcanic island nature and convenient flight access from world. Jeju, as the southernmost and largest island of Korea, was created by volcanic eruptions that occurred millions of years ago. The island has significant academic values as well as marvelous natural landscapes. Dominated by Halla Mountain (1,950 m) at its center, Jeju also has many parasitic cones (called "Oreum"), vast open pastures, and lovely trekking courses (called 'Olle-gil'). I believe that you will like Jeju if you are either the first-time visitor or have visited the island several times before.

All members of the ISOCC 2019 organizing committee look forward to seeing you and your valuable research works in Jeju, Korea, in ISOCC 2019.

Hyunchol Shin
General Chair
ISOCC 2019

Organizing Committee

General Chair
Kwang Hyun Baek *(Chung Ang University, Korea)*

General Co-Chairs
Jun Jin Kong *(Samsung Electronics, Korea)*

Mohamad Sawan *(Polytechnique Montréal, Canada)*

Yoshifumi Nishio *(Tokushima University, Japan)*

General Vice Chairs
Hyunchol Shin *(Kwangwoon University, Korea)*

Sung Weon Kang *(ETRI, Korea)*

Conference Secretary
Youngmin Kim *(Kwangwoon University, Korea)*

Special Session Chairs
Byeong-Gyu Nam *(Chungnam National University, Korea)*

Kyeong-Sik Min *(Kookmin University, Korea)*

Minkyu Je *(KAIST, Korea)*

Hanho Lee *(Inha University, Korea)*

Byeongho Choi *(KETI, Korea)*

Finance Chairs
Kuk Tae Hong *(LG Electronics, Korea)*

Seung Ho Lee *(Hanbat University, Korea)*

Soo Youn Kim *(Dongguk University, Korea)*

IEEE Liaison Chairs
Myung Hun Sunwoo *(Ajou University, Korea)*

Yong Moon *(Soongsil University, Korea)*

Jin Sang Kim *(Kyung Hee University, Korea)*

Organizing Committee

Publication Chairs

Seung Eun Lee *(Seoul National University of Science and Technology, Korea)*

Gyungsu Byun *(Inha University, Korea)*

Joonho Kong *(Kyungpook National University, Korea)*

Publicity Chairs

Changsik Yoo *(Hanyang University, Korea)*

Yong Ho Song *(Hanyang University, Korea)*

Hyung-Min Lee *(Korea University, Korea)*

Woojoo Lee *(Myongji University, Korea)*

Registration Chair

Seokhyeong Kang *(POSTECH, Korea)*

Local Arrangement Chairs

Byungin Moon *(Kyungpook National University, Korea)*

Junghyup Lee *(DGIST, Korea)*

Youngjoo Lee *(POSTECH, Korea)*

Poster Session & Tutorial Chairs

Jong Sun Kim *(Hongik University, Korea)*

Hyung Tak Kim *(Hongik University, Korea)*

Jaehyuk Choi *(Sungkyunkwan University, Korea)*

Chip Design Contest Chairs

Kyoung Rok Cho *(Chungbuk National University, Korea)*

Tae Wook Kim *(Yonsei University, Korea)*

Steering Committee

Steering Committee Chair

Yun Sik Lee *(UNIST, Korea)*

Steering Committee Co-Chairs

Shiho Kim *(Yonsei University, Korea)*

Jinwook Burm *(Sogang University, Korea)*

Jun Rim Choi *(Kyungpook National University, Korea)*

Kwang Sub Yoon *(Inha University, Korea)*

Kyeongsoon Cho *(Hankuk University of Foreign Studies, Korea)*

Young Hwan Kim *(POSTECH, Korea)*

Hang Geun Jeong *(Chonbuk National University, Korea)*

Hi Seok Kim *(Cheongju University, Korea)*

Shin Il Lim *(Seokyeong University, Korea)*

Kwang Hyun Baek *(Chung Ang University, Korea)*

Advisory Committee

Advisory Committee

Chanho Lee *(Soongsil University, Korea)*

Yeon Mo Chung *(Kyung Hee University, Korea)*

Sang Bock Cho *(University of Ulsan, Korea)*

Jin Ku Kang *(Inha University, Korea)*

Joongho Choi *(University of Seoul, Korea)*

Chi Ho In *(Semyung University, Korea)*

Kwang-Yeob Lee *(Seokyeong University, Korea)*

Jin-Gyun Chung *(Chonbuk National University, Korea)*

Min-Kyu Song *(Dongguk University, Korea)*

Technical Program Committee

Technical Program Chair

Kyung Ki Kim *(Daegu University, Korea)*

Technical Program Co-Chairs

Meng-Fan(Marvin) Chang *(National Tsing Hua University, Taiwan)*

Chip Hong Chang *(Nanyang Technological University, Singapore)*

Technical Program Vice Chairs

Kang-Yoon Lee *(Sungkyunkwan University, Korea)*

Jongsun Park *(Korea University, Korea)*

Technical Program Track Chairs

Analog Circuits

Franklin Bien *(UNIST, Korea)*

Young Ho Jung *(Daegu University, Korea)*

Hyungil Chae *(Kookmin University, Korea)*

Data Converters

Seung-Tak Ryu *(KAIST, Korea)*

Minjae Lee *(Gwangju Institute of Science and Technology, Korea)*

RF/Microwave/Wireless

Ilku Nam *(Pusan National University, Korea)*

Donggu Im *(Chonbuk National University, Korea)*

Kuduck Kwon *(Kangwon National University, Korea)*

Technical Program Track Chairs (Cont.)

Wireline

Young-Chan Jang *(Kumoh National Institute of Technology, Korea)*

Dong-Woo Jee *(Ajou University, Korea)*

Digital Architecture and Systems

Youngjoo Lee *(POSTECH, Korea)*

Hoyoung Yoo *(Chungnam National University, Korea)*

Sumedh Dhabu *(Ericsson, Sweden)*

Leonel Sousa *(Universidade de Lisboa, Portugal)*

Digital Circuits and Memories

Joonho Kong *(Kyungpook National University, Korea)*

Jaeha Kung *(DGIST, Korea)*

Yuan Cao *(Hohai University, China)*

Fan (Terry) Zhang *(Zhejiang University, China)*

SoC Design Methodology

Seokhyeong Kang *(POSTECH, Korea)*

Jaeyong Chung *(Incheon University, Korea)*

Emerging Technologies

Hyouk-Kyu Cha *(Seoul National University of Science and Technology, Korea)*

Hyung-Min Lee *(Korea University, Korea)*

Jiajia Chen *(Nanjing University of Aeronautics and Astronautics (NUAA), China)*

Chao Wang *(Singapore University of Technology and Development (SUTD), Singapore)*

CONFERENCE INFORMATION

Time Table

		Lobby	Convention Hall	Joyful Hall (Clavel)	Happy Hall (Camelia)	Amante Hall		Ladies Hall	
MONDAY_ NOVEMBER 12, 2018									
From	Till		2F	1F	1F	2F		2F	
13:00	14:00	On-site Registration		Industrial Talk 1					
14:00	15:00			Industrial Talk 2					
15:00	16:00			Industrial Talk 3					
16:00	16:15		Break Time (15min.)						
16:15	17:45		Tutorial 1	Tutorial 2					
17:45	18:00	Break Time (30min.)							
18:30	21:30	**Welcome Reception**							

		Lobby	Convention Hall	Joyful Hall(Clavel)	Happy Hall(Camelia)	Amante Hall		Ladies Hall	
TUESDAY_NOVEMBER 13, 2018									
From	Till		2F	1F	1F	2F		2F	
08:45	09:00					CDC 1	O01	CDC 2	O01
09:00	09:15						O02		O02
09:15	09:30						O03		O03
09:30	09:45		CDC Demo & CDC Poster 1				O04		O04
09:45	10:00			Break Time (15min.)					
10:00	10:15			**Opening Ceremony** (Joyfull Hall_1F)					
10:15	10:55			**Keynote Speech-1** (Joyfull Hall_1F)					
10:55	11:35			**Keynote Speech-2** (Joyfull Hall_1F)					
11:35	11:50			Break Time (15min.)					
11:50	12:30			**Keynote Speech-3** (Joyfull Hall_1F)					
12:30	13:45		**Lunch** (Convention Hall_2F)						
13:45	14:00	On-site Registration		111	53	DCM 1	4	SS 1	166
14:00	14:15			35	130		81		167
14:15	14:30		Analog 1	94	SoC 1 / 102		82		188
14:30	14:45		CDC Demo & CDC Poster 2	218	140		84		216
14:45	15:00				87		105		
15:00	15:15			Break Time (15min.)					
15:15	15:30			24	8	DCM 2	34	SS 2	164
15:30	15:45			33	41		162		168
15:45	16:00		RF 1	49	SoC 2 / 46		139		175
16:00	16:15			47	101		142		185
16:15	16:30			133			144		183
16:30	16:45		Chip Design Contest(CDC) Poster Exhibition						184
16:45	17:00		(Convention Hall_2F)						
17:00	17:15			27	45	DAS 1	9	SS 3	163
17:15	17:30			113	52		177		169
17:30	17:45		Analog 2	128	ET / 52		83		170
17:45	18:00			131	85		127		171
18:00	18:15			178	17		153		173
18:15	18:30		Break Time (15min.)						
18:30	20:30		**Banquet** (Convention Hall_2F)						

International SoC Design Conference 2018

CONFERENCE INFORMATION

WEDNESDAY_NOVEMBER 14, 2018

From	Till	Lobby	Convention Hall 2F	Joyful Hall (Clavel) 1F		Happy Hall (Camelia) 1F		Ladies Hall 2F	
09:00	09:15	On-site Registration		DCM 3	19	Analog 3	221	RF 2	25
09:15	09:30				42		23		29
09:30	09:45				106		165		14
09:45	10:00				121		36		160
10:00	10:15				100		143		219
10:15	10:30						151		220
10:30	10:45		Break Time (15min.)						
10:45	11:30		Short Tutorial 1 (Happy Hall_1F)						
11:30	12:15		Short Tutorial 2 (Happy Hall_1F)						
12:15	13:45		Lunch (Convention Hall_2F)						
13:45	14:00		Intro. of Research Lab. (Poster)	DAS 2	7	SoC 3	18	SS 4	176
14:00	14:15				15		37		179
14:15	14:30				62		63		180
14:30	14:45				118		73		182
14:45	15:00				119		149		191
15:00	15:15		Break Time (15min.)						
15:15	15:30		Poster Session			Analog 4	6	SS 5	187
15:30	15:45						44		190
15:45	16:00						186		192
16:00	16:15						222		217
16:15	16:30						135		
16:30	16:45						159		
16:45	17:00		Break Time (15min.)						
17:00	17:30		**Closing Ceremony** (Joyfull Hall_1F)						

THURSDAY_NOVERMBER 15, 2018

10:00	11:00	Committee Meeting & Discussion

Regualr Session	Title
Analog 1	Analog Circuits 1
Analog 2	Analog Circuits 2
Analog 3	Analog Circuits 3
Analog 4	Analog Circuits 4
RF 1	RF and Wireless Circuits 1
RF 2	RF and Wireless Circuits 2
ET	Emerging Technologes: Emerging Circuits and Systems
DAS 1	Digital Architectures for Multimedia Systems
DAS 2	Digital Architectures for Communication Systems
DCM 1	Digital Circuits & Memories : Memory Reliability
DCM 2	Digital Circuits & Memories :Memory and Hardware Security
DCM 3	Digital Circuits & Memories : Digital Integrated Circuits
SoC 1	SoC: SoC Testing & Low-Power Design
SoC 2	SoC: Efficient System Design
SoC 3	SoC: Efficient Deep Learning Systems

Special Session	Title
SS 1	Low Power Design Sensors
SS 2	Design and Analysis of Nonlinear Circuits
SS 3	Design, Analysis and Tools for Integrated Circuits and Systems
SS 4	Bio sensing, bio mimicking, and bio inspired circuits and systems
SS 5	RF and Analog IC Design for IoT

International SoC Design Conference 2018

CONFERENCE INFORMATION

Industrial Talk

Industrial Talk-1

Chair: Woojoo Lee (Myongji University, Korea)

13:00~14:00, MONDAY_NOVEMBER 12, 2018

Joyful (Clavel) Hall (Main Building 1F)

Arm Cortex CPU cores and technology portfolio

Shean Chung
Senior FAE Manager
ARM, Korea

Biography

Shean Chung has been working in Arm as an FAE and an FAE manager since 2006, in order to deploy Arm technologies as well as to enable many kinds of Arm products. Prior to Arm, he had taken rich experiences for over 10 years in software engineering on several architectures and systems, and he possesses a wide range of technical experience on heterogeneous platforms crossing from embedded devices to enterprise machines. He has recently been interested in heterogeneous system design-in for best efficient performance at edge devices.

Abstract

Computing has now become a central part of our everyday life, smartphones have enabled computing everywhere from smallest devices such as smartwatches and VR goggles to smart homes and smart cars. In the presentation, we can see that Arm Cortex processors are from a single Arm architecture but have been differently implemented and widely used for a diverse range of applications, and Arm compute technology is deployed throughout the ecosystem – beyond mobile, from sensor to cloud.

International SoC Design Conference 2018

CONFERENCE INFORMATION

Industrial Talk-2

Chair: Woojoo Lee (Myongji University, Korea)

14:00~15:00, MONDAY_NOVEMBER 12, 2018

Joyful (Clavel) Hall (Main Building 1F)

Efficient inference and training of deep neural networks in limited precision

Jun Haeng Lee
Research Master
Samsung Advanced Institute of Technology (SAIT), Korea

Biography

Jun Haeng Lee is Research Master at Samsung Advanced Institute of Technology (SAIT), where he has been working on neuromorphic engineering, deep learning, and neural processors since 2009. Lee received his undergraduate degree (1999), master degree (2001), and Ph.D. (2005) in Electrical Engineering from Korea Advanced Institute of Science and Technology (KAIST). Before he joined SAIT, he was a research staff at KDDI R&D Labs. He was a visiting researcher at Institute of Neuroinformatics (INI) from Feb. 2015 to Feb. 2017. His work primarily focuses on designing efficient deep neural networks for low precision inference engines.

Abstract

It is a challenging task to deploy state-of-the-art deep neural networks (DNNs) to embedded systems due to the inherent nature of huge number of computations and large memory requirements. These impediments are partly caused by the considerable amount of the redundancies found in the network parameters intended for ease of training. Thus, there are abundant opportunities for trimming strategies, such as pruning and quantizing to low precision. I will introduce techniques for converting DNNs trained in full precision into low-precision accelerator friendly formats. I will also review ideas enabling training of DNNs in low precision computing units

International SoC Design Conference 2018

CONFERENCE INFORMATION

Industrial Talk-3

Chair: Jong Sun Kim (Hongik University, Korea)

15:00~16:00, MONDAY_NOVEMBER 12, 2018

Joyful (Clavel) Hall (Main Building 1F)

Memory Subsystem design for Consumer Electronics

Ken Kyuseok Cho
DDR PHY Professional, MID Team, SIC Center, CTO
LG Electronics, Korea

Biography

Ken Kyuseok Cho has been working on high performance and low power memory interfacing IP design for SoC, memory circuit design engineering since 1995. Experienced deep sub-micron CMOS technology including 16/14/12/10nm finfet technology, full custom analog-digital mixed signal designs and ASIC FE/BE design flows. Before joining LG Electronics as an IP design engineer, he worked as DRAM circuit designer and program manager for 18 years in several major DRAM companies in 3 continents. Ken Kyuseok Cho holds B.S. in ECE and M.S. in Semiconductor Engineering from Korea University.

Abstract

Consumer electronic devices with UHD and 8K displays, Artificial Intelligence enabled intelligent applications require very high performance, cost effective and low power memory subsystem design. Various challenges on memory subsystem design of these Smart TV SoC and Intelligent Consumer devices will be introduced and ideas to solve these problems to be explained.

CONFERENCE INFORMATION

Tutorial-1

Chair: Jaeha Kung (Daegu Gyeongbuk Institute of Science and Technology(DGIST), Korea)

16:15~17:45, MONDAY_NOVEMBER 12, 2018

Joyful (Clavel) Hall (Main Building 1F)

AI Processor with Nano Core-In-Memory Architecture for Function-Safe Autonomous Driving

Youngsu Kwon

Group Leader, AI Processor Research Group
Electronics and Telecommunications Research Institute (ETRI), Korea

Biography

Youngsu Kwon received B.S., M.S., and Ph.D. degrees from Korea Advanced Institute of Science and Technology (KAIST), Republic of Korea at 1997, 1999, and 2004, respectively. He had been with Microsystems Technology Laboratory (MTL), Massachusetts Institute of Technology as a Postdoctoral Associate from 2004 to 2005 for designing 3-Dimensional FPGA. He is now Group Leader of AI Processor Research Group, Intelligent SoC Research Department, Electronics and Telecommunications Research Institute (ETRI), Republic of Korea since 2005. In ETRI, he is leading the design of Korean AI Processor, Aldebaran. He has authored over 30 internal journal and conference papers with special interest in low-power processor core design, many-core architecture, CAD and algorithmic optimizations of circuits and systems. He received Government Recognition Award for Science and Technology in 2016, Excellent Researcher Award from Korea Research Council in 2013, Industrial Contributor Award from Korean Federation of SMEs in 2013, and medals from Samsung Humantech Thesis Prize. The Aldebaran CPU core and Application processor for which he acts as a leading architect received the Presidential Award of Korean Semiconductor Design Competition in 2016.

International SoC Design Conference 2018

CONFERENCE INFORMATION

Abstract

State-of-the-art neural network accelerators consist of arithmetic engines organized in a mesh structure datapath surrounded by memory blocks that provide neural data to the datapath. While server-based accelerators coupled with server-class processors are accommodated with large silicon area and consume large amounts of power, electronic control units in autonomous driving vehicles require power-optimized AI processors with a small footprint. An AI processor for mobile applications that integrates general-purpose processor cores with mesh-structured neural network accelerators and high speed memory while achieving high-performance with low-power and compact area constraints necessitates designing a novel AI processor architecture. In this tutorial, we present the design of an AI processor for electronic systems in autonomous driving vehicles targeting not only CNN-based object recognition but also MLP-based in-vehicle voice recognition. The AI processor integrates Super-Thread-Cores (STC) for neural network acceleration with function-safe general purpose cores that satisfy vehicular electronics safety requirements. The STC is composed of tens of thousands of programmable nano-cores organized in a mesh-grid structured datapath network. Designed based on thorough analysis of neural network computations, the nano-core-in-memory architecture enhances computation intensity of STC with efficient feeding of multi-dimensional activation and kernel data into the nano-cores. The quad function-safe general purpose cores ensure functional safety of Super-Thread-Core to comply with road vehicle safety standard ISO 26262. The designed AI processor exhibits 32 Tera FLOPS, enabling hyper real-time execution of CNN, RNN, and FCN.

CONFERENCE INFORMATION

Tutorial-2

Chair: Jong Sun Kim (Hongik University, Korea)

16:15~17:45, MONDAY_NOVEMBER 12, 2018

Happy (Camelia) Hall (Main Building 1F)

Benchmarking Advanced CMOS and Beyond-CMOS Technologies

Andrew Marshall

Research Professor, Department of Electrical and Computer Engineering
The University of Texas at Dallas, USA

Biography

Dr. Andrew Marshall is a research professor at the University of Texas at Dallas, where he specializes in advanced CMOS, Analog Security and beyond CMOS benchmarking. He was with Texas Instruments for 27 years, leading teams developing high voltage and current devices, analog IC design, and power integrated circuits, at technology nodes from 10µm to 20nm. Dr. Marshall also worked on benchmarking of semiconductors IC processes, including performance characteristics of MOS and passive devices. During this time he attained the Texas Instruments Fellow (TI Fellow) technical rank.

He has authored or co-authored over 100 papers in conferences, peer reviewed journals and proceedings, and holds 85 issued patents. Dr. Marshall is a Fellow of the United Kingdom Institute of Physics, and a Fellow of the IEEE.

Abstract

Benchmarking and performance closure have become increasingly important with continued feature size reduction of ICs. The aim of this tutorial is to explain the history of benchmarking, beginning with CMOS logic, and describing the evolution of changes and additions to the methodology as CMOS density has increased.

Development of each advanced CMOS is more difficult and expensive than the prior one, and to extend the performance characteristics of planar CMOS technology there have been many efforts to create new technologies. Some of these are CMOS extensions, such as Finfet devices. Others are so-called beyond CMOS devices, which include charge-based logic such as Tunnel FET based systems, others are non-charge based, which include nano-magnetic structures, spintronics device and quantum structures.

All these need to be benchmarked and evaluated against each other, as it is important to determine which technologies offer advantages over conventional CMOS. Therefore there has developed a need for comparative benchmarking across logic families. We here detail benchmarking of CMOS and of some of the newer beyond CMOS technologies, and consider how benchmarking standards have changed by the addition of beyond CMOS capability.

International SoC Design Conference 2018

CONFERENCE INFORMATION

Short Tutorials

Short Tutorial-1

Chair: Soo Youn Kim (Dongguk University, Korea)

10:45~11:30, WEDNESDAY_NOVEMBER 14, 2018

Happy (Camelia) Hall (Main Building 1F)

Basics of Jitter Analysis

Jae-Yoon Sim
Professor, Department of Electrical Engineering
Pohang University of Science and Technology (POSTECH), Korea

Biography

Jae-Yoon Sim received the B.S., M.S., and Ph.D. degrees in electrical engineering from Pohang University of Science and Technology (POSTECH), Korea, in 1993, 1995, and 1999, respectively. From 1999 to 2005, he was a Senior Engineer in the Samsung Electronics, Korea. From 2003 to 2005, he was a Postdoctoral Researcher at the University of Southern California, USA. From 2011 to 2012, he was a Visiting Scholar at the University of Michigan, Ann Arbor, MI, USA. In 2005, he joined POSTECH, where he is currently a Professor. His research interests include clock generation, serial and parallel links, data converters, neuromorphic circuits and sensor interface circuits. He has served in the Technical Program Committees of the IEEE International Solid-State Circuits Conference, Symposium on VLSI Circuits, and Asian Solid-State Circuits Conference. He is a Distinguished Professor nominated by Korea Institute of Science and Technology. He is an IEEE Distinguished Lecturer from 2018. He was a recipient of the Takuo Sugano Award and Special Author-Recognition Award at ISSCC 2001 and 2013, respectively.

Abstract

Jitter, as the temporal noise, is a general indicator for quality of timing generation. There are a number of definitions of measuring the amount of jitter, and they differently affect performance depending on applications. This tutorial reviews basics of jitter analysis and design considerations for each case of frequency generators such as oscillator, phase-locked loop and clock/data recovery loop in serial link.

International SoC Design Conference 2018

CONFERENCE INFORMATION

Short Tutorial-2

Chair: Soo Youn Kim (Dongguk University, Korea)

11:30~12:15, WEDNESDAY_NOVEMBER 14, 2018

Happy (Camelia) Hall (Main Building 1F)

Minimum-Energy-Driven Integrated Circuits Design for Green Electronics

Tony Tae-Hyoung Kim
Associate Professor, School of Electrical and Electronic Engineering
Nanyang Technological University, Singapore

Biography

Tony Tae-Hyoung Kim received the B.S. and M.S. degrees in electrical engineering from Korea University, Seoul, Korea, in 1999 and 2001, respectively. He received the Ph.D. degree in electrical and computer engineering from University of Minnesota, Minneapolis, MN, USA in 2009. From 2001 to 2005, he worked for Samsung Electronics where he performed research on the design of high-speed SRAM memories, clock generators, and IO interface circuits. In 2007 ~ 2009 summer, he was with IBM T. J. Watson Research Center and Broadcom Corporation where he performed research on circuit reliability, low power SRAM, and battery backed memory design, respectively. On November 2009, he joined Nanyang Technological University where he is currently an associate professor.

He received "Best Demo Award" ay APCCAS2016, "Low Power Design Contest Award" at ISLPED2016, best paper awards at 2014 and 2011 ISOCC, "AMD/CICC Student Scholarship Award" at IEEE CICC2008, Departmental Research Fellowship from Univ. of Minnesota in 2008, "DAC/ISSCC Student Design Contest Award" in 2008, "Samsung Humantec Thesis Award" in 2008, 2001, and 1999, and "ETRI Journal Paper of the Year Award" in 2005. He is an author/co-author of +140 journal and conference papers and has 17 US and Korean patents registered. His current research interests include low power and high performance digital, mixed-mode, and memory circuit design, ultra-low voltage circuits and systems design, variation and aging tolerant circuits and systems, and circuit techniques for 3D ICs. He serves as an Associate Editor of IEEE Transactions on VLSI Systems. He is an IEEE senior member and the Chair of IEEE Solid-State Circuits Society Singapore Chapter. He has served numerous conferences as a committee member.

Abstract

Recently, various ultra-low power applications such as Internet-of-Things (IoT), wearable devices, and biomedical devices have emerged opening up a new domain of integrated circuits design. In these applications, ultra-low voltage circuit techniques for improving the power and energy efficiencies have been the main research focus. While supply voltage scaling has been considered as the most effective way of achieving high energy efficiency, it generates many challenging design issues such as significantly degraded parametric margins, large variations, etc. This tutorial will provide a brief introduction of various integrated circuits design techniques that are essential for minimum-energy-driven.

CONFERENCE INFORMATION

10:00~10:15 TUESDAY, NOVEMBER 13, 2018
Joyful (Clavel) Hall, Main Building 1F

Welcome Address

Kwang Hyun Baek, General Chair (Chung Ang University, Korea)

Conference Statistics

Kyungki Kim, TPC Chair (Daegu University, Korea)

Announcements

Kyungki Kim, TPC Chair (Daegu University, Korea)

International SoC Design Conference 2018

CONFERENCE INFORMATION

Keynote - 1

10:15~10:55, TUESDAY_NOVEMBER 13, 2018

Joyful (Clavel) Hall (Main Building 1F)

Shift left - Disrupting the Current Automotive Development Process

Burkhard Huhnke
Vice President, Automotive Strategy
Synopsys, USA

Biography

Dr. Burkhard Huhnke is the Vice President of Automotive Strategy at Synopsys.   He joined Synopsys earlier this year.

Prior to Synopsys, he was SVP of Product Innovation & E-Mobility at VW, based in Silicon Valley.   He was responsible for synchronizing VW's innovation activities and alliances to identify new concept ideas, business models and partners in the US and had end-to-end ownership of the electric vehicle platform in North America.
Prior to that, he held several positions both in the US and Germany, including Senior GM, Electronics System Integration and Whole Vehicle Integration.

Dr. Huhnke studied electrical engineering, at the University of Braunschweig. His dissertation about optical distance measurement was awarded with the International Measurement Prize.

Dr. Huhnke serves as Research Fellow the Hult Business School in San Francisco, and is a member of the Board of Advisors at the College of Engineering at University of Tennessee Knoxville and at the College of Engineering and Computer Science at University of Tennessee Chattanooga.

International SoC Design Conference 2018

CONFERENCE INFORMATION

Abstract

In the automotive world, recalls for electronics affects about five percent of the vehicles on the road. That means 5 out of every 100 vehicles today have a problem with their electronics. If we want to see more autonomous driving vehicles, that number must be improved. There must be more robustness into the development process.

Unlike a blue screen of death that may occur on a desktop or laptop computer, a chip or software failure within a vehicle traveling at highways speeds could result in significant injury or death to the driver, their passengers, or others.

Preventing this possibility requires building in functional safety. Specifically, ISO 26262 sets out definition for various Automotive Safety Integrity (ASIL) levels with Level D (ASIL-D) being the highest automotive software integrity level, and often the hardest to achieve. To guarantee the safety of any vehicle, ASIL-D must be a part of even the smallest pieces of electronics.

Robust design can and should begin at the SOC level by integrating functional safety from automotive industry. In other words, functional safety doesn't have to wait until later -- in the system world, in the ECU world – it can be built into the chip design itself. This can be done with a virtual prototype, a model of the hardware in development, which can allow software development on those SOCs to begin at a very early phase.

Implicit is the continuous testing of these hardware and software designs throughout the automotive development lifecycle. If the chips and the software can be secured, step by step, within the automotive development process, this creates clear milestones that build in safety and security and quality along the way. It also "shifts left" the mitigation of errors that traditionally compound as the vehicle gets closer and closer to production, the traditional error curve begins to flatten. This, then, accelerates the overall time to production.

The keynote talks about tools to optimize the quality, security, and safety throughout the vehicle lifecycle. For example, safety-island IP and dual-core lockstep processors and the new ASIL-D-ready-certified processors come with a self-checking safety monitor as well as hardware safety features, such as error-correcting code and a programmable watchdog timer to help detect system failures and runtime faults.

The virtual prototype technology allows the entire automotive ecosystem to begin software development in parallel with hardware development – accelerating time to production, which means disrupting the current development process and a shift left by co designing hardware / software for automotive electronics is possible.

CONFERENCE INFORMATION

Keynote - 2

10:55~11:35, TUESDAY_NOVEMBER 13, 2018

Joyful (Clavel) Hall (Main Building 1F)

Intelligent SoC Solutions for the Innovative Products

Seung-Jong Choi
Senior Vice President and Head, System IC Center
LG Electronics, Korea

Biography

Dr. Seung-Jong Choi was born in Seoul, Korea, in 1964. He received the B.S. degree from Seoul National University, Seoul, Korea, in 1987, the M.S. degree from Korea Advanced Institute of Science and Technology, Taejon, Korea, in 1989, and Ph.D. degree from Rensselaer Polytechnic Institute, Troy, NY, in 1996, all in electrical engineering. His doctoral research was on video compression and video signal processing.

Since 1989, he has been at LG Electronics Inc. and is presently a Senior Vice President and the Head of the System IC Center, Seoul, Korea. His main research activities have been related with TV including HDTV, 3DTV, UHDTV and OLED TV. He has been developing multiple generation of TV SoC solutions, and successfully applied them to the LG TV product lineups, so far. Currently, he is leading the system IC R&D activities including TV SoCs for the OLED TV/8KTV, intelligent SoCs for home appliance products, and LSI chips for camera module/power electronics

Abstract

The transition to the era of artificial intelligence is rapidly going on, and IT industry is making great efforts to apply the artificial intelligence to the virtually most of services and products. This keynote emphasizes the big wave behind, that is, to meet demands for the intelligent user experiences along with the customer-tailored evolving products.

On the one hand, speech recognition becomes ubiquitous driven by Google and Amazon focusing to maintain and widen their platforms. Even more sophisticated image recognition and context understanding are also emerging to provide much convenient user experiences. On the other hand, the concept of the intelligence has a great potential to revolutionize the performance of the systems, that is, to enhance the fundamental characteristics of products. For example, the picture quality of TV can be enhanced with AI, and the energy efficiency of home appliance products can be improved with AI.

LG has been successfully launching various market-leading products such as OLED TV, and home appliances. As a system design/manufacturing company, LG seems to have more advantages for making innovative customer-tailored products with AI utilizing huge amount of accumulated data. Intelligent SoC plays a crucial role for the innovative products. This talk introduces LG intelligent SoC R&D activities, and achievements.

International SoC Design Conference 2018

CONFERENCE INFORMATION

Keynote - 3

11:50~12:30, TUESDAY_NOVEMBER 13, 2018

Joyful (Clavel) Hall (Main Building 1F)

Energy Harvesting for Smart Wireless Sensor Nodes

Dong Sam Ha

Professor,
Multifunctional Integrated Circuits and Systems (MICS) Group,
The Bradley Department of Electrical and Computer Engineering
Virginia Polytechnic Institute and State University, USA

Biography

Dong Sam Ha received a B.S. degree in electrical engineering from Seoul National University, Korea, and M.S. and Ph.D. degrees in electrical and computer engineering from the University of Iowa. Since Fall 1986, he has been a faculty member of the Department of Electrical and Computer Engineering, Virginia Tech. He is Director of Multifunctional Integrated Circuits and Systems (MICS) group with four faculty members and a member of Center for Energy Harvesting Materials and Systems (CEHMS) of Virginia Tech.

He specializes in low-power IC design for analog/mixed-signal and RF ICs targeting for embedded system applications. His research interests include energy harvesting from various sources such as solar, thermal, vibration, and RF. He is a Fellow of the IEEE.

Abstract

A smart wireless sensor node (WSN) built on SoC processes sensed data locally and transmits it wirelessly. A smart WSN will be more pervasive in the era of IoT (Internet of Things). A major design issue for WSNs is autonomous power, and energy harvesting is a promising solution. Energy harvested from ambient sources aims to recharge the battery of a WSN or even remove the battery perpetually. Typical energy harvesting sources include solar, thermal, kinetic, and radio frequency. The operating environment of ambient energy sources changes and the energy level of typical ambient energy sources is low. They poss major design issues for energy harvesting circuits, which aim to maximize the energy flowing into storage devices such that rechargeable batteries or supercapacitors. This talk covers design issues of energy harvesting circuits for various energy sources and major design schemes to address the problems. It also covers recent advances and research issues in energy harvesting circuits in the context of smart WSNs built on SoC.

International SoC Design Conference 2018

12-bit 20M-S/s SAR ADC using C-R DAC and Capacitor Calibration

Eunji Youn
Department of Electronic Engineering
Kumoh National Institute of Technology
Gumi, Korea
ejyoun@kumoh.ac.kr

Young-Chan Jang
Department of Electronic Engineering
Kumoh National Institute of Technology
Gumi, Korea
ycjang@kumoh.ac.kr

Abstract—**A successive approximation register (SAR) analog-to-digital converter (ADC) using a capacitor-resistor(C-R) digital-to-analog-converter (DAC) is proposed to implement the resolution of 12 bits maintaining the area for 10 bits. A calibration for upper-bit capacitors of the C-R DAC is proposed to increase the performance of the static and dynamic performances. To evaluate the proposed ADC, a 12-bit 20M-S/s SAR ADC is implemented using a 110-nm CMOS process with a supply of 1.2 V. The area and power consumption of the proposed ADC are 0.204 mm² and 1.24 mW, respectively.**

Keywords: SAR ADC, Capacitor-Resistor DAC, Capacitor-DAC calibration,

I. INTRODUCTION

With the rapid development of mobile systems in recent years, the resolution of analog-to-digital converters (ADCs) required by related systems is continuously increasing without increasing area and power consumption. A successive approximation register (SAR) ADC using a capacitor digital-to-analog converter (CDAC) is generally used to minimize the power consumption reducing the static current in mobile applications [1][2]. However, whatever the resolution of the SAR ADC is increased by one bit, the area of the CDAC used in the SAR ADC is doubled. In addition, unlike other ADCs, SAR ADCs require more control and reduction of the noise generated by a comparator and a DAC, since there is no amplification process for the analog signal to reduce the power consumption even if the resolution is increased. A capacitor-resistor DAC (C-R DAC) with capacitors and resistors combined instead of a CDAC can be used to reduce the area increase as the resolution increases in the SAR ADC [3]. In this case, it may be a design burden to drive the reference voltage divided by the resistance, that is, the output voltage of the RDAC, with the CDAC. In addition, the noise due to the mismatch of the CDAC must be managed more tightly considering the resolution increase due to RDAC. A digital-domain calibration reported in the prior literature [4] reduces the noise generated due to the mismatch of the CDAC. However, this calibration technique requires a lot of hardware and makes it difficult to perform the high-speed data conversion of several tens of MS/s.

In this paper, a SAR ADC using a calibration for the upper bit capacitors of C-R DAC and C-R DAC is proposed to realize ADC with 12-bit resolution and 20-MS/s sampling rate.

II. PROPOSED SAR ADC USING C-R DAC

A. Configuration and Operarion of Proposed SAR ADC

The proposed 12-bit 20MS/s SAR ADC consists of a C-R DAC, a comparator, and a SAR logic with calibration logic, as shown in Fig. 1(a). The C-R DAC has an architecture that combines a VCM-based 9-bit CDAC with a 2-bit RDAC. Since the 2-bit RDAC drives only the LSB capacitor (C_0) of CDAC, it does not consume much power to supply the reference voltage. The C-R DAC has capacitor arrays and a resistor string for generation of the reference voltages with the resolution of 11 bits. In addition, additional calibration capacitors are used to eliminate the mismatch of the 9-bit CDAC. Figure 1(b) shows the timing diagram of the proposed SAR ADC. The sample process for the analog signal input to the SAR ADC begins at the last rising edge of the signal *VALID*, indicating that the comparison of the previous data in the comparator is completed. This sampling process continues until the rising edge of *EXCLK*. After the sample process, the data conversion process is achieved by the basic operation of an asynchronous SAR ADC using the signals *VALID* and *CLKC*. The comparator has an architecture of a sense-amplifier based voltage comparator and includes a meta-stability detector for stable asynchronous SAR operation [5].

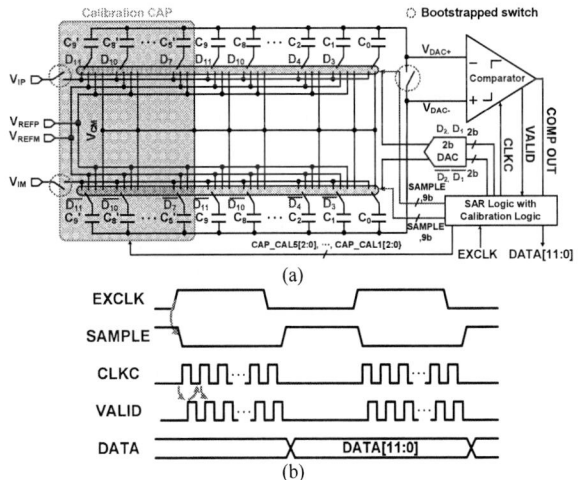

Fig. 1. Proposed SAR ADC (a) block diagram (b) timing diagram

978-1-5386-7961-6/18 $31.00 © 2018 IEEE

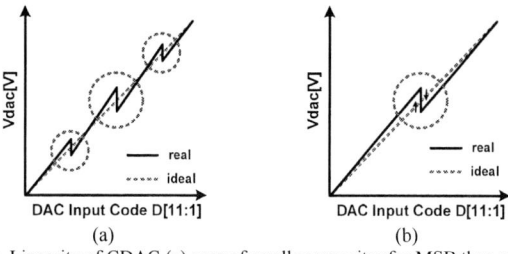

(a)   (b)

Fig. 2. Linearity of CDAC (a) case of smaller capacitor for MSB than sum of capacitors for LSBs (b) concept for improvement of linearity

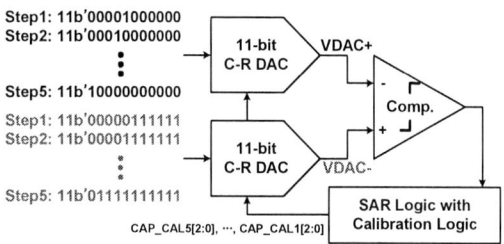

Fig. 3. Concept of calibration for upper 5-bit capacitors of C-R DAC

B. Calibration for Upper-bit Capacitors of C-R DAC

Basically, a CDAC used for a SAR ADC should have the capacitor for the MSB equal to the sum of the capacitors for the remaining LSBs. The sizes of the capacitors for the upper 5 bits of the C-R DAC used in the proposed SAR ADC are designed to be smaller than the ideal values. Thus, the initial transfer curve of the C-R DAC is distorted as shown in Fig. 2(a) because the MSB capacitor is smaller than the sum of the LSB capacitors. However, the linearity of the C-R DAC is improved as shown in Fig. 2(b) by increasing the value of the capacitor for the MSB. To improve the linearity of the C-R DAC by the method mentioned above, the calibration capacitors shown in Fig. 1(a) are added and controlled by making the value of the capacitor of each upper bit equal to the sum of the capacitors of the lower bits. The proposed capacitor calibration simplifies the control logic by only performing the process of increasing the capacitor for the upper 5 bits. The upper 5-bit capacitors of the proposed SAR ADC are C_9, C_8, C_7, C_6 and C_5, respectively, and the corresponding calibration capacitors are defined as C_9', C_8', C_7', C_6' and C_5'. In the first step for the proposed capacitor calibration, two adjacent codes shown in Fig. 3 are fed to two C-R DACs for the differential structure, respectively, to calibrate the value of C_5. Then, the value of C_5' is controlled using the code of $CAP\_CAL1[2:0]$ so that the output of the two C-R DACs, VDAC+ and VDAC-, are equal to each other. Similarly, the values of the capacitors C_9, C_8, C_7, and C_6 are also calibrated by performing all the steps from step 2 to step 5.

III. IMPLEMENTATION AND SIMULATION RESULTS

The proposed 12-bit 20-MS/s SAR ADC shown in Fig. 4 was designed using a 110-nm CMOS process with a 1.2-V supply voltage. Its active area and power consumption are 540 × 380 μm² and 1.24 mW, respectively. Figure 5 shows the power spectrums of the ADC output measured at sampling rate of 20-MS/s about a 2.4 V_{PP} differential sinusoidal input with frequency of 9.453 MHz. The proposed capacitor calibration used in the SAR ADC improved the effective number of bits (ENOBs) to 11.85 bits.

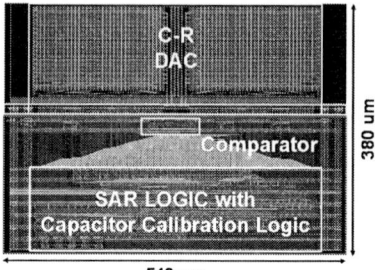

Fig. 4. Layout of designed SAR ADC

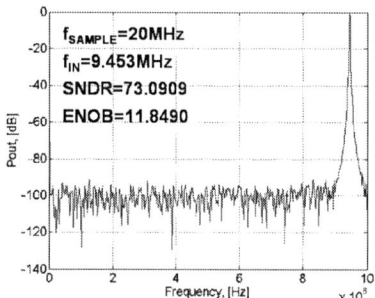

Fig. 5. Simulated power spectrum of ADC output

IV. CONCLUSION

The 12-bit 20-MS/s SAR ADC using the C-R DAC with the capacitor calibration for the upper 5 bits was proposed for mobile applications requiring small area and low power consumption. The capacitor calibration for the upper 5 bits of the C-R DAC simplifies the control logic by only performing the process of increasing the capacitor for the upper 5 bits. The proposed SAR ADC was designed using a 110-nm CMOS process with a supply of 1.2 V. The area and power consumption of the proposed ADC are consumption are 540 × 380 μm² and 1.24 mW, respectively.

ACKNOWLEDGMENT

This research was supported by the MOTIE (No. N0001883, HRD Program for Intelligent semiconductor Industry), the Basic Science Research Program through NRF funded by the Ministry of Education under Grant 2016R1D1A3B03934487, and in part by IDEC.

REFERENCES

[1] N. Verma, *et. al.*, "An Ultra Low Energy 12-bit Rate-Resolution Scalable SAR ADC for Wireless Sensor Nodes," *IEEE J. Solid-State Circuit*, vol. 42, no. 6, pp.1996-1205, June 2007

[2] G.-Y. Huang, *et. al.*, "A 10 b 200MS/s 0.82mW SAR ADC in 40nm CMOS," *IEEE ASSCC*, pp.289-292, Nov 2013

[3] H. Zhou, *et. al.*, "Design of a 12-bit 0.83 MS/s SAR ADC for an IPMI SoC," *IEEE ISOCC*, pp.175-179, Sept. 2015

[4] J. Um, *et. al.*, "A Digital-Domain Calibration of Split-Capacitor DAC for a Differential SAR ADC Without Additional Analog Circuits," *IEEE Trans. Circuits Sys. I*, vol. 60, no. 11, Nov. 2013, pp. 2845-2856.

[5] S.-M. Park, *et. al.*, "A 10-bit 20-MS/s Asynchronous SAR ADC with Meta-stability Detector using Replica Comparators," *IEICE transaction on Electronics*, vol. E99-C, no. 6, pp. 651-654, Jun., 2016.

Current Feedback-Based High Load Current Low Drop-Out Voltage Regulator in 65nm CMOS Technology

Melody Teoh
Department of Electrical Engineering
San Jose State University
San Jose, CA, USA
melody.teoh@sjsu.edu

Sotoudeh Hamedi-Hagh
Department of Electrical Engineering
San Jose State University
San Jose, CA, USA
sotoudeh.hamedi-hagh@sjsu.edu

Abstract— This paper features a current feedback-based capacitor-less low drop-out (LDO) voltage regulator with a bandgap voltage reference (BGR) performing an average temperature coefficient (TC) of 13.34 ppm/°C in the range of -40 to 125 °C in accordance with military standard and thus gain higher stability and power supply rejection ratio (PSRR). The proposed capacitor-less LDO achieves a 200 mA load current with an error percentage of 0.246 % and a -21.47 dB PSRR at 100 KHz with current based structure implemented in TSMC 65-nm CMOS technology for internet of things (IoT) system on chip (SoC) applications.

Keywords— *BGR; LDO; capacitor-less; load current; IoT*

I. INTRODUCTION

Internet of Things (IoT) is the successor of wired communication amongst devices. It is a new generation technology that opens up gateway to new opportunities. In the recent years, many sensing devices have been conducted for IoT application [1][2]. These devices have led to the inclination of industry in IoT connection which requires the performance of low power consumption, stabilized power supply, small chip area, and wireless features to improve many aspects on health systems.

Although an LDO with high PSRR has an ability to achieve ultra-low output noise, the PSRR and load current are inversely proportional to each other. The high load current gives the LDO an ability to support a high fanout to several branches. This paper proposes an LDO with a current feedback-based that injects current into the second stage of the error amplifier. In addition, the capacitor-less structure reduces chip cost and area with better loop-gain bandwidth and slew rate performance.

II. BANDGAP VOLTAGE REFERENCE

A high gain operational amplifier (OPAMP) balances the current in the BGR circuit, thereby enabling the BGR to provide a stable voltage source to the LDO. Although it is a rule of thumb that the number of the stages and its gain is directly proportional, there is a significant drawback— poles that can lead to problems in the phase margin [3].

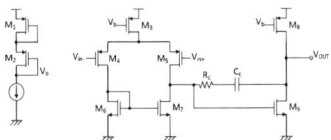

Figure 1.   Proposed Two-Stage Common Source OPAMP

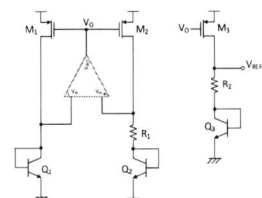

Figure 2.   Proposed BGR Circuit

Even though a folded cascaded OPAMP is able to provide a gain of approximately 45 dB within one stage, it is much easier to implement a common source structure OPAMP, presented in Fig. 1, which takes less area on the chip due to its simplicity. The proposed BGR circuit, shown in Fig. 2, consisting of a two-stage common source differential amplifier that gives out a gain of approximately 60 dB, is able to provide a consistent average output voltage of 1 V for the proposed LDO. An RC compensated circuit is added to complement the poles created when frequency is near 100 MHz. Fig. 4 shows the OPAMP and BGR simulation results, whereas Table I shows the comparison of folded cascaded and common source OPAMPs.

III. PROPOSED LDO WITH BGR

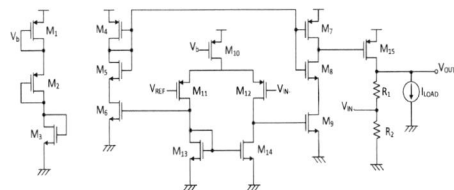

Figure 3.   Proposed LDO

978-1-5386-7961-6/18 $31.00 © 2018 IEEE

The proposed LDO, demonstrated in Fig. 3, injects current into the second stage of the amplifier with a replica circuit. This design doesn't require the usage of capacitors to reduce the noise, as the capacitor-less LDO is adopted for power management of the Always-On Domain, which is a better fit for SoC compared with traditional LDO [4].

IV.   SIMULATION RESULTS

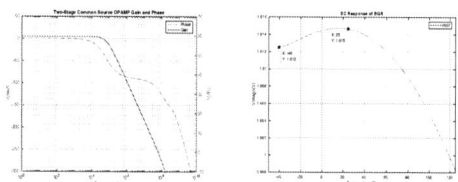

Figure 4.   Gain and Phase of OPAMP (left) and DC Response of BGR (right)

TABLE I.   THE PERFORMANCE OF OPAMP DESIGN

OPAMP Type	[5] Folded Cascaded	[6] Common Source	This Work Common Source
Technology	0.18μm	0.35μm	65nm
Power Supply	1.8 V	3.3 V	2.5 V
Gain	>80 dB	78 dB	~60 dB
Phase Margin	62.3 °	63.9°	66.72°
GBW	1.437 GHz	5.82 MHz	3.79 GHz

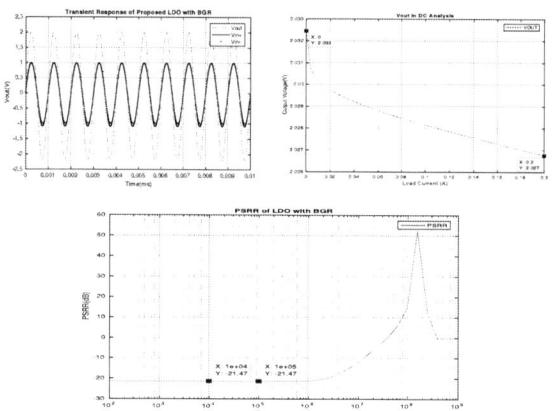

Figure 5.   Transient Response (top left), DC Analysis (top right), and PSRR (bottom) of the proposed LDO with BGR

With a power supply voltage of 2.5 V, the proposed LDO yields a V_{out} of 2.032 V, as shown in Fig. 5, making the $V_{drop-out}$ ≈ 470 mV with an error percentage of 0.246 %. Although the LDO is able to handle a 200 mA load current, the tradeoffs are a huge pass transistor and relatively large feedback network, which in turn increases the total area of the chip layout.

The performance comparison of this work with a few other recent works is demonstrated in Table II. The full chip layout area (BGR and LDO) shown in Fig. 6 is 0.051mm².

Figure 6.   Proposed LDO and BGR full chip layout

TABLE II.   PERFORMANCE COMPARISON TABLE

		[7]	[8]	[9]	This Work
Technology (nm)		180	350	350	65
V_{OUT} (V)		1.5	2.8	2.8	2.03
$V_{DROP-OUT}$(mV)		300	200	500	470
Max. I_{LOAD} (mA)		25	100	50	200
On-chip Capacitor		100pF	Cap-Less	Cap-Less	Cap-Less
PSRR (dB)	10 KHz	-62	-56	N/A	-21.47
	100 KHz	-62	-56	-45	-21.47
LDO Die Area (mm²)		0.069	N/A	N/A	0.0317

V.   CONCLUSION AND FUTURE WORK

The OPAMP, BGR and LDO have distinct features that are applicable to the IoT technology in SoC and medical fields. This work proposes the capacitor-less LDO with high load current and stable reference voltage, which is cost efficient in medical device development. The BGR and LDO are integrated and fabricated in TSMC 65nm CMOS technology.

REFERENCES

[1] Jaroonrut Prinyakupt and Thanakorn Yootho, "Multichannel temperature monitor on IoT," 2016 9th Biomedical Engineering International Conference (BMEiCON), pp. 1–4, 2016.

[2] Sunil Kumar Maddikatla and Srivatsava Jandhyala, "An accurate all CMOS temperature sensor for IoT applications," 2016 IEEE Computer Society Annual Symposium on VLSI (ISVLSI), pp. 349–354, 2016.

[3] D. Saari, "Op amp design in nanoscale processes using fixed-length devices," Master, University of Waterloo, 2014.

[4] C. Wu, J. Lou and Z. Deng, "An ultra-low power capacitor-less ldo for always-on domain in nb-iot applications," in Proc. IEEE International Conference or Applied System Innovation, pp. 138, 2018.

[5] G. Wei, "Design of OTA with common drain and folded cascade used in ADC," World Academy of Science, Engineering and Technology International Journal of Electronics and Communication Engineering, vol. 6, no. 7, pp. 632, 2012.

[6] S. K. Rajput and B. K. Hemant, "Two-stage high gain low power opamp with current buffer compensation," in IEEE Global High Tech Congress on Electronics, pp. 1–4, 2011.

[7] B. Yang, B. Drost, S. Rao and P. K. Hanumolu, "A high-PSR LDO using a feedforward supply-noise cancellation technique," in Proc. IEEE Custom Integrated Circuits Conf. (CICC), pp. 1, 2013.

[8] A. Saberkari, E. Alarcon and S. B. Shokouhi, "Fast transient current-steering CMOS LDO regulator based on current feedback amplifier," VLSI Journal of Integration, Vol. 46, pp. 165–171, 2013.

[9] M. Khan, M. H. Chowdhury, "Capacitor-less low-dropout regulator (LDO) with improved psrr and enhanced slew-rate," in IEEE International Symposium on Circuits and Systems (ISCAS), pp. 4, 2018.

Gap in pagination due to formatting issues.

Pages 5-6

Dynamic voltage Drop induced Path Delay Analysis for STV and NTV Circuits during At-speed Scan Test

[1]Hyunggoy Oh, Heetae Kim, Sangjun Lee, and [2]Sungho Kang
Departments of Electrical & Electronic Engineering
Yonsei University
Seoul, Korea
[1]{kyob508, kht2161, lsj920807}@soc.yonsei.ac.kr, [2]shkang@yonsei.ac.kr

Abstract— The NTV circuit has been introduced as a new low power design concept, which increases energy efficiency significantly. However, delay sensitivity of the NTV circuit is a major challenge. In addition, this problem can be more critical during at-speed scan test because of the dynamic voltage drop issue. In this paper, we propose a comparison of dynamic voltage drop induced path delay between STV and NTV circuits during at-speed scan test. To the best knowledge of the authors, it is the first time to analyze the voltage drop induced path delay during the NTV circuit scan test. Experimental results show that the path delay increment of NTV is larger than that of STV although the dynamic voltage drop of NTV is smaller than that of STV.

Keywords; near threshold voltage; dynamic voltage drop; at-speed scan test; path delay;

I. INTRODUCTION

Recent trends of various electronic devices become smaller such as internet of things (IOT), the energy efficiency of VLSI has become the most important issue. According to [1], supply voltage at near threshold voltage (NTV) region is the optimal range for the energy efficiency because operating frequency decreases linearly but power consumption is decreased exponentially. Nonetheless, it also brings the critical design challenge, which is the delay sensitivity. In the NTV region, the performance variation increases significantly compared to the super-near voltage (STV) region, and therefore, it is very important to handle the power supply noise (PSN) to mitigate the variation in the NTV circuit [2]. From this point of view, the delay sensitivity of NTV can be more serious during at-speed scan test. This is because test power is typically much higher than functional power and excessive voltage drop is still a major concern during at-speed scan test (e.g., yield loss) [3]. Hence, it is required to understand power, dynamic voltage drop, and voltage-drop induced path delay during NTV circuit scan test. For this purpose, this paper presents comparative analysis for STV and NTV circuit. The experimental results show that the path delay increment of NTV (~50%) is larger than that of STV (~5%) during launch-off cycle although test power and dynamic voltage drop of NTV is much smaller than that of STV.

II. EXPERIMENAL FLOW

To analysis the power, dynamic voltage drop, and voltage drop induced path delay, the automated experimental flow is

This work was supported by the IT R&D program of MOTIE/KEIT. [10052716, Design technology development of ultra-low voltage operating circuit and IP for smart sensor SoC].

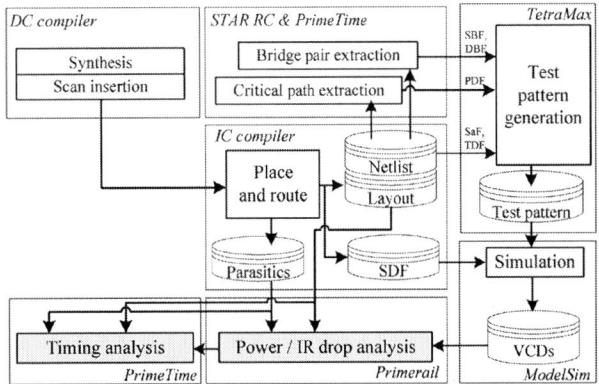

Fig. 1. Flow of the power, dynamic voltage drop and timing analysis

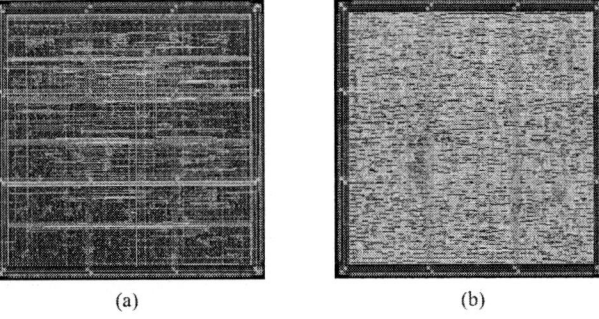

Fig. 2. Layout design of ISCAS s38584 using (a) STV library (b) NTV library

implemented as described in Figure 1. In this experiment, ISCAS s38584 benchmark circuit is synthesized by two libraries (STV = 1.2V and NTV = 0.5V) with 10 scan chains. The operating frequency of STV is 250Mhz and that of NTV is 25 Mhz. It is noted that the NTV circuit frequency is typically smaller 10 times than STV [4]. And then they are placed and routed with 400x400um^2 core size. The power distribution network is designed with four virtual power/ground pads, a power ring and 2x2 power straps. Each layout design is described in Figure 2. After the layout design, transition delay fault (TDF) test patterns are generated by the ATPG tool. Using these patterns, the logic simulator is performed with *standard delay format* (SDF) and the switching information is stored in the *value change dump* (VCD) format. With these generated files, power, dynamic voltage drop and voltage drop induced path delay can be analyzed.

TABLE I
COMPARATIVE ANALYSIS OF POWER AND DYNAMIC VOLTAGE DROP BETWEEN STV AND NTV ISCAS s38584 CIRCUITS

	Functional pattern (100 cycles, random patterns)				TDF test pattern (launch cycle)			
	Total cell power (mW)				Total cell power (mW)			
	TSP	TIP	TLP	TP	TSP	TIP	TLP	TP
STV(1.2v)	1.5257	3.4048	0.0013	4.9319	4.3720	9.7262	0.0015	14.0998
NTV(0.5v)	0.0277	0.0572	0.0012	0.0862	0.0619	0.0809	0.0010	0.1440
	VDD network				VDD network			
	Peak transient voltage (mV)		AVG transient voltage (mV)		Peak transient voltage (mV)		AVG transient voltage (mV)	
	MAX voltage drop		MAX voltage drop		MAX voltage drop		MAX voltage drop	
STV(1.2v)	81.892		8.725		110.594 (35%↑)		18.089 (107%↑)	
NTV(0.5v)	5.296		0.376		6.247 (17%↑)		0.526 (43%↑)	
	VSS network				VSS network			
	Peak transient voltage (mV)		AVG transient voltage (mV)		Peak transient voltage (mV)		AVG transient voltage (mV)	
	MAX voltage rise		MAX voltage rise		MAX voltage rise		MAX voltage rise	
STV(1.2v)	182.31		7.826		183.006 (0.5%↑)		16.198 (106%↑)	
NTV(0.5v)	7.138		0.299		7.571 (6%↑)		0.553 (84%↑)	

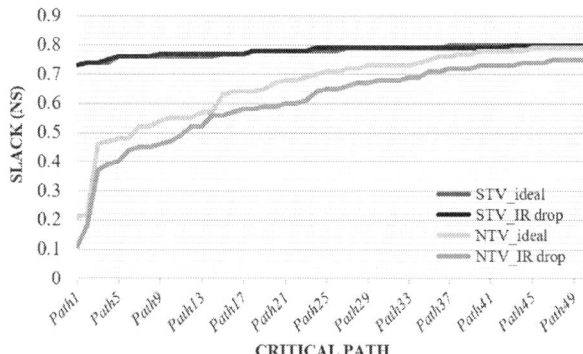

Fig. 3. Comparative analysis of ideal path delay and voltage drop induced path delay between STV and NTV ISCAS s38584 circuits

III. POWER, VOLTAGE DROP, PATH DELAY ANALYSIS

Table 1 shows the comparative analysis of power and dynamic voltage drop between STV and NTV circuits. First, a functional pattern is adapted to measure total cell power and voltage drop. Total switching power (TSP), total internal power (TIP), total leakage power (TLP) and total power (TP) are calculated in the STV and NTV circuit. Because dynamic power is strongly related to V_{DD}, total power of STV is much larger than that of NTV. During 100 cycles, dynamic voltage drop and voltage rise appears on VDD and VSS network because of the switching activity of each circuit. And then, a TDF pattern is adapted to compare to the functional pattern. In this case, the only one at speed launch cycle is observed because the excessive voltage drop issue during at speed scan test is typically related to this launch cycle [4]. The TDF test patterns of STV and NTV are same, and therefore, the total switching activity is almost same. Nonetheless, the rate of increase of total power, voltage drop and voltage rise is totally different between STV and NTV. Voltage drop of at speed scan testing in STV is larger than that of functional operation. However, in the NTV case, the rate of increase of voltage drop during at speed scan testing is around 43%. In addition to voltage rise on VSS network, the effect of PSN increment during at speed testing of NTV is much less than that of STV.

However, the rate of increase of dynamic voltage drop induced path delay of NTV is much higher than that of STV as described in Figure 3. This timing analysis result is achieved among 50 critical paths using timing analysis. The maximum slack reduction rate is around 50% in NTV but it is almost under 5% in STV. Therefore, the delay sensitivity of NTV is still stringent compared to STV during at-speed test.

IV. DISCUSSION AND CONCLUSION

It is imperative to test NTV circuit to increase reliability and detect structural faults. In this paper, the automated experimental flow is introduced and the comparative analysis of power, voltage drop and path delay between STV and NTV. Although the PSN of NTV during at-speed scan test is much less than that of STV, the rate of increase of path delay of NTV is still larger than that of STV. This is because the delay of the NTV circuit is more sensitive. In addition, the power or timing margin of NTV is stricter than that of STV [5]. If the size of the benchmark circuit is very large (>100K gates), this trend is strongly exacerbated, and it can be concluded to yield loss or test cost increase. Therefore, the NTV-aware test methodology is strongly required to deal with this challenge.

ACKNOWLEDGMENT

This research was partially supported by the Graduate School of YONSEI University Research Scholarship Grants in 2018.

REFERENCES

[1] H. Kaul et al., "Near-threshold voltage (NTV) design—Opportunities and challenges," in Proc. DAC, San Francisco, 2012, pp. 1149–1154.

[2] Zhang, X et al., "Characterizing and evaluating voltage noise in multi-core near-threshold processors," in Proc. ISLPED, 2013, pp. 82-87.

[3] A. Bosio et al., "Why and How Controlling Power Consumption During Test : A Survey," in Proc. ATS, 2012, pp. 221-226

[4] Wey.I-C, et al., "Near-threshold-voltage circuit design : The design challenges and chances," in Proc ISOCC, 2014, pp. 138-141.

[5] R. Dreslinski et al., "Near-threshold computing: Reclaiming Moore's law through energy efficient integrated circuits," in Proc. IEEE, vol. 98, no. 2, pp. 253–266, Feb. 2010

A Multi-Fault Dynamic Compaction Technique for Test Pattern Count Reduction

Bo-Yi Li[1] and Jiun-Lang Huang[1,2]
[1]Graduate Institute of Electronics Engineering
[2]Department of Electrical Engineering
National Taiwan University

Abstract—With the advance of IC fabrication technology, the manufacturing test data volume keeps growing because more fault models must be considered to ensure the product quality. In this paper, we develop a multi-fault dynamic compaction (MFDC) technique to reduce the number of test patterns by improving the pattern compaction efficiency. Compared to conventional single-fault dynamic compaction, the resulting test set is more compact, which lowers the test application time and cost. Simulation results on benchmark and industry circuits show an average of 18% pattern count reduction without fault coverage degradation.

Index Terms—automatic test pattern generation; ATPG; dynamic compaction; test data volume

I. INTRODUCTION

As the IC fabrication technology continues advancing, the device feature sizes keep shrinking, which allows the designers to pack more functionalities into a single die. However, this also poses challenges to manufacturing test — in addition to the sheer number of transistors, there are more defect mechanisms to consider [1]. For example, multiple-detect test [2] and cell-aware fault models [4] are known to improve product quality because they detect defects that traditional fault models may not cover.

The ever-growing transistor count and list of fault models lead to test pattern set explosion; as a result, test pattern compaction techniques, static or dynamic, have been utilized in automatic test pattern generation (ATPG) to reduce the test pattern set [6]. This work is related to dynamic compaction which tries to detect secondary faults by making more assignments to a partially specified test pattern. Fault ordering significantly affects the dynamic compaction efficiency; the common approach is to process faults in descending order of testability because most easy-to-detect faults can be detected during fault simulation. [5] shows that structure based fault ordering improves dynamic compaction efficiency.

In this paper, we propose a multi-fault dynamic compaction (MFDC) technique to reduce the test pattern count by improving the dynamic compaction efficiency. Compared to conventional dynamic compaction that processes one secondary fault at a time, i.e., single-fault dynamic compaction (SFDC), according to the pre-determined order, MFDC considers multiple secondary faults at the same time, which allows it to utilize the best local ordering that maximizes test pattern compaction. To avoid CPU overhead, a multi-threading approach is adopted

This work was partially supported by the Ministry of Science and Technology of Taiwan, under Grant No. MOST 105-2221-E-002-214-MY3.

to generate tests for multiple secondary faults at the same time. Care was taken to ensure determinism so that MFDC produces the same result regardless of the thread timing and thread count.

The proposed technique is validated with benchmark and industry circuits. Without sacrificing fault coverage and CPU times, MFDC ATPG achieves an average of 18% pattern count reduction compared to SFDC ATPG.

II. PRELIMINARIES

Fig. 1a depicts the conventional SFDC ATPG flow. Each pattern generation iteration (the outer loop) starts with selecting a target fault f, called the primary fault. $p = \text{TPG}(\phi, f)$ indicates that test generation to detect f starts with a totally unspecified initial pattern, ϕ. Then, in each dynamic compaction iteration (the inner loop), a target fault f, called the secondary fault, is selected and test generator tries to make more assignments to p to detect f. The compaction process continues until the stopping criteria is met. The following fault dropping process performs fault simulation to remove faults detected by the newly generated pattern.

The proposed MFDC ATPG uses PODEM [3], a branch-and-bound search algorithm, as the test generation engine. In general, to avoid spending too much time on a particular fault or pattern, per-fault and per-pattern backtrack limits are set to restrict the numbers of "bounds" associated with a fault and a pattern, respectively.

III. PROPOSED MFDC ATPG

The proposed MFDC ATPG replaces the SFDC step (the shaded block in Fig. 1a) with MFDC. Fig. 1b illustrates the MFDC flow. At the beginning, the secondary fault pattern set $\mathbf{P}$ is empty. Then, the next target fault f is selected and targeted with the given initial pattern p. Instead of replacing p as in SFDC, p' is saved in $\mathbf{P}$. This repeats until N secondary fault patterns are generated and saved in $\mathbf{P}$. Finally, the N patterns are merged into one single pattern for next MFDC run.

A. Deterministic Parallel MFDC

MFDC may incur CPU overhead because some patterns are discarded during the merging process. This is addressed by a multi-thread TPG that generates test patterns for multiple secondary faults concurrently. To ensure determinism, a secondary fault pattern will not be saved in $\mathbf{P}$ until all the faults in front of it have been processed.

978-1-5386-7961-6/18 $31.00 © 2018 IEEE

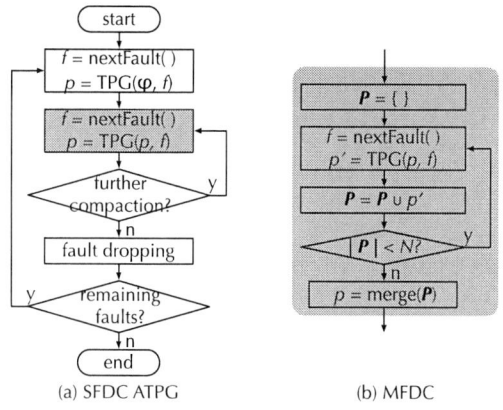

Fig. 1. The SFDC ATPG flow (a) and the proposed MFDC flow (b).

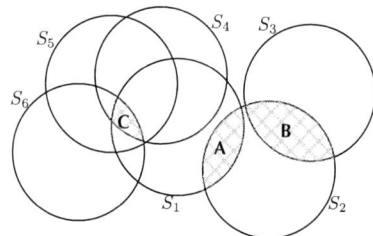

Fig. 2. A dynamic compaction example.

B. Pattern Merging

The merge process first sorts the patterns in $\mathbf{P}$ according to the number of specified bits in ascending order; the reason is to increase the number of patterns that can be merged. The first pattern p is selected as the seed; the other patterns are merged with p one at a time according to the order — patterns that cannot be merged are discarded.

C. Backtrack Limit

Compared to SFDC ATPG, MFDC ATPG tends to incur more backtracks due to redundant assignments; thus, it requires larger backtrack limits.

For example, let $p_1 = \texttt{10xx0xxxxx1}$ be the initial pattern that has f_1 as the primary fault. With MFDC, we may have $p_{1,2} = \texttt{10110xxxx01}$ that further detects f_2 and $p_{1,3} = \texttt{10x10x10x01}$ that further detects f_3. After merging $p_{1,2}$ and $p_{1,3}$, we have $p_{1,2,3} = \texttt{10110x10x01}$ where the the two underlined bits are assigned twice.

D. An MFDC Example

Fig. 2 shows an SFDC vs. MFDC example. In this example, there are six target faults, from f_1 to f_6, and S_i represents the solution space of f_i. Note that the faults are processed according to their indices.

With SFDC, the first generated pattern is $p_{1,2}$ which detects f_1 (primary fault) and f_2 (secondary fault). Dynamic compaction with f_3 to f_6 fails because S_3 to S_6 have no intersection with $\mathbf{A} = S_1 \cap S_2$. Two more patterns p_3 and $p_{4,5,6}$ are generated to detect the remaining faults.

Assume that $N = 4$. With MFDC, we first have p_1 that detects f_1. Then, four patterns, $p_{1,2}$, $p_{1,4}$, $p_{1,5}$, and $p_{1,6}$ are generated and included in $\mathbf{P}$. Since larger space corresponds

to less specified bits, the order to merge the four patterns is: $p_{1,4}$, $p_{1,5}$, $p_{1,2}$, and $p_{1,6}$. Except for $p_{1,2}$, the other three are merged to obtain $p_{1,4,5,6}$. One more pattern $p_{2,3}$ is generated to detect f_2 and f_3. At the end, MFDC produces one less pattern than SFDC to detect the six faults.

IV. EXPERIMENT RESULTS

Experiment results of MFDC ATPG ($N = 8$) on benchmark and industry circuits are shown in Table I. In column 2 ("gate #"), the circuit gate counts are listed. In columns 3 to 5, SFDC pattern counts ("conv.") and the pattern count reduction of [5] and MFDC ("this") are shown. MFDC performs very well in the two largest circuits, netcard and industry, but shows no improvement in b17, which is being investigated. Compared to the available results from [5] (8-core configuration), MFDC performs better except for s35932 which has a small test pattern set. At the same time, MFDC achieves very similar fault coverage (columns 6 and 7) and CPU times (columns 8 and 9) as the conventional SFDC approach.

V. CONCLUSION

We have presented a multi-fault dynamic compaction (MFDC) technique for test pattern count reduction; the idea is to consider multiple secondary faults concurrently. Compared to conventional single-fault dynamic compaction, it achieves an average of 18% pattern count reduction without sacrificing fault coverage and CPU times. In the future, we will integrate MFDC with parallel ATPG to further reduce the CPU times.

TABLE I
EXPERIMENT RESULTS

ckt	gate #	PC reduction (%)			FC (%)		CPU times (s)	
		conv.	[5]*	this	conv.	this	conv.	this
b15	9,343	423		5.91	89.598	89.594	16	20
b17	25,610	506		0.00	86.639	86.642	66	95
b22	15,807	720		3.75	96.269	96.276	21	16
s38584	22,450	266	17.68	35.71	99.331	99.331	4	2
s38417	25,588	188	9.97	14.36	99.582	99.588	1	1
s35932	19,879	38	15.12	2.63	94.146	94.146	9	10
netcard	712,763	11,448		41.03	99.668	99.668	19,541	11,844
industry	634,311	2,582		38.77	96.429	96.429	23,419	22,289

REFERENCES

[1] R. Aitken. New defect behavior at 130nm and beyond. In *Proceedings of IEEE European Test Symposium (Emerging Ideas Contribution)*, pages 279–284, 2004.

[2] B. Benware, C. Schuermyer, N. Tamarapalli, and K.-H. Tsai. Impact of multiple-detect test patterns on product quality. In *Proceedings of IEEE International Test Conference*, pages 1031–1040, 2003.

[3] P. Goel. An implicit enumeration algorithm to generate tests for combinational logic circuits. *IEEE Transactions on Computers*, C-30(3):215–222, March 1981.

[4] F. Hapke, R. Krenz-Baath, A. Glowatz, J. Schloeffel, H. Hashempour, S. Eichenberger, C. Hora, and D. Adolfsson. Defect-oriented cell-aware ATPG and fault simulation for industry cell libraries and designs. In *Proceedings of IEEE International Test Conference*, 2009.

[5] J. C. Y. Ku, R. H.-M. Huang, L. Y.-Z. Lin, and C. H.-P. Wen. Suppressing test inflation in shared memory parallel automatic test pattern generation. In *Proceedings of the Asia and South Pacific Design Automation Conference*, pages 664–669, 2014.

[6] I. Pomeranz, L. N. Reddy, and S. M. Reddy. COMPACTEST: A method to generate compact test sets for combinational circuits. *IEEE Transactions on Computer-Aided Design of Integrated Circuits and Systems*, 12(7):1040–1049, July 1993.

A Test Methodology for Neural Computing Unit

Minho Cheong, Ingeol Lee and Sungho Kang
Electrical and electronic engineering
Yonsei University
Seoul, Korea
{cmh9292, keor}@soc.yonsei.ac.kr, and shkang@yonsei.ac.kr

Abstract— **As convolutional neural networks (CNN) has been widely employed in deep learning applications, the accelerator for CNN has been proposed. Neural computing unit (NCU), which is an accelerator for CNN, includes thousands of identical cores named multiplier and accumulate (MAC), so testing NCU with the conventional methods are inefficient. This paper proposes a novel method to test NCU by applying test patterns for a MAC to all MACs in NCU. The experimental results indicate that the new method reduces test time to 1.38% and test data volume to 0.03%.**

Keywords; neural computing unit (NCU); convolutional neural networks (CNN); identical cores; testing;

I. INTRODUCTION

Recently, applications based on deep learning algorithms have rapidly grew in various field such as image recognition, machine translation and speech recognition. Especially, deep convolutional neural networks (CNN) has been widely employed because it can achieve high accuracy. CNN includes specific computation pattern such as matrix multiplication, so general purpose processors are not efficient for CNN. Thus, various accelerators have been proposed to improve the performance of CNN [1].

An accelerator for CNN named neural computing unit (NCU) includes multiple identical cores named multiplier and accumulate (MAC) [2]. CNN includes thousands of identical cores, so the general test methods are not efficient in testing application time and test data volume. Testing a MAC is simple because it is a simple circuit which performs multiplying, adding and storing, but testing NCU is difficult because it includes thousands of MACs.

This paper proposes an efficient test method for NCU to reduce testing application time and test data volume. By applying the test pattern for a MAC to all MACs in NCU, they can be tested, then the test results can be achieved by comparing the test results of MACs.

II. BACKGROUND

Fig. 1 shows NCU which includes $N \times N$ MACs to calculate $N \times N$ matrix multiplication. $A_1 - A_N$ represent input signals of one side and $B_1 - B_N$ represent input signals of the other side in NCU. The structure of NCU is a form of systolic array which means that the MACs in the first row and column which receive signal directly from input signals pass signals to the MACs in the second row and column at next clock and so on. As a result,

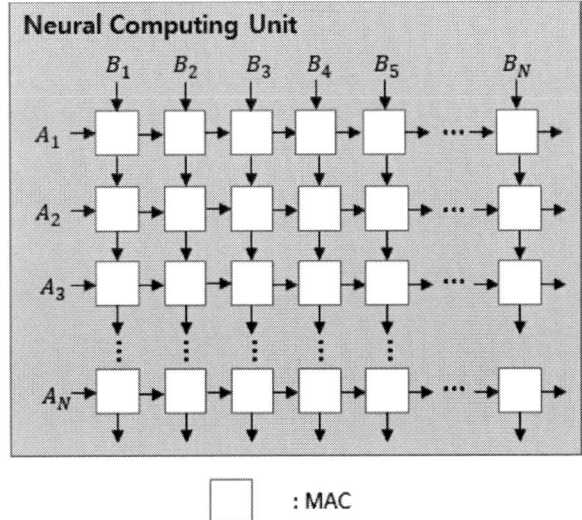

Figure 1. NCU for $N \times N$ matrix multiplication

the input signal can be transmitted to the MACs in south-side and east-side. The input A_i and B_i need to be 0 for i-1 clocks because the input of other side reaches to the MAC which receives the input signal directly from A_i or B_i signal at i-1 clocks, so it takes 3N clocks for NCU to end the calculation.

III. PROPOSED TEST METHOD

Fig. 2 shows proposed test method for NCU which contains 4×4 MACs. In the proposed method, the test pattern for only one MAC is applied to all MACs in NCU. The primary inputs of test patterns are assigned to MACs through $A_1 - A_4$ and $B_1 - B_4$. The test patterns assigned to MAC through $A_2 - A_4$ and $B_2 - B_4$ are delayed as the functional mode of NCU. Also, the scan inputs of test patterns are broadcasted to all MACs and the scan inputs of MACs located in the same dotted diagonal line are assigned simultaneously. As a result, the MACs located in the same diagonal line are tested simultaneously where the MACs in different diagonal line are tested gradually.

The outputs of MACs are compared with two other neighboring MACs in the same diagonal line except for the MACs in the first row, the first column, the last row and the last column. If the outputs from one MAC are different with two other MACs, the MAC can be assumed to be defected. The MACs in first row and first column can compared with one other

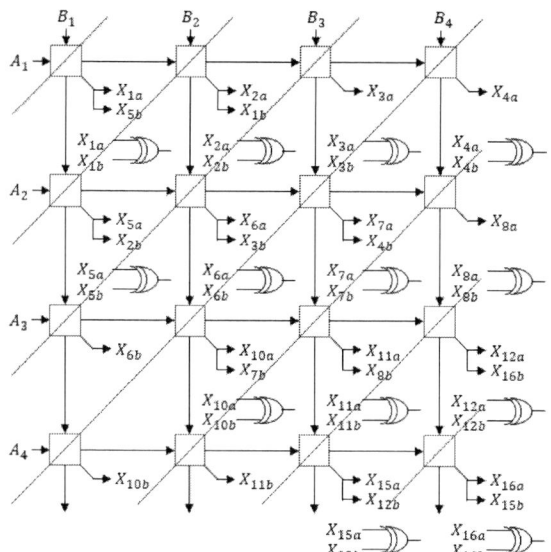

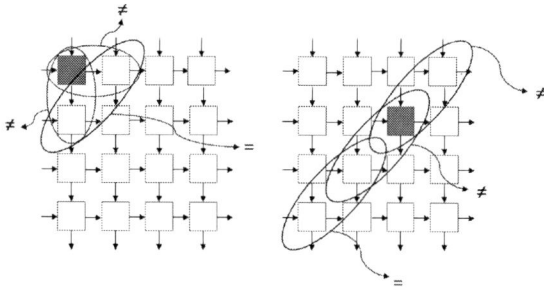

Figure 2. Proposed test structure for NCU

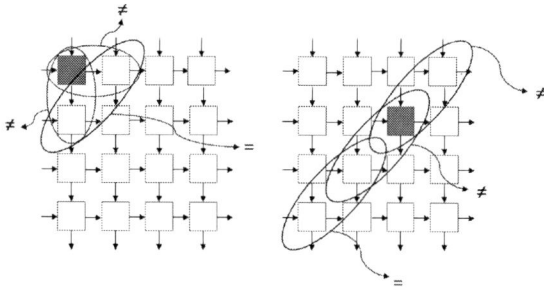

Figure 3. Example of comparison result when a MAC is failed

neighboring MAC in the same diagonal line except for the MAC in the MAC in left upper side and the MAC in right bottom side. The MAC in left upper side is compared with two neighboring MACs in the next diagonal line and the MAC in the right bottom side is compared with the neighboring MACs in the previous diagonal line with timing delay. Those MACs are assumed to be defected if the comparison result between two other MACs are different, too. The other MACs in first row, first column, last row and last column are considered as a defected MAC if the comparison result between itself and neighboring MAC is different, and the comparison result between neighboring MAC and the next MAC is the same.

Fig. 3 shows an example of comparison results when there is a fault in a MAC. When there is a fault in the MAC located at the first row and the first column, the comparison results between neighboring MACs are different while the comparison result with two other MACs is the same. Similarly, when there is a fault in the MAC located not in the first row, the first column, the last row and the last column, the comparison results between two other neighboring MACs are different.

IV. EXPERIMENTAL RESULT

Table I shows test clock and test data volume of the conventional method and the proposed method. The

TABLE I.   TEST CLOCK AND TEST DATA VOLUME OF CONVENTIONAL METHOD SOLUTION AND PROPOSED METHOD

N	Input bit	Conventional method		Proposed method	
		Test clock	*Test data volume*	*Test clock*	*Test data volume*
4	8	4,543	83,074	1,027	3,572
8	8	10,650	410,402	1,035	3,572
16	8	31,510	2,480,322	1,051	3,572
32	8	78,548	12,502,274	1,083	3,572

conventional method assumes that there are $2N - 1$ scan chains in NCU. The test pattern can test stuck-at faults and transition faults. Test clock of the conventional method increases very fast while that of the proposed method increases linearly. This is because the scan chain length of conventional method increases when N increases, while that of proposed method does not increase. Also, the test data volume of the conventional method increases proportional to the square of N while that of the proposed method stays the same. This is because the circuit to be tested of the conventional method becomes larger when N increases, while that of the proposed method is fixed to a MAC. As a result, the test clock of the proposed method is 1.38% of that of conventional method and the test data volume of the proposed method is 0.03% of that of the conventional method.

V. CONCLUSION

This paper proposes a novel test method for NCU. In the proposed technique, test pattern of only one MAC is used by utilizing the characteristic that the input signal of NCU transmitted to MAC without modifying. The test results of MACs are obtained by comparing with two other MACs which output the same value at the same time, or with the neighboring MACs with some timing delays. As a result, the test clock of proposed method achieves only 1.38% clock and the test data volume shows 0.03% data compared with conventional method.

ACKNOWLEDGMENT

This work was supported by Samsung Electronics Company, Ltd., Hwasung, Korea.

REFERENCES

[1] S. Ji, W. Xu, M. Yang, and K. Yu. "3D convolutional neural networks for human action recognition," *IEEE Trans. Pattern Anal. Mach. Intell.*, 35(1):221–231, Jan. 2013.

[2] C. Zhang, P. Li, G. Sun, Y. Guan, B. Xiao, J. Cong, "Optimizing FPGA-based accelerator design for deep convolutional neural networks", *Proc. ACM/SIGDA Int. Symp. Field-Program. Gate Arrays*, pp. 161-170, Feb. 2015.

[3] T. T. Hoang, M. Sjalander and P. L. Edefors, "A High-Speed, Energy-Efficient Two-Cycle Multiply-Accumulate (MAC) Architecture and Its Application to a Double-Throughput MAC Unit," *Circuits and Systems I: Regular Papers, IEEE Transactions on*, vol.57, no.12, pp.3073,3081, Dec. 2010.

An Effective Approach for Building Low-Power General Activity-Driven Clock Trees

Chen-Hsien Lin[†], Shih-Hsu Huang[†], Wei-Kai Cheng[‡]

[†] Department of Electronic Engineering, Chung Yuan Christian University, Taoyuan, Taiwan
[‡] Department of Information & Computer Engineering, Chung Yuan Christian University, Taoyuan, Taiwan
g10576015@cycu.edu.tw, shhuang@cycu.edu.tw, wkcheng@cycu.edu.tw

Abstract—**It is known that clock gating is a useful technique to reduce power consumption. Based on activity patterns of modules, previous works utilized AND gates to construct activity-driven gated clock trees. Recently, it was pointed out OR gates can be used at the bottom level for further power saving. In this paper, we present a general activity-driven clock tree structure in which both AND gate and OR gate can be utilized at any node. Based on this general structure, an effective synthesis algorithm is proposed. Benchmark data show that the proposed approach can reduce 11.3% clock power consumption.**

Keywords—Activity Patterns, Clock Gating, Clock Tree, Low Power, Design Methodology)

I. INTRODUCTION

It is recognized that clock gating [1,2] is one of the most useful techniques to reduce clock power consumption. Based on activity patterns of modules, activity-driven clock trees [3-6] are used to represent gated clock trees. Several works [3-6] have tried to construct low-power activity-driven clock trees by using only AND gates to merge nodes with similar activity patterns to disable the clock signal as possible. Different from previous works [3-6] that only used AND gates, Lin et al. [7] utilized OR gates at the bottom level for further power saving. However, they [7] did not use OR gates at other levels of gated clock tree. Furthermore, they [7] restrict that the gates at the same level must use in the same logical type. In this paper, we present a general activity-driven clock tree structure in which both AND gate and OR gate can be utilized at any node of gated clock tree. Based on this general structure, we propose an effective synthesis algorithm for building a low-power activity-driven clock tree. Since the trade-off between AND gates and OR gates are made during the stage of activity-driven clock tree construction, clock power consumption can be greatly minimized. Compared with previous work [7], benchmark data show that our approach can reduce 11.3% clock power consumption.

II. MOTIVATION

In this section, we use an example to demonstrate the motivation of general activity-driven clock tree structure. This circuit has fifteen modules M1 ~ M15. The activity patterns of modules M1, M2, M3, M4, M5, M6, M7, M8, M9, M10, M11, M12, M13, M14 and M15 are 0111, 0110, 0100, 1000, 1100, 1110, 1111, 1010, 0010, 1011, 0011, 0101, 1101, 0001 and

1001, respectively. In other words, module M1 is idle at cycle 1, active at cycle 2, active at cycle 3, and active at cycle 4; module M2 is idle at cycle 1, active at cycle 2, active at cycle 3, and idle at cycle 4; and so on.

Fig. 1 gives the activity-driven clock tree built by the previous work [3], which only use AND gates. As shown in Fig. 1, the total active cycles at level 1 (i.e., the bottom level), level 2, level 3, level 4, and level 5 are 32, 21, 13, 8, and 4, respectively. The summation of active cycles of all the clock gates is 78. Suppose that the power consumption of each clock gate at an active cycle is 10 µW. Then, clock power consumption is 780 µW.

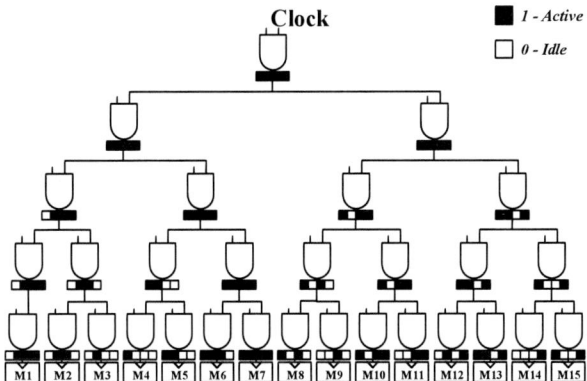

Fig. 1. Clock tree obtained by previous work [3].

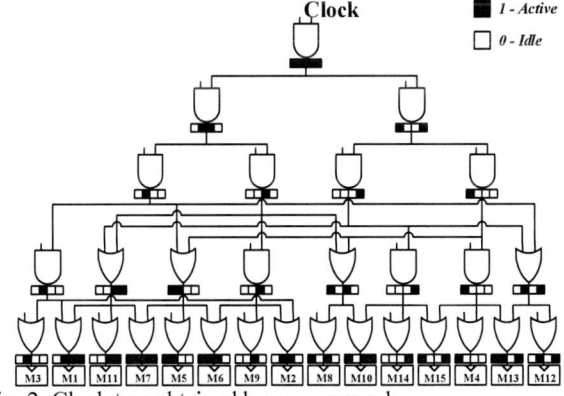

Fig. 2. Clock tree obtained by our approach.

Fig. 2 gives the activity-driven clock tree constructed by

our approach (i.e., both AND gates and OR gates can be considered to utilized at each node). As shown in Fig. 2, the total active cycles at level 1, level 2, level 3, level 4, and level 5 are 32, 12, 4, 4, and 4, respectively. The total active cycles of all the clock gates is 56. Then, clock power consumption becomes only 560 μW.

III. PROPOSED APPROACH

In this section, we propose an approach to synthesize a low-power general activity-driven clock tree (i.e., both AND gate and OR gate can be utilized at any node of gated clock tree). Based on activity patterns of modules, the gated clock tree is constructed in the bottom-up manner. Here, due to page limit, the detailed pseudo codes are omitted. In the following, we highlight the basic ideas.

For each level of gated clock tree, we try to apply two strategies. In the first strategy, the activity patterns of parents are directly propagated from their children. We select the children whose total active cycles is 50% of the total cycles. Take Fig. 3 as an example. In Fig. 3(a), children B_3 and B_4 are selected since their total active cycles is 50% of the total cycle. Then, the activity patterns of B_9 and B_{10} are directly propagated from B_3 and B_4, respectively. Moreover, we can use the AND-ing of parent B_9 to generate the activity pattern of child B_2.

If the first strategy is not applicable for some children, we use the second strategy. In Fiug. 3, the activity patterns of B_1, B_5, B_6, B_7 and B_8 cannot be generated by only B_3 and B_4. In Fig. 3(b), we add parent B_{11} for generating activity patterns of B_1, B_5, and B_8. Similarly, in Fig. 3(c), we add B12 for generating activity patterns of B_6 and B_7. Note that the activity patterns of B_{11} and B_{12} are determined by a greey (heuristic) algorithm.

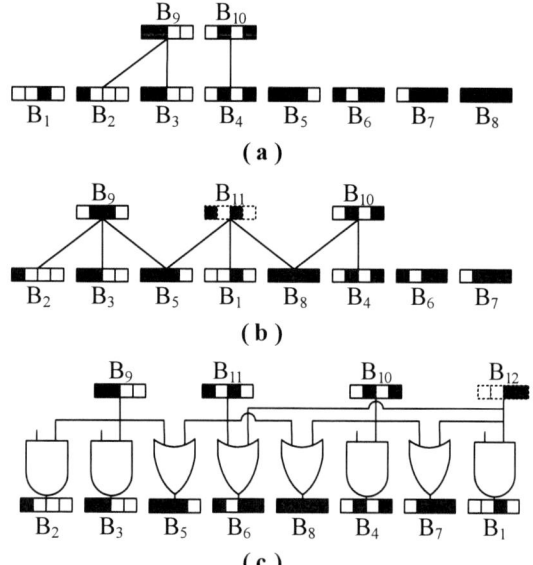

Fig. 3. Snapshots of the second strategy.

IV. EXPERIMENTAL RESULTS

In the experiments, we adopt seven circuits from ISCAS'89 benchmark suite to test the effectiveness of the proposed approach. These test circuits are targeted to TSMC 90nm process technology. In each test circuit, we implement our approach, the previous work [3] (i.e., AND-gate-only clock tree), the previous work [7] (i.e., OR gates at the bottom level) for comparisons. For each activity-driven clock tree, we set 0.1 ns as the clock skew constraint. Table I tabulates the comparisons on their clock power consumptions. We find that, in each test circuit, our approach has the smallest clock power consumption. Compared with previous work [3], in average, the proposed approach can reduce 29.2% clock power consumption. Compared with previous work [7], in average, the proposed approach can reduce 11.3% clock power consumption.

TABLE I. COMPARISONS ON CLOCK POWER CONSUMPTION.

Circuit	Power Consumption (nW)			Reduction	
	[3]	[7]	Ours	Ours vs [3]	Ours vs [7]
s208.1	178.6	170.1	150.3	15.8%	11.6%
s349	387.0	278.9	252.7	34.7%	9.4%
s400	570.7	461.2	406.6	28.7%	11.8%
s1269	893.2	653.5	584.0	34.6%	10.6%
s3271	2560.4	1992.9	1754.7	31.5%	12.0%
s6669	5319.4	4222.8	3721.6	3.00%	11.9%
s15850	12866.7	10398.8	9143.3	29.0%	12.1%
Average				29.2%	11.3%

V. CONCLUSIONS

In this paper, we present a novel approach to construct a general low-power activity-driven clock tree, in which both AND gate and OR gate can be utilized at any node of gated clock tree. Compared with the conventional activity-driven clock tree, our approach can reduce 29.2% clock power consumption. Compared with the activity-driven clock tree that uses OR gates at the bottom level, our approach also can reduce 11.3% clock power consumption.

ACKNOWLEDGMENT

This work was supported in part by Ministry of Science and Technology, Taiwan, under grant number 107-2218-E-033-007.

REFERENCES

[1] J. S, M. Rao, et al., "Clock Gating for Power Optimization in ASIC Design Cycle: Theory & Practice", Proc. of ISLPED, 2008.

[2] J. Shinde and S. S. Salankar, "Clock gating - A Power Optimizing Technique for VLSI Circuits", Proc. of Annual IEEE India Conference, 2011.

[3] A. Farrahi, et al., "Activity Driven Clock Design", IEEE TCAD, 2001.

[4] W. Shen, et al., "An Effective Gated Clock Tree Design Based on Activity and Register Aware Placement", IEEE Trans. on VLSI Systems, 2010.

[5] L. Li, et al., "Activity-Driven Fine-Grained Clock Gating and Run Time Power Gating Integration", IEEE Trans. on VLSI Systems, 2013.

[6] S.H. Huang, et al., "Low-Power Anti-Aging Zero Skew Clock Gating", ACM TODAES, Vo. 18, No. 2, Article 27, 2013.

[7] C.H. Lin, S.H. Huang, et al., "New Activity-Driven Clock Tree Design Methodology for Low Power Clock Gating", Proc. of ISNE, 2017.

Fast Steady-State Thermal Analysis

Lih-Yih Chiou[1], Chun-Hao Chang[1], Liang-Ying Lu[1,2], Wei-Hsuan Yang[1], Yeong-Jar Chang[2] and Juin-Ming Lu[2]

[1]Department of Electrical Engineering, National Cheng Kung University, Tainan, Taiwan, R.O.C.

[2]Information and Communications Research Laboratories, Industrial Technology Research Institute, Hsinchu, Taiwan

{lihyih, n26064896, n2897132, n26044927}@mail.ncku.edu.tw , {ot, jm}@itri.org.tw

Abstract—**Thermal issues are gaining critical in system-on-a-chip. We propose a two-stage thermal analysis approach using big and little matrix operations with a matrix compression method in general successive over-relaxation (SOR) to speed up the analysis of a steady-state thermal profile, referred to as TSA. Experimental results show that the analysis performance of TSA is faster than the general SOR by 2145× in average and faster than the popular Supernodal LU (SuperLU) at least by 1.55× for large size of sparse matrices.**

Keywords- successive over-relaxation approach (SOR), steady-state thermal analysis, Supernodal LU (SuperLU)

I. INTRODUCTION

With the increase of the number of processors in multiprocessor systems and the advancement of stacking process technologies, power density of system-on-a-chip (SoC) is raise significantly. However, high power density may cause serious thermal effects and thus heat issues of chips cannot be ignored [1]. Chip/system designers require the thermal analysis to predict temperature distribution of chips. The mesh-based steady-state thermal analysis can fast predict the distribution of hot spots in a chip when compared with the mesh-base transient-state thermal analysis at the early design stages.

For the mesh-based steady-state thermal analysis, the heat transfer equation [2] can be transferred as follows:

$$GT = P . \qquad (1)$$

where G represents the thermal conductance, T is the temperature, P is the power consumption. G is an $m \times m$ matrix where m is the number of cells in a mesh; T and P are $m \times 1$ vectors. Then, G and P are given, but T is unknown.

T in (1) can be solved by two kinds of approaches. One is to solve directly, and the other is to solve iteratively.

Direct approach: The method can obtain an exact solution by finite number of operations, such as Gaussian elimination method, Cholesky decomposition method, lower-upper (LU) decomposition method [3], or Supernodal LU (SuperLU) [4].

Iterative approach: The method uses an initial solution to obtain final approximate solution by iterative operations, like Gauss-Seidel method, Jacobi method, or successive over relaxation (SOR) method [5].

Two popular thermal simulators by academia are introduced as follows. Skadron *et al.* [6] proposed a thermal analysis simulator called HotSpot for analyzing architecture-level chip temperature. And, SuperLU was used in the steady-state thermal analysis. Sridhar *et al.* [2] presented a compact transient thermal model (CTTM) for the thermal simulation of three-dimensional (3D) ICs considering multiple inter-tier microchannel liquid, called 3D-ICE. 3D-ICE adapted the SuperLU method for the steady-state thermal analysis.

SuperLU is commonly used in several popular thermal simulators. SuperLU improves LU decomposition by using the matrix compression method with compressed sparse row (CSR) format [7]. Although SuperLU is fast in the steady-state thermal analysis, the proposed two-stage analysis approach using big and little matrix operations with matrix compression method based on CSR format in SOR can even run faster.

II. TWO-STAGE ANALYSIS METHOD

To calculate the temperatures of a $2^n \times 2^n$ mesh in a chip, the two-stage analysis method (called TSA) using a iterative approach is proposed as shown in Fig. 1 and depicted as follows. In the first stage (Stage 1), we use a $2^{n-2} \times 2^{n-2}$ mesh to calculate the steady-state temperature, and thus there are $2^{n-2} \times 2^{n-2}$ cells in the mesh. Then, each cell includes a temperature value. Here, n must be large than 2. To calculate the temperature T of the $2^{n-2} \times 2^{n-2}$ cells, we can use a little matrix operation in $(2^{n-2} \times 2^{n-2}) \times (2^{n-2} \times 2^{n-2})$ matrix G according to (1). The calculated temperature value as a red point shown in Fig. 1(a). In the second stage (Stage 2), a big matrix operation in $(2^n \times 2^n) \times (2^n \times 2^n)$ matrix G is utilized to calculate the target steady-state temperature. The steady-state temperature value per cell in the $2^{n-2} \times 2^{n-2}$ mesh is assigned to the initial temperature values of the 16 cells in the $2^n \times 2^n$ mesh as red block shown in Fig. 1(b). Therefore, the initial temperature values of the 16 cells are set to the same value.

For example, Fig. 2(a) shows an 8×8 ($2^3 \times 2^3$) mesh in the floorplan of a chip. According to TSA, we can use a 4×4 little matrix operation in (1) to obtain the temperatures T of 4 cells in the 2×2 mesh in the first stage as shown in Fig. 2(b). In the second stage, the temperature value per cell as red block shown in Fig. 2(b) is assigned to the initial temperature values in 16 cells as the red block shown in Fig. 2(a). Then, a 64×64 big matrix operation in (1) is used to obtain the temperatures T of 64 cells in the 8×8 mesh. By two-stage analysis, TSA using matrix compression method can effectively reduce the computational complexity than general iterative approaches.

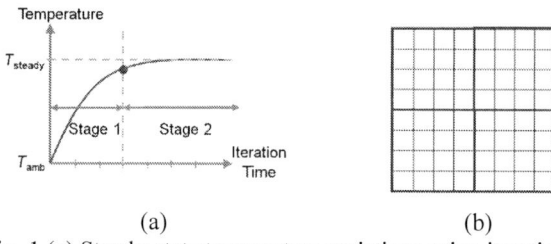

(a)   (b)

Fig. 1 (a) Steady-state temperature variations using iterative approach in TSA. (b) Mesh size for TSA

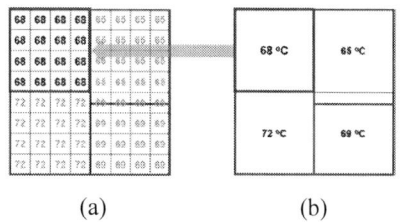

(a)   (b)

Fig. 2 Floorplan of a chip in (a) 8 × 8 mesh and (b) 2 × 2 mesh

III. EXPERIMENTAL RESULTS

The experiments are conducted on a 64-bit Linux workstation with an Intel Xeon E5-2640 2GHz CPU and 98GB memory.

We compare the analysis performance of LU decomposition, SuperLU (Version 5.2.1), SOR and TSA using four cases. In addition, the error is compared between SurperLU and TSA. TABLE I shows the comparison results between the methods for four cases.

In the first comparison, LU decomposition and SuperLU are chosen. As the size of matrix increases, the difference of the time cost increases significantly. When the matrix size is 65536 × 65536, SuperLU is faster 105938.67× when compared with LU decomposition. The result is expected as mentioned due to using matrix compression method. The sign "—" in TABLE I means that the memory is not enough to store the sparse matrix.

In the next comparison, SOR and TSA are compared. The result shows similar trend like those in the first comparison. When the matrix size is 65536 × 65536, TSA is faster 5454.71× when compared with SOR, which demonstrates that TSA effectively improves the analysis performance of SOR.

In the last comparison, we compare SuperLU and TSA since SuperLU and TSA take less time in the previous comparisons. The results indicate that TSA is faster than SuperLU when the size of matrix is more than 16384 × 16384. In addition, the maximum error of TSA is 6.4×10^{-3} % when compared with SuperLU in case 4.

IV. CONCULSION

We proposed a two-stage analysis (TSA) method for steady-state thermal analysis to accelerate analysis performance. The experimental results indicated that TSA outperformed the popular SuperLU for large size of meshes.

ACKNOWLEDGMENT

The authors would like to thank the Taiwan Ministry of Science and Technology for research grant (MOST 107-2634-F-006-002 -).

REFERENCES

[1] Assem Gupta, Sudeep Pasricha, Nikil Dutt, Fadi Kurdahi, Kamal Khouri and Magdy Abadir, "On Chip Communication-Architecture Based Thermal Management for SoCs," in *Proc. International Symposium on VLSI Design, Automation and Test*, 2009, pp. 76-79.

[2] Arvind Sridhar, Alessandro Vincenzi, David Atienza, and Thomas Brunschwiler, "3D-ICE: A Compact Thermal Model for Early-Stage Design of Liquid-Cooled ICs," *IEEE Transactions on Computers*, vol. 63, no. 10, pp. 2576-2589, Oct. 2014.

[3] Juliet N. Gaithuru, Mazleena Salleh, and Ismail Mohamad, "NTRU Inverse Polynomial Algorithm Based on the LU Decomposition Method of Matrix Inversion," in *Proc. IEEE Conference on Application, Information and Network Security*, 2017, pp. 1–6.

[4] James W. Demmel, Stanley C. Eisenstat, John R. Gilbert, Xiaoye S. Li, and Joseph W. H. Liu, "A supernodal approach to sparse partial pivoting," *SIAM Journal on Matrix Analysis and Applications*, vol. 20, no. 3, pp. 720–755, July 2006.

[5] Noreen Jamil, Johannes Muller, Christof Lutteroth, and Gerald Weber, "Speeding up SOR Solvers for Constraint-Based GUIs with a Warm-Start Strategy," in *Proc. International Conference on Digital Information Management*, 2013, pp. 268–273.

[6] Kevin Skadron, Mircea R. Stan, Wei Huang, Sivakumar Velusamy, Karthik Sankaranarayanan, and David Tarjan, "Temperature-Aware Microarchitecture," in *Proc. International Symposium on Computer Architecture*, 2003, pp. 2-13.

[7] Nathan Bell and Michael Garland, "Implementing Sparse Matrix-Vector Multiplication on Throughput-Oriented Processors," in *Proceedings of the Conference on High Performance Computing Networking, Storage and Analysis*, November 14 - 20, 2009, Article No. 18.

TABLE I.   ANALYSIS PERFORMANCE AND ERROR COMPARISON

Case	Sparse Matrix Size (#Cells in a mesh for X, Y, and Z)	Number / Ratio (%) for Nonzero Items in Sparse Matrix	Times (s)				Error (%)	Speedup (×)			
			LU [3]	SOR [5]	SuperLU [4]	TSA	TSA vs. SuperLU	TSA vs. LU	TSA vs. SOR	TSA vs. SuperLU
1	4096 × 4096 (64 × 64 × 1)	28672 / 0.17	68.4	24.12	0.12	0.16	4.0×10^{-5}	427.5	150.75	0.75
2	16384 × 16384 (128 × 128 × 1)	114688 / 0.0427	3957.45	365.3	0.68	0.44	7.3×10^{-4}	8994.20	830.23	1.55
3	65536 × 65536 (256 × 256 × 1)	458752 / 0.01068	289212.57	7691.14	2.73	1.41	5.1×10^{-3}	205115.30	5454.71	1.94
4	262144 × 262144 (512 × 512 × 1)	1835008 / 0.00267	—	—	14.02	7.26	6.4×10^{-3}	—	—	1.93

A Rule-of-thumb Condition to Avoid Large HRS Current in ReRAM Crossbar Array Design

Yelim Youn
Center of System Integrated Chip
LG Electronics
Seoul, Korea
lim9538@gmail.com

Kwangmin Kim and Byungsub Kim
Department of Electronics Engineering
POSTECH
Pohang, Korea
byungsub@postech.ac.kr

Abstract— **This paper presents a rule-of-thumb condition on the size of an ReRAM crossbar array to avoid large HRS current in read operation. Although HRS current must be small for large read margin, the worst HRS current can be too large significantly reducing the read margin if the array size is too large. According to our analysis, the worst HRS current starts steeply increasing as the array size increases beyond a certain size limit. We derived an approximate condition on the size limit in terms of design parameters to avoid large HRS current. The derived formula is verified with SPICE simulation demonstrating that engineers can nicely estimate the maximum array size without causing large HRS current.**

Keywords; resistive memory; crossbar array; reverse leakage current; read current; HRS current; read margin

I. INTRODUCTION

A crossbar array of resistive random access memory (ReRAM) is a promising candidate for the next generation nonvolatile memory solution [1]. An ReRAM crossbar array consists of two sets of parallel electrode lines called wordline (WL) and bitline (BL). At all crossing points, hysteretic switching material is sandwiched between WLs and BLs [1]. The switching material can indicate one bit with two different resistance states called low resistance state (LRS) and high resistance state (HRS). Because of its simple structure, the ReRAM crossbar array is a strong candidate of the next generation low-cost and high density memory [2].

To utilize an ReRAM crossbar in memory application, a designer should guarantee large read margin. When a cell is selected for read operation, read voltage V_{read} is typically applied to the selected WL while the other lines are grounded (Fig. 1). The sense amplifier at the end of the selected BL determines the state of the selected cell from *the read current* I_{read}. The read margin is the difference between the minimum *LRS read current* $I_{readLRS}$ (I_{read} of an LRS cell) and the maximum *HRS read current* $I_{readHRS}$ (I_{read} of a HRS cell), and must be large for reliable read operation.

One of the mechanisms which reduce the read margin is reverse leakage current [3]. A large array is preferred for high memory density. However, if a designer increases the array size too much, *a reverse leakage current* $I_{reverse}$ can flow through the selected bit line raising $I_{readHRS}$ [3]. This mechanism may cause large $I_{readHRS}$ that significantly reduces the read margin if design parameters are poorly chosen [3].

This work was supported in part by POSTECH-Samsung Electronics Future Persistent Memory Cluster Research Project and also in part by the National Research Foundation of Korean (NRF) grant funded by the Korean government (MSIP) (No. 2018R1A2A2A16022248 and No. 2018R1A4A1025679).

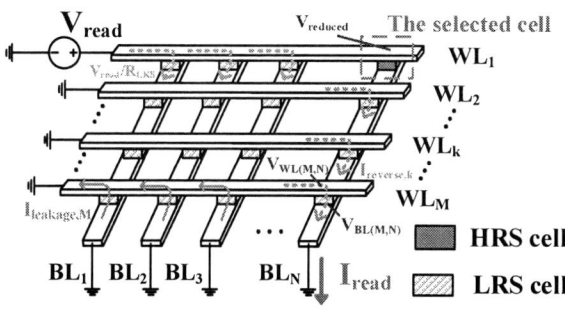

Figure. 1. Read operation of a resistive memory crossbar array in ground scheme [3]. M is the number of WLs and N is the number of BLs.

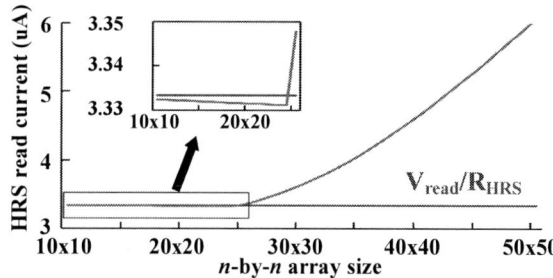

Figure. 2. The HRS read current in the worst-case scenario versus the size n of an n-by-n crossbar array.

Youn suggested and verified a rule-of-thumb guideline on how to avoid this $I_{reverse}$ [3].

In this paper, we propose a rule-of-thumb guideline to avoid large $I_{readHRS}$ of an ReRAM crossbar array. By extending the previous analysis on $I_{reverse}$ [3], we analyze the dependency of $I_{readHRS}$ on design parameters (*line resistance R_{line}, LRS resistance R_{LRS}, HRS resistance R_{HRS}, and an array size*). According to our analysis, the worst $I_{readHRS}$ starts steeply increasing as the array size increases beyond the range given by the guideline. We derive an approximate simple formula of the condition to avoid this steep ramp of the worst $I_{readHRS}$.

II. THE DEPENDENCY OF THE WORST HRS READ CURRENT ON THE ARRAY SIZE

For the analysis of the worst $I_{readHRS}$, the worst-case scenario of HRS reading in [4] (the HRS cell at the nearest

978-1-5386-7961-6/18 $31.00 © 2018 IEEE

from the driving sources and the sense amplifiers is selected and all unselected cells are in LRS) is used (Fig. 1).

The worst $I_{readHRS}$ slightly decreases with the increasing array size if the array size is small, but the worst $I_{readHRS}$ starts steeply increasing as the array size increases beyond a certain size limit (Fig. 2). The HRS read current is reduced by R_{line} of the selected WL and raised by $I_{reverse}$ [3]. As the array size increases, $I_{reverse}$ becomes dominant raising the worst $I_{readHRS}$ [3]. Therefore, the size limit can be approximated as the size where the effects of these two factors on the read current are balanced.

III. RULE-OF-THUMB GUIDELINE TO AVOID LARGE HRS CURRENT

In this section, based on the previous analysis [3], we suggest a rule-of-thumb guideline on the design parameters to avoid large $I_{readHRS}$ in an ReRAM crossbar design.

The voltage drop along the selected WL is caused by the large current flowing into the unselected BLs (Fig. 1). When reading an HRS cell, the largest current flowing into each unselected BL from the selected WL is approximately V_{read}/R_{LRS}. Because these currents are aggregated and flowing through the WL, the voltage drop across of kth R_{line} is approximately $(N-k)R_{line}V_{read}/R_{LRS}$ for M-by-N array size. By adding the voltage drops along the selected WL for $k=1,...,N-1$, the reduced voltage $V_{reduced}$ (Fig. 1) of the top electrode of the selected HRS cell can be approximately derived as

$$V_{reduce} = V_{read} - \frac{V_{read}}{R_{LRS}} \frac{N(N-1)R_{line}}{2}. \quad (1)$$

Therefore, the amount of the current reduction $I_{reduced}$ by the line resistance of the selected WL can be approximated as

$$I_{reduce} = \frac{V_{read}}{R_{HRS}} - \frac{V_{reduced}}{R_{HRS}} = \frac{V_{read}}{R_{HRS}R_{LRS}} \frac{N(N-1)R_{line}}{2}. \quad (2)$$

The reverse leakage current $I_{reverse,k}$ flowing from the unselected kth WL into the selected Nth BL (Fig. 1) equals $\{V_{WL(k,N)} - V_{BL(k,N)}\}/R_{LRS}$ [3]. Therefore, using equations of $V_{WL(k,N)}$ and $V_{BL(k,N)}$ in [3], we can derive $I_{reverse,k}$ as

$$I_{reverse,k} = (M-k+1)\left\{\frac{V_{read}}{R_{LRS}^3}\frac{N(N-1)R_{line}^2}{2} - \frac{V_{read}R_{line}}{R_{LRS}R_{HRS}}\right\}. \quad (3)$$

The total amount of the reverse leakage current $I_{reverse,tot}$ flowing into the selected BL from M-1 unselected WLs can be calculated as the summation of $I_{reverse,k}$ where k = 2, 3, 4, ..., M:

$$I_{reverse,tot} = \frac{V_{read}R_{line}^2}{R_{LRS}^3}\frac{N(N-1)M(M-1)}{4} - \frac{V_{read}R_{line}}{R_{LRS}R_{HRS}}\frac{M(M-1)}{2}. \quad (4)$$

The HRS read current starts increasing beyond its nominal value V_{read}/R_{HRS} as the array size increases if $I_{reverse,tot}$ becomes larger than $I_{reduced}$. By comparing (2) and (4), we can derive the approximate condition to avoid the steep increase of HRS read current in a simple formula as

$$\frac{M(M-1)N(N-1)}{M(M-1)+N(N-1)} < \frac{2R_{LRS}^2}{R_{HRS}R_{line}}. \quad (5)$$

Equation (5) implies that small array size, large R_{LRS}, small R_{HRS}, and small R_{line} are preferred in ReRAM crossbar design

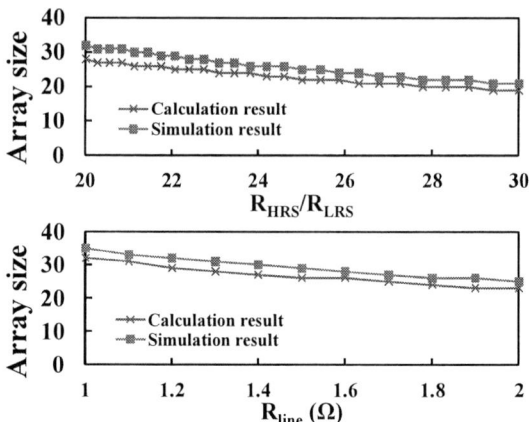

Figure. 3. The calculations and simulations of the size n of an n-by-n ReRAM crossbar array where the worst HRS read current starts increasing beyond its nominal value versus on-off ratio (R_{HRS}/R_{LRS}) and R_{line}.

to avoid large $I_{readHRS}$.

To verify the accuracy of (5), we calculated the array size where the worst $I_{readHRS}$ starts increasing beyond its nominal value and compared the result with the SPICE simulation. Fig. 3 shows plots of calculated and simulated size n of an n-by-n ReRAM crossbar versus on-off ratio (R_{HRS}/R_{LRS}) and R_{line}. The comparison results show that our formula accurately estimates the size limit to avoid large $I_{readHRS}$.

IV. CONCLUSION

This paper proposes a rule-of-thumb guideline on the design parameters to avoid large HRS read current in ReRAM crossbar design. We approximately derive the formula of the condition to avoid the steep increase of the HRS read current in terms of the design parameters. The derived formula is verified with SPICE simulation, and the results showed that our formula accurately estimates the array size beyond which the worst HRS read current starts steeply increasing as the array size increases. This formula can be a useful rule-of-thumb guideline in ReRAM crossbar design to avoid large HRS read current.

ACKNOWLEDGMENT

The author would like to thank IC Design Education Center for CAD tool support.

REFERENCES

[1] A. Flocke and T. G. Noll, "Fundamental analysis of resistive nanocrossbars for the use in hybrid nano/CMOS-memory," in *IEEE European Solid-State Circuits Conference*, Jan., 2007, pp. 328-331.

[2] H. Akinaga, and H. Shima, "Resistive random access memory (ReRAM) based on metal oxides," Proc. *IEEE*, vol. 98. No. 12, pp. 2237-2251, Dec. 2010.

[3] Y. Youn, J.-Y. Sim, H.-J. Park, and B. Kim, "An approximate condition to avoid the reverse leakage current in ReRAM crossbar design," *IEEE international SOC design conference*, 2015.

[4] Y. Deng, et al. "RRAM crossbar array with cell selection device: A device and circuit interation study." *IEEE transactions on Electron Devices*, vol. 60, no. 2, pp. 719-726, Feb. 2013.

Spin Orbit Torque-RAM Write Energy Reduction with Self-Verification Scheme

Taehwan Kim and Jongsun Park

School of Electrical Engineering, Korea University.

Abstract— As a promising candidate for replacing CMOS-based memories, non-volatile magnetic memory has been on a rise. While Spin transfer torque random access memory (STT-RAM) is considered as most promising candidate, it still suffers from various shortcomings concerning write operation. As a result, spin orbit torque random access memory (SOT-RAM) is considered as next generation non-volatile magnetic memory, for it offers relatively better performance and lower power. Even though SOT-RAM shows various advantages over STT-RAM, to meet the power level of CMOS-based memories, significant reduction of write power is highly required. Therefore, in this paper, we propose novel technique for reducing write power of SOT-RAM with redundant write prevention and early write termination. For application of two techniques, self-verification scheme is exploited. Simulation results using 65nm CMOS technology show that up to 69.5% of write energy can be saved compared to the conventional write operation.

Keywords; spin-orbit torque(SOT); write power reduction;

I. INTRODUCTION

As the amount of data usage in modern world has increased exponentially in the last decades, the demand for memories has grown explosively as well. Furthermore, for better memory performance, CMOS has been excessively downscaled, resulting in leakage power within the memory. As a result, modern electronic systems exploiting CMOS-based memory suffers severely from excessive standby power. As a solution, non-volatile magnetic memories have been widely researched due to its numerous advantages complementing CMOS-based memories. Among various non-volatile magnetic memories, spin transfer torque random access memory (STT-RAM) is considered as the most promising candidate. However, STT-RAM shows poor performance and efficiency in write operation, hindering it from becoming viable substitute for CMOS-based memories. Alleviating the drawbacks of STT-RAM, spin orbit torque random access memory (SOT-RAM) is considered as the next generation non-volatile magnetic memory. By exploiting spin-hall-effect (SHE), SOT-RAM outperforms STT-RAM in aspect of performance and efficiency. However, in comparison to the CMOS-based memories, SOT-RAM still has its limitation. Therefore, variety of techniques for improving SOT-RAM is being widely researched [1]. As a part of improving efficiency of the SOT-RAM write operation, in this paper we propose novel approach for reducing write energy of SOT-RAM. The rest of paper is organized as follows. Section II presents basics of SOT-RAM. In Section III, the proposed energy reduction technique will be demonstrated. Section IV shows numerical and simulation results, and the conclusions are presented in section V.

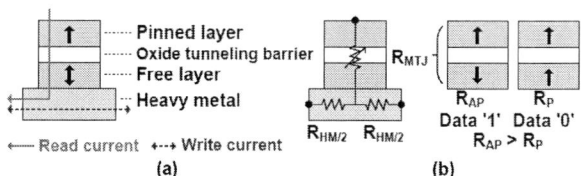

Figure 1. (a) SOT-RAM device. (b) Equivalent resistor model.

II. SPIN ORBIT TORQUE RANDOM ACCESS MEMORY

SOT-RAM [2] is 3-terminal device and is composed of magnetic tunnel junction (MTJ) and heavy metal (HM) as shown in Figure 1 (a). MTJ is composed of two ferromagnetic layers; the pinned layer and the free layer, which are separated by an oxide tunneling barrier. The magnetization of pinned layer cannot be changed while magnetization of free layer can be changed. The write operation of the SOT-RAM, often referred as switching operation, is done by flowing current through HM to change the magnetization of the free layer. As shown in Figure 1 (b), if the magnetization of free layer is parallel to that of pinned layer's, the resistance of MTJ is small. On the other hand, if the magnetization of the free layer is anti-parallel of pinned layer's, the resistance of MTJ is large. By exploiting this characteristic, the data is stored inside of MTJ in form of resistance. Therefore, the read operation is done by flowing current through MTJ and sensing the resistance value it contains. The unique nature of SOT-RAM is that the write operation is done stochastically [3]. The Probability of switching operation is proportional to the magnitude and duration of the write current flowing through HM.

III. PROPOSED WRITE ENERGY REDUCTION TECHNIQUE

In this section, a novel approach for reducing write energy consumption of SOT-RAM will be presented. Figure 2 demonstrates schematic of SOT-RAM array with verification circuit for write energy reduction technique and its timing diagram [4]. To reduce the energy consumed in write operation, we focused on two major cause of energy waste; stochastic nature of the SOT-RAM and redundant write operation. The stochastic nature of SOT-RAM challenges efficient write operation. While in most of the switching activity occurs in much earlier stage of write cycle, because of the stochastic nature, some switching operation occurs in the latter stage of the cycle. Therefore, for reliable operation, current should be applied for entire write cycle, which causes huge waste of energy. Furthermore, redundant write causes energy waste since the energy is wasted with no actual result.

978-1-5386-7961-6/18 $31.00 © 2018 IEEE

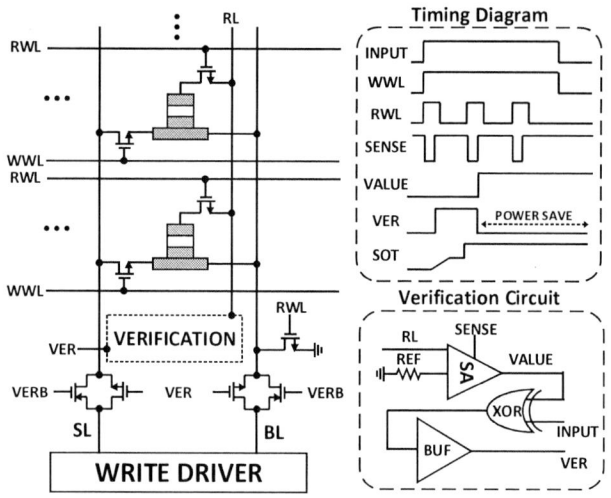

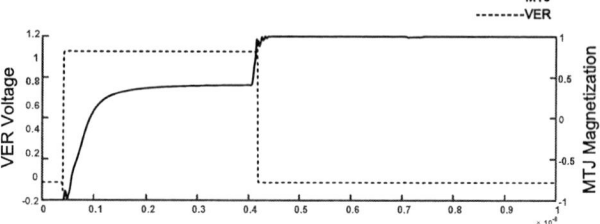

Figure 3. Simulation waveform of write operation

TABLE I. ENERGY COMPARISON OF WRITE OPERATION

	Energy Reduction Techniques		
	Conventional	Early Write Termination	Redundant Write Prevention
Energy (fJ)	167.325	61.464	38.8303

Figure 2. Schematic of SOT-RAM array with proposed technqiue and its timing diagram.

With consideration of these issues concerning write operation of SOT-RAM, our first technique starts by monitoring the data stored in SOT-RAM when new data is fed to prevent redundant write operation. The verification circuit decides whether the write operation should be done or not by comparing the input data and the sensed data. If the input is identical to the stored data, the write operation is considered as redundant write operation and the operation will not be done. Only if the input is different from the stored data the write operation is performed. To explain in detail, XOR gate in verification circuit will generate logic output 1 to shut down the transmission gate connected to the BL and SL, preventing unnecessary current from flowing. If the read data is different from the input data, the XOR gate in verification will result logic value 0, enabling transmission gate to begin write operation.

Our second technique is employed in the middle of the write operation. Since switching operation of SOT-RAM is stochastic, it can be done in any time of the write cycle. Therefore, in the middle of the write operation, we sensed the data stored in SOT-RAM to monitor if the switching operation has been done. If not, the sensed data will be different from input data and verification circuit will enable the transmission gate, starting the write operation again. On the other hand, if the switching has been done, the sensed data and the input data will be identical. Then, verification circuit will shut down transmission gate, saving energy for the rest of the cycle. This self-verification for early write termination is done twice in the middle of the write cycle.

For reliable operation, transmission gate connected to the BL and SL are shut down during read operation, since line driven to VDD taints the sensing operation. To provide discharge path of the circuit during read operation, 1 transistor controlled by RWL is used. Amount of transistor overhead used to employ proposed technique is negligible since they are added on the peripheral circuits for entire column.

IV. SIMULATION RESULTS

The proposed write energy reduction approach has been implemented using 65nm CMOS technology. The SOT-RAM device model is based on [5], with modification to include giant SHE. Figure 3 shows simulation waveform with proposed technique. Table I shows the energy measurement result, with write cycle set to 10ns. In our early termination technique, wbe sensed the switching activity at 4ns and 7ns. With only early write termination, 62.7% of the energy can be saved on average. 76.8% of the energy was reduced with only redundant write prevention technique. With both techniques applied, 69.5% of the energy can be saved assuming 50% probability of redundant write operation.

V. CONCLUSION

In this paper, we propose a low power write technique for the SOT-RAM. By sensing the switching operation in the middle of the write operation and sorting out redundant write operation, write energy is significantly reduced. Our proposed approach achieves about 69.5% of energy reduction, with negligible amount of transistor overhead.

REFERENCES

[1] G. Kang, Y. Jang, and J. Park, "Charge-Recycling based Redundant Write Prevention Technique for Low Power SOT-MRAM," *2018 IEEE Int. Symp. Circuits Syst.*, vol. 2, pp. 1–4, 2018.

[2] R. A. Buhrman, "Spin-Torque Switching with the Giant Spin Hall Effect of Tantalum," *Science (80-.).*, vol. 336, no. May, p. 555, 2012.

[3] K. S. Lee, S. W. Lee, B. C. Min, and K. J. Lee, "Thermally activated switching of perpendicular magnet by spin-orbit spin torque," *Appl. Phys. Lett.*, vol. 104, no. 7, 2014.

[4] L. Zhang *et al.*, "A 16 Kb Spin-Transfer Torque Random Access Memory with Self-Enable Switching and Precharge Sensing Schemes," *IEEE Trans. Magn.*, vol. 50, no. 4, 2014.

[5] J. Kim, A. Chen, B. Behin-Aein, S. Kumar, J. P. Wang, and C. H. Kim, "A technology-agnostic MTJ SPICE model with user-defined dimensions for STT-MRAM scalability studies," *Proc. Cust. Integr. Circuits Conf.*, vol. 2015–November, pp. 8–11, 2015.

2-D Failure Bitmap Compression Using Line Fault Marking Method

Keewon Cho, Young-woo Lee, Sungyoul Seo and Sungho Kang

Dept. of Electrical and Electronic Engineering
Yonsei University
Seoul, Korea
{ckw1505, roberto, sungyoul}@soc.yonsei.ac.kr, and shkang@yonsei.ac.kr

Abstract—As memory densities have rapidly increased, memory testing and repairing processes are the major keys to prevent the decline in the yield. For redundancy analysis (RA), fail addresses should be extracted by the external automatic test equipment (ATE) and stored into the failure bitmap. However, full size of the failure bitmap can be a huge burden on the ATE costs. In order to reduce the storage size, this paper presents a new failure bitmap compression method. The proposed method marks all of the addresses of the line fault, so that repairing solutions can be easily decided. Experimental results show that the proposed compression method greatly reduces the size of failure bitmap while minimizing the failure data loss.

Keywords- redundancy analysis (RA); faillure bitmap; compression; automatic test equipment (ATE); yield

I. INTRODUCTION

Advances in technology have forced to increase the capacities and densities of semiconductor memories [1]. Memories are becoming more vulnerable to faults so that the yield degradation issue is constantly raising the total cost. To deal with the yield problem, a redundancy analysis (RA) technique which repairs memory faulty cells with prepared redundant cells is generally used. For the repairing process, repair solutions are analyzed by the external automatic test equipment (ATE) based on the stored faulty information in the failure bitmap. However, the full size of the failure bitmap becomes a huge burden to the external ATE considering its high cost.

Various studies have suggested failure data compression methods. In case of the code based compression, run-length methods are frequently used in order to get high compression rate [2]. However, these methods have many constraints for implementation and cannot save the characteristics of faults. For saving the characteristics of faults, the method which only stores actual faulty addresses is presented [3]. But this method has lower compression rate because of storing the raw data of faulty addresses.

This paper presents a new failure bitmap compression method which marks all of addresses of the line fault in order to help repair decisions easily. By doing so, the failure bitmap can be greatly compressed while only a little part of single faulty information is omitting. The rest of the paper describes details of the compression method with a simple example.

This research was supported by the MOTIE(Ministry of Trade, Industry & Energy(10067813) and KSRC(Korea Semiconductor Research Consortium) support program for the development of the future semiconductor device.

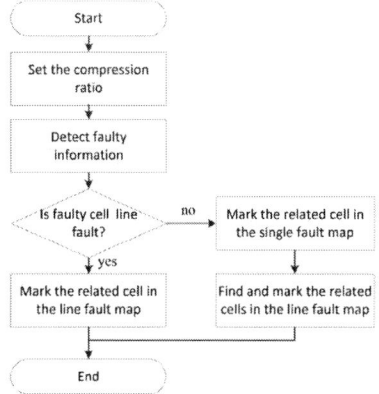

Fig. 1. Flow chart of the proposed compression method.

II. PROPOSED METHOD

Faults can be classified into two types, line faults and single faults. A line fault consists of faults which share either the same row address or the same column address. If faults share the same row address, they can be repaired by one row spare or multiple column spares. On the other hand, if faults share the same column address, they can be repaired by one column spare or multiple row spares. Since the number of spares is limited, repairing a row line fault with a row spare is an effective way for RA in general cases and vice versa. A single fault is a faulty cell which shares no address with any other faulty cell. To repair a single fault, either a row spare or a column spare can be used. Therefore, its repair decisions can be postponed until all repair decisions of line faults are decided.

Fig. 1 shows the flow chart of the proposed compression method. First, compression ratio should be selected considering the failure data loss. Let's suppose that the compression ratio is $1/n$. Then, first n columns of the failure bitmap are utilized as a line fault map and the rest of the failure bitmap is utilized as a single fault map. Since the size of the line fault map depends on the compression ratio, addresses of faults in the failure bitmap are easy to decode. When a line fault is detected, its address information is marked in the line fault map. When a single fault is detected, the single fault map is enabled. However, its address information cannot be expressed by the single fault map alone. So, the address information of a single fault is always expressed as grouped cells in the failure bitmap.

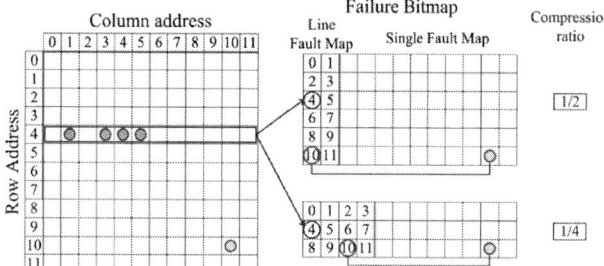

Fig. 2. Row address compression examples of a line fault and a single fault.

A specific example of the proposed compression method is described in Fig. 2. For the convenience of understanding, only a row address compression method is depicted. Compression ratios are set to 1/2 and 1/4. Then, the first two and four columns of the failure bitmap are utilized as the line fault map, respectively. The line fault map and the single fault map of each case are divided with red lines in Fig. 2. In the example, the row address 4 is determined as the line fault. Therefore, the relevant cell in the line fault map is marked. In addition, a faulty cell (10, 10) is determined as a single fault. Then, the related cell in the line fault map and 10th column cell in the single fault map are marked concurrently so that the address information of the single fault can be saved in the reduced failure bitmap.

Fig. 3 shows an example of the proposed compression method when the total compression ratio is 1/4. The faulty memory block in Fig. 3 has a 16 by 16 memory array. In the faulty memory block, row address 4 and 11 and column address 8 are line faults. In addition, faulty cells (8, 10) and (15, 15) are defined as single faults. Since the compression ratio of row address is 1/2, row address 4 is marked in the cell (2, 0) in the failure bitmap. Row address 11 is marked in the cell (5, 1) in the same manner. The compression ratio of column address is also 1/2, so that column address 8 is marked in the cell (0, 4). The faulty cell (8, 10) consists of row address 8 and column address 10. Row address 8 can be expressed as (4, 0) in the failure bitmap, and column address 10 can be expressed as (0, 5). Therefore, the faulty cell (8, 10) can be expressed as {(4, 0), (0, 5), (4, 5)}. Similarly, the faulty cell (15, 15) can be expressed as {(7, 1), (1, 7), (7, 7)}. After the compression, the failure bitmap size is reduced while preserving the characteristics of faulty cells.

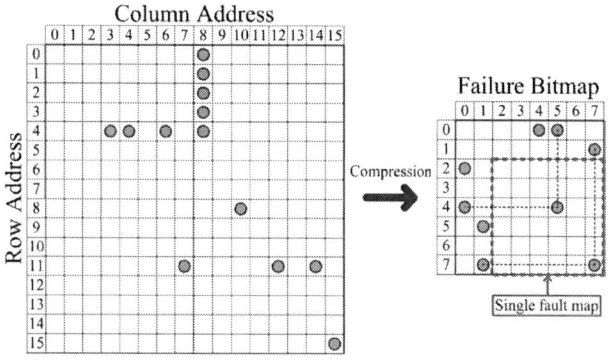

Fig. 3. Example of the proposed compression method (compression ratio=1/4).

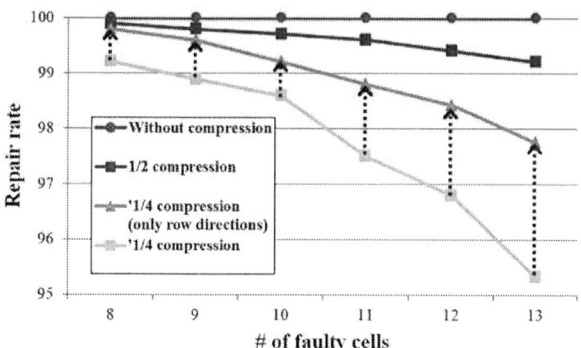

Fig. 4. Repair rate of various compression ratios.

In the experiments, 1024 by 1024 memory arrays are utilized. Fig. 4 shows repair rates of various compression ratios when a memory block has 4 row spares and 4 column spares. Each case was repeated 100,000 times with a c-code based simulator. Repair rates with a 1/2 compressed failure bitmap are higher than 99% at any circumstances as shown in the graph. Repair rates with a 1/4 compressed failure bitmap rapidly decrease as the number of faulty cells increases. However, repair rates can be improved if the failure bit map is compressed in a rectangular shape. For example, if the failure bitmap is only compressed in a row direction, repair rates with a 1/4 compressed failure bitmap are improved by 98% as shown in the graph. In addition, repair rates can be improved further by proper RA algorithms and additional spares. Therefore, experimental results show that the proposed compression method can reduce the size of failure bitmap while maintaining the reasonable repair rate.

III. CONCLUSION

A novel failure bit map compression method utilizing the line fault marking is presented in this paper. To perform the RA process more easily, all faults are classified with two categories, line faults and single faults. And then, faulty information is compressed depending on the types of faults. Therefore, the proposed compression method greatly reduces the failure bitmap size while preserving most of faulty information. As the result, the proposed idea can lower the burden of the total test cost.

ACKNOWLEDGMENT

This research was supported by the MOTIE(Ministry of Trade, Industry & Energy(10067813) and KSRC(Korea Semiconductor Research Consortium) support program for the development of the future semiconductor device.

REFERENCES

[1] S. Soin, Y. Kishore and G. Singh, "Memory testing and fail bitmap generation for fault correlation," International Conference on Microelectronic Devices, Circuits and Systems, Vellore, Aug. 2017.

[2] L. Lee, W. Tseng, R. Lin and C. Chang, "2^n pattern run-length for test data compression," IEEE Transactions on Computer-Aided Design of Integrated Circuits and Systems, vol. 31, no. 4, pp. 644-648, 2012.

[3] J. Vollrath, U. Lederer and T. Hladschik, "Compressed bit fail maps for memory fail pattern classification," in Proc. of IEEE Conference on European Test Workshop, Cascais, May. 2000.

Diagnosis of Resistive Nonvolatile-8T SRAMs

Yu-Ting Li and Jin-Fu Li
Department of Electrical Engineering
National Central University
Taoyuan, Taiwan 320

Chun-Lung Hsu and Chi-Tien Sun
Information and Communications Research Laboratories
Industrial Technology Research Institute
Hsinchu, Taiwan 310

Abstract—**Resistive nonvolatile static random access memory (SRAM) can preserve data in power down mode and provide fast power-on speed. A resistive nonvolatile 8T (Rnv8T) SRAM cell consists of a 6T SRAM cell, two memristive devices, and two transistors. RAM faults and memristor-related faults should be considered for the testing and diagnosis of Rnv8T SRAMs. In this paper, a diagnosis methodology is proposed to distinguish RAM faults and memristor-related faults.**

Index Terms—**Memristor, nonvolatile SRAM, diagnosis, March test**

I. INTRODUCTION

Resistive nonvolatile static random access memory (SRAM) is a good candidate for the low-power applications. It can preserve data in power down mode and provide fast power-on speed. Furthermore, its fabrication process is compatible with the CMOS process [1], [2]. The resistive nonvolatile-8T (Rnv8T) SRAM has many advantages over other existing resistive nonvolatile SRAMs [1]. A Rnv8T SRAM cell is composed of a 6T SRAM cell, two memristors (R_L and R_R), and two access transistors (M_2 and M_3) for the memristors as shown in Fig. 1(a) [1]. It has four functional operations: read, write, store, and restore operations. The read/write operations are similar to those of a 6T SRAM cell. The store operation programs the data stored in the 6T SRAM cell into the memristors. On the contrary, the restore operation writes the data of the memristors back to the 6T SRAM cell.

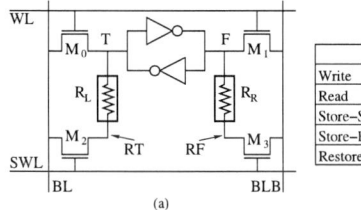

	WL	SWL	BL	BLB
Write	V_{DD}	V_{DD}	—	—
Read	V_{DD}	V_{SS}	—	—
Store–Set	V_{SS}	V_{DD}	V_{DD}	V_{DD}
Store–Reset	V_{SS}	V_{DD}	V_{SS}	V_{SS}
Restore	V_{SS}	V_{DD}	V_{SS}	V_{SS}

(a)   (b)

Fig. 1.   (a) Rnv8T SRAM cell structure (b) Voltage bias for different operations [1]

Fig. 1(b) shows the bias conditions of the Rnv8T SRAM cell for different operations [1]. The word line (WL) and the switch line (SWL) are set to V_{DD} for the execution of write operation. Since SWL=V_{DD}, the access transistors (M_2 and M_3) of memristors are turned on. That provides a parallel write path such that the write margin is increased. The WL is set to V_{DD} and SWL is set to V_{SS} for the executing of read operation to increase the read stability.

For performing the store operation, the WL is set to low and SWL is set to high. The store operation is divided into two phases: the store-set and the store-reset operations. If the store-set operation is executed, both BL and BLB are set to V_{DD}. On the contrary, both BL and BLB are set to V_{SS} for the store-reset operation. For executing the restore operation, the WL is set to V_{SS} and SWL is set to V_{DD}. Also, both BL and BLB are set to low to clear residual charges at the storage nodes T/F such that the states of the two nodes

are equalized. Subsequently, the supply voltage is increased and the nodes T/F are charged through the PMOS transistors. The resistance values of R_R and R_L are different, which creates a voltage difference at the nodes T/F during the power-on process. The memristor with low (high) resistance has a large (small) discharge current which results in a low (high) voltage state at the corresponding storage node.

Testing and diagnosis of resistive nonvolatile SRAMs should consider the RAM faults and memristor-related faults [3], [4]. Diagnosis is a key technique for the yield improvement of memories. In this paper, we propose a diagnosis methodology for distinguishing the RAM faults and memristor-related faults.

II. DIAGNOSIS METHODOLOGY FOR RNV8T SRAM

A. RAM Faults and Memristor-Related Faults

The following typical functional fault models for RAMs defined in [5] are considered in this work. Those functional faults include stuck-at fault (SAF), state coupling fault (CFst), idempotent coupling fault (CFid), and inversion coupling fault (CFin). Memristor-related faults defined in [3], [4] are considered. Those faults include memristor stuck-at fault (MASF), slow store fault (SSF), store destructive fault (SDF), and memristor disturb read fault (MDRF). Subsequently, we describe those memristor-related faults in more details. In a Rnv8T SRAM cell, if the data x of a SRAM cell after a restore-y operation is always k, where $k \in \{0, 1\}$, it has a $MSAF$. The $MSAF$ has two sub-types: $MSA0F$ and $MSA1F$ for $k = 0$ and $k = 1$, respectively. If a Rnv8T SRAM cell has a SSF, the store-x operation cannot program the memristors to expected resistances within the defined store operation time. In a Rnv8T SRAM cell, if the data of SRAM cell is flipped when a store-x operation is executed, a SSF exists. A Rnv8T SRAM cell in a column has a $MDRF$ if it executes a read-x operation and returns unexpected data when the SRAM state and memristor state of another Rnv8T SRAM cell in the same column both are $\bar{x}$. The $MDRF$ can further be divided into $MDR0F$ and $MDR1F$ for $x = 0$ and $x = 1$, respectively.

B. Two-Phase Diagnosis Algorithm

To distinguish all the simple RAM faults and memristor-related faults, a two-phase diagnosis methodology is proposed. In the first phase, the March-17N diagnosis algorithm is applied to distinguish the simple RAM faults including stuck-at faults, state coupling faults, idempotent coupling faults, and inversion coupling faults [6]. In this phase, only the Read/Write operations are executed. The memristor-related faults are not activated. Once the RAM faults are identified, a proposed diagnosis algorithm March-MD for distinguishing memristor-related faults is applied in the second phase. The diagnosis algorithm March-MD can distinguish the MSAF, SSF, SDF, and MDRF.

978-1-5386-7961-6/18 $31.00 © 2018 IEEE   23   ISOCC 2018

The proposed diagnosis algorithm March-MD is as below.

$$
\left\{
\begin{array}{c}
(\Updownarrow_{even}\ (w0); \Updownarrow_{odd}\ (w1)(store);\\
\Updownarrow_{even}\ (w1); \Updownarrow_{odd}\ (w0); (store);)^n\\
\Updownarrow_{even}\ (w0); \Updownarrow_{odd}\ (w1); (store); \Updownarrow_{even}\ (r0);\\
\Updownarrow_{odd}\ (r1); (poff); (restore); \Updownarrow_{even}\ (r0);\\
\Updownarrow_{odd}\ (r1); \Updownarrow_{even}\ (w0); \Updownarrow_{odd}\ (w1); \Updownarrow_{even}\ (r0); \Updownarrow_{odd}\ (r1);\\
(store); (poff); (restore); \Updownarrow_{even}\ (r0); \Updownarrow_{odd}\ (r1);\\
\Updownarrow_{even}\ (w1); \Updownarrow_{odd}\ (w0); (store); \Updownarrow_{even}\ (r1);\\
\Updownarrow_{odd}\ (r0); (poff); (restore); \Updownarrow_{even}\ (r1);\\
\Updownarrow_{odd}\ (r0); \Updownarrow_{even}\ (w1); \Updownarrow_{odd}\ (w0); \Updownarrow_{even}\ (r1); \Updownarrow_{odd}\ (r0);\\
(store); (poff); (restore); \Updownarrow_{even}\ (r1); \Updownarrow_{odd}\ (r0);
\end{array}
\right\}
$$

Here $\Updownarrow_{even}$ and $\Updownarrow_{odd}$ denote the addressing sequence can be ascending or descending with even and odd address sequence. Also, rx and wx denotes a read operation with expected data x and a write operation with data x, respectively.

March-MD begins with a self-forming sequence consisting of n times of write-and-store operations, which can fix the over-forming problems and keep the resistance of memristors within the working region. Table I shows the fault dictionary of the March-MD algorithm for the memristor-related faults, where $E_i=0$ denotes the ith Read operation cannot detect the corresponding faults. If $E_i=$ even or odd, the ith Read operation can detect the corresponding faults at even addresses or odd addresses, respectively. To distinguish MSAF and SSF, the store operations of the same value have to execute two times for each value. SS0F and SS1F cause a stuck value during the first store operation which can be detected by E_2, E_3, E_{10} and E_{11}. After the second store operation is executed, the faulty cell of SS0F and SS1F can be programmed to the corrected value while the faulty cell of the MSAF still stuck at the wrong state. The store destructive fault can be detected by the read operations which is after the store operations (E_0, E_1, E_8 and E_9). The MDRF can be activated since applying the checkerboard data background, so each read operation applied with the checkerboard data background can observe the fault effect of MDRF. As Table I shows, all the faults have different syndromes. Thus, the March-MD can distinguish all the memristor-related faults.

TABLE I
FAULT DICTIONARY OF MARCH-MD

Fault Types	E_0	E_1	E_2	E_3	E_4	E_5	E_6	E_7
MSA0F	0	0	0	odd	0	0	0	odd
MSA1F	0	0	even	0	0	0	even	0
SS0F	0	0	even	0	0	0	0	0
SS1F	0	0	0	odd	0	0	0	0
SD0F	0	odd	0	odd	0	0	0	odd
SD1F	even	0	even	0	0	0	even	0
MDR0F	even	0	even	0	even	0	even	0
MDR0F	0	odd	0	odd	0	odd	0	odd

Fault Types	E_8	E_9	E_{10}	E_{11}	E_{12}	E_{13}	E_{14}	E_{15}
MSA0F	0	0	even	0	0	0	even	0
MSA1F	0	0	0	odd	0	0	0	odd
SS0F	0	0	0	odd	0	0	0	0
SS1F	0	0	even	0	0	0	0	0
SD0F	even	0	even	0	0	0	even	0
SD1F	0	odd	0	odd	0	0	0	odd
MDR0F	0	odd	0	odd	0	odd	0	odd
MDR1F	even	0	even	0	even	0	even	0

III. SIMULATION AND ANALYSIS RESULTS

A fault simulator has been implemented to verify the diagnostic resolution of diagnosis algorithms for nvSRAMs. Fig. 2 shows the comparison result of diagnostic resolution of the proposed March-MD and the test algorithm March-CM reported in [3]. We see that the proposed two-phase diagnosis methodology (March-17N+March-MD) can achieve 100% diagnosis resolution for the RAM faults and memristor-related faults. However, the March-CM only can achieve 86.67% diagnosis resolution since it is developed for the purpose of testing.

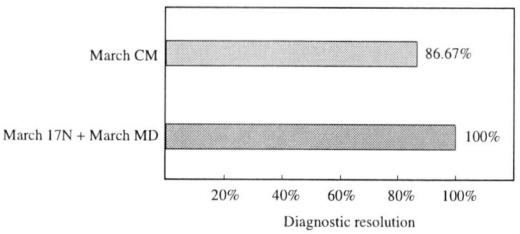

Fig. 2. Diagnostic resolution of different diagnosis algorithms

The test time of the March-MD diagnosis algorithm can be estimated by the following equation.

$$t_f + 4t_w RC + 8t_r RC + 4(t_s + t_{poff} + t_{re})$$

t_f denotes the required time of self-forming operation which is determined by the physical characteristic of memristors. t_w and t_r denote the required time of a read and a write operation, respectively. t_s and t_{re} denote the required time of a store and a restore operation, respectively. t_{poff} denotes the required time of the power off process. R and C denote the number of rows and columns of the Rnv8T SRAM under test.

IV. CONCLUSION

In this paper, we have proposed a two-phase diagnosis methodology for distinguishing RAM faults and memristor-related faults of Rnv8T SRAMs. The March-17N algorithm is used to distinguish the RAM faults first and then a March-MD diagnosis algorithm is proposed to distinguish memristor-related faults.

ACKNOWLEDGMENT

This work was supported in part by the Ministry of Science and Technology (MOST), R.O.C., under Contract NSC 102-2221-E-008-108-MY3 and MOST 104-2220-E-008-009.

REFERENCES

[1] P.-F. Chiu, M.-F. Chang, C.-W. Wu, C.-H. Chuang, S.-S. Sheu, Y.-S. Chen, and M.-J. Tsai, "Low store energy, low VDDmin, 8T2R nonvolatile latch and SRAM with vertical-stacked resistive memory (memristor) devices for low power mobile applications," *IEEE Jour. of Solid-State Circuits*, vol. 47, no. 6, pp. 1483–1496, June 2012.

[2] W. Wei, K. Namba, J. Han, and F. Lombardi, "Design of a nonvolatile 7T1R SRAM cell for instant-on operation," *IEEE Transactions on Nanotechnology*, vol. 13, no. 5, pp. 905–916, Sept. 2014.

[3] Y.-T. Li, Y.-X. Chen, and J.-F. Li, "Fault modeling and testing of resistive nonvolatile 8T-SRAMs," in *Proc. IEEE VLSI Test Symp. (VTS)*, Las Vegas, April 2016, pp. 1–6.

[4] B.-C. Bai, K.-L. Luo, C.-A. Chen, Y.-W. Chen, M.-H. Wu, C.-L. Hsu, L.-C. Cheng, and J. C.-M. Li, "Back-end-of-line-defect analysis for Rnv8T nonvolatile SRAM," in *Proc. IEEE Asian Test Symp. (ATS)*, 2013, pp. 123–127.

[5] R. Dekker, F. Beenker, and L. Thijssen, "A realistic fault model and test algorithm for static random access memories," *IEEE Trans. on Computer-Aided Design of Integrated Circuits and Systems*, vol. 9, no. 6, pp. 567–572, June 1990.

[6] J.-F. Li, K.-L. Cheng, C.-T. Huang, and C.-W. Wu, "March-based RAM diagnosis algorithms for stuck-at and coupling faults," in *Proc. Int'l Test Conf. (ITC)*, Baltmore, Oct. 2001, pp. 758–767.

Reliability Optimization of ReRAM Architecture using Heterogeneous Error Correcting Code Scheme

Kwangjin Lee

Department of Semiconductor and Display Engineering
Sungkyunkwan University, Suwon, Korea
lee931126@skku.edu

Tae Hee Han

Department of Semiconductor Systems Engineering
Sungkyunkwan University, Suwon, Korea
than@skku.edu

Abstract— **The memristor-based resistive random access memory (ReRAM) has been emerged because of its beneficial properties such as non-volatility, high density, and process scalability. However, ReRAM also suffers from process variation caused by down scaling of device size and operating voltage level. In this paper, we propose a new reliable ReRAM architecture with heterogeneous ECC-based subblock architecture focusing on reducing implementation overhead. Experimental results show that the proposed approach reduces the number of parity bits by 18.80 %, area by 20.43 %, respectively, and increases the code rate by 3.90 % compared to those of conventional homogeneous ECC scheme.**

Keywords : ReRAM, Error Correcting Code, Error Pattern Modeling, Data Reliability

I. INTRODUCTION

ReRAM is a resistive-based nonvolatile memory technology with several advantages such as excellent scalability (< 10nm), low programming voltage (< 3V), short read / write time (< 10ns) and excellent compatibility with existing CMOS process [1], [3].

For higher density, ReRAM is implemented in 1-selector-1-resistor (1S1R) crosspoint structure as shown in Fig. 1. The 1S1R crosspoint structure is superior to 1T1C DRAM architecture in terms of the area because it does not require access transistors. However, 1S1R structure is affected adversely by sneak current and voltage drop depending on the distance from the drivers. Specifically, due to the absence of access transistors, sneak current leaks into the half-selected cell as shown in Fig. 2.

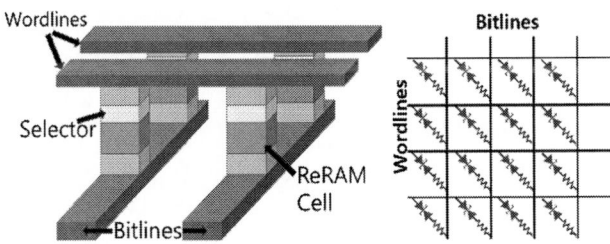

Figure 1. ReRAM crosspoint architecture with selector

As the distance of the cell from the driver is increasing, the cell is becoming more vulnerable to voltage drop due to the reduced noise margin. For example, the voltage level at bitline b_5, far from the driver, can be lower than the required level. For this reason, the BER difference between the leftmost and the rightmost of 512-bit data block is approximately 100 times as demonstrated by Mao *et al.* [2].

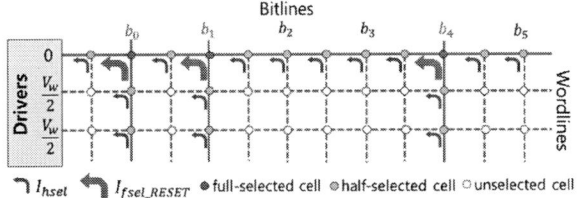

Figure 2. Currents in reset phase of ReRAM subblock

Thus, we propose a low overhead ECC scheme, considering the non-uniform BER distribution of the ReRAM crosspoint architecture, which applies the different strength ECC using the analytic block failure rate (BFR) model. The contribution of this paper is: (i) an analytic model for the correctability of the ECC logic required for data that has a difference in error rate for each bit, (ii) based on the model in (i) we design a heterogeneous ECC scheme for 512-bit 1S1R ReRAM data block for higher memory density.

II. PROPOSED MATHEMATICAL MODEL

Equation (1) is an existing model for BFR when BER is uniform among the overall data block [2], [4].

$$BFR = 1 - \sum_{i=0}^{c} \binom{N}{i} BER^i (1 - BER)^{N-i}, \qquad (1)$$

where c indicates the correctability of ECC and N indicates the whole block size. However, in the case of ReRAM, modeling by the above equation is not appropriate because the error rate in the block is nonuniform. Focusing on this point, we introduce a modified BFR model to take into account the error pattern of ReRAM crosspoint architecture.

$$\Lambda = \{\lambda_i | i = 1, 2, \cdots, D\}, \qquad R = \left\{ r_i = \frac{\lambda_i}{1 - \lambda_i} \middle| i = 1, 2, \cdots, D \right\}$$

$$P(n) = \sum_{\substack{\Lambda_n \subset \Lambda \\ |\Lambda_n| = n}} \prod_{\lambda \in \Lambda_n} \lambda \cdot \prod_{\lambda \notin \Lambda_n} (1 - \lambda)$$

978-1-5386-7961-6/18 $31.00 © 2018 IEEE

$$= \prod_{\substack{\lambda \in \Lambda}} (1-\lambda) \sum_{\substack{\Lambda_n \subset \Lambda \\ |\Lambda_n|=n}} \prod_{\substack{\lambda \in \Lambda_n}} \frac{\lambda}{1-\lambda} = P(0) \sum_{\substack{\Lambda_n \subset \Lambda \\ |\Lambda_n|=n}} \prod_{\substack{\lambda \in \Lambda_n}} \frac{\lambda}{1-\lambda}$$

$$T_n(i) = \sum_{\substack{R_n \subset R - \{r_i\} \\ |R_n|=n}} \prod_{r \in R_n} r$$

$$S_n = \sum_{\substack{\Lambda_n \subset \Lambda \\ |\Lambda_n|=n}} \prod_{\lambda \in \Lambda_n} \frac{\lambda}{1-\lambda} = \sum_{\substack{R_n \subset R \\ |R_n|=n}} \prod_{r \in R_n} r = \frac{1}{n} \cdot \sum_{i=1}^{D} r_i T_{n-1}(i)$$

$$T_n(i) = S_n - a_i T_{n-1}(i) \quad (n \geq 1), \quad T_0(i) = 1$$

$$P(n) = P(0) \cdot S_n = P(0) \cdot \frac{1}{n} \cdot \sum_{i=1}^{D} r_i T_{n-1}(i).$$

D : The total number of data bits in the block

λ_i : The probability of error at the i-th bit

$P(n)$: The probability of n errors occurring in the data block

Finally, if the correctability of the used ECC is c, the error rate after ECC is able to be calculated as:

$$BFR = 1 - \sum_{n=0}^{c} P(n). \qquad (2)$$

Eq. (2) produces more accurate BFR compared to that of Eq. (1) because of using different λ_i values which reflects BER distribution of ReRAM. Therefore, model (2) allows to achieve required reliability with minimum ECC strength.

The baseline system used Bose-Chaudhuri-Hocquenghem (BCH) code after dividing the 512-bit data block into four subblocks of 128-bit units [4]. To achieve the BFR of 10^{-10}, the same four-bit correctable BCH code is applied to all 4 subblocks in the baseline system. However, the proposed method applies ECC of different strength according to the BER of subblocks calculated through the model of (2). With our new scheme, the ECC correctability required for BFR of 10^{-10} are 2, 3, 4, and 4 for subblock 1, 2, 3, and 4, respectively. As a result, the total required correctability for BFR of 10^{-10} is only 13 as shown in Fig. 3.

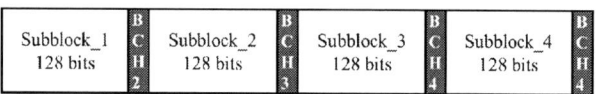

Figure 3. Heterogeneous ECC scheme

III. EVALUATION

We calculated the ECC overhead by analytical C++ simulation based on the model of (2) as well as the experimental results of the previous work in [4]. Table 1 shows a comparison of the proposed method with a homogeneous ECC scheme that applies the same ECC to four parts of 512-bit ReRAM data block. It is shown that the number of parity bits is reduced by

18.8 % from 128 to 104 using a heterogeneous ECC scheme which reflects the BER differences between the subblocks depending on the distances from the driver. Moreover, the area overhead is also reduced by 20.43 % in the proposed method because the complexity of the heterogeneous ECC logic is decreased when compared to that of the homogeneous ECC logic. Consequently, required memory reliability is accomplished with lower overhead by the heterogeneous ECC scheme.

Table 1. Comparison of the number of parity bits and area overhead

	Conventional ECC scheme					Heterogeneous ECC scheme				
Subblock	1	2	3	4	Total	1	2	3	4	Total
No. of parity bits	32	32	32	32	128	16	24	32	32	104
Area overhead (%)	4.83	4.83	4.83	4.83	19.33	2.19	3.52	4.83	4.83	15.38

IV. CONCLUSION

In this paper, we propose a new error model considering the BER difference between cells of ReRAM crosspoint architecture to achieve better efficiency in terms of implementation complexity without harming the required reliability. Based on this point, we derive the ECC correctability need to achieve the target BFR of the 512-bit data block and implement the heterogeneous ECC scheme that has lower ECC overhead. Our experimental results show that the proposed scheme achieves the required BFR with 18.8 % fewer parity bits, 3.9 % increased code rate and 20.43 % smaller area compared to previous studies.

ACKNOWLEDGMENT

This work was supported by the MOTIE (Ministry of Trade, Industry & Energy) (10080594) and KSRC (Korea Semiconductor Research Consortium) support program for the development of the future semiconductor device.

REFERENCES

[1] C. Xu, D. Niu, N. Muralimanohar, R. Balasubramanian, T. Zhang, S. Yu, Y. Xie, "Overcoming the Challenges of Crossbar Resistive Memory Architectures," High Performance Computer Architecture (HPCA), 2015.

[2] M. Mao, P. Y. Chen, S. Yu, C. Chakrabarti, "A Multilayer Approach to Designing Energy-Efficient and Reliable ReRAM Cross-Point Array System," IEEE Transactions on Very Large Scale Integration (VLSI) Systems, vol.25, No 5, 2017.

[3] S. Yu and P. Y. Chen, "Emerging memory technologies: Recent trends and prospects," IEEE Solid-State Circuits Mag., vol. 8, no. 2, pp. 43–56, Mar. 2016..

[4] D. Niu, Y. Xiao, Y. Xie "Low Power Memristor-Based ReRAM Design with Error Correcting Code," in 17th ASP-DAC, Jan 2012, pp. 79–84.

[5] D. Strukov, "The area and latency tradeoffs of binary bit-parallel bch decoders for prospective nanoelectronic memories," in 2006 Fortieth Asilomar Conference on Signals, Systems and Computers, Oct 2006, pp. 1183–1187

A Digital ΣΔ Modulated Class-S Transmitter with Two-Step Up-conversion and Filter-less Front End

Pan Xue[1], Haijun Shao[1], Dan Fang[1], Gan Guo[2], Wei Che[2], Zhiliang Hong[1*]

[1]State Key Laboratory of ASIC and System, Fudan University, Shanghai, China

[2]Analog Devices, Inc., Shanghai/Beijing, China

*Email: zlhong@fudan.edu.cn

Abstract—**A fully-digital Class-S transmitter implemented in 65-nm CMOS is proposed. In the filter-less digital front end (DFE) of the transmitter, a 40 MHz bandwidth time-interleaved (T-I) ΣΔ modulator operating up to 1.92 GHz sampling rate at center frequency of 480 MHz is performed. And two-step mixing technique is adopted to achieve the gigahertz carrier frequency at implementable operating frequency. The measured power consumption of the DFE is 13mW under a 1.2 V supply voltage when a 10 MHz LTE signal is applied and the carrier frequency of the transmitter is up to 2.4 GHz.**

Keywords—Class-S transmitter; ΣΔ Modulator; power amplifier (PA); two-step up-conversion;

I. INTRODUCTION

In recent years, many scholars have focused on the research of fully-digital radio frequency (RF) transmitters for their advantages in design flexibility and power efficiency. Although digital transmitters with DACs, such as [1], have achieved good performance on out-of-band noise compression, these transmitters need additional digital pre-distortion techniques to ensure the in-band linearity. To alleviate this issue, intensive Class-S transmitters are studied [2]-[6]. In the Class-S transmitters, ΣΔ modulator is utilized to generate a single-bit switching signal which conveys both phase and amplitude information to drive a digital PA (DPA), providing a high in-band linearity.

In [2] and [3], two fully-custom designed low-pass ΣΔ modulators are employed to deal with I/Q baseband signals. And the sampling rate of ΣΔ modulator in [2] can be up to 4 GHz, but the large power consumption and design intricacy make it costly to be integrated. In [3], ΣΔ and pulse-width modulators are integrated with switched capacitor PAs, however, sophisticated off-chip passive components are required to combine I/Q signals. Reference [4] employs a band-pass ΣΔ modulator to realize combination of I/Q signals achieving the monolithic integration with DFE and PA, but its maximal carrier frequency is 1.6 GHz.

This paper presents a fully-digital Class-S transmitter with 2.4 GHz carrier frequency. A time-interleaved band-pass ΣΔ modulator with 40 MHz bandwidth is performed to convert the multi-bit input to 1-bit signal. A single-bit DPA with on-chip transformer is integrated to amplify the signal.

II. SYSTEM ARCHITECTURE AND CIRCUIT DESIGN

The architecture of the proposed Class-S transmitter is presented in Fig. 1, which is mainly partitioned into the DFE and

PA. The 11-bit digital I/Q baseband signals are converted and mixed into 1-bit signal in the DFE. And then the mixed 1-bit signal is amplified by a DPA and transmitted at RF.

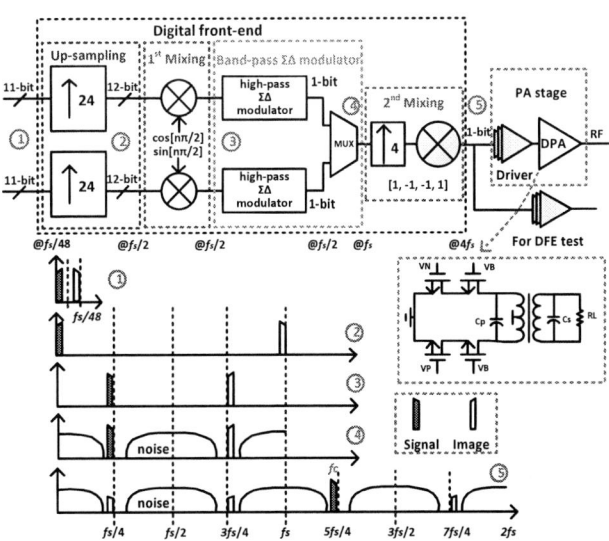

Fig. 1. Proposed digital ΣΔ modulated Class-S transmitter architecture and its corresponding spectrum illustration.

To implement the DFE by automatic design synthesis in current CMOS technologies, a two-step up-conversion architecture is proposed, in which two mixing stages convert the baseband signal to RF as Fig.1 demonstrates. This approach utilizes the spectrum repeat property of sampling and takes the third order harmonic at carrier frequency as the final RF output, then the timing constraint is relaxed. For the mixing operations, when the carrier frequency is chosen as a quarter of the sampling rate, the carrier waves will only have of [1, 0, -1, 0] and [0, 1, 0, -1], then the mixer can be implemented by changing the sign of the digital signal, eliminating multipliers entirely.

ΣΔ modulator is the most power hungry block in the DFE due to its high sampling rate and massive computations. To alleviate the large power consumption, time-interleaving algorithm is applied to lower the practical sampling rate of the ΣΔ modulator. Fig. 2 illustrates the designed time-interleaved band-pass ΣΔ modulator. Two identical 6th-order high-pass ΣΔ modulators are performed on the half sampling rate of the DFE, and their outputs are combined and up-sampled by a multiplexer (MUX), making a band-pass characteristic. Benefiting from

978-1-5386-7961-6/18 $31.00 © 2018 IEEE   27   ISOCC 2018

the band-pass filtering feature of the $\Sigma\Delta$ modulator, filters for images compression in the up-sampling block are removed.

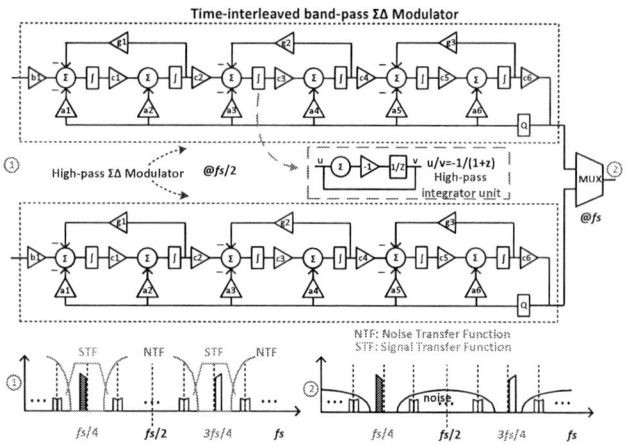

Fig. 2. Proposed time-interleaved band-pass $\Sigma\Delta$ modulator and its corresponding spectrum illustration.

The single-bit DPA is in a current-mode Class-D (CMCD) topology with on-chip transformer-based matching network, and integrated with the DFE using the same supply voltage. Moreover, there is no pre-distortion calibration needed due to the inherent high linearity of single-bit metric.

III. Experimental Results

Fig. 5 (b) presents the chip photograph of the proposed Class-S transmitter. It is fabricated in 65 nm GP CMOS. The chip occupies a 1.5 mm × 1.4 mm area with a core area of 0.64 mm². Fig. 4 shows the measured output spectrums when a 1 MHz single tone input signal is applied. The operating frequency of the proposed $\Sigma\Delta$ modulator can reach to 1.92 GHz and the RF carrier frequency is 2.4 GHz. A maximum 48 dB in-band SNR is achieved at an oversampling ratio of 24. It can be seen that there are no harmonic distortions in the desired signal band. Fig. 5(a) shows the measured output spectrum of the transmitter with a carrier frequency of 2.4 GHz when a 10 MHz LTE signal is applied to it. The total power consumption is 13 mW in DFE under a 1.2 V supply voltage. Table 1 summarizes the main performance metrics of proposed transmitter and gives the comparison with other works.

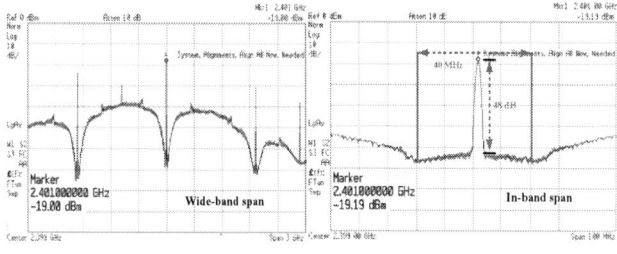

Fig. 4. Measured transmitter output spectrum with 1 MHz single tone input.

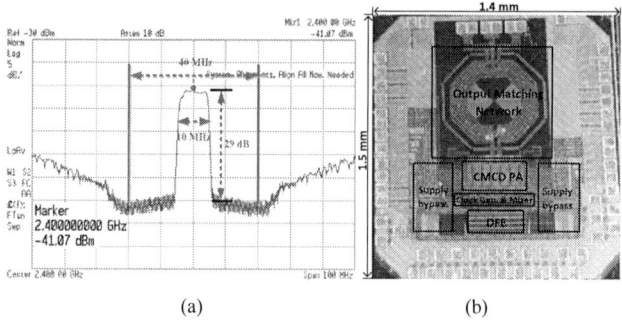

Fig.5 (a) Measured transmitter output spectrum for 10 MHz LTE input (b) Chip photograph of the transmitter.

TABLE I.   COMPARISON WITH OTHER WORKS

	[2]	[4]	**This Work**
Process	90 nm CMOS	65 nm CMOS	65 nm CMOS
Architecture	3rd ord. LP	T-I 10th ord. BP	T-I 6th ord. BP+Filter-less
Integration	DFE only	Fully	Fully
$\Sigma\Delta$ modulator output	1 bit	1 bit	1 bit
$\Sigma\Delta$ modulator BW	30 MHz	40 MHz	40 MHz
Carrier Frequency	1.95 GHz	1.6 GHz	2.4 GHz
Sampling Frequency	1.95 GHz	1.28 GHz	1.92 GHz
Output Stage	CML buffer	CMCD PA +On-chip balun	CMCD PA +On-chip balun
Power Consumption	30 mW ($\Sigma\Delta$M)	15 mW (DFE)	13 mW (DFE)
Area (mm²)	0.15 (w.o.PA)	0.252(DFE) 0.56 (PA)	0.034 (DFE) 0.52 (PA)

IV. Conclusion

A monolithic fully-digital $\Sigma\Delta$ modulated Class-S transmitter with two-step up-conversion and digital filter-less DFE is implemented in 65 nm CMOS, and a time-interleaved $\Sigma\Delta$ modulator is presented. The transmitter can support a 10 MHz LTE signal at 2.4 GHz carrier frequency.

Acknowledgment

This work is supported by the RFTX program of Analog Devices, Inc.

References

[1] M. Alavi, et al., "A 2 x 13-bit All-Digital I/Q RF-DAC in 65 nm CMOS," in *Proc. of IEEE RFIC Symp.*, 2013, pp.167–170.

[2] A. Frappe, et al., "An All-Digital RF Signal Generator Using High-Speed $\Sigma\Delta$ Modulators," *IEEE J. Solid-State Circuits*, vol. 44, no. 10, pp. 2722-2732, Oct. 2009.

[3] R. Hezar, et al., "A 23dBm Fully Digital Transmitter using $\Sigma\Delta$ Modulator and Pulse-width Modulation for LTE and WLAN Applications in 45nm CMOS," in *Proc. IEEE RF Integr. Circuits (RFIC) Symp.*, 2014, pp.217-220.

[4] Yang Zhao, et al, "An All-Digital Gigahertz Class-S Transmitter in a 65-nm CMOS," *IEEE Trans. Very Large Scale Integr. (VLSI) Syst.*, vol. 24, no. 4, pp. 1402-1411, Apr. 2016.

[5] J. Keyzer, et al, "Digital Generation of RF Signals for Wireless Communications with Band-pass Delta-Sigma Modulation," in *IEEE MTT-S Int. Microwave Symp.. Dig.*, vol.3, pp.2127-2130, 2001.

[6] S. Hatami, et al., "Single-Bit Pseudo-parallel Processing Low-Oversampling Delta-Sigma Modulator Suitable for SDR Wireless Transmitters," *IEEE Trans. Very Large Scale Integr. (VLSI) Syst.*, vol. 22, no. 4, pp. 922-931, Apr. 2014.

Gap in pagination due to formatting issues.

Pages 29-33

A novel high precision SDR-based positioning system

G. Piccinni, G. Avitabile, G. Coviello
Electrical and Information Engineering Department
Polytechnic University of Bari
Italy
{g.piccinni, gfa, g.coviello}@poliba.it

C. Talarico
Electrical and Computer Engineering Department
Gonzaga University, Spokane, WA
USA
talarico@gonzaga.edu

Abstract— **This paper introduces a novel positioning system based on a Software Defined Radio (SDR) architecture that evaluates with a high level of accuracy and precision the position of an active target placed in a harsh environment. The use of an SDR architecture allows the properties of the transmitted signal to be varied according to the accuracy and precision required by diverse applications and operating conditions. The positions are extracted using a fully-digital algorithm implemented with an FPGA. Transmitting a 64-length OFDM symbol of only 100 MHz of bandwidth, the proposed system achieves an accuracy that is within one cm.**

Keywords; Indoor Positioning System, Localization, OFDM, Software Define Radio; Zadoff-Chu sequence

I. INTRODUCTION

The aim of this work is to introduce a software reconfigurable indoor positioning system able to evaluate the position of an active device with a high-level of precision and an accuracy that is suitable for medical monitoring. The proposed system relies on the use of a cluster of SDR receivers and a non-coherent demodulation scheme. The SDR architecture guarantees system re-configurability as a function of design specifications, and it requires a very basic analog front-end, composed of a down/up-conversion mixer, a low-noise power-amplifier, an A/D converter, and a D/A converter. The system operates well in presence of severe multipath conditions and relies on a full digital signal whose characteristics, including bandwidth, period and gain, can be modified via software.

II. SYSTEM ARCHITECTURE

The proposed system measures the 2-D and 3-D positions of an active target exploiting a GPS-like scheme composed by four receivers. One of them is selected as reference of the system and so the positions are computed evaluating the Time-Difference-of-Arrival (TDOA) of the received signal. The transmitter sends an OFDM-like signal to the four receivers. The receivers (RX) compute their relative distance from the transmitter (TX) and then send the information to a central server that has the function of triangulating the TX position. The OFDM symbol comprises only pilot subcarriers which represent a coefficient of a Zadoff-Chu sequences (ZC) [1]. The use of these sequence allows to exploit a Software Define Radio (SDR)-based architecture for both the transmitter and the receivers (Fig. 1). Once generated the coefficients of a ZC sequence can be stored in a lookup table and converted in an I/Q analog baseband signal by means of a

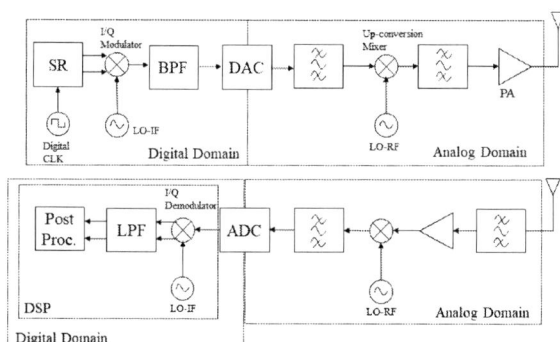

Figure 1. System Architecture (a) Transmitter (b) Receiver.

Digital to Analog converter. At the receiver side, the analog front-end translates the RF signal into an IF signal and then feed it to an Analog to Digital converter that transforms the analog signal into digital samples. The distances and the target positions are computed by means of a fully digital algorithm implemented with an FPGA. The resulting system architecture is easily and widely scalable as function of the operating spatial resolution and the precision required. The signal period (T), the bandwidth (B), and the number of subcarriers can be varied through a very modest software modification. Moreover, the proposed system requires the same analog front-end for both the transmitter and the receivers, and, as long as it can be designed to cover a wide range of bandwidths, allows to leave to the digital section of the system the tasks of filtering, demodulating and modulating. Finally, since the I/Q demodulation and modulation is performed in the digital domain, any offset or mismatch in amplitude, frequency, and phase between the I/Q paths can be easily avoided or recovered via the software algorithm.

III. SYSTEM ANALYSIS

As mentioned, the distances and the positions of the transmitter and the receivers are extracted with a TDOA algorithm. Hence, no synchronization mechanism is implemented between the transmitter and the receivers. However, the receivers share the same clock to generate the time acquisition signal and the same local oscillator (LO) source needed by the down-conversion process. This way, any temporal shift or carrier frequency offset between the transmitter and the receivers appear as constant terms in the distance computation and therefore they can be removed using multiple TDOA measurements. The distance ranging algorithm was presented in

978-1-5386-7961-6/18 $31.00 © 2018 IEEE   34   ISOCC 2018

[2] and requires four steps. In the first step the starting point of the sequence, is correctly computed and selected, for the reference receiver. In the second step a coarse difference of distances is evaluated in the time domain by performing the cross-correlation function between the receivers' signals and a clean copy of the transmitted signal. In the third step the previous values are finely adjusted computing the carrier offset between the subcarriers in the frequency domain and finally in the last step the position is extracted considering a geometrical model of the system. However, the presence of multipaths can affect the distance values extracted with the proposed algorithm [2]. Thus, the third step is slightly modified to include the generation of a Finite Impulse Response (FIR) filter that allows to remove the multipath echoes [3]. The coefficients of the FIR filter are extracted considering the properties of the cross-correlation function of the ZC sequences. The cross-correlation function presents a series of peaks that represents the direct signal and its copies generated by the obstacles of the indoor environment (Fig. 2). The presence of multipaths, in time domain, modifies the amplitude and the position of the main peak and the FIR coefficients can be computed and adjusted by means of well-known gradient methods. The criterion used relies on the error in time-shift and amplitude estimation:

$$F_1 = \hat{\tau}_i - \tau_{i-1}$$
$$F_2 = \hat{\alpha}_i - \alpha_{i-1}$$

(1)

where $\hat{\tau}_i$ and $\hat{\alpha}_i$ are the time-shift and the amplitude of the FIR coefficients estimated at i^{th} iteration cycle step. Using the method of steepest descent to minimize the cost function (1) the coefficients are updated according to:

$$\tau_i = \hat{\tau}_i - \gamma_1 \cdot F_1$$
$$\alpha_i = \hat{\alpha}_i - \gamma_2 \cdot F_2$$

(2)

The coefficients γ_1 and γ_2 determine the convergence speed of the algorithm and are dynamically adjusted in each iteration cycle. The value of γ_1 varies in range from 0.1 to 1.5 while γ_2 is chosen in the range from 0.1 to 1.1. The increment of these coefficients is inversely proportional to the standard deviation of the functions F1 and F2 and determine the stop condition of the algorithm. The algorithm is terminated, i.e. it has minimized the expression (1) and estimated the multipath peaks when there are no longer variations in the amplitude and the time-shift of the multipath peaks. Once the termination condition is achieved, the coefficients of the FIR filter are set, and therefore the equalized signal $x_{eq}(z)$ can be computed (by convoluting the FIR filter with a clean copy of the transmitted signal), and the received signal $y_r(z)$ extracted.

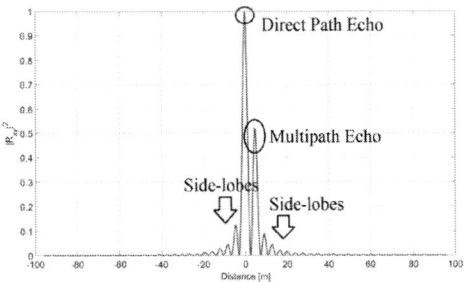

Figure 2. Cross-Correlation in time domain in presence of multipath echo

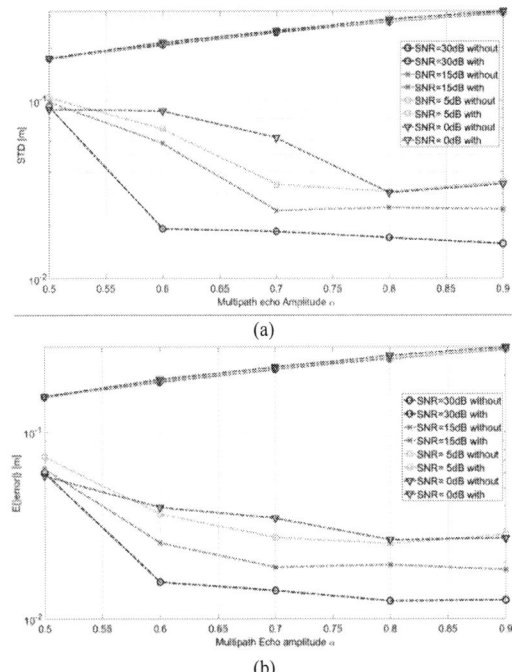

(a)

(b)

Figure 3. (a) System precision and (b) System accuracy as function of multipath amplitude and SNR.

The system was validated considering a channel impulse response comprises of a fixed direct path and a single multipath echo with a variable amplitude. The distance of the multipath echo is varied, and the performances are extracted as function of the SNR of the system exploiting a 64-length OFDM symbol with 100 MHz bandwidth. Fig. 3 shows how the use of the echo cancellation algorithm improves the overall distance accuracy. In facts, the system is able to compute the TDOA with an accuracy that is within one cm even in presence of severe multipath conditions and for low SNR.

CONCLUSIONS

A novel solution is presented to implement an indoor positioning system that exploits an SDR-based architecture for both the transmitter and the receiver. The resulting system is easily and widely scalable as function of the operating spatial resolution and the precision requirements. Exploiting the properties of the transmitted signal a simple algorithm is implemented to estimate and remove the multipath echo. With the use of a 100 MHz 64-length OFDM symbol and a fully-digital distance ranging algorithm, the proposed system achieves a precision providing an accuracy that is within one cm even with low SNR and in presence of severe multipath conditions.

REFERENCES

[1] G. Piccinni, G. Avitabile, G. Coviello, "An Improved Technique based on Zadoff-Chu Sequences for Distance Measurements", IEEE 4th Radio and Antenna Days of the Indian Ocean", 10-13 October 2016, Reunion Island.

[2] G. Piccinni, G. Avitabile, C. Talarico, G. Coviello "Analysis and Modeling of a Novel SDR-based High-Precision Positioning System", Proc. 15th IEEE SMACD, July 2018.

[3] K.-D. Kammeyer, Mann R. and W. Tobergte, "A Modified Adaptive FIR Equalizer for Multipath Echo Cancellation in FM Transmission", IEEE J. Select. Areas Commun., vol. 5, no. 2, pp. 226–237, 1987.

Implementation of Systolic Co-processor for Deep Neural Network Inference based on SoC

Erwin Setiawan[1], Trio Adiono[2]

University Center of Excellence on Microelectronics, Institut Teknologi Bandung
IC Design Laboratory, PAU Building, ITB Campus, Jl. Tamansari No. 126, Bandung 40132, Indonesia
Tel. +62-22-2506280/Fax. +62-22-2508763
Email: erwin.ouyang@gmail.com[1]

Abstract— **In this paper, we present an implementation of systolic co-processor for Deep Neural Network (DNN) inference. The co-processor is used in matrix multiplication between input on every DNN layer and weight values for corresponding DNN layer. The co-processor is implemented on FPGA inside the programmable System-on-Chip (SoC). The co-processor can be accessed from the ARM Cortex-A9 processor through the AXI4 bus. The DNN inference result from the co-processor has been verified by comparing to the MATLAB simulation. The co-processor has been implemented on Xilinx Zynq-7000 SoC. The computation result has been verified by comparing to the MATLAB simulation.**

Keywords— *systolic co-processor; deep neural network; SoC; AXI4 bus*

I. INTRODUCTION

The Deep Neural Network (DNN) training and inference process are generally done in GPU. GPU is faster and more efficient than CPU because of its parallelism [1]. There is a custom ASIC solution for computing DNN inference called Tensor Processing Unit™ (TPU), as in [2]. The custom ASIC solution is faster and more power efficient than GPU. TPU is designed to be used for servers in datacenters. The DNN computations consist of matrix multiplication on every DNN layer. The hardware implementation of matrix multiplication can be done by using systolic array architecture [3]. The systolic array is a network of processing elements (PEs) that are uniform and fully pipelined. The data is passed and computed through the PEs. The systolic array architecture of matrix multiplication can reduce area of the chip, while it decreases speed.

This paper presents the implementation of systolic co-processor for DNN inference. The co-processor is integrated to the ARM Cortex-A9 processor through the standard AXI4 bus. The implementation is done on Xilinx Zynq-7000 SoC. This paper has correlation with our work as presented in [4]. In [4], the methodology for mapping the DNN matrix multiplication into a systolic architecture is presented. In this paper, we use the same DNN model as in [4], but the batch size is eight, while in [4], the batch size is only one. If the batch size is only one, then it is a vector-matrix multiplication and the systolic array architecture is one-dimensional. In this paper the batch size is eight, hence it is a matrix-matrix multiplication and the systolic array architecture is two-dimensional.

This paper is organized as follows. In section II, the DNN model and proposed SoC architecture are explained. In section III, the systolic array architecture is explained. In section IV, the result and conclusion for this work is explained.

II. DNN MODEL AND SOC ARCHITECTURE

A. DNN Model

The DNN model in this paper has five layers, as in [4]. It has one input layer, three hidden layers, and one output layer. The number of neurons in input layer, first hidden layer, second hidden layer, third hidden layer, and output layer is shown in Fig. 1. The batch size for the DNN is eight, so it can process up to eight input samples at a time.

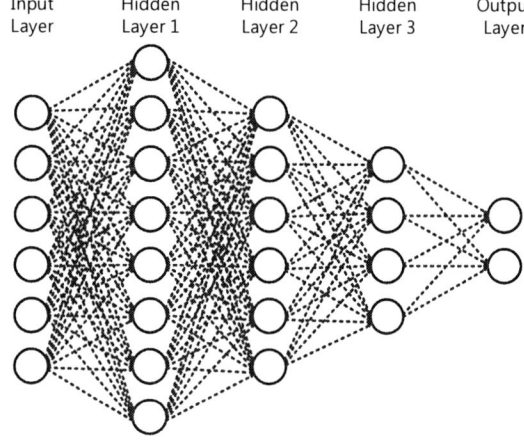

Fig. 1. DNN model

B. SoC Block Diagram

The SoC block diagram for the proposed design is shown in Fig. 2. The Zynq-7000 SoC consists of Processing System (PS) (hard core) and Programmable Logic (PL) (FPGA). In PS, the main processor is Dual Core ARM Cortex-A9 and the memory is 512 MB DDR3. There are several peripherals that are connected to the processor through the AHB and APB bus. The peripherals are SDIO, Ethernet, USB, GPIO, and UART. The SDIO is used for accessing SD card that is used to store Linux OS. The USB can be used for peripheral such as keyboard and mouse. The Ethernet and UART can be used for communication. In PL, there are VGA controller and DNN co-processor. VGA can be used for displaying Linux GUI, so this system is like single board computer. The DNN co-processor is connected to the main processor through the AXI4 bus.

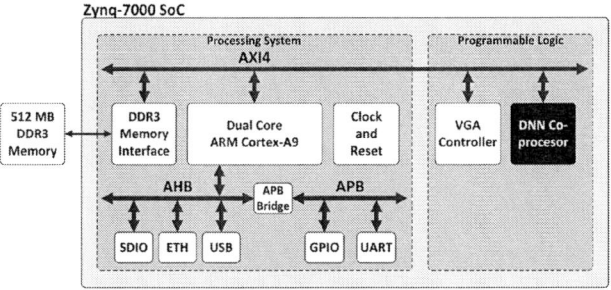

Fig. 2. SoC block diagram

III. SYSTOLIC ARRAY ARCHITECTURE AND IMPLEMENTATION RESULT

A. Systolic Block Diagram

The systolic array architecture for DNN matrix multiplication is shown in Fig. 3. The architecture consists of 8×8 PEs. The input of the PEs are neuron values and weights. The output of the PEs is the multiplication between neuron values and weights. Each of the PE consists of one multiplier and one adder. The data format is 16-bit fixed point number. It has six two's complement integer bits and ten fractional bits.

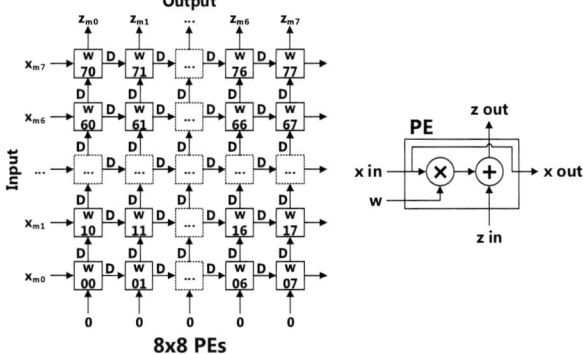

Fig. 3. Systolic array architecture for DNN matrix multiplication

B. Co-processor Block Diagram

The systolic array is used for one matrix multiplication, i.e. the matrix multiplication in one DNN layer. To compute the whole DNN layer, we need four matrix multiplication process which is implemented by time multiplexing method in DNN controller as shown in Fig. 4. The output of PEs is connected to the sigmoid activation block, and the result is stored in a buffer. The data in the buffer can be transferred to the I/O data registers. The I/O data registers can be accessed by the main processor through the AXI4 bus. Weight registers are used for storing the weight values.

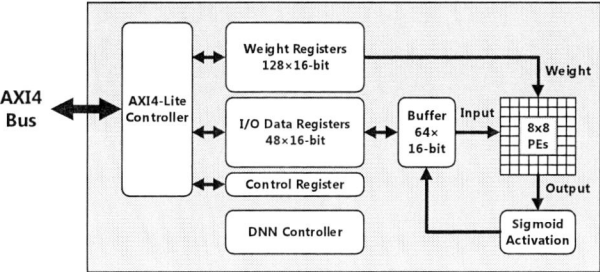

Fig. 4. Co-processor block diagram

C. Implementation Result

The functional verification is done by comparing the DNN inference result from co-processor with the result from MATLAB simulation. The result is same for all DNN layers. The latency for one matrix multiplication is eight clocks. The computation time for one inference process is 97 clocks. Form the synthesis report using Xilinx Vivado®, the maximum frequency clock of 196.35 MHz is obtained. The PL resource utilization on Zynq-7000 is shown on Table 1. The total on-chip power estimation is 1.708 W.

TABLE I. PL RESOURCES UTILIZATION

Resource	Utilization	Utilization %
LUT	6752	38.36
LUTRAM	62	1.03
FF	5515	15.67
BRAM	4	6.67
DSP	64	80

IV. CONCLUSION

The proposed DNN co-processor has been successfully designed and verified on Zynq-7000 SoC. The main processor can access the co-processor by using memory-mapped method through the AXI4 bus.

REFERENCES

[1] NVIDIA, "GPU-based deep learning inference: a performance and power analysis," NVIDIA Whitepaper, November, 2015. [Online]. Available: https://www.nvidia.com/content/tegra/embedded-systems/pdf /jetson_tx1_whitepaper.pdf. [Accessed May 9, 2018].

[2] N. P. Jouppi, et al., "In-datacenter performance analysis of a tensor processing unit," in 44th International Symposium on Computer Architecture (ISCA), Toronto, Canada, June 26, 2017. [Online]. Available: https://arxiv.org/ftp/arxiv/papers/1704/1704.04760.pdf. [Accessed May 9, 2018].

[3] S. Y. Kung, "VLSI array processors," IEEE ASSP Magazine, pp. 4-22, July 1985.

[4] E. Setiawan, S. Fuada, and T. Adiono, "A methodology for mapping the deep neural networks inference into systolic array architecture," unpublished.

Approximate Adder Generation for Image Processing Using Convolutional Neural Network

Ryuta Ishida
Graduate School of ECS
Fukuoka University
td172002@cis.fukuoka-u.ac.jp

Toshinori Sato
Dept. of EECS
Fukuoka University
toshinori.sato@computer.org

Tomoaki Ukezono
Dept. of EECS
Fukuoka University
tukezo@fukuoka-u.ac.jp

Abstract— **This paper proposes a design methodology for configurable approximate arithmetic circuits. It considers the processed data of the target circuits. A prototype system, which relies on deep neural network, is built to confirm the practicability of the methodology.**

Keywords; approximate circuit; design methodology; CNN

I. INTRODUCTION

Approximate computing [3] is a promising technique that achieve low power, high performance, and small footprint at the same time. There are a lot of studies on approximate arithmetic circuits. Configurable approximate circuits [6,11,12] recently interest researchers. Unfortunately, it is a difficult and tedious task to determine parameters for the configurations. There are some previous studies that generate approximate circuits. SALSA [13] modifies the given RTL to its approximate version. Similarly, ABACUS [7] synthesizes approximate circuits based on its behavioral descriptions. SABER [9] determines the length of approximate bits in the trailing bits of operands through an analytical expression of the function. All of these previous studies consider only circuit structure but their target data are out of considerations. Hence, any optimizations regarding specific data cannot be conducted. In contrast, this paper studies a design methodology that is oriented to processed data.

II. DATA-DIRECTED CIRCUIT GENERATION

Different data require different circuits, which are optimized for several design constraints, even if their functionality is the same. Based on the observations, data-directed design methodology shown in Fig. 1 is proposed. The approximate circuit generator considers its target data as well as some design constraints and user's requirements. Parameterized approximate circuits are handled and their parameters are automatically determined. This relieves designers from the tedious task of determining the parameters.

III. A PROOF-OF-CONCEPT

The image sharpening circuit [5] is used to build a proof-of-concept prototype. First, it performs a Gaussian smoothing:

$$R(i,j) = \frac{1}{273} \sum_{k=-2}^{2} \sum_{l=-2}^{2} G(k+2, l+2) \cdot I(i+k, j+l)$$

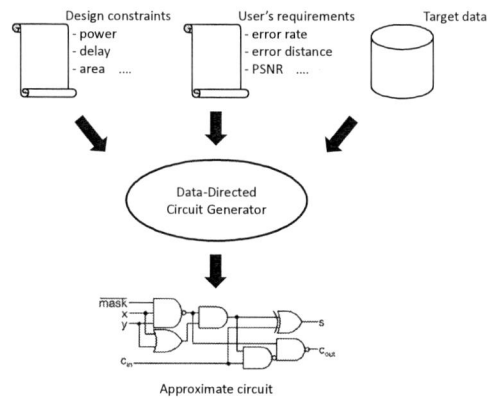

Figure 1. Data-Directed Circuit Generation

$$G = \begin{bmatrix} 1 & 4 & 7 & 4 & 1 \\ 4 & 16 & 26 & 16 & 4 \\ 7 & 26 & 41 & 26 & 7 \\ 4 & 16 & 26 & 16 & 4 \\ 1 & 4 & 7 & 4 & 1 \end{bmatrix}$$

where $I(i, j)$ and $R(i, j)$ are each pixel of the input and the smoothed images, and G is its Gaussian kernel. Second, it obtains the sharpened image S by $S=2I-R$. Multiplications are replaced by additions and shift operations. After that, only additions are approximated. Lower-part OR adder (LOA) [6] and Carry-Maskable Adder (CMA) [10,11] are chosen as the target of circuit generation in this prototype.

The LOA consists of the precise adder in the leading bits and OR gates in the trailing bits. This is because the lower bits are less important in accuracy than the upper ones. Although it is very simple, it is moderately accurate and is efficient both in power and in area. The LOA does not have dynamic configurability, and thus its configuration should be statically determined at the design time. The CMA consists of multiple Carry-Maskable Full Adders (CMFAs), which are connected in cascaded with the carry output from each CMFA connected to the carry input of the CMFA in the chain, just like a ripple carry adder (RCA). It is depicted in Fig. 2. The signal ¬mask configures the CMFA. When it is 0 (mask is 1), it works as an OR gate. At this time, C_{out} as well as C_{in} is fixed to be 0. Otherwise (mask is 0), it works as a precise full adder (FA). Hence, if CMFAs in the leading bits are configured as FAs and those in the trailing bits are configured as OR gates, it works similarly as to the LOA. The number of OR gates in the LOA

978-1-5386-7961-6/18 $31.00 © 2018 IEEE

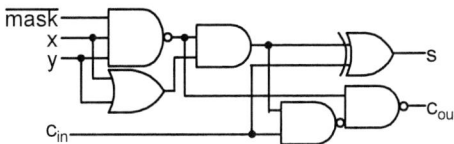

Figure 2.   CMFA

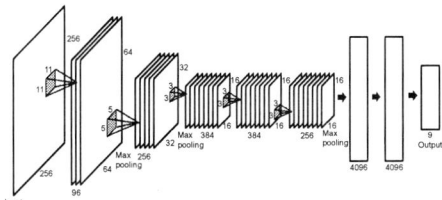

Figure 3.   AlexNet-based CNN

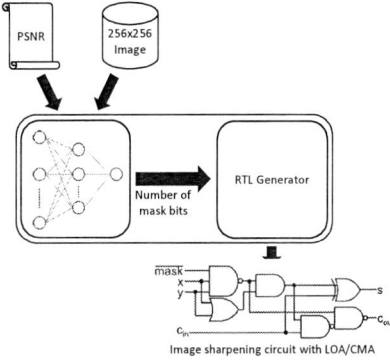

Figure 4.   Prototype Tool

and the number of the masked bits in the CMA can be determined in a same way. The noticeable difference is that the CMA has dynamic configurability.

A slight variation of AlexNet [4] is used to determine the bitwidth, which is the number of OR gates in the LOA or that of the masked bits in the CMA. The trailing eight bits are candidates of this choice. As shown in Fig. 3, it has five convolutional layers, three max pooling layers, and three fully-connected layers. TensorFlow [1] with Keras [14] is used to implement it. The input is a 256x256 grayscale bitmap image with 8-bit pixels. The output is the number of the mask bit: 0 to 8. Its training and the test sets are 10,000 and 2,000 images, which are randomly selected from ILSVRC2012 dataset [8]. The batch and the epoch sizes are 250 and 50, respectively.

The prototype design tool is depicted in Fig. 4. The representative images and user's requirement provided as PSNR are entered into the tool, which mainly consists of AlexNet and RTL generator. PSNR is defined as follows [2]:

$$PSNR = 10 \log_{10}(\frac{As_{max}^2}{MSE})$$

$$MSE = \frac{1}{X \cdot Y} \sum_{x=0}^{X-1} \sum_{y=0}^{Y-1} (s[x,y] - \tilde{s}[x,y])^2$$

where As_{max}, $s[x,y]$, and $\tilde{s}[x,y]$ are the maximum, the precise, and the approximate values of each pixel, respectively, and X and Y are the image dimensions. Due to the immaturity of the prototype, only 40dB of PSNR is currently selectable. AlexNet predicts the LSB width of the approximate adder. By receiving the bitwidth, the approximate adder, which satisfies the PSNR, is provided by the RTL generator.

In the experiments, the LSB width is correctly determined for 98.75% of 2,000 images and their dedicated approximate circuits are generated. Power, delay, and area of the sharpening circuit are evaluated by using Mentor Graphics ModelSim and Synopsys Design Compiler with NanGate 45nm library, In the case of Lena, which is not included in the training nor test sets, the optimal number of five is successfully selected for the LOA. The power, delay, and area are improved by 33.28%, 5.67%, and 21.86%, respectively. Fig. 5 shows the images processed

by the precise and by the LOA, which are almost nondistinguishable visually.

(a) Precise adder   (b) LOA (5-bit ORs)

Figure 5.   Sharpened Images

IV.   CONCLUSIONS

This paper proposes a data-directed methodology for designing approximate arithmetic circuits. Exploiting the characteristics of processed data, parameters configuring the target circuit are decided. A proof-of-concept prototype is build and the practicability of the methodology is confirmed.

ACKNOWLEDGMENT

This work was supported by JSPS KAKENHI (JP17K00088), by funds (No.175007 and No.177005) from the Central Research Institute of Fukuoka Univ., and by VDEC, the University of Tokyo in collaboration with Synopsys, Inc..

REFERENCES

[1]   M. Abadi, et al., arXiv:1603.04467, 2015.

[2]   D. Bull, "Communicating pictures," Academic Press, 2014.

[3]   J. Han, et al., doi:10.1109/ETS.2013.6569370, ETS, 2013.

[4]   A. Krizhevsky, doi:10.1145/3065386, CACM, 2017.

[5]   M. S. Lau, et al., doi:10.1145/1629395.1629434, CASES. 2009.

[6]   H. R. Mahdiani, et al., doi:10.1109/TCSI.2009.2027626, TCAS I, 2010.

[7]   K. Nepal, et al., doi: 10.7873/DATE.2014.374, DATE, 2014.

[8]   O. Russakovsky, et al., arXiv:1409.0575, 2014.

[9]   D. Sengupta, et al., doi:10.1145/3061639.3062314, DAC, 2017.

[10]   K. Tajima, et al., http://id.nii.ac.jp/1001/00185028/, 2018.

[11]   T. Yang, et al., doi:10.1109/ISQED.2018.8357311, ISQED, 2018.

[12]   T. Yang, et al., doi:10.1109/ASPDAC.2018.8297389, ASP-DAC, 2018.

[13]   S. Venkataramani, doi:10.1145/2228360.2228504, DAC, 2012.

[14]   https://github.com/keras-team/keras

Design of a Generic Security Interface for RISC-V Processors and its Applications

Hyunyoung Oh, Junmo Park, Myonghoon Yang, Dongil Hwang and Yunheung Paek

ECE and ISRC, Seoul National University
Seoul, Republic of Korea
{hyoh, jmpark, mhyang, dihwang}@sor.snu.ac.kr, ypaek@snu.ac.kr

Abstract— **In this paper, we propose the design of a generic security interface for RISC-V. This interface increases flexibility of security modules by creating an environment that can operate independently on a host processor. We also present an application using this interface for the memory protection. To check the feasibility of our idea, we implement an early prototype where a RISC-V processor is connected with the proposed hardware components using our interface. The empirical results show that our security interface has enabled a security module operated independently of the processor with no performance and low area overhead.**

Keywords; RISC-V, security interface, memory protection

I. INTRODUCTION

As the security threats become increasing steadily, several studies have proposed various security solutions that protect the underlying system against security attacks. Most of the early security solutions were implemented in software, and often suffered from a substantial performance overhead. To reduce the overhead, many researchers have proposed hardware-based security solutions. Conventional hardware-based solutions, however, have a drawback in that they cannot be utilized extensively since they require a permanent modification to the microarchitecture of the host processor [1]. Thus, without a permanent modification, it becomes necessary to have an interface that enables security modules to be used independently of a host processor. In this sense, we have developed a *security interface* that can extract the internal information of the host processor needed for security modules. This interface is connected to components of the pipeline structure inside the RISC-V Rocket core, which can obtain the internal information to recognize the control flow and data access patterns of the host. In addition, we present an application that can effectively perform memory protection by using the security interface.

II. ASSUMPTION AND THREAT MODEL

In this work, we target embedded systems employing processors available in today's market, which cannot afford MMUs for sophisticated virtual memory managements. We assume that the target system comes in the form of an SoC and therefore any external hardware can be attached during implementation.

As our adversaries, we assume that untrusted software modules can be installed in the target devices. These untrusted

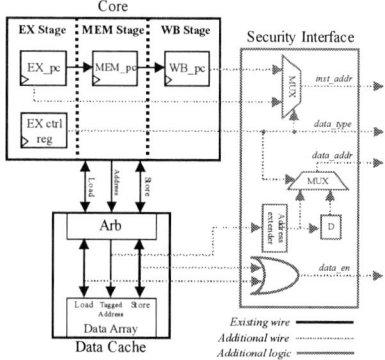

Figure 1. Information extraction through Security Interface

modules can lead to compromising an entire program when (1) the modules have vulnerabilities which can be exploited by an attacker or (2) the modules themselves are malicious. This is due to the fact that, without address protection between modules, a software module would be able to corrupt the data used by another module.

III. SECURITY INTERFACE

Security interface, as shown in Fig. 1, extracts the RISC-V processor's internal information such as instruction/data address and data access pattern. The security interface probes pipeline stages to get an instruction address and internal signals of data cache to get a data address. But synchronizing between those extracted addresses is not trivial. We observed that there exist some delay relations between them and founded that those relations are correlated to control signal of pipeline stages. Thus, we carefully multiplex each delayed address by using the control signal as a selector. And we also create the enable signal to indicate the validity of the output.

IV. EXEMPLARY APPLICATION FOR SECURITY: MEMORY PROTECTION

In this section, we propose a security module integrated together the security interface. This module, called *memory region protector* (MRP), supports the security function that is to establish the software isolation environment by checking the occurrence of illegal memory accesses. Fig. 2 shows the overall architecture of our MRP. To detect an illegal memory access, MRP utilizes a register file, called *access permission*

978-1-5386-7961-6/18 $31.00 © 2018 IEEE 40 ISOCC 2018

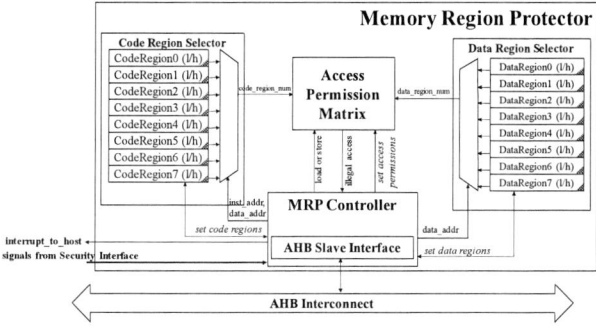

Figure 2.   Memory Region Protector architecture

matrix that indicates legal access permission. The access permission matrix maintains two types of relationship: code versus code region and code versus data regions, as described in TABLE I. After receiving an instruction address and a data address via the security interface, code region selector and data region selector find each region index respectively. And then as considering those region indexes with the received access type, MRP checks whether or not the access can be found in the access permission matrix. If not, the MRP controller interrupts the host CPU to report the illegal memory access.

During the boot-up, the code stored in PROM initializes the access permission matrix. When the code stored in PROM is executed, the processor launches an instruction. Over the system bus, the instruction is encoded by the bus protocol. On receiving the command, the *APB slave interface* in the *MRP controller* module decodes the command and sends the decoded information to the access permission matrix for configuring the registers. The MRP controller also updates the registers to define protected code and data regions in the *code region selector* and *data region selector*, respectively. To define the range of each code or data region, two registers are provided to set the low and high addresses of the region.

V.   EXPERMINETAL RESULTS

In our prototype, we use the Xilinx Zynq-7000 board and use a version 1.7 of RISC-V Rocket core parameterized by FPGA configuration DefaultFPGAConfig, as the host processor. In our implemented MRP, the number of code and data region are both configured to be eight. The bus compliant with AMBA AHB2 protocol [2] is used to interconnect all the modules in our prototype system.

We quantified the resources necessary for our hardware components in terms of lookup tables for logic (LUTs) and FFs with synthesis constraint of 33MHz clock rate. The design statistics show that, compared to the baseline Rocket core, our components incur the resource overhead of 6.07% and 2.96% for FFs and LUTs, respectively. We also estimated the gate count of our hardware components using Synopsys Design Compiler. With a commercial 45nm process library, the total gate-count of the proposed modules is 16,586(1,712 gates for the security interface, 12,828 gates for MRP and 2,046 gates for the access permission matrix).

The security interface incurs zero performance overhead because it extracts the internal information without changing the critical path of the host CPU. Using the interface, our MRP runs in parallel with the functional execution of the host. Hence, the access permission check of MRP also does not impact the performance of the target system.

TABLE I.   ACCESS PERMISSION MATRIX

SUBJECT \ OBJECT	Code Region0	Code Region1	Code Region2	Data Region0	Data Region1	Data Region2
Code Region0	RX	-	R	RW	-	RW
Code Region1	-	RX	-	-	R	-
Code Region2	-	R	RX	RW	R	RW

Access Permissions — R : Readable, W : Writable, X : eXecutable - : No access is permitted

As a hardware component that plays a role similar to MRP designed in this paper, there are MPU [3] of ARM Cortex series and PMP [4] which can be optionally attached to RISC-V processor recently. However, MRP can arbitrarily set the size of the area compared to these, providing more flexibility in setting up the system. Also, MRP provides at least equal or less overhead because it requires the same amount of registers to set the lower and upper bounds of an area.

VI.   CONCLUSION

We present a security interface that makes it possible to extract the internal information of RISC-V needed for the security modules. We have also demonstrated an exemplary application for security called MRP that can effectively implement a memory protection mechanism using a security interface. MRP plays the role of deciding the existence of invalid memory accesses by observing the pattern of data transfers outside the host. The experimental results showed that MRP which is our hardware-based solution successfully operated using a security interface and provides enhanced memory protection with low area and virtually no performance overhead. As a result, the proposed generic security interface has the potential to design more advanced security modules based on the internal information of RISC-V that has been successfully extracted.

ACKNOWLEDGMENT

This work was partly supported by Institute for Information & communications Technology Promotion(IITP) grant funded by the Korea government(MSIT) (No.2018-0-00230, Development on Autonomous Trust Enhancement Technology of IoT Device and Study on Adaptive IoT Security Open Architecture based on Global Standardization [TrusThingz Project]) and supported by Institute for Information & communications Technology Promotion(IITP) grant funded by the Korea government(MSIT) (No.2017-0-00213, Development of Cyber Self Mutation Technologies for Proactive Cyber Defense) and supported by the National Research Foundation of Korea(NRF) grant funded by the Korea government(MSIT) (NRF-2017R1A2A1A17069478)." and supported by IDEC.

REFERENCES

[1] Patrick Koeberl, Steffen Schulz, Ahmad-Reza Sadeghi, and Vijay Varadharajan. 2014. TrustLite: a security architecture for tiny embedded devices. In Proceedings of the Ninth European Conference on Computer Systems. ACM, 10.

[2] ARM 1999. AMBA Specification. ARM.

[3] ARM 2014. ARM Cortex-M7 Processor. ARM.

[4] Electrical Engineering and Computer Sciences University of California 2016. The RISC-V Instruction Set Manual Volume II: Privileged Architecture. Electrical Engineering and Computer Sciences University of California.

Convolutional Neural Network Accelerator with Reconfigurable Dataflow

Myungwoo Oh, Chaeeun Lee, Sanghun Lee, Youngho Seo, Sunwoo Kim, Jooho Wang and Chester Sungchung Park

Department of Electrical Engineering, Konkuk University, Seoul, South Korea

E-mail: {woper363, dlcodms123, sanghuniya, ddongseo11, sunwkim, joohowang, chester}@konkuk.ac.kr

Abstract— **Convolutional-Neural-Network (CNN) is used in broad applications. There are dataflows for convolutional layers in CNN such as row-stationary and weight-stationary. However, these dataflows have strengths and weaknesses. This paper analyzed two representative dataflows and introduce the dataflow-reconfigurable CNN accelerator that takes advantage of both dataflows.**

Keywords- CNN, dataflow, row-stationary, weight-stationary.

I. INTRODUCTION

Convolutional neural network has been applied to various applications with unprecedented accuracy. To handle this energy-consuming computation in an embedded environment, a highly energy efficient accelerator architecture is required. In this paper, we investigated the two dataflows, row-stationary (RS) and weight-stationary (WS), which are applied to several state-of-the-art CNN accelerator [1,2,3] and analyzed advantages and disadvantages of them. In computing multiple layers of convolution, the performance results are deeply affected by which dataflow is selected. Operating with appropriate dataflow according to the conditions leads to performance improvements and energy efficiency.

II. ANALYSIS OF DATAFLOWS AND SYSTEM

The RS dataflow is introduced in Eyeriss [1]. Each row of filter matrix and input feature map is transferred to the corresponding row and diagonal of PE array respectively. Each processing element (PE) executes a 1-D convolution. Partial sums are accumulated from bottom to top of the array. For the WS dataflow [2], each value of input feature map is broadcasted to multiple PEs and filter value is staying in each PE. Unlike RS, WS PE executes a single multiplication and accumulation (MAC) from top to bottom of the array. The main difference between these dataflows is data reuse. RS dataflow have reuse patterns of all data types (input, filter and output) by having scratch pad (spad) inside each PE. WS maximizes filter reuse by performing every MAC operations that require currently loaded filters. Therefore, WS needs large global buffer size in order to broadcast all inputs required. On the other hand, WS has a relatively simple PE structure and control logic compared to RS.

Fig. 1 shows PE array and memory hierarchy of the system. It is assumed that the CNN accelerator, like in Eyeriss [1], consists of a spatial array of 168 PEs, a 108KB global buffer, X-Y bus and local link. The memory level includes an off-chip DRAM, on-chip global buffer, X-Y bus with local link (array), and 520B spads per PE. (Register file) The number of PEs which are utilized by RS convolutional layer is determined by

filter size, output size and number of PE sets. The PE utilization in a WS convolutional layer of the PE array is affected by the number of filters with the number of columns and by filter size with the number of rows. Remaining PEs are not used.

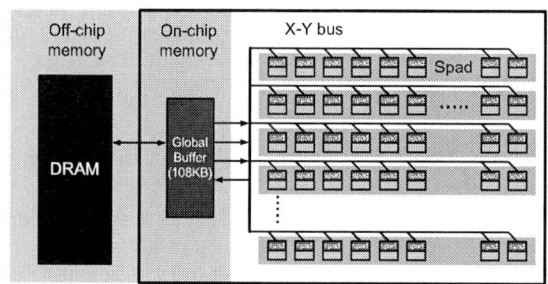

Fig 1. 12x14 PE array and memory hierarchy of the system

In a previous analysis [3], it is assumed that there is insufficient memory for WS, so additional DRAM access occurs. Another analysis [2] does include full-dimensional but two-dimensional convolution. We compared full-dimensional simulation result of RS and WS, (including batch, channels and filters) in the circumstance with enough global buffer size for WS.

TABLE 1   Number of memory access in the two dataflows

RS dataflow	Input	Filter	Output
DRAM	H^2C	R^2CM/n	E^2M
Global buffer	$H^2CM/(pt)$	R^2CM/n	$2E^2CM/(qr)$
Array	H^2RCM/p	ER^2CM/n	$E^2RCM/(qr)$
Register file	E^2R^2CM/p	E^2R^2CM	E^2R^2CM

WS dataflow	Input	Filter	Output
DRAM	H^2C	R^2CM	E^2M
Global buffer	H^2CM/x	R^2CM	$2E^2CM$
Array	H^2R^2CM	R^2CM	E^2R^2CM
Register file	H^2CM/x	R^2CM	E^2RCM

x :Number of different filters in PE array

Table 1 shows memory access counts of RS and WS on the four levels of memory hierarchy illustrated in Fig. 1. The CNN layer parameters follow Eyeriss [1]. The values shown in this table represent the major terms for memory access of each cases. Other terms have been omitted because they are an order of magnitude smaller than major terms.

RS shows low access count for DRAM, global buffer and array, because RS pre-load data from large memories to spad. RS can interleave MACs with q channels and p filters to increase data reuse inside PEs. It makes smaller access to

978-1-5386-7961-6/18 $31.00 © 2018 IEEE   42   ISOCC 2018

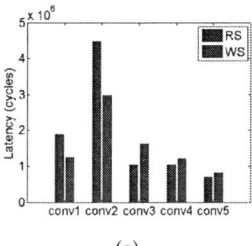
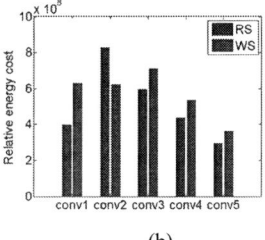

(a)   (b)

Fig. 2. Comparison of two dataflows about (a) latency and (b) relative energy cost

global buffer than WS. However, RS re-loads data from spads frequently, RF read access count becomes large. Relatively, WS have low access count for RF and filter data. WS must re-load input features from global buffer many times, but filter data reuse rate is maximized with very small RF access.

III. PROPOSED ARCHITECTURE AND SIMULATION

Based on Table 1 and relative energy cost table [3], energy consumption according to memory hierarchy can be obtained by multiplying Table 1 and memory level relative energy costs. We also designed cycle-accurate model of RS and WS to simulate latency of two dataflows.

In Fig. 2(a), WS is faster in layer 1 and 2, RS is faster in the others. Using spads, RS shows high overall data reusability. WS does not have spad, but it maximizes filter reusability. Weakness of WS is input reusability, but layer 1 and 2 have small channel parameter, so disadvantage of WS is reduced. In Fig. 2(b), WS is energy efficient in layer 2. RS can extend batch size, which reduce filter data DRAM access. Layer 1 and 2 needs a lot of global buffer memory, batch size is determined to be 1. However, stride of layer 1 causes large amount of unutilized input broadcast on WS dataflow. Therefore, RS is energy efficient in layer 1,3,4 and 5.

Energy costs and latency of two dataflows, RS and WS, are affected by layer parameters, so if the layer parameters change, the dataflows efficiency also changes. Layer parameter changes even in a single CNN network, but existing dataflow-fixed accelerators cannot select advantageous dataflow

In PE of Eyeriss [1], MAC is performed with input, filter and stored psum. However, WS does not reuse psum in PE, it need to perform MAC with incoming psum. Since existing RS PE structure does not support WS datapath, we implemented PE hardware that support both RS and WS dataflows. First, as illustrated in Fig.3 (a), modified RS PE can supports WS. The difference with existing RS PE is dataflow selection mux. In RS, it runs along the black line and WS runs as the red line. Second, we designed PE controller that shifts control policy for both dataflows on line.

Fig. 3(b) is the structure of RS PE controller. It need to generate the index to store and load data for spads and mux selection signal for receiving the input psum transmitted between PEs. Ready gen block generates input port ready signals by counting the handshake of the valid & ready signal of each incoming data type. Based on the number of each handshake from counters, MAC counter calculates current number of MACs and Mux select enables/initializes MAC. According to the MAC count, W index and R index indicate write and read indices of spads respectively.

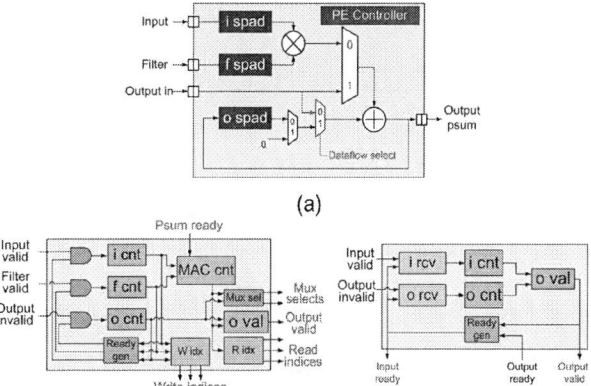

(a)

(b)   (c)

Fig. 3. (a) Reconfigurable PE. (b) RS and (c)WS PE controllers

Fig. 3(c) illustrates WS PE controller. With input receiver and output receiver blocks, PE receives only necessary inputs/psums. Counters provide numbers of enabled input data, output valid is generated with counter out. Ready gen makes ready signals for input port and receiver blocks. To support these dataflows efficiently, Counters, Ready gen, Output val blocks are shared with blocks of RS. As a result, additional hardware resource for dataflow-reconfigurable PE is minimized

As shown in Table 2, it is shown that the resource difference between proposed architecture and the Eyeriss architecture is negligible.

TABLE 2.   RTL synthesis result of Eyeriss and proposed PE

(TSMC 0.18um)	Eyeriss [1]	Proposed
Single PE (mm²)	0.66	0.67
PE array (mm²)	114.53	114.77

IV. CONCLUSION

In this paper, the dataflow-reconfigurable PE architecture is proposed. It can be utilized to take advantage of two dataflows, RS and WS selectively, with little additional resource. It switches appropriate dataflow while performing single network, online.

ACKNOWLEDGMENT

This research was supported by the MSIP (Ministry of Science, ICT & Future Planning), Korea, under the National Program for Excellence in SW (2018-0-00213) supervised by the IITP (Institute for Information & communications Technology Promotion).

REFERENCES

[1] Y.-H. Chen, T. Krishna, J. S. Emer, and V. Sze, "Eyeriss: An energyefficient reconfigurable accelerator for deep convolutional neural networks," *IEEE J. Solid-State Circuits*, vol. 52, no. 1, pp. 127–138, Jan. 2017.

[2] J. Jo, S. Kim, and I.-C. Park, "Energy-Efficient Convolution Architecture Based on Rescheduled Dataflow." IEEE Transactions on Circuits and Systems I: Regular Papers 99 (2018): 1-12.

[3] Y.-H. Chen, J. Emer, and V. Sze, "Eyeriss: A spatial architecture for energy-efficient dataflow for convolutional neural networks," in Proc. ACM/IEEE 43rd Annu. Int. Symp. Comput. Archit. (ISCA), Jun. 2016, pp. 367–379.

Novel Low Power FinFET SRAM Cell Design With Better Read and Writabilty For Cache Memory

Shreyash Patel

Electrical and Computer Engineering Department
Illinois Institute of Technology
Chicago, USA
spatel115@hawk.iit.edu

YoungBae Kim, Dr.Ken Choi

Electrical and Computer Engineering Department
Illinois Institute of Technology
Chicago, USA
ykim102@hawk.iit.edu, kchoi12@iit.edu

Abstract— **In this paper, we have proposed a novel low power Finfet based Sram cell design which has different read access path and write access path. The proposed cell achieves better performance compared to conventional sram cell. Proposed cell achieves 78% for hold 1 and 67% for hold 0 static power reduction than conventional 8T and 6T Sram cell. The proposed sram cell achieves 69% and 58% for write performance in terms of power and delay. Read delay is also reduced by approximately 89% and read power is reduced by 87%. Compared to conventional 8T Sram Cell. The proposed cell is read SNM free and has better SNM for write compared to Conventional Sram Cell.**

Keywords; *Static Power reduction; Read SNM free; FinFet (Short Gate); Novel 8T-Sram cell.*

I. INTRODUCTION

Conferring to ITRS road map, the SRAM arrays is occupying most of memory chip area. In VLSI design power consumption has become cynical in Sram cell. As Most of the cache area in a SOC is taken by Sram. i.e. Approximately 70% of the cache is filled with Srams. As the size of the die is reducing, so power supply has a major impact on power consumption of chip. After a certain point size of MOSFET cannot be reduced further because of gate leakage, supply voltage cannot be lowered, rise in cost per transistor as it was getting difficult as we move to lower technologies. That give rise to Finfet. The term "FIN" as it had fin shape gate which helps in overcoming the problems of DIBL effect, Ion/Ioff ratio, Leakage, Need of more current through the device. At Present, the FinFET has become primary IC technology because of its major reduction in leakage and performance improvement compared with flat plane CMOS. However, in the FinFET devices, Fin limits design space of the circuits that demand appropriate WL ratio transistor, especially SRAM cells, where the number of fins must be very refined to insure steady read and write. It was first used by Intel Corporation in 3rd Generation i7 series of processors from 22nm technology node. Major part of dynamic power is consumed during write operation as the write bit lines are confined to have full voltage swing between Vdd and 0. Main goal of this is paper is to offer enhanced write stability and low power altogether.

This work is supported by the Industrial Core Technology Development Program of MOTIE/KEIT, KOREA. [#10083639, Development of Camera-based Real-time Artificial Intelligence System for Detecting Driving Environment & Recognizing Objects on Road Simultaneously]

II. PROPOSED 8-T SRAM CELL DESIGN

To construct FinFet SRAM circuit, we used a BSIM-IMG model using HSPICE. To reduce the switching power consumption, we separated reading part and write part as shown in Figure 1. Unlike conventional 6T Sram cell, where proposed 8-T Sram cell uses only one driver transistor(X3). We have two transistor X7 and X8 which are access transistor's. Transistor X1 and X2 are pull-up transistor which are weaker than access transistor X7 and X8. The driver transistor is stronger than access transistor. This constraint is called read stability and writability. The proposed circuit is shown in Figure 1. Size of the FinFet can be determined by number of fins it has. The sizing for FinFet is shown in following Table 1.

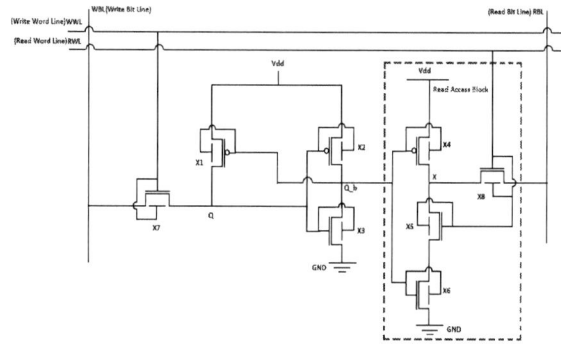

Figure 1. Proposed novel 8-T Sram Cell

Table I. SIZE OF THE FINFET

No	Name of transistor	Size (No. of fins)
1	X1	1
2	X2	1
3	X3	4
4	X4	1
5	X5	4
6	X6	4
7	X7	2
8	X8	2

A. Write Operation

In the proposed circuit, we have different read and write paths. Because of which write speed and write margin increases. It has only one write bit line (WBL) which avoids full voltage swing on bit line during write operation, to reduce the dynamic power. For write operation, if the data to be written is 1 then write bit line is charged to vdd. If the data to be written is 0 then write bit line is discharged to gnd.

B. Read Operation

For read mode, we have different read access block. During read mode X8 and X5 are switched on by asserting the read word line (RWL). Relying upon the value of Q_b, node x is either 0 or 1. For read 0, read bit line is pre-charged to vdd. Q_b is one because the Q is 0. So, when Q_b is 1, X6 get activated and then my read bit-line gets discharged through X8, X5, X6. Zero is read on read bit line. For read 1, Q is holding 1 and Q_B is holding 0 because of which node x is one and read bit line would not discharge and one can read on read bit line. Since X5 and X6 are series we have static power reduction.

III. SIMULATION AND RESULTS

For simulation, we have used a BSIM-IMG HSPICE model. Hence read and write have different access path we can show that our sram cell is faster than the conventional sram cell. The waveform and results are compared with the conventional sram. The static power is also calculated and compared with conventional sram.

Table 2. Static Power Consumption

Operations	Conventional (nw)	Proposed (nw)	Improvement
Hold 0	1.98	0.49	67%
Hold 1	1.80	0.39	78%

As seen in Table 2, due to X5 and X6 are connected in series during off condition we get the benefit of stack effect. Because of which we get improvement in static power consumption during when Sram is in standby mode or they are off. Advantage of different read and write part is that there is not much stress on bit-lines.

Table 3. Dynamic Power Consumption

Operation	Conventional (nw)	Proposed (nw)	Improvement
Write 1	457.2	188.8	58.1%
Write 0	327.5	102.7	69.3%
Read 0	41.7	4.9	88.1%
Read 1	39.4	4.2	89.4%

Table 4. Delay Comparison

Operation	Conventional (ps)	Proposed (ps)	Improvement
Write 0	36.8	13.97	62%
Write 1	28.47	8.94	68.8%
Read 0	21.4	3.12	85.1%
Read 1	18.2	1.83	90.15%

From the above table we see that the dynamic power for write and read is improved significantly. Dynamic power for read is improved by approximately 88% and for write is improved by approximately 65%. The wave forms of read and write operation our shown below in (Figure 2). We have also given pseudo data sequence (1100) to show read and write operation. Now with this sequence we first write 1 to cell holding 0, then we read 1. After that we write 0 to cell holding

1 and then we read 0. This way we have checked all the read and write operations. Separate table and wave form is shown below.

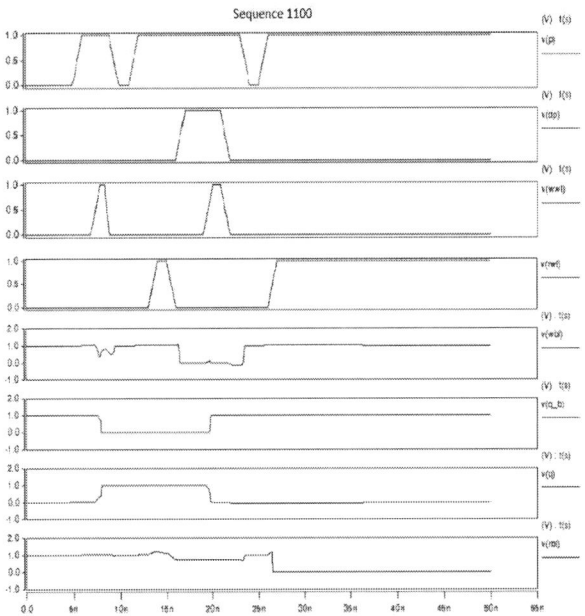

Figure 2. SRAM Read and Write Operation

IV. CONCLUSION

In this paper, we have proposed a low power and better read and write ability 8-T SRAM cell based on FinFet (SG) technology. The proposed cell achieves lower dynamic and static(leakage) power due to single-ended write bit line scheme, because of only one driver transistor and stack effect. The proposed cell also achieves better read stability, writability and its read SNM free because of different read and write paths and greater pull up ratio. So, because of individual read and write Sram speed increased such as on an average read speed and write speed is 65% and 88%. We see major improvement in terms of power also where static and dynamic power with 67% for hold zero and 79% for hold 1. 65% and 88% for write and read operations.

Acknowledgment

We thank our colleagues from KETI and KEIT who provided insight and expertise that greatly assisted the research and greatly improved the manuscript.

References

[1] Dr Brad D. Gaynor and Soha Hassoun "Fin Shape Impact on FinFET Leakage with Application to Multi-threshold and Ultra low-leakage FinFET Design", *Senior Member,* IEEE, 2014.

[2] Sanjana S R, Balaji Ramakrishna S, Samiksha, Roohila Banu, Prateek Shubham "DESIGN AND PERFORMANCE ANALYSIS OF 6T SRAM CELL IN 22nm CMOS AND FINFET TECHNOLOGY NODES",2017 International Conference on Recent Advances in Electronics and Communication Technology.

[3] Rajiv Joshi, Fellow, IEEE, Keunwoo Kim, Senior Member, IEEE, 1 Rouwaida Kanj "FinFET SRAM Design", Member, IEEE IBM TJ Watson Research Center, Yorktown Heights NY 10598 1 IBM Austin Research Labs, Austin Texas 78758

Low Cost Hardware Implementation of LEA-128 Encryption using Bit-Serial Technique

Byungjun Choi, Bohun Kim and Jongsun Park
School of Electrical Engineering, Korea University.

Abstract— **The Lightweight Encryption Algorithm (LEA) is a block cipher algorithm developed to provide confidentiality in high-speed environments, big data, cloud, and mobile devices. There is an issue that these security algorithms' modules must be small in area and power in order to be loaded onto resource constraint devices such as passive RFID. In this paper, for the first time in LEA which is one of the block ciphers, Bit-Serial technique is used to realize small area and power. Using Samsung 65nm process, we could obtain 2240GE which is 49.8% less in terms of area compared to the state-of-the-art area-optimized LEA-128. The IP designed in this paper can be applied as a security module to resource constraint devices.**

Keywords—LEA; Bit-Serial

I. INTRODUCTION

Currently, we are getting closer to the Internet of Things (IoT), where devices connected to the Internet can exchange information without human intervention. This connectivity is useful, however it cause the personal private information to be exposed to the outside, thus increasing the security risk. Therefore, in order to protect the information from this risk, the encryption must be a core technology essential for implementing the IoT system. In order to implement a cryptographic module in small devices such as IoT, smart devices and a passive RFID, the cryptographic module should have low area and power consumption while the throughput has sufficient margin [2]. According to this trend, the LEA algorithm, one of the lightweight block ciphers, has been proposed. LEA [1] is classified into LEA-128, LEA-192, and LEA-256 according to the length of the key, and can be selected according to the stability criteria. Currently, the LEA is being applied to the IoT field. However, previously implemented hardware has a lot of area to get into specific tiny devices [1], [2]. We have implemented the LEA-128 encryption algorithm, which is the most light and widely used, in a bit-serial technique to solve this limitation. Compared to the area-optimized hardware for existing LEA, area is reduced by 50.2%. In this paper, we describe the LEA-128 background algorithm in Section 2, and in Section 3, we describe the LEA-128 with the bit-serial method. In section 4, we compare the area with existing architectures of LEA and Section 5 is the conclusion of this paper.

II. LEA-128 ENCRYPTION ALGORITHM

LEA-128 receives 128-bit plaintext and 128-bit keys as input and generates 128-bit ciphertext. The encryption of LEA is performed in following orders. First, 128-bit plaintext is divided into four 32-bit words, represented by X as shown in Fig. 1 where confusion and diffusion are performed. Confusion is the

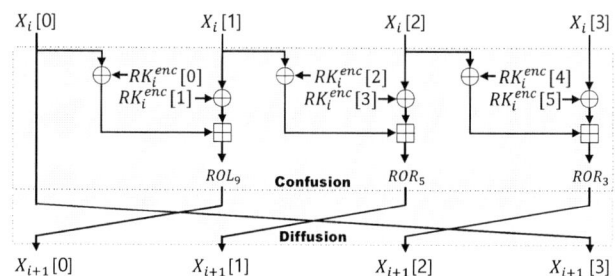

Fig 1. i-th round function.

property that hides the relationship between keys and ciphertexts, while diffusion is the property which hides the relationship between plaintext and ciphertext. This is done once, and when a new X is created, it is said that it has progressed a round, and the X after the total 24 rounds becomes the final ciphertext. Where ROL_i and ROR_i represent the left and right i-bit rotations, respectively.

The key used to perform confusion is shown in equation (1).

$$T[0] \leftarrow ROL_1(T[0] + ROL_i(\delta[i \bmod 4])),$$
$$T[1] \leftarrow ROL_3(T[1] + ROL_{i+1}(\delta[i \bmod 4])),$$
$$T[2] \leftarrow ROL_6(T[2] + ROL_{i+2}(\delta[i \bmod 4])),$$
$$T[3] \leftarrow ROL_{11}(T[3] + ROL_{i+3}(\delta[i \bmod 4])),$$
$$RK_i \leftarrow (T[0], T[1], T[2], T[1], T[3], T[1]). \qquad (1)$$

The delta values are as follows.

$$\delta[0] = 0\text{xc3efe9db}, \qquad \delta[1] = 0\text{x44626b02},$$
$$\delta[2] = 0\text{x79e27c8a}, \qquad \delta[3] = 0\text{x78df30ec}$$

III. PROPOSED BIT-SERIAL ARCHITECTURE OF LEA-128

In this section, a bit-serial architecture of LEA is proposed. Compared to other implementations, the bit-serial architecture can achieve the smallest area and lower power consumption trading-off against throughput. This architecture performs a bit-wise operation and all data path is in one bit. Moreover, the registers to store text and key are converted to shift registers which are always shifted without any exception, to eliminate additional multiplexer(MUX) of the flip-flops. This scheme can reduce the area of filp-flops by approximately 60%, which occupied the largest portion of the total area in conventional implementation methods [2]. However, straightforward implementation of serial architecture of LEA is interrupted because of the rotation operation. The problem occurs since the rotation operation extracts data from an intermediate position in

978-1-5386-7961-6/18 $31.00 © 2018 IEEE

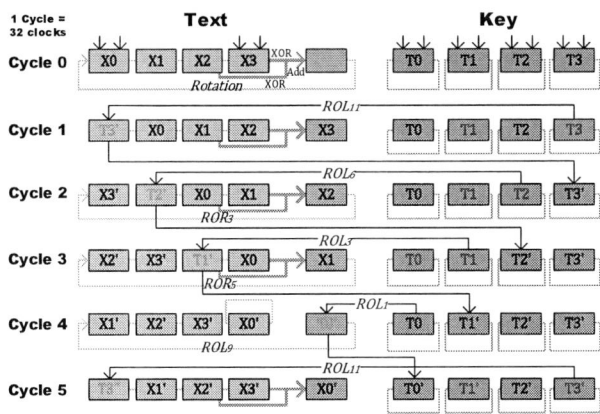

Fig 2. Dataflow of bit-serial LEA-128

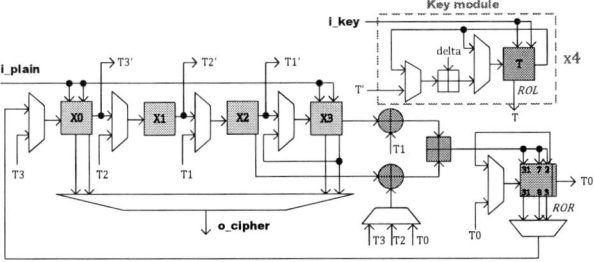

Fig 3. Hardware architecture of bit-serial LEA-128

a register, whereas the addition should be done at LSB of register. Otherwise, we have to hold the data stored in the registers to adjust the timing, resulting in area overhead for controlling registers. As a solution, we introduced additional flop-flops to implement proposed architecture with minimal overhead. It allows all rotations except delta selection with relatively low hardware cost. It also enables diffusion operation by performing 32-bit left rotation, which completes the round function for encryption. Fig. 2 illustrates the dataflow of the proposed architecture. The yellow boxes are text register and the cyan boxes are key register while each unit is 32-bit flip-flops. Every flip-flops are triggered at every clock cycle and shifted to the next flip-flops connected by the lines. On the other hand, the round function is divided into the two steps. First, fundamental computation of LEA encryption, expressed as a violet arrow, is composed of two XOR operations and an addition. The two XOR operations are performed between the extracted texts and designated keys for the cycle. And the results are added to each other. These bit-wise operations are performed from the LSB. The orange additional text register is controlled to perform the rotation of the data after the previous addition and it is forwarded to the first unit. In the fourth cycle, the last 32-bit texts are rotated by itself so that the data stored in the register return to their positions by the end of the cycle. As a result, the round function of encryption including diffusion is completed. Meanwhile, at every cycle, one unit remains unused. While key scheduler provides a key for the round function, the key is stored at the relevant unit each cycle simultaneously. At the next cycle, the stored key is reloaded to the key register from the LSB sequentially. This process allows the rotation of the key module to be performed with a single port. Fig. 3 shows the detail of the proposed bit-serial LEA-128 hardware architecture.

Table 1 AREA COMPARISON OF LEA-128 IMPLEMENTATIONS

	Area (GE)			Tech. (nm)
	FOM-opt.	Area-opt.	Bit-serial	
[1]	5426	3826	-	UMC 130
This work	6323	4458	2240	Samsung 65

The four key modules are in the whole key scheduler. Each key module has its own position of the output port to perform a rotation. In LEA-128, the odd keys (*RK[1], RK[3], RK[5]*) are T1, and the even keys (*RK[0], RK[2], RK[4]*) are T0, T2, T3. Therefore we shared the delta value between T0, T1, T2 to minimize the cost since delta selection is performed by several MUX. In the additional text register, there are also multiple input and output ports at specific positions to perform rotations of the round function. For instance, to perform a ROR_3, outputs are extracted from 3rd bit and the next inputs are entered into 2nd bit at the same time. The register consists of 33-bit flip-flops including 1-bit flip-flop for the key transfer.

IV. IMPLEMENTATION RESULTS AND COMPARISONS

The proposed LEA-128 bit-serial architecture has been implemented using Samsung 65nm CMOS technology at 100MHz. We have used Synopsys Design Compiler from to synthesize the RTL of the designs. Table 1 shows the results comparisons between the previous works and the proposed design. In case of the FOM-optimized implementation, the targeted metric is throughput/area. It processes one round function in one clock cycle. On the other hand, the area-optimized implementation has less area with 7 times less throughput than the FOM-optimized implementation. We implemented the architectures of the previous work using our library for fair comparison. As a result, the proposed bit-serial architecture shows outstanding results in terms of area. We reduced area by 50.2% compared to the area-optimized architecture and by 35.4% compared to the FOM-optimized architecture.

V. CONCLUSION

In this paper, we have proposed a bit-serial architecture of LEA-128 encryption. By reducing data path and MUX of flip-flops, we have achieved 2240GE and 50.2% area of state-of-the-art area-optimized implementation while trading-off against throughput. In conclusion, the proposed bit-serial LEA-128 can be applied to tiny devices like a passive RFID.

ACKNOWLEDGMENT

This work was supported as part of Military Crypto Research Center(UD170109ED) funded by Defense Acquisition Program Administration(DAPA) and Agency for Defense Development(ADD).

REFERENCES

[1] Hong D, Lee JK, Kim DC, Kwon D, Ryu KH, Lee DG, "LEA: A 128-bit block cipher for fast encryption on common processors," InInternational Workshop on Information Security Applications 2013 Aug 19 pp. 3-27.

[2] Jean J, Moradi A, Peyrin T, Sasdrich P. "Bit-sliding: a generic technique for bit-serial implementations of SPN-based primitives-applications to AES, PRESENT and SKINNY," CHES 2017. 2017:687-707.

978-1-5386-7961-6/18 $31.00 © 2018 IEEE

A Cost-Effective High Accuracy Auto-Trimming System without Tester Constraint for Low-End Embedded Flash Memory

Junichi Suzuki, Junichi Yamashita, Masami Hanyu, Masamichi Ido, Tatsuya Saito,
Yasuhiro Nakashima, Masanori Hayashikoshi, Yukiyoshi Kiyota
Renesas Electronics Corporation, Tokyo, Japan
junichi.suzuki.xe@renesas.com

Abstract— As the demand and production volume for embedded flash MCUs increase, their flash memory test time reduction is getting more important. Among them, trimming test occupies to a certain extent. To decrease it, a dedicated on-chip test circuit of a current comparator can be an answer, but such a current comparator tends to consume large area when high precision is needed. In this paper, we proposed a new current judgement circuit for reference current trimming, of small area and high precision. A test chip of 110 nm process embedded flash memory for MCUs with the new trimming circuit has been fabricated and the new circuit effects have been confirmed by the test chip. The 0.59 percent of Flash macro size can be reduced with this proposed auto-trimming circuit.

Keywords—*test time reduction, reference current, embedded flash memory, test circuit*

I. INTRODUCTION

In recent years, the number of MCUs used for electronic devices including IoT has been increasing. To respond such a strong demand and to produce so many MCUs, high production throughput, reducing their test time, is very important. Many MCUs have on-chip flash memory in many cases, and test time for the flash memory often occupy a significant percentage in these embedded flash MCUs test time. To reduce test time, both test time reduction per chip and chip

number increasing in multi-chip testing are necessary. An on-chip auto-trimming system [1] can solve these issues. However, due to high accuracy requirement for analog circuits, the circuit area size gets large. In this paper, we propose a new judgement circuit for reference current (Iref) trimming with small area and high precision for auto-trimming system, of fast test time.

II. IREF TRIMMING CURCUIT

An example of Iref trimming circuit is shown in Fig. 1. Iref is the current flowing through the Ref Tr (SA), and IrefR is the current flowing through Ref Tr (Replica) in the Iref test circuit. The Ref Tr (Replica) is a replica of the Ref Tr (SA), and its shape and size are designed to be the same as those of Ref Tr (SA).

In this circuit, when trimming Iref, the gate voltage of Ref Tr (Source) is set by the voltage ladder circuit using a trimming code signal. In the test, the current IrefR flowing through Ref Tr (Replica) in Iref test circuit is measured by a tester. The proper trimming code for Iref in the target range is searched with the tester measurement and determines Iref for the read operation.

Current measurement by a tester requires settling time for stabilization of the measurement system including the tester itself, test board, etc. It takes more than 100 ms. Since it is quite slower than the signal operation in the chip, if the current is measured a plurality of times, the test time will be increased a lot. Therefore, we propose the Iref test circuit without

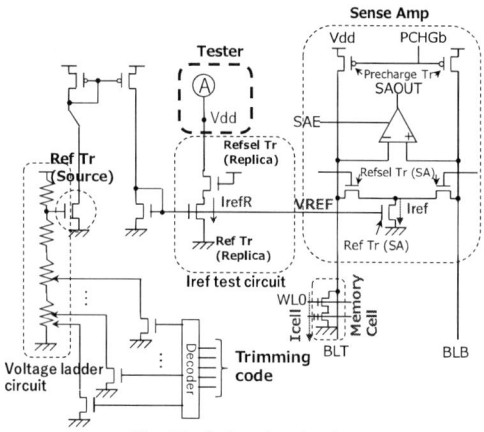

Fig. 1 Iref trimming circuit.

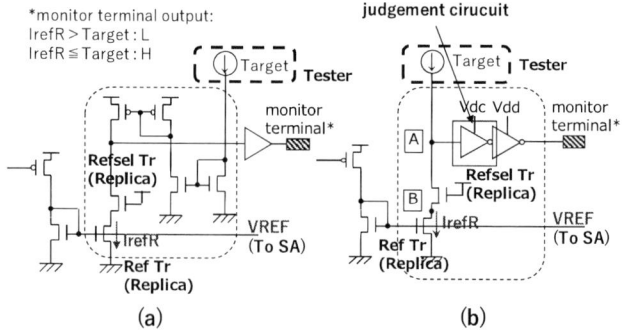

Fig. 2 Current comparator Iref test circuits: (a) Conventional (b) Proposed.

constraint of the tester.

III. ON-CHIP IREF TEST CIRCUIT

A. Conventional Iref test circuit

To achieve fast trimming, we adopted the above-mentioned auto-trimming system basically. In the auto-trimming system, an on-chip voltage comparator is used, and the trimming code is set automatically. Although IrefR is a current value, the system can be adopted with using a current comparator [2]. However, to compensate variations, the size of the analog circuit, particularly the transistor size of the mirror circuit increases, and the total chip size increases. For example, Fig. 2(a) is a current comparator using simple mirror circuits. The target current is copied by the mirror circuit, and to keep mirror errors small enough, using large size transistors is inevitable.

B. Proposed Iref test circuit

To have the auto-trimming system with smaller area size, we contrived and propose a new Iref test circuit for Iref trimming in Fig. 2(b). The target current is applied from the outside chip (a Tester), IrefR is switched on-chip by the trimming code, and the judgement is made by the judgement circuit.

In this circuit, to produce IrefR as equal as possible with Iref in Fig. 1, it is important to set the voltages at point A and B equal to those at BLB and the drain of Ref Tr (SA). Those can be realized by setting the power supply (Vdc) to twice Vdd for judgement circuit. Thereby, the judgement circuit can be made extremely small. The size and gate voltage of Refsel Tr (Replica) is set to same as those of Ref Tr (SA). With those treatment, we can get IrefR almost equal to Iref.

Fig. 3(a) shows the measurement variance in the conventional Iref test circuit using the mirror circuit by Monte Carlo simulation, and the measurement variance in the proposed Iref test circuit by the normal probability distribution diagram. For the conventional circuit, several transistor sizes are plotted. In the proposed test circuit, the influence of transistor variation is very small.

Fig. 3(b) shows the relation between 3σ variation of IrefR and total gate area of each circuit. The target range of IrefR is $\pm$ 0.5 uA from the trimming code width. In order to obtain the variation of IrefR $\pm 3\sigma$ less than $\pm$ 0.5 uA, the total gate size is necessary larger than 1300 μm2 for conventional, but for the proposed circuit only 4 μm2 is required. Thus, the area size is much decreased by using our proposed Iref trimming circuits for auto-trimming system.

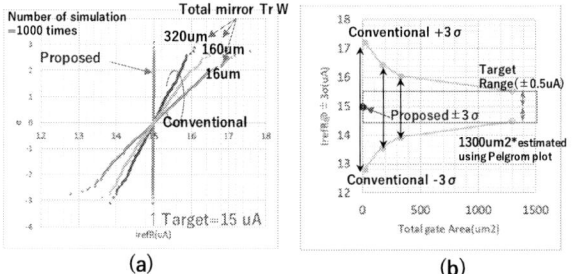

(a)   **(b)**

Fig.3 Simulation Result (a): normal probability distribution of measured IrefR: conventional vs proposed (b): Total Tr gate area and IrefR variation

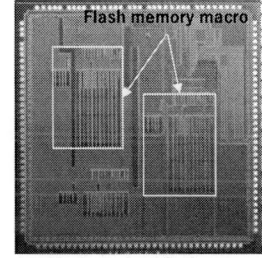

Fig. 4 A microphotograph of flash memory test chip.

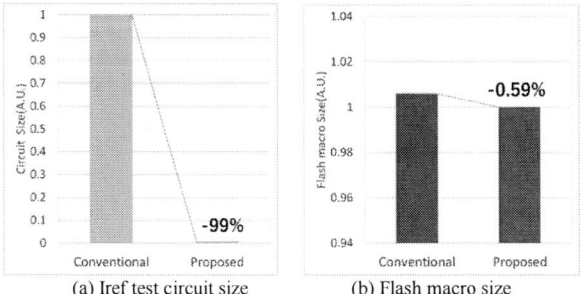

(a) Iref test circuit size   (b) Flash macro size

Fig.5 Conventional and Proposed Auto-trimming size comparison.

IV. DESIGN AND EVALUATION RESULT OF THE TEST CHIP

A microphotograph of the flash memory macro on the test chip and its features are shown in Fig. 4. Fig. 5 compares the size of the flash memory macro and the test time, in case of not using the auto-trimming, using the conventional auto-trimming, and using the proposed auto-trimming. In this comparison, the trimming code is 5 bits, flash memory macros are two of 64 KB and 4 KB, and with binary search. This proposed circuit reduces 99 percent of Iref test circuit area as shown in Fig. 5(a). The effect of the area reduction for flash memory macro with 0.59 percent may look small as shown in Fig. 5(b), however, this cost effect is large enough from the stand point of cost-conscious low-end MCUs.

V. CONCLUSION

We reduced the trimming test time by using auto-trimming system in 110 nm embedded flash memory. We contrived and proposed a new current judgement circuit for Iref trimming of small area size and sufficient accuracy. Furthermore, we confirmed its accuracy and area size. With the proposed system, fast test time like the conventional auto-trimming and small area size almost equal to non-auto-trimming can be achieved.

REFERENCES

[1] Hiroyuki Tanikawa et al., "An Internal Voltage Generation System of Flash Memory Module Embedded in a Microcontroller," IEEE Asian Solid-State Circuits Conference, 2005, pp 121-124.

[2] Dileep Reddy Desai et al., "Design of an accurate min-max current selector," 53rd IEEE International Midwest Symposium on Circuits and Systems, 2010, pp 562-565

Integration of Retention-aware Refresh and BISR Techniques for DRAM Refresh Power Reduction

Wei-Kai Cheng, Jian-Kai Chen
Department of Information and Computer Engineering
Chung Yuan Christian University
Taoyuan City, Taiwan
wkcheng@cycu.edu.tw, jainkaic@gmail.com

Shih-Hsu Huang
Department of Electronic Engineering
Chung Yuan Christian University
Taoyuan City, Taiwan
shhuang@cycu.edu.tw

Abstract—**Due to few cells in memory that have shorter retention time, DRAM controller have to raise the refresh frequency to keep data integrity, and hence produce unnecessary refresh for the other normal cells, which result in large refresh overhead and the delay of memory access. In this paper, we propose an integration scheme to integrate retention-aware refresh and BISR techniques. Based on the RAAR method, our strategy can choose the most appropriate way of weak cell fixing to minimize the waste of non-weak row refresh. A dynamic programming algorithm with a state transition equation is proposed to resolve this problem. Experimental results show that with this BISR integration scheme, we can further reduce refresh power than without applying it.**

Keywords; DRAM; retention time; retention-aware refresh; BISR (built-in self-repair)

I. INTRODUCTION

DRAM cell is composed of a transistor and a capacitor. Because of the capacitor has the characteristic of leakage, it must be periodically refreshed. The time required for this periodic refresh is called the retention time. When DRAM execute refresh operation, memory request must be stopped until the refresh operation finished, bring the delay of memory access. Research [1] showed that in a 32GB DDR3 DRAM, there are about only 30 cells with retention time less than 128ms, and only about 1000 cells require refresh interval shorter than 256ms. However, due to weak data storing ability of these cells, the whole memory cells are forced to refresh together and cause too many unnecessary refresh, result in large refresh power consumption and performance degradation.

In research [2], DRAM refresh weak rows by using RAAR technique, only parts of banks which contain weak rows have to be refreshed frequently. However, because it chose the same refresh scheme all over the RAAR period, this method still waste quite number of refresh time for memory banks that do not contain weak rows but need to be refreshed simultaneously with other memory banks that contain weak rows. Research [3] further improve this method by integrating all the RAAR refresh schemes and quarter-bank refresh technique. Under the premise of given weak rows distribution, and with the consideration of refresh overhead and refresh energy, it chose the optimal refresh scheme in each refresh time to fit the different weak rows distribution.

This work was supported in part by Ministry of Science and Technology, Taiwan. (Grant number MOST 107-2218-E-033-008)

As mentioned in research [4], even if we choose the optimal refresh method corresponding to the refresh bundle [5], it still contain lots of normal memory cells in the refresh, result in unavoidable memory access delay and extra refresh power consumption. In contrast to the bank reordering technique proposed in research [4], we propose an integration scheme to integrate retention-aware refresh and BISR techniques.

As shown in Fig. 1, there exist spare rows and spare columns in DRAM architecture to fix faulty cells. In this paper, we propose that except fixing faulty cells, BISR technique can also be applied to minimize the waste of non-weak row refresh and further degrade the consumption of refresh power. Based on RAAR method, with the given distribution of weak rows, the available number of spare rows and spare columns, our strategy can choose the most appropriate way of weak cell fixing by the dynamic programming algorithm with a state transition equation.

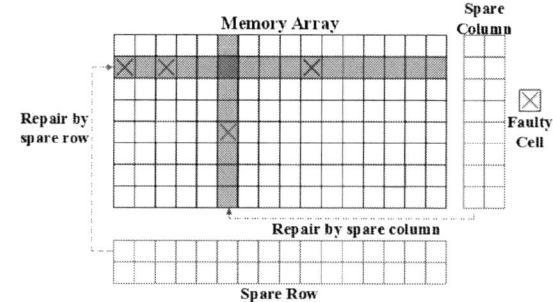

Figure 1. BISR repair schematic.

II. PROROSED METHODOLOGY

A. The best repair scheme for each row group

On DRAM refresh, all rows are evenly partitioned into row groups, and memory controller issue commands to refresh each row group sequentially. Fig. 2 shows all-bank refresh, per-bank refresh and partial-bank refresh proposed in research [2].

In this paper, we would like to reduce refresh overhead by repairing weak cells. Therefore, we firstly need to find out the best repair solution for each row group, based on the tRFC (recharge time of memory cells) ratio of different refresh schemes [2], as shown in Table I.

978-1-5386-7961-6/18 $31.00 © 2018 IEEE

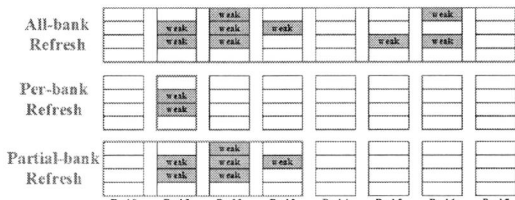

Figure 2. Refresh method schematic.

Table I. tRFC RATIO

Refresh Method	tRFC
All-bank Refresh	K
Partial-bank Refresh	K / 1.32
Per-bank Refresh	K / 2.3

The information in Table I figures out how much tRFC can be reduced after repair. We use the example in Fig. 3 to illustrate how to find the best repair scheme for a row group. Before the repair, the refresh method is all-bank. After we repair the weak row of bank1, bank2 and bank4, its refresh method changed to per-bank. Bring K of Table I to be 1, tRFC of per-bank refresh is 0.435, and tRFC reduction is 0.565. Because it needs five spare rows for this repair scheme, in average one spare row can reduce tRFC by 0.113. Table II shows the results of all repair scheme, the none-refresh scheme is the best one because of highest tRFC reduction per spare row in average..

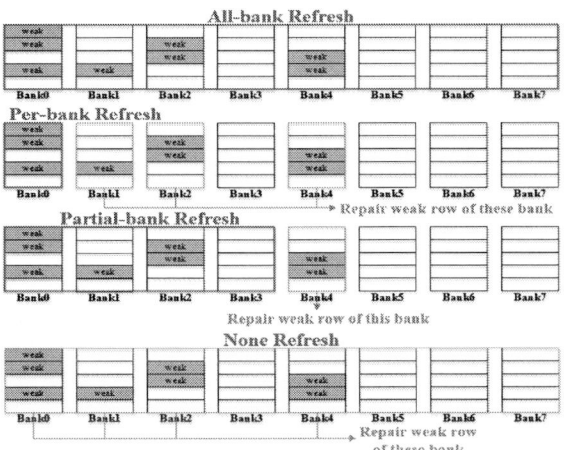

Figure 3. Repair scheme for different refresh methods.

Table II. RESULT OF ALL REPAIR SCHEME

Refresh Method	tRFC Reduction	Spare Row Cost	tRFC_SR avg.
All-bank	0	0	0
Partial-bank	0.242	2	0.121
Per-bank	0.565	5	0.113
None	1	8	0.125

B. tate-transition equation of Dynamic Programming

The state transition equation for dynamic programming is shown as below. Notation n is the number of row groups contain weak rows. Notation **tRFC Reduce** is reduction of tRFC in the best repair scheme for a row group. Notation **needSR** is how many spare rows are needed to repair this row group. Notation **SR** is available number of spare rows. The largest collection of sum of **tRFC Reduce** are the row groups we will repair.

$$f[n, SR] = \begin{cases} f[n-1, SR], & SR < needSR \\ max\{f[n-1, SR], f[n-1, SR-needSR] + tRFCReduce\}, & SR \geq needSR \\ 0, & nSR = 0 \end{cases}$$

III. EVALUATION AND RESULTS

We integrate gem5 and DRAMSim2 as our simulation environment, with benchmarks form Spec2006. The system setting parameters are shown in TABLE III, and there are 60 spare cells for BISR. We compare five refresh schemes with auto refresh baseline, Fig. 4 shows the comparison results, in which "Repair" is the proposed repair scheme, "Integrate" is the refresh scheme of research [3], and the other three are the refresh schemes of research [2]. Experimental results show that our repair technique can further improve integration scheme of research [3], and has the best results.

Table III. System Configurations

Processor	1 core, 3.2GHz
Cache	L1-I cache 128KB, L1-D cache 128KB, L2 cache 4MB
DRAM	Micron MT41K1G8,32GB
Weak row	0.15% weak row, random distribution

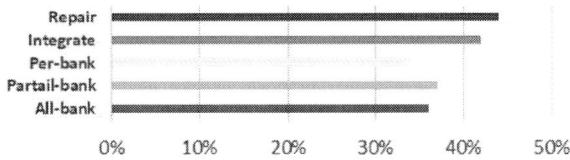

Figure 4. Refresh energy reduction normalized to baseline auto refresh

REFERENCES

[1] Jamie Liu, Ben Jaiyen, Richard Veras, and Onur Mutlu, "RAIDR: Retention-Aware Intelligent DRAM Refresh", International Symposium on Computer Architecture (ISCA), June 2012.

[2] Wei-Kai Cheng, Po-Yuan Shen, "Retention-aware Refresh Techniques for DRAM Refresh Power Reduction", Workshop on Synthesis And System Integration of Mixed Information technologies (SASIMI), October 2016.

[3] Wei-Kai Cheng, Xin-Lun Li, Jian-Kai Chen, "Integration Scheme for Retention-aware DRAM Refresh", International Conference on Electron Devices and Solid-State Circuits (EDSSC), October 2017.

[4] Wei-Kai Cheng, Xin-Lun Li, Jian-Kai Chen, "DRAM Refresh Improvement with Bank Reordering", International Symposium on Next-Generation Electronics (ISNE), May 2018.

[5] Ishan G. Thakkar, Sudeep Pasricha, "Massed Refresh: An Energy-Efficient Technique to Reduce Refresh Overhead in Hybrid Memory Cube Architectures", International Conference on VLSI Design and 15th International Conference on Embedded Systems (VLSID), January 2016.

Study on Intel CPU-FPGA Architecture

Security Perspective

Jinyong Lee

Electronics and Telecommunications Research Institute (ETRI)
Hyper-Connected Communication Research Laboratory,
System Security Research Group, Daejeon, South Korea
jinyonglee@etri.re.kr

Abstract— **To respond to constantly changing workloads and emerging technologies for data centers, Intel decided to maximize performance while minimizing power consumption by providing a hybrid CPU-FPGA system packaged in a single processor. The internal FPGA can be used to implement hardware logics that can accelerate specific algorithms for data centers, but it can also embody hardware modules that can improve the overall security of the system. This paper first examines this newly launched processor and presents some possible use cases of this system for enhanced security.**

Keywords; Intel Xeon processor, FPGA, Security monitor

I. INTRODUCTION

Recently, Intel has announced a new processor line-up, Intel Xeon Scalable Processor 6138P [1]. The distinguishing feature of 6138P compared to former ones is that it includes, in its package, a high performance field programmable logic array (FPGA). The reason for this integration can be attributed to the demands of today's data centers: high performance and power efficiency. The best way to meet these contradicting demands would be to provide application specific hardware accelerators as they are known to have small area and low power consumption while attaining high performance. However, owing to the fact that the data center workloads are quite diverse and tend to change as new applications emerge, Intel has taken another approach. They decided to provide customers with more hardware resource, FPGA in this case, that is flexible enough for users to customize freely. In doing so, customers can create next generation data center systems with flexible workload-optimized performance and power efficiency.

As one can imagine, the internal FPGA can be used to implement hardware modules that speed up certain algorithms frequently used in data centers such as data preprocessing/manipulation or enhance the performance of emerging technologies like inference engine for machine learning. However, there may be another use cases, enhancing the system security, of which importance is recently and rapidly been emphasized. In this paper, we investigate the processor, in the goal of improving security, and present some possible use cases that can be realized with the help of this newly added hardware feature.

II. BACKGROUNDS

A. Intel Xeon Scalable Processor

Figure 1 shows a block diagram of the processor. As can be seen, it includes the Intel Arria 10 GX 1150, which is at the high end of Arria line, providing 1.5 million logic elements and DSP blocks capable of delivering 1.366 single-precision teraflops. There are three Ultra Path Interconnect (UPI) links in the processor, and the two below are used to connect to other processors to form a NUMA system and the remaining one for communicating the FPGA. It has six DDR4 memory controllers, two PCI-Express 3.0 x16 slots and one DMI x4 slot. The remaining PCI-Express 3.0 x8 slots are reserved for connecting the Arria 10 GX FPGA to the CPU. The internal FPGA is known to provide up to 160Gbps of I/O bandwidth, and can implement its own cache memory that is coherent to the CPU, thanks to the cache-coherent interconnect, UPI.

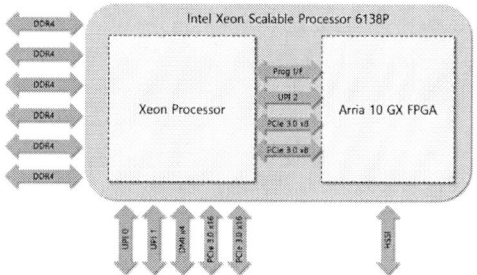

Figure 1. Diagram of Intel Xeon 6138P Processor

III. POSSIBLE USE CASES FOR SECURITY

A. Memory Integrity Checking

As the name suggests, memory integrity checking requires constant acquisition of the content in system memory and verification of its integrity. One prominent example of this, in the field of security, is the kernel integrity monitoring, and there has been rich body of research conducted in academia. Especially, monitoring with an independent hardware component has been proposed, quite recently, in literature. For instance, Hollingworth et al. [2] suggested the use of Symmetric Multi-Processor (SMP) for monitoring. Later, Petroni et al. proposed Copilot system [3], which a kernel runtime integrity monitor is implemented on a PCI-card. J. Wang et al. proposed HyperCheck [4], an integrity monitor for hypervisors, implemented on a PCI card and operated with the help of System Management Mode (SMM) [5]. These works in common suggested a mechanism in which a hardware monitoring module is deployed on some hardware resource located outside the CPU, such as a PCI-card, and periodically monitors the content of DRAM, thus investigating the integrity of the content.

The same mechanism, we suspect, can be realized in the newly launched processor without much trouble. The hardware monitor can now be relocated into the internal FPGA, and yet it would be able to perform the same task of integrity monitoring. The acquisition of memory can be performed via the interconnects the Xeon processor provides; as depicted in Figure 2, the processor possesses numerous channels connected to DRAM, such as UPI and PCIe. Even though the exact latency between the FPGA and the memory controller located on the Xeon processor is still unknown, but the latency would be lower than that of the original implementation since the later requires traveling all over the off-chip PCI-Express bus to access memory.

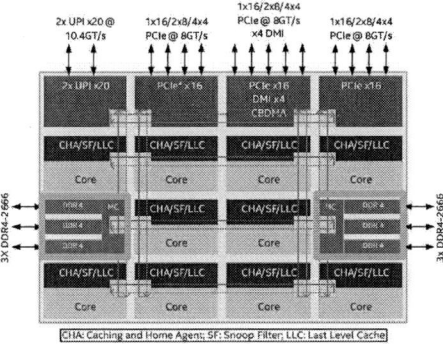

Figure 2. Memory subsystem of Xeon Processor [6]

B. CPU Behavior Monitoring

The main role of CPU behavior monitors is to watch the execution patterns of a target program running on the CPU. There can be various reasons for the monitoring, but our focus lies on schemes that try to find the existence of attacks by analyzing execution patterns. Many researchers suggested hardware-based monitoring schemes, focusing on enforcing control flow integrity (CFI) or tracking data flow information during the code execution [7,8].

In our opinion, such hardware-based monitoring schemes can be implemented with efficiency on the newly proposed processor. Figure 3 shows one example of monitoring schemes that realizes dynamic information flow tracking (DIFT). As seen in the figure, the hardware DIFT engine (dubbed as PAU) is located outside the CPU and, for its operation, it requires constant communication with the CPU and accessibility to the system memory. The researchers in [7] suggested the overall concept, and experimentally implemented it on FPGA-based prototyping system. We suspect that the system can easily be implemented on the Xeon processor, placing the DIFT engine in its internal FPGA, utilizing the abundant interconnect already implemented in the processor.

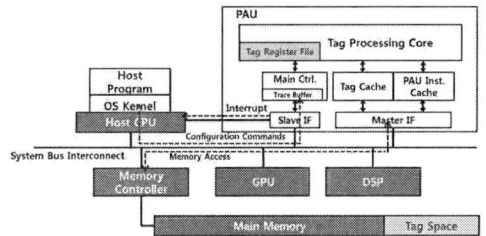

Figure 3. Dynamic information flow tracking system [7]

IV. CONCLUSION

Although this new processor is currently available, for now, to selected few customers and the release of detailed specification is still on the way, this processor opens up opportunities for users who seek to provide best performance/efficiency of their workloads. As demonstrated in this paper, it can also be used to enhance system security, by implementing hardware-based security modules in its internal FPGA, utilizing rich interconnects to communicate with the CPU and the system memory. Our future work will focus on realizing the presented systems on a real processor and examining the implications and difficulties that arise in the process.

ACKNOWLEDGMENT (HEADING 5)

This work was supported by Institute for Information & communications Technology Promotion(IITP) grant funded by the Korea government(MSIT) (No.2018-0-00312, Developing technologies to predict, detect, respond, and automatically diagnose security threats to automotive Ethernet-based vehicle)

REFERENCES

[1] https://ark.intel.com/products/139940/Intel-Xeon-Gold-6138P-Processor-27_5M-Cache-2_00-GHz

[2] D. Hollingworth and T. Redmond, "Enhancing operating system resistance to information warfare," MILCOM 2000. 21st Century Military Communications Conference Proceedings, volume 2, pages 1037 –1041 vol.2, 2000.

[3] N. L. Petroni, Jr., T. Fraser, J. Molina, and W. A. Arbaugh, "Copilot - a coprocessor-based kernel runtime integrity monitor," Proceedings of the 13th conference on USENIX Security Symposium - Volume 13, SSYM'04, pages 13–13, Berkeley, CA, USA, 2004. USENIX Association

[4] Wang, J., Stavrou, A. and Ghosh, A., "Hypercheck: A hardware-assisted integrity monitor," International Workshop on Recent Advances in Intrusion Detection (pp. 158-177). Springer, Berlin, 2010, September.

[5] L. Duflot, D. Etiemble, and O. Grumelard, "Using cpu system management mode to circumvent operating system security functions," In Proceedings of the 7th CanSecWest conference, 2006

[6] https://www.hotchips.org/wp-content/uploads/hc_archives/hc29/HC29.22-Tuesday-Pub/HC29.22.90-Server-Pub/HC29.22.930-Xeon-Skylake-sp-Kumar-Intel.pdf

[7] Heo I, Kim M, Lee Y, Choi C, Lee J, Kang B, and Paek Y, "Implementing an application-specific instruction-set processor for system-level dynamic program analysis engines," ACM Transactions on Design Automation of Electronic Systems . 2015, no. 4 (2015): 53.

[8] Lee Y, Lee J, Heo I, Hwang D, Paek Y, "Using CoreSight PTM to Integrate CRA Monitoring IPs in an ARM-Based SoC," ACM Transactions on Design Automation of Electronic Systems. 2017 May 31;22(3):52

A 13.56 MHz Active Rectifier With Self-Switching Comparator for Wireless Power Transfer Systems

Yingfei Xiang[1], Yu Wang[1] and C.-J. Richard Shi[1,2]

[1]Brain Chip Research Center, Fudan University, Shanghai, PR China, 201203
[2]Department of Electrical Engineering, University of Washington, Seattle, WA, 98195, USA

Abstract—This paper presents a 13.56 MHz active rectifier for wireless power transfer (WPT) systems with self-switching comparator technology. The self-switching scheme reduces the static power consumption of the comparators that gate the power transistors in the active rectifier. The power conversion efficiency (PCE) of the rectifier is enhanced by more than 6.7% using the self-switching technology. The rectifier was fabricated in a 65nm CMOS process. Measured results show that the rectifier achieves a peak PCE of 75.4% under 10dBm input power.

Index Terms–Active rectifier, wireless power transfer, self-switching comparator, power conversion efficiency.

I. Introduction

Wireless power transfer (WPT) has been widely used in implantable devices such as retinal prostheses and neural recorders to eliminate the use of the bulky battery [1]-[3]. In most of these applications, inductive near-field couping is utilized due to its high efficiency. The operation frequency of these devices is usually restricted to the ISM band, in which 13.56 MHz is one of the most commonly used frequency for the well balance between coil size, switching losses and power attenuation through body tissue [2]. Rectifier is the most important block in typical WPT systems to convert AC power to DC. Recently, active rectifier is widely used for the excellent control of the power transistors [3]-[5]. However, the comparators used in the active rectifier consume extra power, which impacts the overall power conversion efficiency (PCE) of the rectifier. In this work, a self-switching comparator technology is proposed to address the problem.

II. Proposed Active Rectifier Design

The schematic of the proposed active rectifier with self-switching comparators is shown in Fig. 1. The rectifier is composed of four power transistors (M_{N1-2}, M_{P1-2}), four comparators (CMP_{N1-2}, CMP_{P1-2}), a bias generator and four self-switching transistors (M_{N3-4}, M_{3-4}) , as shown in Fig. 1. There are two types of the comparators that control the gates of the PMOS and NMOS power transistors, respectively. The comparators are powered by the rectified voltage V_{dc} and gated by the self-switching transistors. The control signals of the power transistors generated by the comparators are reused to control the self-switching transistors in the cross coupling way, which means the N-type comparator is controlled by P-type comparator outputs, and vice verse, as shown in Fig. 1.

The working principle of the active rectifier is as follows. As shown in Fig. 1, V_{gp} (V_{gn}) and V_{p-ctrl} (V_{n-ctrl}), both generated by comparator CMP_{P1} (CMP_{N2}) are the control

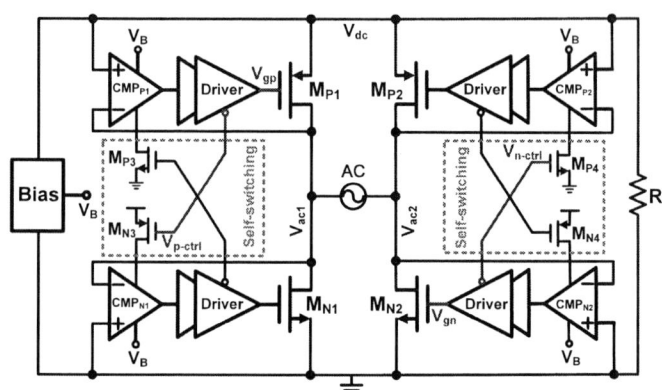

Fig. 1. The schematic of the proposed active rectifier

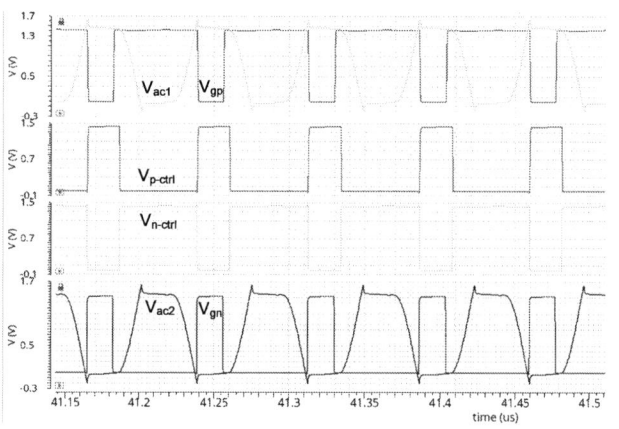

Fig. 2. The voltage waveforms of nodes in the rectifier

signals of the power transistors M_{P1} (M_{N2}) and the self-switching transistors M_{P4} (M_{N3}), respectively. Consider the positive half cycle of the input signal ($V_{ac1} > V_{ac2}$), during which CMP_{P1} is enabled. When V_{ac1} is higher than V_{dc}, the buffered output signal of CMP_{P1} turns on M_{P1} (with V_{gp}), which rectifies V_{ac1} to the DC output, and cut off the M_{N3} as well as CMP_{N1} (with V_{p-trcl}) . Therefore, there will be no power consumption by CMP_{N1}, which will not affect the function of the rectifier since M_{N1} keeps cut off during this period. Given the symmetry of the rectifier circuit, CMP_{N2} turns on M_{N2} with V_{gn} and cuts off CMP_{P2} with V_{n-ctrl} during the same period. Fig. 2 shows the waveforms of these nodes during the whole process. In the same way, during the

978-1-5386-7961-6/18 $31.00 © 2018 IEEE

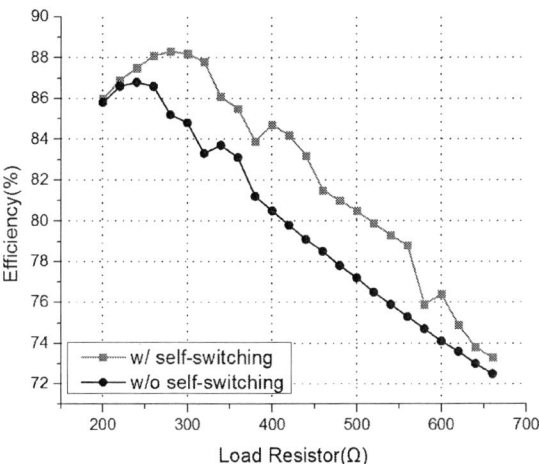

Fig. 3. Simulated PCEs under different load conditions with and without the self-switching method

Fig. 4. Chip microgragh of the proposed rectifier

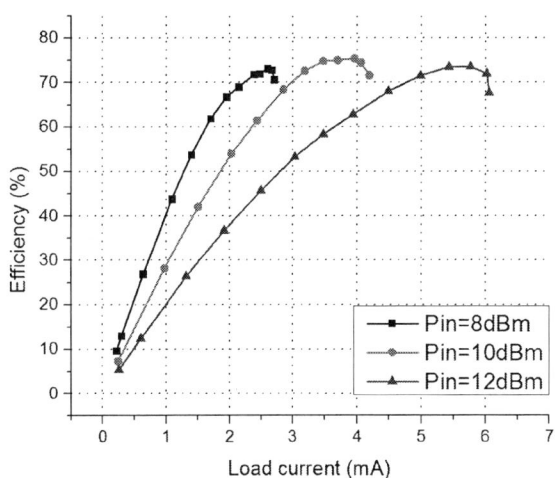

Fig. 5. Measured PCEs under different load current and input power

TABLE I
PERFORMANCE COMPARISON WITH PRIOR ART

Reference	[4]	[5]	**This Work**
Technology	0.18 μm	0.5μm	**65nm**
Area	0.608mm^2	0.144^2	**0.013mm^2**
Freq	10MHz	13.56MHz	**13.56MHz**
Input Amp	0.8-2.7V	1.46V	**1.4-2.2V**
Output Volt	0.3-2.0V	2.39V	**1.3-2V**
Pout(Max)	2mW	5.8mW	**9.2mW**
PCE	37%-80%	68%-79%	**71.9%-75.4%**

next half circle, CMP$_{P2}$ (CMP$_{N1}$) turns on M$_{P2}$ (M$_{N1}$) and cuts off CMP$_{N2}$ (CMP$_{P1}$). Therefore, the power consumption of the comparator is reduced by the on-time duty cycle of the corresponding power transistor, which is thus more effective with heavier loads.

Fig. 3 shows the simulation results of the rectifier's PCE with and without the self-switching method. It is observed that the self-switching technology significantly improves the PCE of the rectifier achieving a peak improvement of 5.4%.

III. MEASUREMENT RESULTS

The proposed active rectifier was fabricated in a 65nm CMOS process. The chip microgragh is shown in Fig. 4. The active rectifier occupies an active area of 0.013mm^2. The PCE of the rectifier is measured under load range from 0.2mA to 6mA and input power range from 8dBm to 12dBm as is shown in Fig. 5. It is observed the PCE of the rectifier increases dramatically with increasing load, and achieves a peak PCE of 75.4% under 4mA load current and 10dBm input power. The performance of the proposed active rectifier is compared to the previous works in Table I.

IV. CONCLUSION

A 13.56 MHz active rectifier with self-Switching comparators for WPT systems is presented in this paper. The proposed self-switching technology effectively reduces the power consumption of the comparators, and improves the rectifier's PCE. The proposed active rectifier was fabricated in a 65nm CMOS process. Simulation and measured results shows the efficiency improvement using the technique and hence a high PCE of the rectifier.

ACKNOWLEDGMENT

This work is supported by Science and Technology Commission of Shanghai Municipality under Grant 16JC1420300 and 18YF1402300.

REFERENCES

[1] X. Li, C. Y. Tsui and W-H. Ki, "A 13.56MHz wireless power transfer system with reconfigurable resonant regulating rectifier and wireless power control for implantable medical devices", in *IEEE J. Solid-State Circuits*, vol.50, no.4, APRIL.2015

[2] L. Cheng, W-H. Ki, Y. Lu and T-S. Yim, "Adaptive On/Off Delay-Compensated Active Rectifier for Wireless Power Transfer Systems", in*IEEE J. Solid-State Circuits*, vol.51, no.3, pp.712-723, Mar.2016

[3] C.Huang, T.Kawajiri and H.Ishikuro, "A Near-Optimum 13.56 MHz CMOS Active Rectifier With Circuit-Delay Real-Time Calibrations for High-Current Biomedical Implants", in*IEEE J.Solid-State Circuits*, vol.51, no.8, pp.1797-1809, Aug.2016

[4] S.S.Hashemi, M.Sawan and Y.Savaria,"A High-Efficiency Low-Voltage CMOS Rectifier for Harvesting Energy in Implantable Devices",in in *IEEE Trans. on Biomedical Circuits and Systems*, vol.6, no.4, pp.326-335, Aug.2012

[5] H.Lee and M.Ghovanloo,"A high frequency active voltage doubler in standard CMOS using offset-controlled comparators for inductive power transmission",in *IEEE Trans. on Biomedical Circuits and Systems*,vol.7,no.3,pp.213-224,Jun.2013

978-1-5386-7961-6/18 $31.00 © 2018 IEEE

Gap in pagination due to formatting issues.

Pages 56-59

1.6-ppm/°C Reference Voltage Generator with PSRR of -93dB based on Threshold Voltage Difference of LVT and SVT Devices

Behnam Samadpoor Rikan, Hamed Abbasizadeh, Reza E. Rad, Arash Hejazi and Kang Yoon Lee

College of Information and Communication Engineering,
Sungkyunkwan University, Suwon, Korea
Email: {behnam, klee}@skku.edu

Abstract— **This paper proposes a reference voltage generator structure based on the difference between the threshold voltages (V_{th}) of the standard threshold voltage (SVT) and low threshold voltage (LVT) transistors. The proposed structure does not apply any passive resistors or bipolar junction transistors (BJT). The reference generator is designed using a 55 nm (shrinked 65 nm) CMOS process and achieved a temperature coefficient (TC) of 1.6-ppm/°C over the -40°C to 85°C temperature ranges. The power supply rejection ratio (PSRR) for this structure is -93dB.**

Keywords— *Voltage Generator, SVT, LVT, PSRR, CMOS*

I. Introduction

These days, power management is a significant function for portable electronic devices such as hand-held computers, cellular phones, and particularly, implanted wireless biomedical chips. Proper power management can reduce power consumption and increase the battery life. Conventional linear low dropout (LDO) regulators generally consist of an error amplifier, a power output stage, and a negative feedback loop where the reference voltage is produced by a bandgap reference (BGR) [1-2]. Precision BGRs are critical building blocks for various analog and mixed–signal electronic devices, such as pulse-width modulation (PWM) controllers, data converters, operational amplifiers, oscillators, linear regulators and PLLs. Conventional BGRs apply BJT transistors to provide a robust voltage or current. However, in ultra-low power applications, these structures have received less attention. Although ultra-low power BGR structures have been recently introduced [2-4], these structures have low PSRR, their TC is not satisfactory, and there is room for further improvements. This work introduces a reference voltage generator structure with a very low current consumption, high PSRR and very low TC based on the threshold voltage difference of SVT and LVT devices.

II. Basic Concept and Proposed Structure

The structure design is intended to apply the difference between the V_{th} voltages to produce a stable reference voltage. To explain the concept, assume we have the circuit shown in Fig. 1 with transistors M1 and M2 having different V_{th} and the currents of two paths being the same. If we assume that M1 and M2 transistors have the same sizes, then:

$$V_A = -V_{GS1} \quad and \quad V_B = V_{ref} - V_{GS2} \tag{1}$$

$$V_A \simeq V_B \quad \longrightarrow \quad V_{ref} = V_{GS2} - V_{GS1} \tag{2}$$

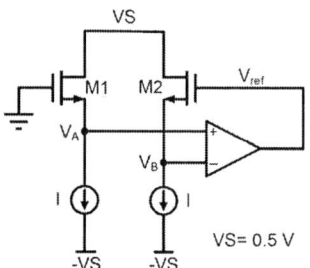

Fig. 1. Conceptual block diagram of the proposed reference voltage generator

$$V_{ref} = \sqrt{\frac{I}{\frac{K_2'}{2}\left(\frac{W}{L}\right)_2}} + V_{th2} - \sqrt{\frac{I}{\frac{K_1'}{2}\left(\frac{W}{L}\right)_1}} - V_{th1} \tag{3}$$

If we assume $\left(\dfrac{W}{L}\right)_1 = \left(\dfrac{W}{L}\right)_2$ *and* $K_1' = K_2'$

$$V_{ref} = V_{th2} - V_{th1} \tag{4}$$

Where V_{ref} is the reference voltage, W is the width and L is the length of the transistors.

As we know, V_{th1} and V_{th2} are temperature and process-dependent parameters. However, in (4) if V_{th1} and V_{th2} have similar TC, their differences can have a very small TC value. Besides, $V_{th2} - V_{th1}$ have much less dependency on the process variation than V_{th2} and V_{th1} do.

Fig. 2 shows the implementation of the proposed reference voltage generator. This structure consists of an error amplifier as well as SVT and LVT transistors. As we can see, M1, M2, M3, M4, M6, M7, and M8 transistors have been selected as LVT devices and M5, M9, M10, and M11 transistors have been selected as SVT devices. We applied the V_{th} difference between the M5 and M6 transistors as the reference voltage. In the applied CMOS process, this value is around 172mV. However, in order to set the reference voltage at 500mV and to overcome the non-idealities that are not considered in (1), and also considering the fact that for low power consumption purposes these transistors have to operate in the subthreshold region, according to the simulations, we have set $(W/L)_1$ to three times of $(W/L)_2$. To avoid the channel length modulation effects, the M1-M8 transistors lengths were selected as 10 μm. The supply voltage for this design is 0.9 to 1.1 V. Each of M1-M8 transistors' current with 1 V supply voltage is 16 nA.

Fig. 2 also includes start-up circuit where after the V_{ref} settling, the current of this path is reduced to only 2 pA.

978-1-5386-7961-6/18 $31.00 © 2018 IEEE

Nevertheless, this is a sample start-up circuit, and it can be replaced with other structures as well. A P-type error amplifier can be applied in this structure. As we can see, this structure has only few devices, and it does not include any BJT, passive resistors or capacitors. This results in a simple structure with a small area. Also, the current consumption of this structure is very low, and it achieves a very high TC and PSRR.

III. SIMULATION RESULTS

This structure was designed and verified using a 55 nm (shrinked 65 nm) CMOS process. The reference voltage of this structure was almost 500 mV. Fig. 3 demonstrates the reference voltage as a function of the temperature in different corner cases. According to this simulation, for the temperature range from -40°C to 85°C, the maximum and minimum output voltages are 0.49945 V and 0.49935 V, respectively, indicating the best TC of 1.6-ppm/°C. Fig. 4 exhibits the Monte-Carlo simulation of the reference voltage generator. For this simulation 100 samples were applied. The mean value for this simulation is 499.66 mV, and the standard deviation (std. dev) is only 145 µV. The coefficient of variation for this simulation is about 0.03%.

Fig. 5 shows the PSRR of the reference voltage generator. PSRR at 100 Hz for FF, TT and SS cases are -90 dB, -93 dB and -98 dB respectively. In addition, the PSRR at 1 GHz for the FF, TT and SS cases are -24 dB, -28 dB and -30 dB, respectively. Excluding the amplifier design, which is not the concern of this work, the current consumption of the rest parts of the reference generator is only 32 nW from a 1 V supply voltage. Table I summarizes and compares the design with other recent papers.

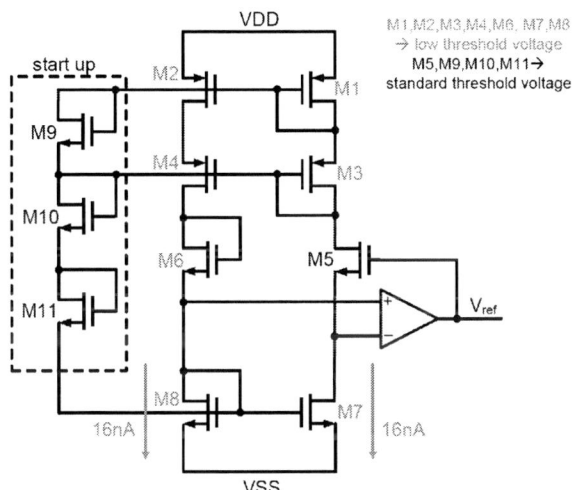

Fig. 2. The designed reference voltage generator with SVT and LVT devices

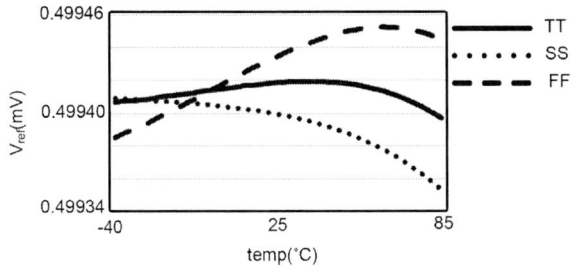

Fig. 3. Reference voltage as a function of temperature in different corner cases

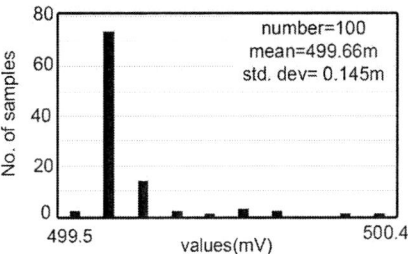

Fig. 4. Distribution of the V_{ref} among 100 Monte-Carlo samples simulation

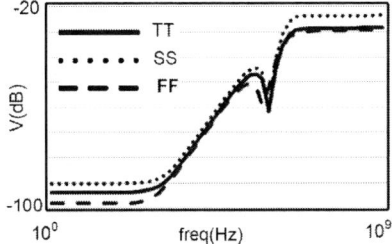

Fig. 5. PSRR of the reference voltage generator for different corner cases

TABLE I.   SUMMARY AND COMPARISON

	This Work	**[3]**	**[4]**
Technology (nm)	55	180	350
Supply Voltage (V)	1	1.2	1.2
Ref. Voltage (V)	0.5	1.09	0.862
TC (ppm/°C)	1.6	147	1
Temp. Range (°C)	-40 to 85	-40 to 120	-40 to 120
PSRR (dB)	-93	-62	-

IV. CONCLUSION

This paper presented a reference voltage generator structure based on the threshold voltage difference between SVT and LVT devices. The structure does not require BJT or passive resistors and achieved the best TC of 1.6-ppm/°C over -40°C to 85°C temperature range. The PSRR of this structure is -93 dB.

ACKNOWLEDGEMENT

This work was supported by the National Research Foundation of Korea (NRF) grant funded by the Korean government (MSIP) (2014R1A5A1011478).

REFERENCES

[1] B. Samadpoor Rikan, H. Abbasizadeh, J.-H Kang and K.-Y Lee "A High Current Efficiency CMOS LDO Regulator with Low Power Consumption and Small Output Voltage Variation," Journal of IKEEE, vol.18, no.1, pp. 37-44, Mar. 2014.

[2] B. Samadpoor Rikan, H. Abbasizadeh, T. T. Kim Nga, S. J. Kim and K. Y. Lee, "A Low Leakage Retention LDO and Leakage-based BGR with 120nA Quiescent Current," ISOCC 2017, Nov. 2017.

[3] Osaki, et. al..: "1.2-V Supply, 100-nW, 1.09-V Bandgap and 0.7-V Supply, 52.5-nW, 0.55-V Subbandgap Reference Circuits for Nanowatt CMOS LSIs", IEEE J. Solid-State Circuits, 2013, 48, pp. 1530-1538.

[4] Zhang, Y., Zhu, J., Sun, W, Yang, B, and Huang, Z.: "1 ppm/°C bandgap with multipoint curvature-compensation technique for HVIC", Microelectronics Journal, 2014, 50, pp. 1908-1910.

Gap in pagination due to formatting issues.

Pages 62-67

A 4.86 µW/Channel Fully Differential Multi-Channel Neural Recording System

Taeju Lee[1], Ji-Hyoung Cha[2], Su-Hyun Han[2], Seong-Jin Kim[2], and Minkyu Je[1]

[1]School of Electrical Engineering, Korea Advanced Institute of Science and Technology, (KAIST), Daejeon, Korea
[2]School of Electrical and Computer Engineering, Ulsan National Institute of Science and Technology (UNIST), Ulsan, Korea
{taeju.lee, mkje}@kaist.ac.kr

Abstract— **This paper presents a fully differential multi-channel neural recording system. The system consists of four key blocks which are a low-noise amplifier (LNA), programmable gain amplifier (PGA), buffer, and successive approximation register ADC (SAR ADC). The input stage of the OTA used in LNA is designed as the inverter-based structure for improving the current efficiency. For an energy efficient system, the dual sample-and-hold (S/H) structure is applied to the SAR ADC. Each channel consumes the power of 4.86 µW/Channel and achieves an input-referred noise of 2.58 µV$_{rms}$. The implemented IC operates under a 1-V supply voltage for core blocks and 1.8-V for output digital buffers. The system is implemented in a standard 1P6M 0.18-µm CMOS process.**

Keywords—Neural recording system, low-noise amplifier, inverter-based input stage, energy efficient system.

I. INTRODUCTION

The low-power, low-noise multi-channel neural recording system is one of the many parts required to perform brain research. Thanks to the development of a multi-channel electrode array, the recorded neurons have steadily increased. Accordingly, the number of recording circuits has to increase. In the case of the implantable biomedical devices, the system must be driven with low power to prevent necrosis caused by heat. At the same time, the system should be designed to have low noise characteristics to measure microscale neural signals. The neural signals are distributed in the frequency band from sub-1 Hz to several kHz. This band is vulnerable to the interference such as power lines and stimulation signals. Therefore, the recording system that is robust against the interference should be designed. In this paper, we present a low-power, low-noise fully differential multi-channel neural recording system for implantable biomedical devices.

II. SYSTEM ARCHITECTURE

Fig. 1 shows the proposed system architecture, which consists of analog front-ends (AFEs), analog multiplexers (AMUXs), dual SAR ADCs, digital multiplexers (DMUXs), a bias circuit, and an output driver. To implement the 64-channel recording system, eight unit blocks (*Set N, N*=1-8) are used. Each unit block consists of LNAs, PGAs, buffers, an AMUX, and a 10-bit dual SAR ADC. The AFE that consists of LNA, PGA, and buffer is designed as a fully differential structure. In the unit block, eight AFEs share an AMUX and an ADC. An AMUX accepts eight channels and sends two channels to the ADC through selection signals *SEL$_A$* and *SEL$_B$*. After the digitization of neural signals, the channel selection is performed by four DMUXs, which are controlled through S_{X1} and S_{Y1}. The final channel selection is conducted by a DMUX controlled by S_{X2} and S_{Y2}. The output signals are 11 bits consisting of one end-of-conversion (EOC) signal and 10-bit data. The overall system is driven with 1-V and 1.8-V supply voltages. The 1-V supply voltage is to operate AFEs, AMUXs, ADCs, DMUXs, and a bias circuit.

This research was supported by the Convergence Technology Development Program for Bionic Arm (2017M3C1B2085296) and the Brain Research Program (2017M3C7A1028859) through the National Research Foundation Korea (NRF) funded by Ministry of Science and ICT.

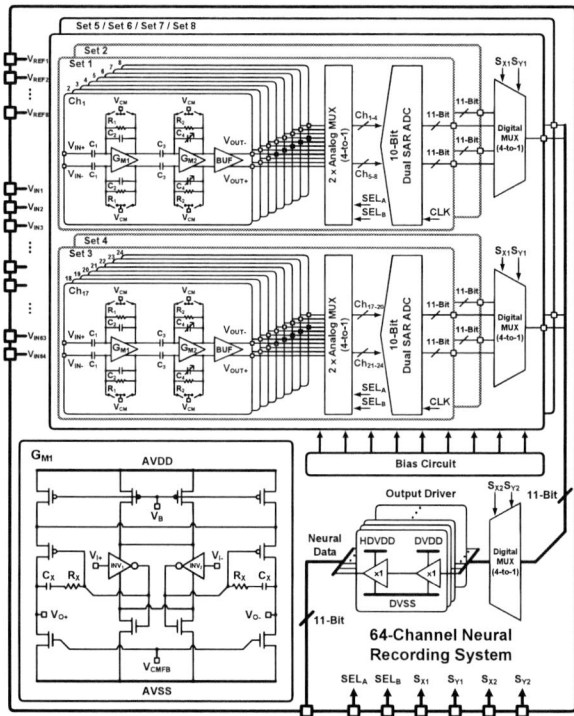

Fig. 1. Architecture of the multi-channel neural recording system.

The 1.8-V supply voltage is used for the output driver, which up-shifts the voltage level of the final output bits to interface with the field-programmable gate array (FPGA). The LNA and PGA are implemented as a capacitively-coupled structure. The OTA G_{M1} used in the LNA is shown in Fig. 1. The input stage of the G_{M1} is implemented as the inverter-based structure, reported in [1], [2], for improving the current efficiency. The output DC bias point of the G_{M1} is set by one common-mode feedback (CMFB) OTA. The input stage of the OTA G_{M2} used in the PGA is designed as only PMOS pair. The voltage gains of the LNA and PGA are set by the ratio of C_1/C_2 and C_3/C_4, respectively. The input DC bias points of G_{M1} and G_{M2} are set by R_1 and R_2, respectively. The R_1 and R_2 are implemented using two diode-connected MOSFETs for area efficient design while achieving the high resistance. The output buffer for driving the ADC input is implemented as a rail-to-rail type based on [2].

In the digitization stage, the dual S/H structure in [2] is employed to increase the sampling time for the A/D conversion, which can reduce the power burden on the buffer. To lower the power consumption in the ADC, the delay chain based time domain comparator, that is reported in [3], is adopted as shown in Fig. 2. In this work, the 4-stage delay chain is used to convert the input signals of V_{DAC+} and V_{DAC-}

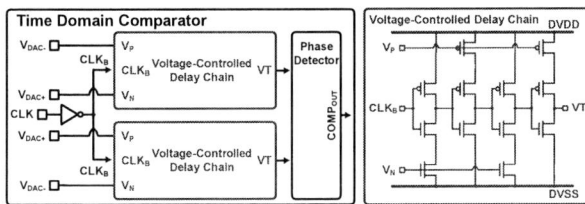

Fig. 2. Block diagram of the time domain comparator [3].

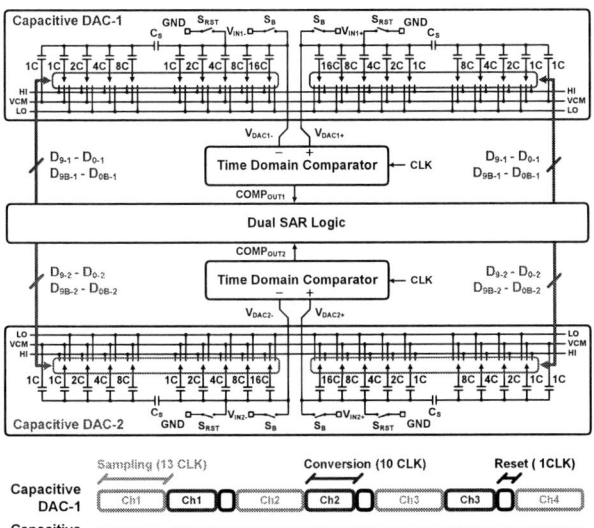

Fig. 3. Schematic of the 10-bit dual SAR ADC and timing diagram.

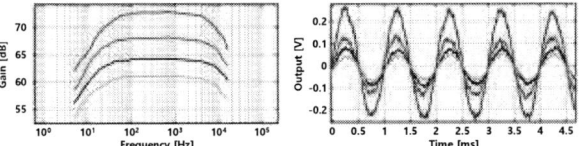

Fig. 4. Measured frequency response (left) and time domain results corresponding to each gain with a 100 μV$_{pp}$ input at 1 kHz (right).

to the delay difference based on the reference clock CLK_B. The dual operation of the ADC is performed using two capacitive DACs (CDACs) that are implemented as the split array for area efficient design. The unit capacitance used in the CDAC is 150 fF with MIM structure occupying the area of 72 μm^2. Fig. 3 shows the schematic of the 10-bit dual SAR ADC and timing diagram for channel conversion.

III. MEASUREMENT RESULTS

The IC is fabricated in a 1P6M 0.18-μm CMOS process. The active area of a 64-channel recording system is 9.4 mm^2. The power consumption of a single channel, including LNA, PGA, buffer, and ADC, is 4.86 μW/Ch. The system gain is measured from 61 dB to 74 dB with the bandwidth of 10 kHz. Fig. 4 shows the measured frequency response and time domain results corresponding to each gain. The input-referred noise (IRN) is measured as 2.58 μV$_{rms}$ when integrated from 10 Hz to 10 kHz as shown in Fig. 5. The noise efficiency factor (NEF) is calculated as 2.19, resulting in the power efficiency factor (NEF^{2}V$_{DD}$) of 4.80. The total harmonic distortion of the AFE is measured as 1.5 %, corresponding to −36.5 dB that is measured using 100 μV$_{pp}$ input at 1 kHz as shown in Fig. 6(a). The output power

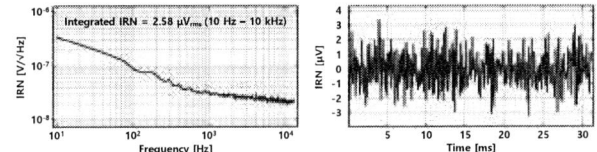

Fig. 5. Measured input-referred noise in the frequency domain (left) and time domain (right).

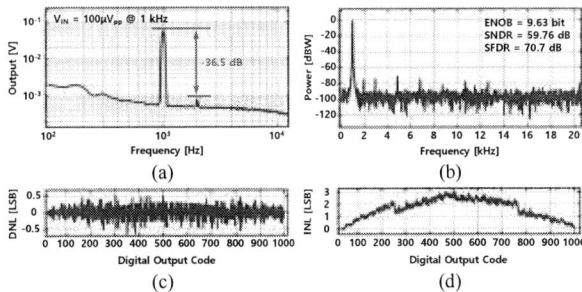

Fig. 6. (a) Measured total harmonic distortion of the analog front-end, (b) Output power spectrum of the ADC, (c) Differential nonlinearity (DNL), and (d) Integral nonlinearity (INL).

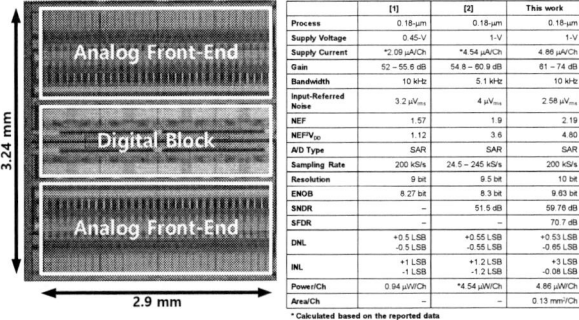

Fig. 7. Die micrograph (left) and performance summary (right).

spectrum of the ADC is measured using 1 V$_{pp}$ input at 1 kHz, resulting in the effective number of bits (ENOB) of 9.63 bit, signal-to-noise and distortion ratio (SNDR) of 59.76 dB, and spurious-free dynamic range (SFDR) of 70.7 dB as shown in Fig. 6(b). The differential nonlinearity (DNL) and integral nonlinearity (INL) are measured as +0.53/−0.65 LSB and +3/−0.08 LSB respectively as shown in Fig. 6(c) and (d). The clock frequency for ADC operation is set to 2.5 MHz. Fig. 7 shows the die micrograph and performance summary.

IV. CONCLUSION

The implemented IC consumes the power of 4.86 μW/Ch and occupies the area of 0.13 mm^2/Ch while achieving the input noise of 2.58 μV$_{rms}$ by employing the inverter-based structure in the LNA input stage. The power burden on the buffer is reduced by using the dual S/H technique while maintaining the sampling rate of 200 kS/s.

REFERENCES

[1] D. Han et al., "A 0.45 V 100-Channel Neural-Recording IC With Sub-μW/Channel Consumption in 0.18 μm CMOS," *IEEE TBioCAS*, vol. 7, no. 6, pp. 735-746, Dec. 2013.

[2] X. Zou et al., "A 100-Channel 1-mW Implantable Neural Recording IC," *IEEE TCAS–I*, vol. 60, no. 10, pp. 2584-2596, Oct. 2013.

[3] S.-K. Lee et al., "A 21fJ/Conversion-Step 100 kS/s 10-bit ADC With a Low-Noise Time-Domain Comparator for Low-Power Sensor Interface," *IEEE JSSC*, vol. 46, no. 3, pp. 651-659, Mar. 2011.

Gap in pagination due to formatting issues.

Pages 70-71

Tapered-Ratio Compression for Residual Network

Sungbum Kang, Joonsang Yu, and Kiyoung Choi

Dept. of Electrical and Computer Engineering, Neural Processing Research Center (NPRC)
Seoul National University, Seoul, Korea
{sb05kang, joonsang.yu}@dal.snu.ac.kr, kchoi@snu.ac.kr

Abstract— **Residual network is a commonly used convolutional neural network, but as the network becomes wider and deeper, it is becoming more difficult for IoT devices or embedded systems to adopt it due to their hardware constraints. Network compression using knowledge distillation is helpful to overcome this problem. In this paper, we propose a tapered compression ratio for residual network compression, which achieves higher accuracy with fewer multiplications on CIFAR datasets.**

Keywords; deep learning; deep neural network; knowledge distillation; network compression.

I. INTRODUCTION

Deep Neural Networks (DNNs) are widely used for many applications such as image classification and voice recognition due to their outstanding performance [1, 2]. However, the deeper and wider networks require the higher computing power and larger memory space and bandwidth. It is not easy for IoT devices or embedded system to adopt such networks due to their small memory space and low computing power. Thus, if we can obtain a smaller network even with higher accuracy, it will be very useful for integrating DNNs into such a system.

Previous work shows that trained networks in general have redundant filters [3], which can be removed to reduce network size [4]. One of the methods for network compression is knowledge distillation (KD) [4] introduced to train a small network called *student network* by using a larger trained network called *teacher network*. During the training phase, the student network mimics the behavior of the teacher network. Through knowledge distillation, the student network can achieve higher accuracy than that obtained by training the student network directly using the conventional back propagation algorithm.

In this paper, we propose tapered-ratio compression technique for residual network (ResNet) [1]. The technique uses the idea of knowledge distillation, but gives higher accuracy at even smaller multiplication points compared with the original knowledge distillation technique.

II. TAPERED-RATIO COMPRESSION

When compressing a network using knowledge distillation, we need to decide the size of the student network. After the target size of the student network is fixed, that single compression ratio is applied equally to all layers [5]; we call it *uniform ratio compression* in this paper. For example, if we want to obtain half size of the teacher network, uniform compression ratio is set to 0.50, which is the same for all layers. Note, in this paper, that compression ratio of 0.90 means the number of channels in student network becomes 0.90 (90%) of the teacher network.

This work was supported by Samsung Advanced Institute of Technology.

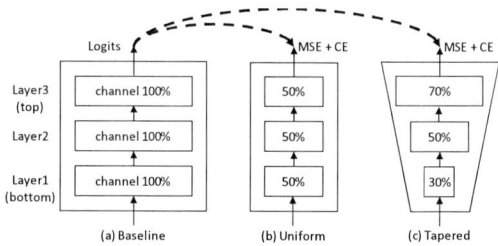

Figure 1. An example of network compression. (a) A teacher network (b) A student network with uniform compression ratio (c) A student network with tapered compression ratio. (b) and (c) are trained by minimizing both Mean Squared Error (MSE) for logits of (a) and Cross Entropy (CE) loss.

We suggest a tapered-ratio for residual network compression. The idea is based on the observation that gradually increasing the compression ratio when moving toward the top layer (output side) gives a better result for residual network compression. We call it *tapered-ratio compression*. Generally, simple features like horizontal line or vertical edge are extracted in lower layers (close to the bottom layer) and more complex features are extracted in upper layers (close to the top layer). Simple features are more likely to have redundancies than complex features. Thus we expect that more filters in lower layers can be removed than those in upper layers. Moreover, since there are more activations in lower layers, if we apply tapered-ratio compression, we can obtain more reduction in memory access.

Figure 1 shows a simple example of compressing a network with three layers. Figure 1 (a) is a teacher network. Figure 1 (b) is a student network with uniform compression ratio, and Figure 1 (c) is a student network with the proposed tapered compression ratio. Figure 1 (b) applies 0.50 compression ratio equally to all layers, and Figure 1 (c) applies [0.30 / 0.50 / 0.70] from bottom to top. (actually, the original teacher network already has a tapered shape in terms of number of filters, but in terms of compression ratio, all of its layers have the same width of 100%).

To obtain higher accuracy, we use knowledge distillation with both logits of teacher and label information for given input similar to hint-training [6]. The student networks are trained to optimize the following loss function:

$$L_{kd} = L_{mse}(logit_T, logit_S) + L_{ce}(y_{true}, logit_S) \quad (1)$$

where L_{mse} is MSE loss for logits, and L_{ce} is cross-entropy loss with true label y_{true} for a given input image). And $logit_T$ and $logit_S$ denote logits of teacher and student, respectively.

978-1-5386-7961-6/18 $31.00 © 2018 IEEE

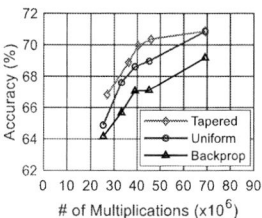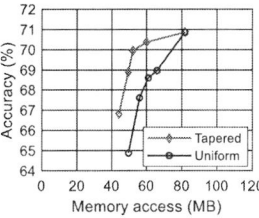

Figure 2. Comparison of accuracy for a given number of multiplications (left) and amont of memory access (right) between tapered and uniform compression for ResNet-32 on CIFAR-100. Batch size is set to 64. *Backprop* denotes accuracy using back propagation from scratch without KD. *Tapered* has higher accuracy and less amount of memory access compared to *Uniform*.

TABLE I.   COMPRESSION RESULTS FOR SEVERAL RATIOS FOR RESNET-32 ON CIFAR-100

Accuracy	Compression Ratio	Multiplications
68.73	0.10 / 0.30 × 5 / 0.85 × 5 / 1.00 × 5	40,039,424
69.05	0.10 / 0.45 × 5 / 0.80 × 5 / 1.00 × 5	41,051,392
69.96	**0.25 / 0.50 × 5 / 0.75 × 5 / 1.00 × 5**	**40,405,248**
69.25	0.45 / 0.60 × 5 / 0.75 × 5 / 0.90 × 5	40,099,328
68.75	0.60 / 0.75 × 5 / 0.75 × 5 / 0.80 × 5	40,403,776

III. EXPERIMENTAL RESULTS

We conducted several experiments for ResNet-32, ResNet-56, and WideResNet-28-10 (WRN-28-10) [2] on CIFAR-10 and CIFAR-100 datasets with PyTorch framework. We used Xavier initializer and Stochastic Gradient Descent (SGD) optimizer with Nesterov momentum to train teacher networks and student networks.

Figure 2 (left) shows compression results (accuracy vs. number of multiplications) of ResNet-34 on CIFAR-100. In the figure, *Backprop* denotes accuracy using back propagation from scratch without KD. *Uniform* denotes accuracy using KD with uniform compression ratio. And *Tapered* denotes accuracy using KD with tapered compression ratio. The baseline accuracy of the teacher network is 69.18%. Tapered shows higher accuracy than others. When uniform compression ratio is set to 0.80, there are 45,155,904 multiplications and the accuracy is 68.97%. In contrast, when tapered compression ratio is set to [0.25 / 0.50×5 / 0.75×5 / 1.00×5] (there are 16 residual blocks in ResNet-32 including first convolution layer.), there are 40,405,248 multiplications and the accuracy is 69.97%. The tapered case has fewer multiplications and 1.00%p higher accuracy than the uniform case, and the accuracy is even higher than the baseline accuracy in spite of smaller network size.

Figure 2 (right) shows the amount of memory access during inference phase for the same points in Figure 2 (left). The amount of memory access is calculated as "batch size (64) × activations + weight parameters". Tapered-ratio compression has much fewer memory accesses at points having similar number of multiplications.

Since tapered-ratio compression has multiple candidates, we compare several candidates changing its ratio in Table I. Compressing the bottom layers while increasing the size of the top layers (to keep about the similar amount of computations), gives a higher accuracy. However, once the size of the top layer

TABLE II.   COMPRESSION RESULTS OF RESNET-32, RESNET-56 AND WRN-28-10 ON CIFAR-10 AND CIFAR-100. (M: MILLION / MB: MEGA BYTE)

CIFAR-10			
	ResNet-32	ResNet-56	WRN-28-10
	Acc (Mult / Mem)	Acc (Mult / Mem)	Acc (Mult / Mem)
BL	92.44 (69.1M / 81.7MB)	93.48 (125.8M / 139.3MB)	95.54 (5243.3M / 757.8MB)
BP	91.71 (45.2M / 66.0MB)	92.67 (82.1M / 112.4MB)	94.40 (477.8M / 204.3MB)
UNI	92.04 (45.2M / 66.0MB)	93.04 (82.1M / 112.4MB)	94.84 (477.8M / 204.3MB)
TPRD	92.24 (40.4M / 52.1MB)	93.14 (74.7M / 90.6MB)	94.96 (4753M / 143.7MB)
CIFAR-100			
	ResNet-32	ResNet-56	WRN-28-10
	Acc (Mult / Mem)	Acc (Mult / Mem)	Acc (Mult / Mem)
BL	69.18 (69.1M / 81.7MB)	70.32 (125.8M / 139.3MB)	81.11 (5243.3M / 757.8MB)
BP	67.10 (45.2M / 66.0MB)	67.71 (82.1M / 112.4MB)	75.81 (477.8M / 204.3MB)
UNI	68.97 (45.2M / 66.0MB)	71.05 (82.1M / 112.4MB)	78.79 (477.8M / 204.3MB)
TPRD	69.96 (40.3M / 52.1MB)	71.91 (74.7M / 90.6MB)	78.87 (4753M / 143.7MB)

Note that BL, BP, UNI, and TPRD stand for baseline, back propagation, uniform, and tapered, respectively.

reaches that of the teacher, the accuracy no longer improves, thus we choose the best tapered compression ratio (the one in bold face in Table I) for our experiments. In case of WRN-28-10, since the size of the top layer cannot be increased to that of the teacher since the number of multiplications increases too much, tapered compression ratio is set to [0.1 / 0.1×4 / 0.25×4 / 0.45×4], which gives the best result among the options that can keep a similar number of multiplications.

Table II shows compression results for ResNet-32, ResNet-56, and WRN-28-10 on CIFAR-10 and CIFAR-100. The results show 0.99%p, 0.86%p, and 0.08%p higher accuracies for ResNet-32, ResNet-56 and WRN-28-10, respectively, with fewer multiplications than uniform results on CIFAR-100. The amount of memory access is only 78.9%, 80.6%, and 70.3%, respectively, compared to uniform cases. ResNet-32, ResNet-56, and WRN-28-10 have 0.20%p, 0.10%p, and 0.12%p higher accuracies, respectively, with fewer multiplications on CIFAR-10 compared to uniform cases for the same networks. Every result shows higher accuracy in spite of fewer multiplications and smaller amount of memory access.

IV. CONCLUSION

We propose tapered-ratio compression using knowledge distillation to obtain higher accuracy for residual network. Tested with ResNet-32, ResNet-50, and WRN-28-10 on CIFAR-10 and CIFAR-100, it reduces number of multiplications and memory accesses while improving the accuracy.

REFERENCES

[1] K. He, X. Zhang, S. Ren, and J. Sun, "Deep residual learning for image recognition," in *CVPR*, June 2016.

[2] S. Zagoruyko and N. Komodakis, "Wide residual networks," in *BMVC*, Sep. 2016.

[3] H. Li, A. Kadav, I. Durdanovic, H. Samet, and H. P. Graf, "Pruning filters for efficient convnets," in *ICLR*, May 2016.

[4] G. Hinton, O. Vinyals, and J. Dean, "Distilling the knowledge in a neural network," in *NIPS Workshop*, Dec. 2014.

[5] J.-H. Luo, J. Wu, and W. Lin, "Thinet: A filter level pruning method for deep neural network compression," in *ICCV*, Oct. 2017.

[6] A. Romero, N. Ballas, S. E. Kahou, A. Chassang, C. Gatta, and Y. Bengio, "Fitnets: Hints for thin deep nets," in *ICLR*, May 2015.

Segmentation-based disparity refinement

GyuJin Bae and Young Hwan Kim
Department of Electrical Engineering
Pohang University of Science and Technology
Pohang, Republic of Korea
youngk@postech.ac.kr

Abstract—Stereo matching consists of four steps: cost computation, cost aggregation, disparity optimization, and disparity refinement. While the first three steps have been actively researched, few efforts have been taken on disparity refinement. In this paper, we propose a novel segmentation-based disparity refinement method. The proposed method uses the property that the pixels belonging to the same segment have similar disparity values. Using this property, the proposed method can detect the error regions with various shapes and sizes. For the detected error regions, the proposed method applies the weighted median filter to remove the disparity error. In the experimental results, the proposed method improves the quality of disparity maps by increasing average bad pixel rate up to 17.5 % compared to the benchmark methods.

Keywords; Stereo matching, Segmentation, Disparity refinement

I. INTRODUCTION

Stereo matching has been actively studied in the field of computer vision. Most of the stereo matching methods consist of four steps: cost computation, cost aggregation, disparity optimization, and disparity refinement. Through the former three steps, the disparity value is calculated. Then, block matching-based method [1] is widely used because of computation efficiency. However, the method often causes the disparity error in the regions such as object boundary and low texture regions.

To remove the disparity error, several refinement methods have been proposed [2]-[3]. Traditional refinement method typically consists of left-right consistency checking, hole-filling, and a noise-removing filter. In recent years, some new methods were proposed. Zhang *et al.* [2] refined the disparities by a weighted median filter. Jiao *et al.* [3] proposed the color image-guided refinement method. The method detects the boundary-inconsistent regions considering the shape of the error regions in the boundary regions. Next, it removes the disparity error in the boundary-inconsistent regions. The method effectively detects the error region if the error exists near the boundary and its size is small. Otherwise, it does not detect the error region.

In this paper, we propose a novel disparity refinement method. The proposed method first detects the error region using the segmentation. For the detected error region, the proposed method applies the weighted median filter to remove the disparity error. This paper is organized as follows. In Section II, we describe the proposed method. In Section III, we

This research was supported by LG Display Co., Ltd. and the MSIT(Ministry of Science and ICT), Korea, under the ICT Consilience Creative program(IITP-2018-2011-1- 00783) supervised by the IITP(Institute for Information & communications Technology Promotion)

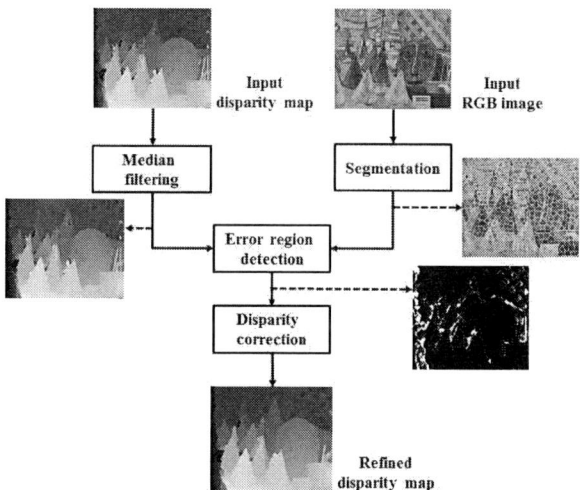

Figure 1. Flowchart of the proposed method.

present the experimental results. Finally, we conclude this study in Section IV.

II. PROPOSED METHOD

The block diagram of the proposed method is illustrated in Fig. 1. The proposed method consists of four steps: median filtering, segmentation, error region detection, and disparity correction. First, the proposed method performs median filtering to remove the small errors in the input disparity map. Then, the search window size is set to 3×3.

Second, we perform segmentation based on the color information. As a segmentation method, we use the SLIC method proposed in [4]. The SLIC method adheres to boundaries, and it is fast and memory efficient.

Third, the proposed method detects the error regions using the segmentation result. To detect the error regions, we have used the assumption that pixels belonging to the same segment have similar disparity values. The error region is detected as follows:

$$Flag_{error} = \begin{cases} 1, & if\ d_i > Threshold \\ 0, & otherwise \end{cases}, \qquad (1)$$

TABLE I.   AVERAGE ERROR PERCENTAGE (%) COMPARISON

Method	Tsukuba			Venus			Teddy			Cones			Avg.
	nocc	all	disc	nocc	all	disc	nocc	all	disc	nocc	all	disc	
Input disparity map	4.066	4.901	<u>10.513</u>	2.116	3.288	16.328	8.879	17.400	<u>20.028</u>	4.065	12.907	11.174	9.638
Joint WM filter [2]	8.640	9.216	22.888	20.157	20.803	25.399	32.92	38.848	37.086	32.335	38.093	35.159	26.795
Jiao's method [3]	4.354	5.047	11.374	4.307	5.275	16.490	9.713	17.981	21.361	5.795	14.277	13.277	10.771
Proposed method	<u>3.965</u>	<u>4.455</u>	13.299	<u>1.291</u>	<u>2.271</u>	<u>12.998</u>	<u>8.575</u>	<u>16.808</u>	20.623	<u>3.941</u>	<u>12.640</u>	<u>10.231</u>	<u>9.258</u>

 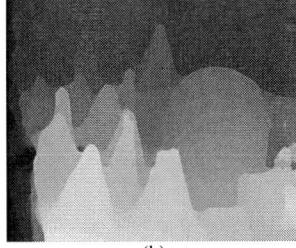  

(a)   (b)   (c)   (d)

Figure 2. Comparison with other refinement methods: (a) Input disparity map, (b) Joint WM filter [2], (c) Jiao's method [3], and (d) proposed method.

where, d_i is the representative disparity value of segment i. Then, the representative value is the median disparity value of the pixels belonging to the segment. The threshold value is set to 1.

Finally, the proposed method corrects the disparity error by performing weighted median filtering on the detected error regions. The weight value is calculated as follows:

$$w(p_C, p_N) = e^{-diff_{color}} \cdot Flag_{error} \cdot Flag_{seg},$$
$$diff_{color} = |R_C - R_N| + |G_C - G_N| + |B_C - B_N|,$$
$$Flag_{seg} = \begin{cases} 1, & if\ Label_C = Label_N \\ 0, & otherwise \end{cases}, \qquad (2)$$

where, p_C and p_N represent the center pixel and neighboring pixel, respectively. The search window size is set to 21×5. $diff_{color}$ is the color difference value between the center pixel and neighboring pixel. R_C, G_C, B_C represent the red, green, and blue value of the center pixel, respectively. R_N, G_N, B_N represent the red, green, and blue value of the neighboring pixel, respectively. $Flag_{seg}$ is the flag signal related to segmentation. $Label_C$ and $Label_N$ represent the label value of the center pixel and neighboring pixel, respectively.

III.   EXPERIMENTAL RESULTS

We used the Middlebury dataset [5]. The dataset is widely used to evaluate the stereo matching algorithms. The dataset contains various regions such as occlusion and low textured regions. Two refinement methods were used as benchmarks: joint WM (weighted median) filter [2] and Jiao's method [3].

To generate the input disparity maps, we used Hosni's method [1].

The accuracy was evaluated as the average percentage of bad pixels, which have an absolute disparity error of greater than one. For each image, bad pixel rates were evaluated for non-occlusion (nocc), all (all), and discontinuous (disc) regions and then averaged.

As shown in Table I, the proposed method shows a lower bad pixel rate, when compared to the benchmark methods. Fig. 2 showed the proposed method removed the disparity error better than other methods.

IV.   CONCLUSION

In this paper, we propose a segmentation-based disparity refinement method. First, the proposed method detects the error regions considering the distribution of disparity value in the segment. Next, the proposed method removes the disparity error by applying the weighted median filter on the detected error regions. As a result, the proposed method the proposed method shows the best average bad pixel rate compared to the benchmark methods.

REFERENCES

[1] A. Hosni, C. Rhemann, M. Bieyer, C. Rother, and M. Gelautz, "Fast cost-volume filtering for visual correspondence and beyond," *The IEEE Conference on Computer Vision and Pattern Recognition*, 2011.

[2] Q. Zhang, L, Xu, J. Jia, "100+ times faster weighted median filter," *The IEEE Conference on Computer Vision and Pattern Recognition*, 2014.

[3] J. Jiao, R. Wang, W. Wang, D. Li, and W. Gao, "Color image-guided boundary-inconsistent region refinement for stereo matching," *IEEE Trans. on circuits and systems for video technology*, vol. 27, no. 5, 2017.

[4] R. Achanta, A. Shaji, K. Smith, A. Lucchi, P. Fua, and S. Susstrunk, "SLIC superpixels compared to state-of-the-art superpixel methods," *IEEE Transactions on Pattern Analysis and Machine Intelligence*, vol. 34, no. 11, 2012.

[5] http://vision.middlebury.edu/stereo/data/

Human Visual Attention Analysis-based Image Segmentation using Color Histogram

Ho Sub Lee and Young Hwan Kim

Electrical Engineering
POSTECH, Pohang, 37673, Republic of Korea
{hslee5210, youngk}@postech.ac.kr

Abstract—**This paper proposes a new image segmentation method which uses 3D color histogram and human visual attention analysis. Existing histogram-based methods find the cluster centers by analyzing the distribution of the 3D color histogram values. It is difficult to consider the overall characteristics of an image by analyzing the distribution of the 3D color histogram values. Thus, Existing histogram-based methods have difficulty to finding the optimal cluster centers. To overcome this drawback, the proposed method uses both 3D color histogram and human visual attention analysis to find the optimal cluster centers by considering the overall characteristic of the image. Compared with the benchmark methods, the experimental results show that the proposed method improved the segmentation accuracy.**

Keywords—image segmentation; human visual attention.

I. INTRODUCTION

Image segmentation is a technique to partition an image into homogeneous regions. It is widely used in computer vision and image processing as pre-processing [1]-[3]. Image segmentation can be divided into two categories: feature space-based methods [4]-[5], and graph-based methods [6]-[7].

Histogram-based method is a popular feature space-based segmentation algorithm, which divides the histogram into several groups using clustering algorithms [8]-[11]. The histogram approach describes a given image as a color or luminance histogram and finds the optimal cluster centers for the pixel grouping. Although the histogram-based method has lower computational complexity, compared with graph-based methods, it provided lower segmentation quality than the graph-based methods. This is because it is difficult to find the optimal cluster centers accurately for a give image. In order to find the optimal cluster centers, it is necessary to analyze the distribution of the 3D color histogram values and saliency information based on the human visual attention analysis.

In this paper, we propose a new image segmentation method by using the 3D histogram and human visual attention analysis. The proposed method clusters pixels in a non-iterative manner by generating the 3D color and the saliency histograms, which considers important human vision information. In addition, connected-component labeling which performs merging by measuring the color similarity is performed to remove the over-segmentation regions.

This research was supported by LG Display Co., Ltd. and the MSIT(Ministry of Science and ICT), Korea, under the ICT Consilience Creative program(IITP-2018-2011-1- 00783) supervised by the IITP(Institute for Information & communications Technology Promotion)

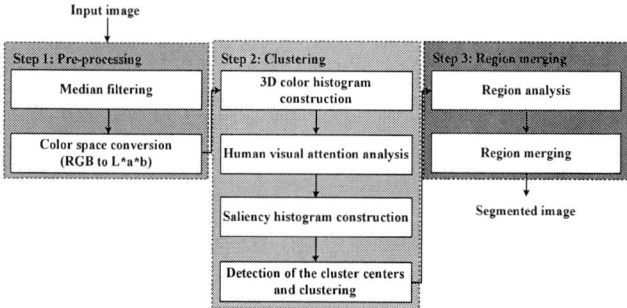

Fig. 1. Flowchart of the proposed method.

II. PROPOSED METHOD

The flowchart of the proposed method is shown in Fig. 1. First, median filtering is performed on the given image to remove the noise (Fig. 2 (b)). Then, the color space of the given image is converted from RGB color space to CIE $L*a*b$ color space. Next, each color component is quantized as follows.

$$L_q(z) = Q\left(\frac{(L(z) - \min(L))}{\max(L)}, N_{level}^L\right)$$
$$a_q(z) = Q\left(\frac{(a(z) - \min(a))}{\max(a)}, N_{level}^a\right) \qquad (1)$$
$$b_q(z) = Q\left(\frac{(b(z) - \min(b))}{\max(b)}, N_{level}^b\right)$$

where $L_q(z)$, $a_q(z)$, and $b_q(z)$ are the quantized integer values of each L, a, and b component at pixel z, respectively. N_{level}^L, N_{level}^a, and N_{level}^b denote the number of quantization levels for L, a, and b component. $Q(x, N_{level})$ is given by the following equation.

$$Q(x, N_{level}) = \lfloor (N_{level} - 1) \cdot x \rfloor + 1, x \in [0,1] \qquad (2)$$

Then, saliency histogram is also generated by accumulating the saliency values for each pixel in the corresponding 3D color histogram bins. The saliency values can be obtained by the image signature algorithm [12]. Peak bin with a histogram value larger than its adjacent histogram bins is selected as a cluster center. Additional peak bin with a saliency histogram

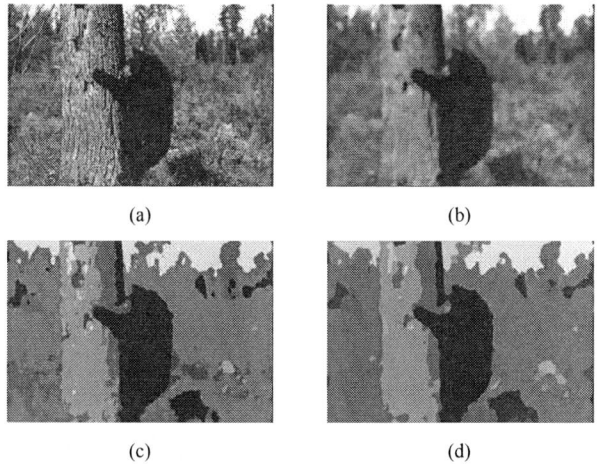

| (a) | (b) |

| (c) | (d) |

Fig. 2. Results of each step of the proposed method. (a) Original image. (b) Result of the median filtering. (c) Result of the clustering. (d) Result of the region merging.

value larger than its adjacent saliency histogram bins is also selected as a cluster center (Fig. 2 (c)).

As post-processing, the proposed method determines the connectivity by considering the color similarity. If two adjacent clusters have the similar color characteristic, we merge these two segments by assigning them with the same label (Fig. 2 (d)).

$$C(z_c, z_n) = \begin{cases} 1, & if \ I_c(z_c) = I_c(z_n) \ or \ CD(z_c, z_n) < TH \\ 0, & otherwise \end{cases} \quad (3)$$

where $C(z_c, z_n)$ is the connectivity between the current pixel z_c and its neighboring pixel z_n and $I_c(z_c)$ and $I_c(z_n)$ are the color indices of z_c and z_n, respectively. $CD(z_c, z_n)$ denotes the color difference of a block centered at pixel z_c and z_n. *TH* denotes the threshold value for the merging process.

III. EXPERIMENTAL RESULTS

For the performance evaluation of the proposed and the benchmark methods, probability rand index (PRI) [13], segmentation covering (SC) [14], boundary displacement error (BDE) [15], and global consistency error (GCE) [16] were used. For the test images, we used Berkeley Segmentation Dataset (BSDS 500) [14]. For the benchmark methods, HTFCM [8], RFHA [9], HAFCM [10], and HPCQ [11] were used.

The values of the evaluation metrics are shown in Table I. The proposed method promoted segmentation accuracy by applying the human visual attention analysis and component labeling which performs merging by measuring the color similarity; SC decreased 0.0022, but PRI, BDE, and GCE were promoted by 0.0079, 0.3068, and 0.027, respectively, compared to HPCQ [12]. It also showed the superior segmentation accuracy to the other histogram-based benchmark methods.

IV. CONCLUSION

This paper proposed a new approach to histogram-based image segmentation which considers human visual attention

TABLE I. AVERAGE PRIs, SCs, BDEs, AND GCEs, FOR THE BENCHMARK AND THE PROPOSED METHODS

Methods	PRI ↑	SC ↑	BDE ↓	GCE ↓
HTFCM [8]	0.6745	0.3546	14.4682	0.4165
RFHA [9]	0.7511	0.4224	14.3887	0.2350
HAFCM [10]	0.7435	0.4420	15.4481	0.2515
HPCQ [11]	0.7787	**0.5356**	12.8726	0.2104
Proposed	**0.7866**	0.5334	**12.5658**	**0.2077**

model. In the proposed method, the peak bins from the 3D color histogram are selected. In addition, our proposed method finds the other peak bins from the saliency histogram. After clustering is performed, component labeling is performed to remove the over-segmentation regions. As a result, the proposed method successfully improved segmentation accuracy than the histogram-based benchmark methods.

REFERENCES

[1] C.-C. Cheng, C.-T. Li, and L.-G. Chen, "A novel 2D-to-3D conversion system using edge information," *IEEE Trans. Consum. Electron.*, vol. 56, no. 3, pp. 1739-1745, Aug. 2010.

[2] S.-J. Kang, and S. Bae, "Fast segmentation-based backlight dimming," *J. Display Tech.*, vol. 11, no. 5, pp. 399-402, May 2015.

[3] F. Yang, H. Lu, and M-H. Yang, "Robust Superpixel Tracking," *IEEE Trans. Image Processing*, vol. 23, no. 4, pp. 1639-1651, Apr. 2014.

[4] D. Comaniciu and P. Meer, "Mean shift: a robust approach toward feature space analysis," *IEEE Trans. Pattern Anal. Mach. Intell.*, vol. 24, no. 5, pp. 603-619, May 2002.

[5] H. Cho, S.-J. Kang, and Y. H. Kim,"Image Segmentation using Linked Mean-Shift Vectors and Global/Local Attributes," *IEEE Trans. Circuits and Systems for Video Tech.*, vol.27, no.10, pp.2132-2140, Oct. 2017.

[6] P. F. Felzenszwalb and D. P. Huttenlocher, "Efficient graph-based image segmentation," *Int. J. Comput. Vis.*, vol. 59, no. 2, pp. 167-181, Sep. 2004.

[7] J. Shi and Malik. J, "Normalized cuts and image segmentation," *IEEE Trans. Pattern Anal. Mach. Intell.*, vol. 22, no. 8, pp. 888-905, Aug. 2000.

[8] K. S. Tan and N. A. M. Isa, "Color image segmentation using histogram thresholding – Fuzzy C-means hybrid approach," *Pattern Recognit.*, vol. 44, no. 1, pp. 1-15, Jan. 2011.

[9] K. S. Tan, N. A. M. Isa, and W. H. Lim, "Color image segmentation using adaptive unsupervised clustering approach," *Appl. Soft Comput.*, vol. 13, no. 4, pp. 2017-2036, Apr. 2013.

[10] K. S. Tan, W.H. Lim, and N.A.M. Isa, "Novel initialization scheme for fuzzy c-means algorithm on color image segmentation," *Appl. Soft Comput.*, vol. 13, no. 4, pp. 1832-1852, Apr. 2013.

[11] S. I. Cho, S.-J. Kang, and Y. H. Kim, "Human perception-based image segmentation using optimizing of colour quantisation," *IET Image Process.*, vol. 8, no. 12, pp. 761-770, Dec. 2014.

[12] X. Hou, J. Harel, and C. Koch, "Image signature: Highlighting sparse salient regions," *IEEE Trans. Pattern Anal. Mach. Intell.*, vol. 34, no. 1, pp. 194–201, Jan. 2012.

[13] R. Unnikrishnan, C. Pantofaru, and M. Hebert, "Toward objective evaluation of image segmentation algorithms," *IEEE Trans. Pattern Anal Mach. Intell.*, vol. 26, no. 6, pp. 929-944, June 2007.

[14] P. Arbelaez, M. Maire, C. Fowlkes, and J. Malik, "Contour detection and hierarchical image segmentation," *IEEE Trans. Pattern Anal. Mach. Intell.*, vol. 33, no. 5, pp. 898-916, May 2011.

[15] J. Freixenet, X. Muñoz, D. Raba, J. Martí, and X. Cufí, "Yet another survey on image segmentation: Region and boundary information integration," in *Proc. 7th Eur. Conf. Comput. Vis.*, 2002, pp. 408–422.

[16] D. Martin, C. Fowlkes, D. Tal, and J. Malik, "A database of human segmented natural images and its application to evaluating segmentation algorithms and measuring ecological statistics," in *Proc. 8th IEEE Int. Conf. Comput. Vis.*, Jul. 2001, pp. 416–425.

A Real-time Tracking Algorithm for Human Following Mobile Robot

Tsung-Han Tsai
Department of Electronic Engineering
National Central University
Chung-Li, Taiwan
han@ee.ncu.edu.tw

Chia-Hsiang Yao
Department of Electronic Engineering
National Central University
Chung-Li, Taiwan
yao13942075@dsp.ee.ncu.edu.tw

Abstract—**The ability to detect and track specific people is considered a key prerequisite for serving mobile robots. This paper presents a tracker using a binocular camera capable of tracking a specific human in both indoor and outdoor environments. The proposed tracker detects the target with the combination of human detection, color histogram, and block matching algorithm (BMA). When the color of the object is similar, the position of the object is determined by the predictor of the Kalman filter. Finally, the mobile robot is controlled based on the depth information of the target object. The effectiveness of the system is verified by human experiments in indoor and outdoor environments.**

Keywords-component; Mobile robot; human-following; visual tracking; binocular camera.

I. INTRODUCTION

In daily life, there are many machines that can make life more convenient. Robots that can follow people at any time has become more trend. As long as the robot can identify the targets that need to be followed, the robot can be operated from the user without having to manipulate the robot.

The main task for the human-tracking robots is to detect and follow the target human. First, the system performs a human detection with Histogram of Oriented Gradient (HOG) features by a Support Vector Machine (SVM) classifier. The background color information is removed by a simple foreground extractor before calculating the color histogram. Then a comparison of the statistics of the color histogram is performed to find the target. Once the color of the object is similar, the location of the target is provided by a predictor based on Kalman filter. Although the human detection system based on HOG and SVM classifier has good detection effect, the execution time is still too high which is a hinder for a real-time system. By adding a block matching algorithm, the tracking execution time can be effectively shortened.

II. TARGET TRACKING ALGORITHM

The proposed method for visual tracking of human is explained in the flowchart provided in Fig.1. In the initialization phase, an object model is defined for the target, which is to be tracked in subsequent frames of the video. Once the object is found by human detection, the color histogram information of each object is compared with that of a target that has previously been registered. The most similar object is the target we want to track. If the object is similar in color to other objects, the location of the target is determined by a predictor based on Kalman filter. In addition, we use block

matching algorithm before human detection. If the target is found, the human detection step is omitted to reduce the execution time of the tracking algorithm.

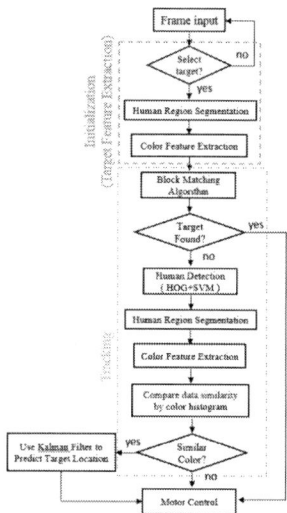

Figure 1. Flowchart of the human tracking algorithm.

A. Human Detection

In [1], Dalal and Triggs proposed the use of the combination of HOG features and SVM classifiers to implement pedestrian detection. They found that this method is a pedestrian detection method with a good balance between speed and effect. Therefore we choose the HOG feature to detect the position of the person with the SVM classifier.

Since the movement of the target between each frame is small, the past positions can be used to predict where the target is in this frame. There is no need to detect the whole image executor after the possible location of the target has been identified. The computation time can be reduced for human detection by defining ROI. In this paper, the width of the ROI is defined as twice as long as the width of the bounding box, and the height is defined as the height of the original image.

B. Target Detection from Color Histogram

In this paper, the color histogram for the HSV color space is built to find out what is required to track in human detection results. In order to adapt to lighting changes, the color histogram of the target is updated at intervals of a few frames.

As shown in Fig. 2, the result of human detection includes the information of background color, which affects the comparison of color histograms. To solve this problem, a simple visual-based foreground extractor is proposed. First, we use depth information to mark pixels larger than 5 meters as background and other pixels as unknown. Whether each unknown pixel is a background or not, it is defined by the pixels around it which have been marked as a background. The method of extracting background is deduced from the modified visual background extractor (ViBe) algorithm [2]. To classify the background, the Euclidean distance is calculated between the pixel value and each model sample, and the number of samples whose Euclidean distance is less than the threshold R is calculated, which is represented as N. We add the concept of confidence score when classifying background, which is decided by the value of N. Background pixels with higher confidence scores are more likely to be sampled in modeling. The results of the human region segmentation are shown in Fig. 2. If human regions are found, the color information of each region is extracted. Next, we need to calculate the similarity between the object and the template color information of the target.

(a) Indoor environment

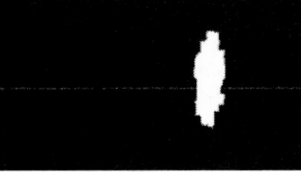
(b) Outdoor environment

Figure 2. The results of the human region segmentation.

C. Block matching algorithm (BMA)

The block matching algorithm (BMA) is used to find the best matching block from the reference frame. In this module, color histogram is used for matching criteria for the object tracking. The search method we use is the CHEXS algorithm [3]. Before human detection, we first search for targets in our defined search window through block matching. The advantage of BMA is that the execution time is short. The disadvantage is that it is more sensitive to object rotation or light changes, so we use the human detection to find the target when the BMA does not find it. Through the combination of the two methods, not only the target can be quickly found, but the accuracy will not decrease also.

III. EXPERIMENTAL RESULTS

The mobile robot is equipped with a Jetson TX1 (Quad ARM A57/2 MB L2), a ZED stereo camera, and the wheeled mobile robot platform. The effectiveness of the proposed system is verified by two evaluation values, Precision and Recall.

A. The experimental results with joining BMA

Table 1 shows that the tracking method combined with BMA can effectively reduce the time of calculation. The average calculation time is only 31.8 milliseconds. The experimental results also show that the precision and recall value is higher than 90%.

TABLE I. THE EXPERIMENTAL RESULTS

	Human detection and color histogram	Proposed
Resolution	640x360	640x360
Execution time	56 (ms)	31.8 (ms)
Precision	96 %	95 %
Recall	95 %	95 %

B. Target tracking experiment in indoor and outdoor environment

First, the target tracking experiment was performed indoors. The brightness of the target's clothes was changed by the sunlight through the windows in the indoor environment. As shown in Fig. 3(a). Next, the experiment of tracking a target was conducted outdoors, a more complicated illumination environment. As shown in Fig. 3(b).

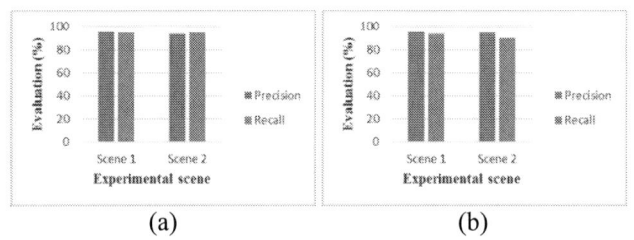

(a)   (b)

Figure 3. Evaluation of results in indoor scenes and outdoor scenes

IV. CONCLUSION

In this paper, we have presented a human tracker using a binocular camera, which can control the robot to follow a specific human. Through these experiments, we confirmed that the method can track specific people in both indoor and outdoor environments. In addition, we have added the block matching algorithm to reduce computation time. From the experimental results, the calculation time of tracking is obviously reduced, and the Precision and recall value is higher than 85%.

REFERENCES

[1] N. Dalal and B. Triggs, "Histograms of oriented gradients for human detection," 2005 IEEE Computer Society Conference on Computer Vision and Pattern Recognition (CVPR'05), San Diego, CA, USA, 2005, pp. 886-893 vol. 1.

[2] O. Barnich and M. Van Droogenbroeck, "ViBe: A Universal Background Subtraction Algorithm for Video Sequences," in IEEE Transactions on Image Processing, vol. 20, no. 6, pp. 1709-1724, June 2011.

[3] K. Belloulata, S. Zhu, J. Tian and X. Shen, "A novel cross-hexagon search algorithm for fast block motion estimation," International Workshop on Systems, Signal Processing and their Applications, WOSSPA, Tipaza, 2011, pp. 1-4.

Drivable Area Detection Method Capable of Distinguishing Vegetation Area on Country Road

Sangjae Lee and Byungin Moon

School of Electronics Engineering, Kyungpook National University, Daegu, Korea
snagjae920@knu.ac.kr, bihmoon@knu.ac.kr

Abstract— The v-disparity method is one of the most widely used algorithms for detecting the drivable area based on stereo vision. However, it mistakes vegetation of county roads as the drivable area. Thus, this paper proposes the drivable area detection method that addresses the problem of treating the roadside vegetation as the drivable area. The experimental results show that the proposed method is approximately 6.86% more accurate than the previous v-disparity method.

Keywords; drivable area detection; v-disparity; free space; vegetation area

I. INTRODUCTION

Drivable area detection is a necessary process for autonomous driving. There are various sensors such as vision, radar, and lidar used for the drivable area detection [1]. In general, stereo vision is used for the drivable area detection because it has relatively low cost and provides various data such as color, texture, and 3D information. The v-disparity method is well known as a stereo vision based algorithm to detect the drivable area [2, 3]. It is a method of detecting in real time accurate drivable area on urban roads. However, because it detects the drivable area using apparent road boundaries such as guardrails or roadside curbs, it is inaccurate on country roads whose boundaries are generally vegetation. Thus, this paper proposes a v-disparity based method to distinguish and eliminate vegetation area from drivable area on country roads.

II. THE PROPOSED METHOD

The proposed method consists of two processes as shown in Fig. 1. One process detects free space using disparity image and the other extracts roadside vegetation in the color image. Then, the results of the two processes are combined to detect the drivable area.

A. Free Space Detection

The first step of free space detection process is the obstacle removal using horizontal edges from a disparity image. In the disparity image, the horizontal edges are not detected in obstacle areas because disparities of the obstacle are almost the same. However, the horizontal edges are detected in road areas where disparities constantly change in vertical direction. Therefore, the obstacle removal process is performed by eliminating the disparities of points where the value of the

This research was supported by Basic Science Research Program through the National Research Foundation of Korea (NRF) funded by the Ministry of Education (NRF-2016R1D1A3B01015379).

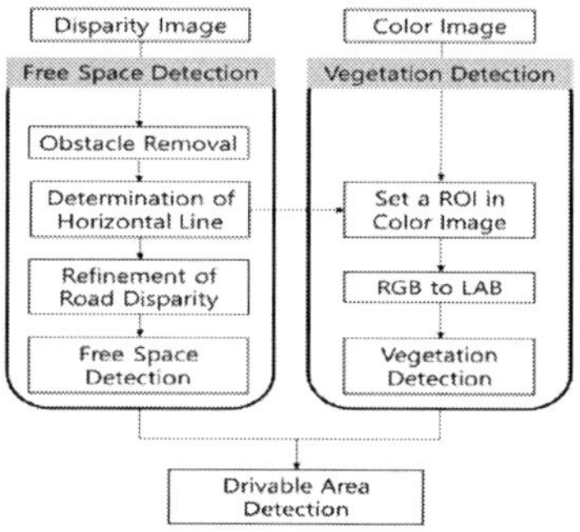

Figure 1. Flowchart of the proposed method.

horizontal edge is not a positive number. The values of horizontal edges are computed using Sobel vertical mask as

$$G_h(y,x) = \{ I_1(y+1,x-1) + 2I_1(y+1,x) + I_1(y+1,x+1) \}$$
$$- \{ I_1(y-1,x-1) + 2I_1(y-1,x) + I_1(y-1,x+1) \}, \quad (1)$$

where y and x are y- and x-axis coordinates in the disparity image, respectively, $I_1(y, x)$ is a disparity at point (y, x), and $G_h(y, x)$ is a value of horizontal edge at point (y, x). Figs. 2(a) and 2(b) show the original disparity image and disparity image after removing obstacles, respectively. After removing obstacles, a v-disparity image is generated by counting the points with the same disparity in each row [2]. In the v-disparity image of Fig. 2(c), points inside the rectangle have count values of disparities at the boundaries of obstacles in Fig. 2(b), so, the points have small values. To remove the points, we simply apply a thresholding operation with a threshold denoted as *threshold_road*. If the value of a point in the v-disparity image is lower than *threshold_road*, the point is removed as Fig. 2(d).

In the second step, the horizontal line is determined to detect real roads, which exist below the horizon [3]. Connected component labeling is performed in Fig. 2(d) and the longest connected component is extracted. Then, the horizontal line is the row of the point that has the smallest column number on the longest connected component. Fig. 2(e) shows the horizontal line below which road area exists. After the horizon detection, the proposed method selects a representative point of each row below the horizontal line in Fig. 2(e). The representative point, whose column number is regarded as road area disparity on the corresponding row, is the one that has the largest value in each

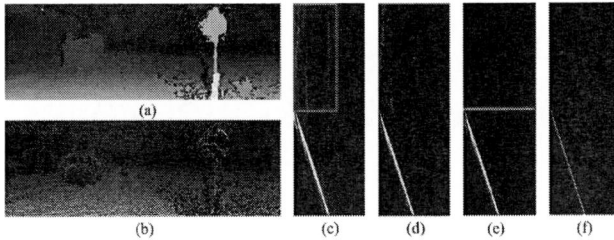

Figure 2. (a) original disparity image, (b) disparity image after removing obstacles, (c) v-disparity image using (b), (d) v-disparity after removing noises, (e) horizontal line, and (f) refined v-disparity image.

row. However, there are some outliers among representative points that should be removed. A representative point is regarded as an outlier and marked as invalid if its column number has a difference larger than a threshold denoted as *offset* from those of neighboring representative points [3]. The column numbers of invalid representative points are interpolated to those of valid neighboring representative points. By doing so, a continuous line is generated as shown in the refined v-disparity image of Fig. 2(f). Points of original image are regarded as belonging to free space if their disparities are smaller than or equal to the column number of the representative point of Fig. 2(f) on each row. The black area of Fig. 3(b) represents the free space detected from Fig. 3(a).

B. Vegetation Detection

We propose a simple pixel-based vegetation detection method because vegetation area, which free space includes on country roads, should be distinguished from road area.

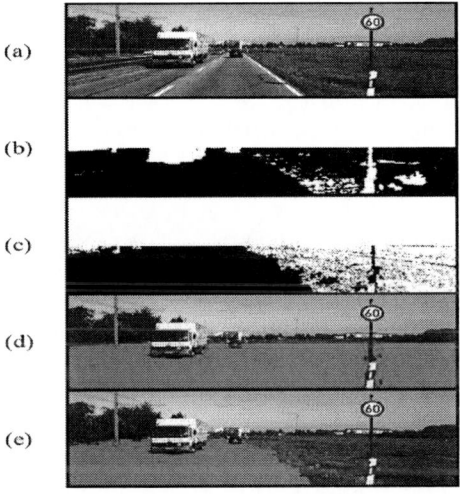

Figure 3. (a) original color image, (b) result of free space detection, (c) result of vegetation detection, (d) result of the drivable area detection using [2], and (e) result of the drivable area detection using the proposed method.

TABLE I.   RESULTS OF PIXEL-BASED DRIVABLE AREA EVALUATION

	Average of precision
Reference [2]	81.47%
The proposed method	88.33%

Experimental results showed that the channel A of LAB can detect vegetation well even in the presence of shadow in vegetation, compared with other color spaces such as RGB, HSV, and YCbCr [4]. So, the proposed vegetation detection considers points as belonging to vegetation if their channel A values are negative, and then applies erosion and dilation operations to remove small holes. The white areas below the horizontal line of Fig. 3(c) are vegetation areas detected from Fig. 3(a).

C. Drivable Area Detection

The proposed method derives the drivable area by combining the result of free space detection and that of vegetation detection, and then fills holes in the drivable area by median filtering.

III.   EXPERIMENTAL RESULTS

We compared the drivable area detection results of [2] and the proposed method. For performance comparison, 92 pairs of left and right country road images were selected from the KITTI vision benchmark road data set, and the accuracy was measured using the labeled images for training [5]. We set *threshold_road* to 15 and *offset* to 3. Results of the experiment are summarized in Table 1. The proposed method is approximately 6.86% more accurate than [2]. This is because the proposed method removes vegetation from the drivable area unlike [2], as shown in Figs. 3(d) and 3(e).

IV.   CONCLUSION

This paper proposes a v-disparity based method to detect the drivable area on country roads. Unlike the previous method, the proposed method distinguishes and eliminates vegetation area from drivable area on country roads, so the proposed method is about 6.86% more accurate than the previous method. In addition, the proposed method has the advantage of simple computation because it is a non-parametric approach using road attributes unlike the conventional method fitting roads to mathematical models.

REFERENCES

[1] A. B. Hillel, R. Lerner, D. Levi, and G. Raz, "Recent Progress in Road and Lane Detection: A Survey," Machine Vision and Applications, vol. 25, no. 3, pp. 727–745, 2014.

[2] R. Labaytade, D. Aubert, and J. P. Tarel, "Real Time Obstacle Detection in Stereovision on Non Flat Road Geometry Through "V-disparity" Representation," IEEE Intelligent Vehicles Symposium, 2002, pp. 646-651.

[3] M. Wu, S.-K. Lam, and T. Srikanthan, "Nonparametric Technique Based High-Speed Road Surface Detection," IEEE Transations on Intelligent Transportation Systems, vol. 16, no. 2, pp. 874–884, Sep. 2015.

[4] I. Harbas and M. Subasic, "Detection of Roadside Vegetation Using Features from the Visible Spectrum," 37th International Convention on Information and Communication Technology, Electronics and Microelectronics(MIPRO), pp. 1204-1209, 2014.

[5] J. Fritsch, T. Kuehnl, and A. Geiger, "A New Performance Measure and Evaluation Benchmark for Road Detection Algorithms," In International Conference on Intelligent Transportation Systems (ITSC), 2013, pp. 1693-1700.

978-1-5386-7961-6/18 $31.00 © 2018 IEEE

On-Chip Memory Optimization of High Efficiency Accelerator for Deep Convolutional Neural Networks

Tzu-Yi Lai
Department of Electronic Engineering
Feng Chia University
Taichung, Taiwan
m0503292@fcu.edu.tw

Kuan-Hung Chen
Department of Electronic Engineering
Feng Chia University
Taichung, Taiwan
kuanhung@fcu.edu.tw

Abstract— **Artificial intelligence (AI) machine often used Deep Convolutional Neural Networks (DCNN). In this paper, we delved the relevance between memory size and convolutional network architectures, i.e., AlexNet[1] and YOLOv2[2]. The high efficiency accelerator for AlexNet uses 134 k Byte memory size. Comparably, YOLOv2 costs 127 k Byte. The type of network architecture must be analyzed to determine how to save space. From this study, we get two design tips, i.e., we should focus on the use of level-2 filter memory when realizing the DCNN models having larger filter size. Meanwhile, when implementing the models having smaller filter size and large image size, we should focus on the use of level-3 image memory and its disassembly.**

Keywords; CNN, hardware accelerator, on-chip memory

I. OVERVIEW OF SYSTEM ARCHITECTURE

Fig. 1 is the overall chip architecture [3]. All of the data are read in and written out of the chip through the system interface. Input Data Decoder decodes and buffs the data that are imported from system interface. COU contains a PE array to execute convolution operations and a 6k-byte storage space to store the reusable data. In addition, COU is dominated by the COU management unit. This management unit helps us to map the PE array to a variety shape of convolution layers. PEs will be enabled if they are selected by COU management unit. At the output of COU, SAU stores and accumulates the partial sum results which produced by COU in convolution process. SAU consists of 28 partial sum buffers to store the partial sums

that are exported from COU and a 112k-byte memory to store the accumulated partial sum results. After all of the convolution operations are finished, data will be activated and export serially through the system bus.

II. ON-CHIP MEMORY ANALYSIS

On-Chip Memory plays an important role in the convolutional hardware architecture. It stores some reusable data and reduces the number of inputs of the same data from system interface. In the convolution operation, having enough memory can improve the operating efficiency of the hardware architecture. According to the different design goals of the convolutional hardware architecture, its On-Chip Memory will be allocated in different locations. According to the distance between Memory and 1-D PE, we divide the On-Chip Memory into three levels "Lv.3 Memory, Lv.2 Memory, and Lv.1 Memory, as shown in Fig.2, Fig.3, and Fig.4. We analyzed the memory and efficiency of the AlexNet and YOLOv2 models.

A. Lv.3 Image Memory

The function of Lv.3 Image Memory is to store the required current IFMap (Input Feature Map). We will cut the IFMap whose storage space requirement is greater than our Image Memory. For AlexNet, we finally chose to clip the Layer 1 IFMap into four Sub-IFMaps with a size of 13.9 k byte. Using an Image Memory with a size of 16 k byte stores the image data. We used the 12-k byte Image Memory for YOLOv2.

B. Lv.1 Filter Memory and Lv.2 Shared Memory

Lv.1 filter memory is used to store several kinds of filter data to the 1-D PE. Table I shows the Analysis of Lv.1 memory. We chose 256 byte and 128-byte memory for each 1-D PE for AlexNet and YOLOv2, respectively.

In the AlexNet model, if using one multiplier in each 1-D PE, the filter memory only needs to store 64 bytes. Larger filter memory does not have better computing efficiency by the hardware architecture. If 8 multipliers are used inside the 1-D PE, it needs 288 bytes for filter memory in CONV. Layer 1 to run at the best computing performance. At this time, there are 18 kinds of filters stored in the filter memory. In order to calculate the 96 filters of layer 1, it requires six operations of the 288-byte filter memory.

In this case, the number of internally stored filters is 16 if using the 256-byte filter memory. Also, the 96 filters of layer 1

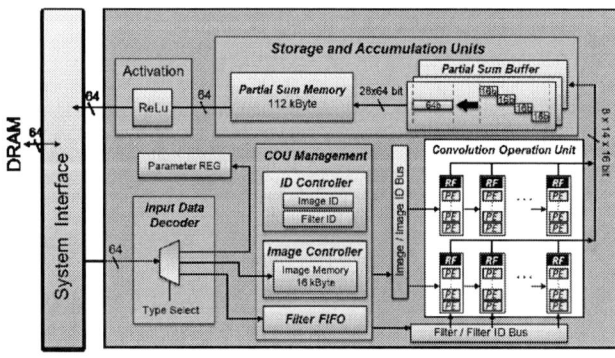

Fig.1 Overall Architecture of Chip

can be calculated by 6 calculations. We decided to adopt 256-bytes filter memory storage space for AlexNet because this size of memory not only makes Layer 1 having performed efficient convolution operations but also keep AlexNet's other convolutional layers at its best. Similarly, we implemented the same way of thinking at YOLOv2 and decided to use 128-byte Lv.1 filter memory for each 1-D PE. The minimum IFMap Size is at least 7x7 according to modern convolution models [1, 2, 4, 5, 6, 7]. In Fig.6, and Fig. 7, we grouped 7 1-D PEs into a 1-D PE Group. The 1-D PEs in this group shared the filter SRAM and MAC. Although this architecture lets down the performance, it can significantly reduce the cost of memory. It only costs 6.15k byte and 3.07k byte for AlexNet and YOLOv2.

C. Lv. 3 Partial Sum Memory

Lv.3 Partial Sum Memory stores the Partial Sum generated during the convolution operation, and is mainly determined by two parts, i.e., the number of filters stored in the chip, and the SAU (Storage and Accumulation Unit). There are 28 Partial Sum Buffers inside the SAU and its maximum output OFMap Pixels is 8x14x16 bit/clock cycle. We use the Partial Sum Memory for a total of 28 banks, and each bank can store 4 k byte size data, a total of 112 k byte for AlexNet. Similarly, it is 112 k byte required for YOLOv2.

III. MEMORY ANALYSIS

Table II is the memory gate-count analysis of AlexNet and YOLOv2. It shows the differences of using shared memory or not and different networks.

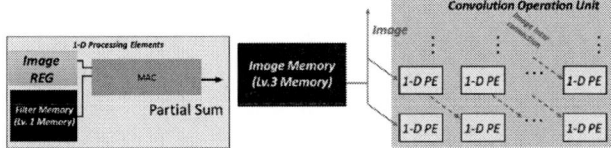

Fig.2 Lv.1 Memory Schematic   Fig.4 Lv.3 Memory Schematic

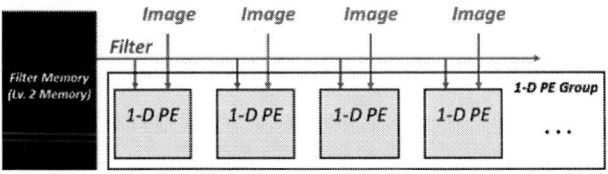

Fig.3 Lv.2 Memory Schematic

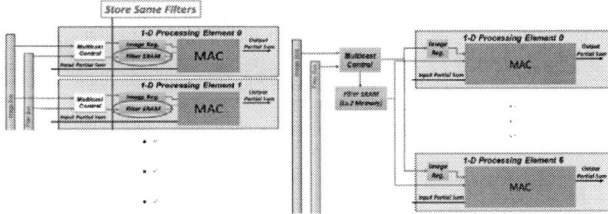

Fig.5. The Original Seven 1-D   Fig.6 The Group of Seven 1-D

PE Architecture   PE Architecture

TABLE I. MAX. FILTER MEMORY SIZE ANALYSIS RESULTS

Networks	AlexNet Layer 1		YOLOv2 Layer 1	
Kernel Width	11		3	
Max. Image ID	35		15	
# of Multipliers in a 1-D PE	1	8	1	8
# of clock cycles can finish one filter	11	2	3	1
the most suitable filter number	4	18	5	15
the most suitable filter memory size	64 Byte	288 Byte	20 Byte	120 Byte
filter memory size we chose	256 Byte		128 Byte	
total filter memory size	42k Byte		21.5k Byte	

TABLE II. MEMORY GATE-COUNT ANALYSIS

Models	AlexNet[1] Gate-Count(k)		AlexNet[1] Capacity (k Byte)		YOLOv2[2] Gate-Count(k)		YOLOv2[2] Capacity (k Byte)	
Lv.3 Image Memory	74		16		55.5		12	
Lv.1 Filter Memory (Lv.2 Shared Memory)	841.7	120.2	42	6	430.9	58.63	21.5	3
Lv.3 Partial Sum Memory	1097.7		112		1097.7		112	
Total	2,013.4	1,291.9	170	134	1,584.1	1,212	145.5	127

IV. CONCLUSION

This paper explored the relevance between memory size and convolutional network architectures, i.e., AlexNet [1] and YOLOv2 [2]. The high efficiency accelerator for AlexNet uses 134 k Byte memory size. On the other hand, YOLOv2 costs 127 k Byte.

REFERENCES

[1] K. Al-ex, I. Sutskever, and G. E. Hinton, "ImageNet Classification with Deep Convolutional Neural Networks," Commun. ACM, vol. 60, no. 6, pp. 84–90, May 2017.

[2] J. Redmon and A. Farhadi, "YOLO9000: Better, Faster, Stronger," arXiv:1612.08242 [cs], Dec. 2016.

[3] J.L. Zeng, J.Y. Wang, Y.S. Ni, S.J. Huang and K.H. Chen, "High Efficiency Accelerator for Deep Convolutional Neural Networks," 2017 VLSI-CAD, Aug.2017

[4] K. Simonyan and A. Zisserman, "Very Deep Convolutional Networks for Large-Scale Image Recognition," arXiv:1409.1556 [cs], Sep. 2014.

[5] C. Szegedy et al., "Going Deeper with Convolutions," arXiv:1409.4842 [cs], Sep. 2014.

[6] K. He, X. Zhang, S. Ren, and J. Sun, "Deep Residual Learning for Image Recognition," arXiv:1512.03385 [cs], Dec. 2015.

[7] G. Huang, Z. Liu, K. Q. Weinberger, and L. van der Maaten, "Densely Connected Convolutional Networks," arXiv:1608.06993 [cs], Aug. 2016.

Performance Metrics of Inexact Multipliers Based on Approximate 5:2 Compressors

Lavanya Maddisetti
C-ACRL, Department of ECE
Vardhaman College of Engineering
Shamshabad, Telangana, India
lavanya.maddisetty@gmail.com

JVR Ravindra
C-ACRL, Department of ECE
Vardhaman College of Engineering
Shamshabad, Telangana, India
jayanthi@ieee.org

Abstract—**Approximate computing has become an important design paradigm in innumerable applications like multimedia processing, signal processing, and machine learning to lower the performance metrics such as power, delay and area of any circuit. In this process approximate multiplier has been demonstrated in this paper, employing inexact 5:2 compressors for partial product reduction stage of multiplication operation. The simulations of all the architectures of this paper are done with Cadence Virtuoso using Spectre ©simulator in 45nm technology node. Error analysis has been performed and Error Distance, Mean Error Distance, Normalized Error Distance are found for the proposed multipliers.**

Index Terms—**Approximate Logic, Error Analysis, Compressor, Multiplier**

I. Proposed 5:2 Compressor and Multiplier

Compressors [1-4] were enrolled to replace adders present in partial product reduction stage of multipliers. This section demonstrates a novel 5:2 compressor and a proposed multiplier employing this state of art compressor. The 5:2 compressor is proposed utilizing a design level technique namely probabilistic pruning [5] in which the unwanted nodes and their corresponding interconnects are deleted from an exact 5:2 compressor. In the exact 5:2 compressor using XOR-XNORs and 2×1 multiplexers [1], the XNOR terminal of the second XOR-XNOR gate has been pruned such that one of the input terminal of the related multiplexer is grounded. The equivalent figure of novel 5:2 compressor is displayed in Fig. 1.

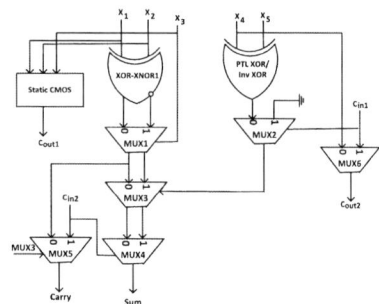

Fig. 1: Proposed inexact 5:2 Compressor

The Sum and Carry expressions for the proposed 5:2 compressor without XNOR of second XOR-XNOR has been listed below from eq. (1) to eq. (5) and there is no change

in C_{out1} and C_{out2} expressions among existing and proposed circuits.

$$Mux1 = (X_1 \oplus X_2) X_3' + (X_1 \oplus X_2)' X_3 \qquad (1)$$

$$Mux2 = (X_4 \oplus X_5) C_{in1}' \qquad (2)$$

$$Mux3 = \left\{ [(X_4 \oplus X_5) C_{in1}']' \begin{bmatrix} (X_1 \oplus X_2) X_3' + \\ (X_1 \oplus X_2)' X_3 \end{bmatrix} \right\} + \\ \left\{ [(X_4 \oplus X_5) C_{in1}'] \begin{bmatrix} (X_1 \oplus X_2) X_3' + \\ (X_1 \oplus X_2)' X_3 \end{bmatrix}' \right\} \qquad (3)$$

$$Sum = C_{in2}' * Mux3 + C_{in2} * Mux3' \qquad (4)$$

$$Carry = Mux3' * Mux1 + Mux3 * C_{in2} \qquad (5)$$

The above novel 5:2 compressor has been placed in an 8×8 Dadda multiplier to make it a proposed multiplier. The architecture of the existing and the proposed multipliers are similar in the count of number of 5:2 compressors, Full Adders and Half Adders but the only dissimilarity is the existing 5:2 compressor has been replaced with proposed 5:2 compressor. The architecture of an 8×8 Dadda multiplier is shown in Fig. 2.

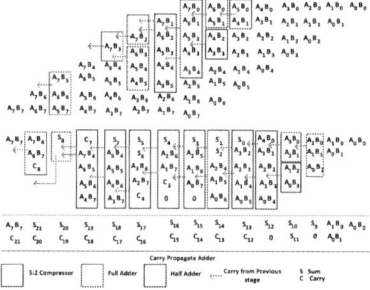

Fig. 2: Proposed inexact Multiplier using 5:2 compressors

II. Simulation and Results of 5:2 Compressor and Multiplier

The simulations of existing and proposed 5:2 compressors and multipliers utilizing these compressors have been performed using Cadence tools. The average power and propagation delay of these structures have been carried out using Cadence Virtuoso in 45nm technology node and the simulation tool employed is Spectre©. Tables I and II exhibit the performance metrics of 5:2 compressors and multipliers respectively.

978-1-5386-7961-6/18 $31.00 © 2018 IEEE   84   ISOCC 2018

TABLE I: Performance Metrics of 5:2 Compressor

Designs	5:2 Compressors with	Transistor Count	Power (nW)	PDP (fJ)
Design 1	42 n 32 TGMUX [4]	68	9000.35	30.34
Design 2	6T XOR-XNOR TG MUX [3]	60	455.1	4.36
Design 3	Full Adders only [1]	54	439.8	4.12
Design 4	Proposed without XNOR2	22	238.7	2.36
Design 5	Proposed without XNOR2 with INV XOR2	20	224.8	2.23

TABLE II: Power, Delay and Energy of Multipliers

Designs	Multipliers using 5:2 Compressors	Power (uW)	Delay (ns)	Energy (uJ)
Mul D1	MulFAsnTGonly [1]	97.77	12.72	275.3
Mul D2	Mul5242n32nTG [4]	76.97	7.685	215.9
Mul D3	Mul526TnTG [3]	30.98	-2.043	186
Mul D4	Prop Mul 52 WO XNOR2	20.79	19.2	96.12
Mul D5	Prop Mul 52 WO XNOR2 with INV XOR2	20.79	37.6	96.12

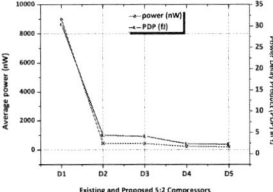

Fig. 3: Average Power, and PDP of 5:2 Compressors

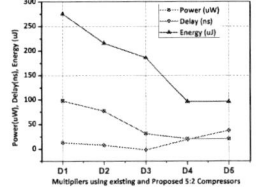

Fig. 4: Average Power, Delay and Energy of Multipliers

From Table I, it is observed that the average power consumption and propagation delay of each 5:2 compressors are according to the transistor count. But Table II has little deviation when compared with the transistor count of compressors. The multiplier which has 5:2 compressor using only Full Adders is consuming more power than the rest of the multipliers. As the difference in number of transistors between proposed 1 and proposed 2 compressors is only two, the power consumption of this pair is same. The delay of all the multipliers are not varying neither with respect to transistor count of 5:2 nor in accordance of average power consumption of multipliers. Fig. 3 and Fig. 4 shows the subsequent graphs of existing, proposed 5:2 compressors and the multipliers employing them.

III. ERROR ANALYSIS

The reliability of inexact compressors and multipliers can be determined using error metrics proposed in [6] which includes Error Distance (ED), Mean Error Distance (MED) and Normalized Error Distance (NED). The definitions and the corresponding equations are mentioned in [6]. These parameters have been calculated for the proposed 5:2 compressor and the corresponding multipliers using Matlab and the same are displayed in Table III.

TABLE III: Error Metrics of Proposed 5:2 Compressor and Multiplier emloying Proposed Compressor

Error Metrics of proposed 5:2 Compressor and Multiplier	Value
Error Rate	50%
Error Distance	1.4035×10^{38}
MED	1.0965×10^{36}
NED	8.5663×10^{33}
Mul D4 & D5	MED= 0.248 NED=3.8146×10^{-16}

Since the exact 5:2 compressors and the multipliers utilizing them have no errors, the error analysis has been shown only to the proposed compressor and multipliers. The first three compressors and multipliers are the exact architectures for which the error metrics are 0's. ED, MED and NED of the proposed compressors and multipliers (Design 4,5 compressors and Mul D4, D5) are same since there is no variation in the logic of the designs.

IV. CONCLUSION

This paper presents a proposed inexact 5:2 compressor and a multiplier utilizing this compressor. Probabilistic pruning technique has been employed to approximate the existing 5:2 compressor, The proposed multiplier used a pass transistor based XOR and Inverter based XOR in place of second XOR-XNOR circuit. with this approximations, the average power has been decreased by 78.73% when compared with Mul D1 and Mul D4. Finally, error analysis has been performed.

V. ACKNOWLEDGMENTS

This research project was carried out at Center for Advanced Computing Research Laboratory (C-ACRL), Vardhaman College of Engineering. The authors would like to thank the management and faculty for their constant support throughout.

REFERENCES

[1] Veeramachaneni et.al, "Novel architectures for high-speed and low-power 3-2, 4-2 and 5-2 compressors," Proc. of Int. Conf.on VLSI Design (VLSID), 2007, pp. 324-329.

[2] Mehdi.G et.al, "A new ultra high speed 5-2 compressor with new structure", 23rd Int. Conf. in MIXDES, Poland, 2016, pp. 151-154.

[3] Balobas.D et.al, "Low-Power high-performance CMOS 5-2 compressor with 58 transistors",Elec. Lett., 54(5), 2018, pp. 278-280.

[4] Lavanya. M et.al, "Low-Power Near-Explicit 5:2 Compressor for Superior Performance Multipliers", International Journal of Engineering Research and Technology, 11(4), 2018, pp.529-545.

[5] Avinash.L et.al, "Energy parsimonious circuit design through probabilistic pruning, in DATE, 2011.

[6] Liang.J et.al, "New Metrics for the Reliability of Approximate and Probabilistic Adders," IEEE Trans. on Comp., 63(9), 2013, pp.1760-1771.

Exploiting Configurability for Correct Sign Calculation in an Approximate Adder

Toshinori Sato
Dept. of Electronics Engineering and Computer Science
Fukuoka University
toshinori.sato@computer.org

Tomoaki Ukezono
Dept. of Electronics Engineering and Computer Science
Fukuoka University
tukezo@fukuoka-u.ac.jp

Abstract— **This paper proposes an error avoidance method for sign bits in an approximate adder. Carry-Maskable Adder (CMA) as well as most existing approximate adder does not target signed numbers and thus sign bit errors frequently occur. The proposed method exploits dynamic configurability of the CMA and turns it to operate precisely when a sign bit error is predicted. To the best of the authors' knowledge, this is the first study exploiting the dynamic configurability of approximate circuits. Simulations of a neural network for handwritten digit recognition, it is found that the proposed method considerably improves recognition rate.**

Keywords; approximate adder;sign bit error; configurability

I. INTRODUCTION

Approximate computing [3] is one the promising technique in the era when power is the first-class constraint, because it can trade accuracy for power. On the other hand, deep neural network is currently one of the most fascinating research topics. It operates huge number of multiply-add operations and thus requires large power [9]. To reduce the power consumed by deep learning devices, approximate computing is expected to be a promising technique and thus an approximate adder, Carry-Maskable Adder (CMA) [11], is adopted.

Unfortunately, a deep neural network [5,6], where the CMA is adopted, seriously fails handwritten digit recognition. One of the reasons is that the approximate adder is not suitable for signed numbers. Using Lower part OR adder (LOA) [7], which consists of the precise adder in the leading bits and OR gates in the trailing bits and works similarly to the CMA, an example of sign bit errors is explained. Consider a 16-bit addition of 64 and -32. It is operated as follows:

	adder	OR	
	00000000	01000000	(64_{10})
+	11111111	11100000	(-32_{10})
	11111111	11100000	(-32_{10})

Although the accurate sum is 32, the approximate one is incorrect and is -32. A sign bit error occurs. This may result in incorrect convolutions and thus in wrong recognitions.

Because target of most existing approximate adders is unsigned numbers, sign bit errors frequently occur when negative values are operated. Some studies [1,2,4,10,12] consider sign error correction. However, three of them [1,2,10] require the additional cycle for correction and the other [4] is dedicated to a specific adder. The scheme proposed in [12] is

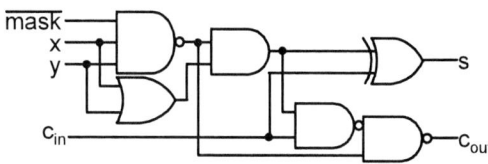

Figure 1. Carry-Maskable Full Adder

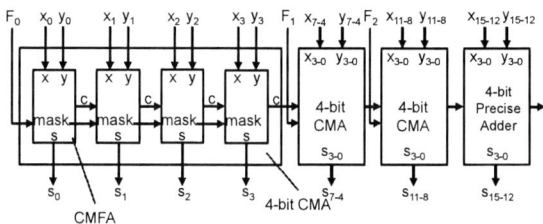

Figure 2. 16-bit Carry-Maskable Adder

the most general one as long as the authors know but it assumes the block-based approximate adders. The LOA and the CMA are not based on that and hence this paper studies another scheme for correct sign calculation.

II. INCORRECT SIGN CALCULATION AVOIDANCE

A. Carry Maskable Adder

Fig. 1 depicts a Carry-Maskable Full Adder (CMFA) [11]. When the signal **mask** equals 0 (¬mask is 1), it works as the precise full adder (FA). Otherwise, C_{out} becomes 0 and it operates as an OR gate. It is assumed that C_{in} is 0. A 4-bit CMA is made of four CMFAs and then is used as a component of a 16-bit CMA. A precise 4-bit adder consisting of four FAs is used in the most significant four bits and three 4-bit CMAs are used in the remaining twelve bits, as shown in Fig. 2. F_2, F_1, and F_0 are mask bits, each of which configures all four CMFAs in its associated sub-adder. For example, when the mask set of $\{F_2,F_1,F_0\}$ is $\{0,1,1\}$, the upper half work as the precise 8-bit adder and the lower half work as eight OR gates. Note that the CMA is the precise 16-bit adder when $\{F_2,F_1,F_0\}$ is $\{0,0,0\}$.

B. Avoidance Method for Incorrect Sign Calculation

It is studied what kind of conditions cause sign bit errors in the CMA. Outputs obtained from $2^{16} \times 2^{16}$ (16-bit x 16-bit) inputs are classified into four categories: correct sign, incorrect

978-1-5386-7961-6/18 $31.00 © 2018 IEEE

TABLE I.   DISTRIBUTION OF FOUR CATEGORIES

Mask	Correct	Incorrect		
		{+} + {+}	{+} + {-}	{-} + {-}
{0,0,1}	99.979%	0.000%	0.013%	0.008%
{0,1,1}	99.613%	0.000%	0.257%	0.130%
{1,1,1}	93.754%	0.000%	4.164%	2.083%

sign from two positive numbers, that from a pair of a positive and a negative ones, and that from two negative ones. The cases where overflow occurs in the precise addition are removed. Table I shows the results. The first column shows the configuration of the CMA. The remaining four columns show the percentages of the categories explained above. Amazingly found, the incorrect situations are very rare and the additions of a pair of a positive and a negative numbers are dominant.

From the observations above, it is expected that the penalty in power would be small even if the possibly incorrect cases were operated precisely. The CMA should be configured to perform precisely when the sign bit error is predicted. A naïve prediction scheme will work well. When the input operands are a pair of a positive and a negative numbers, the sign bit error is predicted and thus the mask is reset to {0,0,0}. The circuit required for the prediction and the mask reset is very simple. An XOR gate operates on two MSBs of the input operands and an OR gate sets each of three ¬mask signals to 1 (mask is 0); only four gates are required for the 16-bit CMA.

III.   EVALUATION

A.   Methodology

A slight variation of LeNet-5 [6], which performs handwritten digit recognition, is selected for evaluation. As shown in Fig. 3, it has seven layers: a convolutional layer with six 5x5 filters, a max pooling layer with six 2x2 filters, a convolutional layer with sixteen 10x10 filters, a max pooling layer with sixteen 5x5 filters, two fully-connected layers with 120 and 84 units, and a softmax layer with ten outputs. The activation function is tanh instead of sigmoid. It is implemented in darkent [8]. The input is a 32x32 pixel image. The training uses 32-bit floating-point operations. Only additions in convolutional and fully-connected layers on inference phase utilize the CMA, which operates 16-bit fixed-point addition. The training and the test sets are 60,000 and 10,000 images from the mnist database [6].

Power consumed by the CMA is estimated by using the results of the previous study. Table II summarizes each configuration's power consumption [11]. The number of additions processed by the CMA is counted and then used to calculate its total power consumption. Regarding the circuit predicting incorrect sign calculation, we roughly estimate its power overhead is 10%.

B.   Results

Table III presents the recognition rate before and after the proposed method is adopted. The second and third columns show the rates when precise 32-bit floating-point and 16-bit

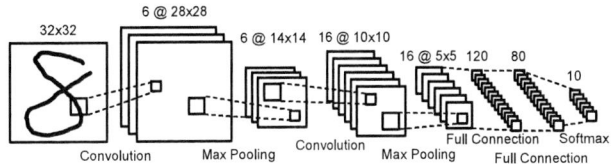

Figure 3.   LENet-5 Convolusional Neural Network

TABLE II.   POWER CONSUMED BY CMA [11]

Mask	{0,0,0}	{0,0,1}	{0,1,1}	{1,1,1}
Power (uW)	95.2	78.3	60.7	42.8

TABLE III.   RECOGNITION RATE

	Floating	Fixed	**{0,0,1}**	{0,1,1}	{1,1,1}
Original	97.37%	97.32%	**70.86%**	10.32%	10.37%
Proposed	n.a.	n.a.	**96.29%**	10.32%	9.80%

fixed-point additions are performed. The remaining three columns show those when the CMA is used with three different configurations. Because the configurations of {0,1,1} and {1,1,1} are out of account, the discussion focuses on the configuration of {0,0,1}. Relying on the proposed method, the recognition rate is improved to 96.3% from 70.9%. The achieved rate is considerably high and will be practical. Power consumed by the CMAs is increased by 5.5% from the original approximate case but is still 5.5% smaller than the precise case. This is because 25.6% of additions are predicted to be incorrect.

IV.   CONCLUSIONS

This paper proposes the error avoidance method for sign bits in the CMA. The detailed simulations unveil that the proposed method works very well on the neural network for handwritten digit recognition.

ACKNOWLEDGMENT

This work was supported by JSPS KAKENHI (JP17K00088), by funds (No.175007 and No.177005) from the Central Research Institute of Fukuoka Univ., and by VDEC, the University of Tokyo in collaboration with Synopsys, Inc..

REFERENCES

[1] A. Cilardo,. doi:10.1109/DATE.2009.5090749, DATE, 2009.
[2] K. Du, et al., doi:10.1109/DATE.2012.6176685, DATE, 2012.
[3] J. Han, et al., doi:10.1109/ETS.2013.6569370, ETS, 2013.
[4] J. Hu, et al., doi:10.7873/DATE.2015.0627, DATE, 2015.
[5] Y. LeCun, et al., doi:10.1109/5.726791, Proc. IEEE, 1998.
[6] Y. LeCun, et al, http://yann.lecun.com/exdb/mnist/
[7] H. R. Mahdiani, et al., doi:10.1109/TCSI.2009.2027626, TCAS I, 2010.
[8] J. Redmon, https:// pjreddie.com/darknet/
[9] V. Sze, et al., doi:10.1109/JPROC.2017.2761740, Proc. IEEE, 2017.
[10] A. K. Verma et al., doi:10.1109/DATE.2008.4484850, DATE, 2008.
[11] T. Yang, et al., doi:10.1109/ISQED.2018.8357311, ISQED, 2018.
[12] R. Zhou, et al., doi:10.1145/2902961.2903012, GLSVLSI, 2016.

Gap in pagination due to formatting issues.

Pages 88-89

A Folded Locking Scheme for the Long-Range Delay Block in a Wide-Range DLL

Yu-Chi Wei   Shi-Yu Huang

Electrical Engineering Department, National Tsing Hua University, Taiwan

Contact E-mail: syhuang@ee.nthu.edu.tw

Abstract—In this work, we present a long-range delay block for a wide range DLL supporting clock rates from 10MHz to 1GHz. The main contribution is a fast-locking scheme that quickly decides the control code of the delay block using a folded scheme. Post-layout simulation using a 90nm CMOS process has demonstrated that the locking time can be slashed dramatically.

Index Terms **— Cell-based DLL, Delay Line, Long-Range Delay Block, Folding Architecture**

I. INTRODUCTION

In today's IC, the Delay-Locked Loop (DLL) circuit is a very common building block for numerous applications, ranging from high-speed multi-phase clock generation, clock synchronization, clock de-skew, timing control for the logic-and-DRAM interface, etc [1][2]. Traditionally, analog circuits are used in building a DLL. However, more and more all-digital solutions have emerged as alternatives [3][4][5][6][7][8]. In some prior works, DLLs constructed by only standard cells have also been proposed [9][10]. Due to their digital nature, compilers for cell-based DLLs or PLLs are also possible [11][12]. These compilers can generate a cell-based DLL or PLL macro based on the users' requirements within a few minutes.

The most important component in a DLL is the **Tunable Delay Line** (TDL). *A TDL is a component with the end-to-end delay from its input to its output depending on some control code(s).* In order to support a super-wide range DLL, a so-called **Long-Range Delay Block** is often needed before a fine-resolution yet short-range TDL. The long-range delay block is quite simple - it consists of a series of delay groups, each consisting of a few delay buffers. For example, we estimate that 512 delay groups (each having a delay of 200ps) are needed to support an operating rang from 10MHz to 1GHz.

The entire operation of a DLL is often divided into two stages - phase-locking and phase-tracking. With a long-range DLL, we have an extra operation before the above two stages - referred to as **long-range delay block locking**, which is *to determine a proper control code value for the long-rang delay block so that its end-to-end delay roughly matches one clock cycle time of the input clock in silicon.* The aim of this work is mainly to propose an area efficient long-range delay block for a wide-range DLL while providing fast locking ability.

II. LOCKING SCHEMES OF LONG-RANGE DELAY BLOCK

A. Basic Architecture of Delay Block

A basic long-range delay block based on the **path selection concept** is shown in **Fig. 1**, the control code for the cascaded 512 *Delay Groups (DG)* is denoted as a 9-bit binary code, $\alpha[8:0]$. When the input clock signal, *Clk_in*, passes through the 512 delay groups, it creates 512 delayed versions along the way, denoted as $\{X_1, X_2, ..., X_{512}\}$, respectively. These X signals are further used to feed a 512-to-1 multiplexer (abbreviated as MUX in the sequel) to produce the output clock signal, *Clk_out*. The α-code is the control signal of the above giant MUX, which

selects one signal out of $\{X_1, X_2, ..., X_{512}\}$ to drive the output clock signal, *Clk_out*, and thereby determining the delay from *Clk_in* to *Clk_out*.

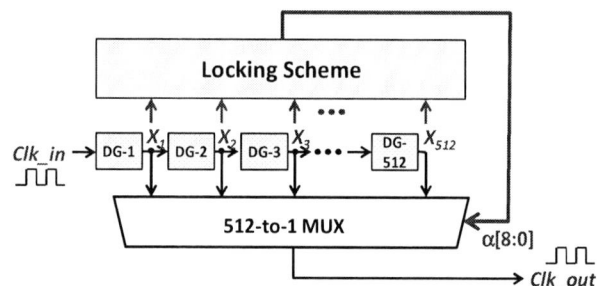

Fig. 1: A path-selection based Long-Range Delay Block, with a locking scheme to generate the control code, $\alpha[8:0]$, of the MUX, so that the delay from *Clk_in* to *Clk_out* is roughly one clock period.

As aforementioned, when a DLL using such a long-range delay block is enabled to operate, the long-range delay block locking process is started. In the following, we first demonstrate how this locking process can be done by a *concurrent scheme* and a *sequential scheme*, respectively, before we introduce a *folded scheme*. A concurrent scheme is the fastest, but requiring an extremely large area overhead. On the other hand, a sequential scheme uses the smallest amount of area, but it may introduce long locking time. The proposed folded locking scheme provides a good trade-off between the concurrent scheme and the sequential scheme.

B. Concurrent and Sequential Locking Scheme

In a concurrent locking scheme, 512 D-type Flip-Flops, $\{FF_1, FF_2, ..., FF_{512}\}$ are incorporated. For i-th Flip-Flop FF_i, signal X_i is used to drive its data input pin, which is sampled at the triggering edge of *Clk_in*. What is produced is denoted Q_i. The 512 resulting signals $\{Q_1, Q_2, ..., Q_{512}\}$ form a Q-code, as illustrated in **Fig. 2**. It is notable that the Q-code is in the form of $<0,0,...,0,1,1,...,\underline{\mathbf{1,0}},0, ..., 0,1,1,..., 1,0,0,...>$, which consists of alternating runs of '0s' and runs of '1s'.

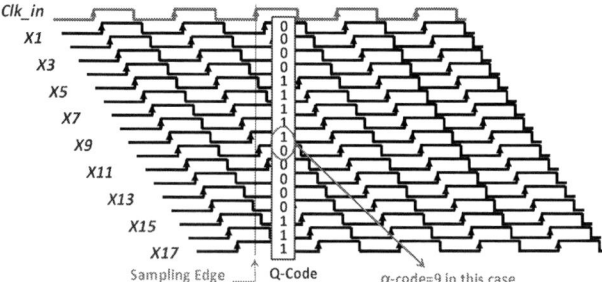

Fig. 2: The samplings of X bits (delayed versions of *Clk_in*) and the resulting Q-code.

Recall that the criterion of long-range locking scheme is to find an index k such that the delay between X_k and *Clk_in* is roughly one clock cycle time. **This criterion is equivalent to finding the location in the Q-code where the 1→0 transition occurs**. This can be realized by some encoder. At the end of this quick locking scheme, the found 1→0 transition location will

---

* This work was supported in part by Ministry of Science and Technology (MOST) of Taiwan under grant MOST-105-2221-E-007-120-MY3. We also acknowledge the help of Chip Implementation Center, Taiwan for their assistance in providing the access to EDA tools.

978-1-5386-7961-6/18 $31.00 © 2018 IEEE

be converted into a binary control code, i.e., α-code, controlling the MUX. A simplified example for the Q-code resulting from the sampling of $\{X_1, X_2, ..., X_{17}\}$ is illustrated in **Fig. 2**. The α-code should be nine in this example.

Obviously, the above concurrent locking scheme will require a large area overhead to accomplish. One alternative is to do it sequentially by only one D-type Flip-Flop. The process will take 512 clock cycles. At clock cycle i, X_i will be sampled at the triggering edge of *Clk_in*. The resulting sequence of Q-bits are also in the form of alternating runs of '0s' and runs of '1s', i.e., $\{0,0,...,0,1,1,...,\mathbf{1,0},0, ..., 0,1,1,..., 1,0,0...\}$. A simple logic can be used to identify the first clock cycle where the resulting Q-bit changes from 1 to 0. At that time, the process is terminated early and the corresponding index is reported as the desired control code. Such a sequential locking scheme may take several hundreds of clock cycles to complete for the cases when the applied input clock frequency is towards the lower end of the supported range of the operating frequency, e.g., 10MHz.

C. Folded Locking Scheme

We propose a **folded locking scheme** to strike a balance between the area overhead and the locking time. The basic idea is to partition the 512 signals $\{X_1, X_2, ..., X_{512}\}$ into 16 segments, with each segment having 32 consecutive X bits. Then, we use 32 D-type Flip-Flips to sample the 32 X bits in one segment at the triggering edge of *Clk_in*, for segments 1, 2, 3, and all the way to 16, in a sequential manner. Like the sequential scheme, **when there is a 1→0 transition in the 32-bit Q-code at certain segment, the process will be terminated early and the control code can be calculated by the following formula:**

*α-code = (segment_id) * 32 + (transition index);*

In the worst case when a very low-frequency input clock is applied, we may need to check every of the 16 segments to find the first 1→0 transition. Nevertheless, this process takes only 16 clock cycles to complete, which is much faster than the 512 clock cycles as in the sequential scheme. On the other, its area overhead mainly consists of 32 D-type Flip-Flops, 32 16-to-1 MUX, and a simple 1→0 transition identification logic with the inputs of 32 input digital bits. Overall, this is much smaller than the concurrent locking scheme.

III. Experimental Results

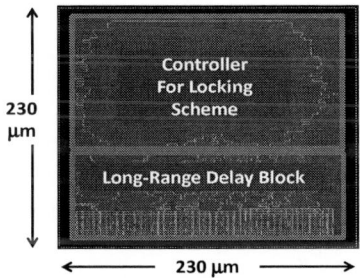

Fig. 3: Layout of our long-range delay block with the proposed folded locking scheme.

We have implemented the proposed long-range delay block using a 90nm CMOS process, with the layout shown in **Fig. 3**. We applied a number representative clock signals to drive our circuit in post-layout simulation. The results of locking time in terms of the number of clock cycles and absolute time in seconds are shown in **Table 1**. Also, we listed the phase difference between the output clock signal *Clk_out* and the input clock signal *Clk_in* after the locking process is complete. One delay group of our delay block is about 200ps. It can be seen that resulting phase difference is smaller than 200ps,

indicating that our rough locking scheme can indeed find the proper control code for the long-rang delay block. It is notable that this amount of phase difference certainly is not final and it can be easily reduced to less than 10ps after the subsequent phase-locking and phase-tracking process in the DLL. It can be seen from the results that for input clocks less than 100MHz, the rough locking times can be reduced significantly by more than 90%.

Table 1. Comparison of locking time and phase difference.

Input Clock Freq.	Sequential Scheme			Folded Scheme			Reduction (%)
	Locking Time (#Cycles)	Locking Time (ns)	Phase Diff. after Locking	Locking Time (#Cycles)	Locking Time (ns)	Phase Diff. after Locking	
10MHz	483	48650	85	20	2250	97	95
20MHz	281	14325	11.4	14	840	37	94
50MHz	112	2349	31.4	9	210	115	92
100MHz	54	595	14.2	7	85	13	87
200MHz	32	167	72	6	38	84	77
500MHz	13	29	20	5	13	48	53

IV. Conclusion

In this work, we proposed a folded locking scheme to facilitate the fast locking of a long-range delay block without introducing too much area overhead. For input clock frequency less than 100MHz, the locking time can be effectively reduced by more than 90%. The proposed circuitry can be easily added to a narrow-range DLL to extend its range of supported input clock frequency.

References

[1] Y.-H. Tu, K.-H. Cheng, H.-Y. Wei, and H.-Y. Huang, "A Low Jitter Delay-Locked-Loop Applied for DDR4", *Proc. of Design and Diagnostics of Electronic Circuits & Systems* (DDECS), pp. 98-101, 2013.

[2] S.-L. Chen, M.-J. Ho, Y.-M. Sun, M.-W. Lin, and J.-C. Lai, "An All-Digital Delay-Locked Loop for High-Speed Memory Interface Applications," *Proc. VLSI Design, Automation and Test* (VLSI-DAT), pp. 1-4, 2014.

[3] H.-H. Chang and S.-I. Liu, "A Wide Range and Fast-Locking All-Digital Cycle-Controlled Delay-Locked Loop," *IEEE J. Solid-State Circuits*, Vol. 40, No. 3, pp. 661-670, Mar. 2005.

[4] R.-J. Yang and S.-I. Liu,"A40-550 MHz harmonic-free all-digital delay-locked loop using a variable SAR algorithm," *IEEE J.Solid-State Circuits*, Vol. 42, No. 2, pp. 361–373, Feb. 2007

[5] J.-A. Tierno, A.-V. Rylyakov, and D.-J. Friedman, "A Wide Power Supply Range, Wide Tuning Range, All Static CMOS All Digital PLL in 65 nm SOI," *IEEE Journal of Solid-State Circuits*, Vol. 43, No. 1, pp. 42–51, 2008.

[6] Y.-S. Kim, S.-K. Lee, H.-J. Park, and J.-Y. Sim, "A 110 MHz to 1.4GHz Locking 40-Phase All-Digital DLL," *IEEE J. Solid-State Circuits*, Vol. 46, No. 2, Feb. 2011.

[7] M.-H. Hsieh, L.-H. Chen, S.-I. Liu, and C. C.-P. Chen, "A 6.7MHz-to-1.24GHz 0.0318 mm² Fast-Locking All-Digital DLL in 90nm CMOS," *IEEE Int. Solid-State Circuits Conf. Dig. Tech. Papers*, pp. 244-245, 2012.

[8] C.-Y. Yao, Y.-H. Ho, Y.-Y. Chiu, and R.-J. Yang, "Design a SAR-Based All-Digital Delay-Locked Loop With Constant Acquisition Cycles Using a Resettable Delay Line," *IEEE Trans. On Very Large Scale Integration Systems*, Vol. 23, No.3, pp. 567-574, 2014.

[9] T. Olsson and P. Nilsson, "A Digitally Controlled PLL for SoC Applications," *IEEE J. Solid-State Circuits*, vol. 39, no. 5, pp. 751-760, May, 2004.

[10] Duo Sheng, Ching-Che Chung and Chen-Yi Lee, "An Ultra-Low-Power and Portable Digitally Controlled Oscillator for SoC Applications," *IEEE Trans. on Circuits Syst., Exp. Briefs*, vol. 54, no. 11, Nov. 2007.

[11] C.-W. Tzeng and S.-Y. Huang, "Parameterized All-Digital PLL Architecture and its Compiler to Support Easy Process Migration", IEEE Trans. on VLSI Systems (TVLSI), Vol. 22, No. 3, pp. 621-630, Mar. 2014.

[12] P.-C. Huang and S.-Y. Huang, "Cell-Based Delay Locked Loop Compiler," *Proc. of Int'l SoC Design Conf.*, pp. 91-92, Oct. 2016.

Temperature Independent Subthreshold Circuits Design

Morteza Nabavi
Polystim Neurotech Lab
Dep. of Elec. Eng.
Polytechnique, Montreal, QC, Canada
morteza.nabavi@polymtl.ca

Maitham Shams
Department of Electronics
Carleton University, Ottawa, ON, Canada
shams@doe.carleton.ca

Mohammad Sawan
Polystim Neurotech Lab
Dep. of Elec. Eng.
Polytechnique, Montreal, QC, Canada
mohamad.sawan@polymtl.ca

Abstract—**Subthreshold circuits are suitable for applications such as bio-medical applications where the amount of energy consumption is critical. However, the drawback of these circuits is their low speed. Parallel transistor stacks technique and designing subthreshold circuits based on optimum PMOS to NMOS width ratio has shown significant speed improvements. In this paper, by providing analytical proof, it is shown that the PMOS to NMOS width ratio is independent of temperature.**

I. INTRODUCTION

Subthreshold circuits are one of the promising methods for applications (e.g., bio-medical applications) where the frequency of operation has lower priority compared to the energy consumption. However, improving the speed of subthreshold circuits can expand the application spectrum of these circuits. In [1]–[4], few attempts are proposed to improve the operational frequency of such circuits with minimum energy cost.

The concept of parallel transistor stacks (PTS) is proposed in [1], [2]. It is shown in [1], that utilizing PTS results in up to three times speed in the subthreshold region. However, it is shown in [3], that utilizing PTS blindly might result in lower speed. Therefore, the appropriate strategy on how to design circuits utilizing PTS is proposed and explained.

The optimum PMOS to NMOS width ratio β_{opt} is explored analytically and by simulation in four CMOS technology in [4].

In this paper, we provide simulations and analytical results on why the temperature have no or negligible effect on the β_{opt}.

Section II presents the parallel transistor stacks technique and how to use this technique appropriately. The temperature effect on β ratio is analyzed in Section III. Section IV concludes this paper.

II. PARALLEL TRANSISTOR STACKS

PTS has been shown to be an effective technique for improving the speed of digital circuits operating in the subthreshold region [1] [2]. However, it is shown in [3] that utilizing PTS blindly does not necessarily result in higher speed. In addition, depending on the technology, the minimum width transistor is not always the optimum width transistor. To utilize PTS appropriately, it is required to plot the current over capacitance

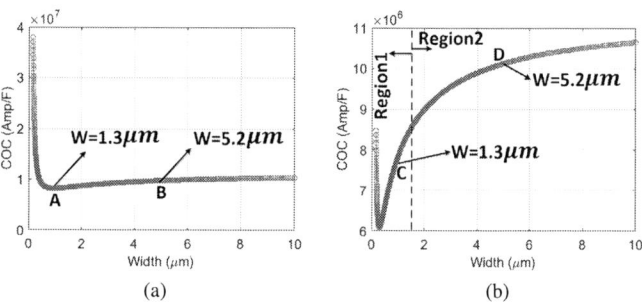

Fig. 1: Current over capacitance ratio of (a) NMOS and (b) PMOS transistors versus width in the 130 nm CMOS technology

(COC) ratio versus width. The COC ratio is a figure of merit of the speed of a logic gate. In other words, the speed of a logic gate is proportional to the current of the transistors and is inversely proportional to the capacitance of the transistors. Therefore, the higher current and less capacitance results in higher speed of the logic gates. The Optimum PMOS to NMOS width ratio (β_{opt}) for maximum speed is obtained in [4].

III. TEMPERATURE EFFECT ON β

The threshold voltage of NMOS and PMOS transistors has strong temperature dependence. This is shown in Figure 2(a).

Figure 2(b) shows $\Delta_{V_{th}}(=|V_{tp}| - V_{tn})$ versus the temperature (T) in the 65 nm CMOS technology. As shown in this figure, $\Delta_{V_{th}}$ is linearly proportional to the temperature (T). Therefore, $\Delta_{V_{th}}$ can be represented by

$$\Delta_{V_{th}} = K_1 T + K_2 \qquad (1)$$

where K_1 and K_2 are fitting parameters dependent to the technology. For example, in the 65 nm CMOS technology $K_1 = -0.0009$ and $K_2 = -0.0001$. Therefore, $\sqrt{\Gamma}$ can expressed by

$$\sqrt{\Gamma} = \sqrt{e^{\frac{|V_{tp}| - V_{tn}}{n\frac{KT}{q}}}} = \sqrt{e^{\frac{K_1 T + K_2}{n\frac{KT}{q}}}} \qquad (2)$$

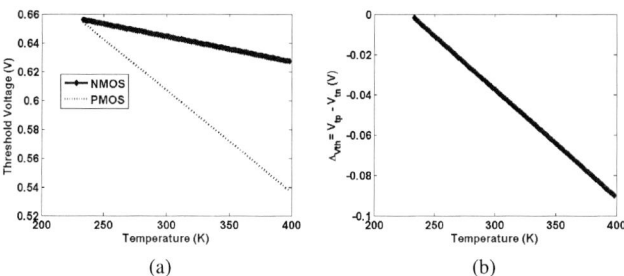

(a)   (b)

Fig. 2: a) Threshold voltage of PMOS and NMOS transistors, and b) the difference of the threshold voltages versus temperature in the 65 nm CMOS technology

The ratio R for two different temperatures T_1 and T_2 in the range of -40°C (233 K) and 120°C (398 K) can be expressed as

$$R = \frac{\sqrt{\Gamma}_{(T2)}}{\sqrt{\Gamma}_{(T1)}} = \frac{\sqrt{e^{\frac{K_1}{n\frac{K}{q}}} e^{\frac{K_2}{n\frac{K T_2}{q}}}}}{\sqrt{e^{\frac{K_1}{n\frac{K}{q}}} e^{\frac{K_2}{n\frac{K T_1}{q}}}}} = \sqrt{e^{\frac{K_2}{n\frac{K}{q}}\left(\frac{1}{T_2} - \frac{1}{T_1}\right)}} \quad (3)$$

In Equation (3), $n = 1.5$, $K_2 = -0.0001$, $q = 1.602 \times 10^{-19}$, and $K = 1.3806488 \times 10^{-23}$. By selecting any value for T_1 and T_2 in the range of -40°C (233 K) and 120°C (398 K), the value of R ($= \sqrt{e^{\frac{K_2}{n\frac{K}{q}}\left(\frac{1}{T_2} - \frac{1}{T_1}\right)}}$) is almost 1 . In other words $\sqrt{\Gamma}$ is almost independent of the temperature. For example, if we assume $T_1 = -40°C$ (233 K) and $T_2 = 0°C$ (273 K), $R = 0.999$. If we assume $T_2 = 70°C$ (343 K), $R = 0.999$, and if we assume $T_2 = 125°C$ (398 K), $R = 0.999$.

In the following, we show that $\sqrt{\Lambda}$ ($= \sqrt{\frac{\mu_n}{\mu_p}}$) is also independent of the temperature variation. As explained in [5], in the range of -40°C (233 K) and 120°C (398 K), μ_n is proportional to $T^{-2.4}$ and μ_p is proportional to $T^{-2.2}$. Therefore, $\sqrt{\Lambda} = \sqrt{\frac{\mu_n}{\mu_p}}$ is proportional to $\sqrt{T^{-0.2}}$. We define P as

$$P = \frac{\sqrt{\Lambda}_{(T_2)}}{\sqrt{\Lambda}_{(T_1)}} = \frac{\sqrt{T_2^{-0.2}}}{\sqrt{T_1^{-0.2}}} = \sqrt{\left(\frac{T_2}{T_1}\right)^{-0.2}} \quad (4)$$

If we assume $T_1 = -40°C$ (233 K) and $T_2 = 0°C$ (273 K), $P = 0.9843$. If we assume $T_2 = 70°C$ (343 K), $P = 0.9621$, and in case of $T_2 = 125°C$ (398 K), $P = 0.9479$. As shown in these examples, P is close to 1 and has negligible variation (at most 5%) to the temperature change.

To support our analysis with simulation results, the operational frequency of a 19-stage inverter ring oscillator versus β is plotted at four different temperatures (-40°C, 0°C, 25°C, 70°C, 125°C) in Figure 3. As shown in this figure the maximum frequency of the ring oscillator occurs at $\beta_{opt} = 1$ for all temperatures.

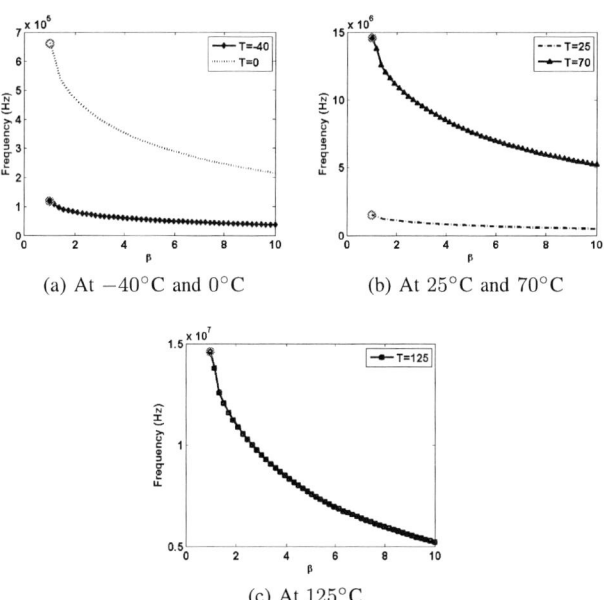

(a) At −40°C and 0°C   (b) At 25°C and 70°C

(c) At 125°C

Fig. 3: Frequency of a 19-stage inverter ring oscillator versus β in the 65 nm CMOS technology at different temperatures

IV. CONCLUSION

Designing subthreshold circuits based on PTS and optimum PMOS to NMOS width ratio shows significant speed improvement with minimum energy cost. In this paper, by providing analytical proof, it is shown that the optimum PMOS to NMOS width ratio is independent of the temperature variation. Simulation results show that the optimum PMOS to NMOS width ratio remains constant for the range of -40°C to 125°C.

REFERENCES

[1] M. Muker and M. Shams, "Designing digital subthreshold cmos circuits using parallel transistor stacks," *Electronics Letters*, vol. 47, no. 6, pp. 372–374, March 2011.

[2] J. Zhou, S. Jayapal, B. Busze, L. Huang, and J. Stuyt, "A 40 nm inverse-narrow-width-effect-aware sub-threshold standard cell library," in *2011 48th ACM/EDAC/IEEE Design Automation Conference (DAC)*, June 2011, pp. 441–446.

[3] M. Nabavi, "Designing Faster CMOS Subthreshold Circuits Using Transistor Sizing and Parallel Transistor Stacks," Ph.D. dissertation, Carleton University Ottawa, 2012.

[4] M. Nabavi, F. Ramezankhani, and M. Shams, "Optimum pmos-to-nmos width ratio for efficient subthreshold cmos circuits," *IEEE Transactions on Electron Devices*, vol. 63, no. 3, pp. 916–924, 2016.

[5] S. M. Sze and K. K. Ng, *Physics of semiconductor devices*. John wiley & sons, 2006.

978-1-5386-7961-6/18 $31.00 © 2018 IEEE

Optimizing the Performance of a Low Power – Area Efficient OTA Design that is Based on Hybrid Current Shunting Technique

Imtinan B. Attili and Soliman A. Mahmoud

Department of Electrical and Computer Engineering
University of Sharjah
Sharjah, UAE
iattili@sharjah.ac.ae

Abstract—**This paper presents a new current mirror operational transconductance amplifier (OTA) design that utilizes a hybrid technique of shunting current from the main differential pair. This technique combines shunting current through two different ways; the first is by using an adaptive biased transistors through a second differential pair, and the other is by using a constant biased current source transistors. Each way separately had been examined in previous studies and have shown an improvement on the performance of the current mirror OTA. Combining these different ways is examined here and the performance have been optimized using 90nm CMOS technology to insure obtaining the best results without compromising the stability of the circuit, therefore keeping the phase margin ≥ 60º.**

Keywords— OTA; current shunting; stability

I. INTRODUCTION

High performing CMOS operational transconductance amplifiers are a vital building block in analog signal processing systems. The appeal of these blocks comes from their simple design, good performance, single-pole characteristics and wide output voltage swing. In integrated circuits, hundreds of amplifiers are built into a single chip, therefore each amplifier must occupy as small a die area as possible to reduce the power consumption while maintaining a high and stable performance [1]. However, in its conventional connection, the OTA lacks in DC gain and output resistance. One way to enhance this drawback is to shunt part of the current passing through the differential pair, which was done in several ways as seen in [2-5]. Combining the two current shunting techniques identified in [2] and [3] is done here and the circuit is optimized for the best performance through altering the current shunt amount and the different current gain factors.

II. THE HYBRID CURRENT SHUNTING OTA DESIGN

The circuit presented here employs two different techniques of shunting current from the main differential pair M1 and M2, as seen in Fig. 1. The first technique is through shunting a fixed amount of current (I_{sh}) using fixed biased current source transistors M15 and M16 [2]. The second is through using an adaptive cross current shunt technique, which is based on using a second differential pair, M11 and M12,which

senesces the amount of input voltage in the main differential pair and accordingly bias the current shunting transistors M9 and M10 [3].

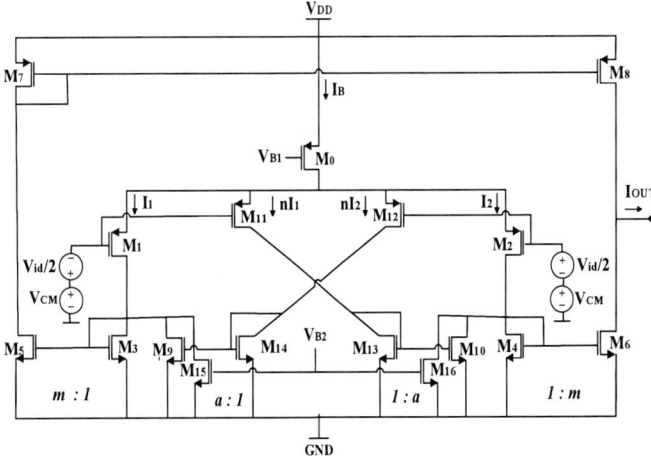

Fig. 1. The CMOS realization of the hybrid current shunting OTA.

Taking into account the different current mirror gains α and m and that the two differential pairs work at different strengths making the current passing through transistors M11 and M12 n times the current in transistors M1 and M2, the effective transconductance, output resistance and DC gain of this OTA can be driven as

$$G_m = m\left[g_{m_1} + a\, g_{m_{12}} \right] \qquad (1)$$

$$R_{out} = \frac{1}{(\lambda_6 + \lambda_8)\, m\left[\frac{I_B}{2} \frac{(1-an)}{(1+n)} - I_{sh} \right]} \qquad (2)$$

$$A_v = \frac{v_{out}}{v_{id}} = \frac{\left[g_{m_1} + a\, g_{m_{12}} \right]}{(\lambda_6 + \lambda_8)\left[\frac{I_B}{2} \frac{(1-an)}{(1+n)} - I_{sh} \right]} \qquad (3)$$

It is evident from the above three equations that the performance of the circuit can be controlled by the tuning of the different factors a, m and n. Understanding the individual effect of both a and n is studied through simulations below.

III. SIMULATION RESULTS

The new OTA is validated through simulations which were carried on LTspice software using 90 nm CMOS model, BSIM4 (level 54) version 4.3 under 1V supply voltage. The bias current used was 2μA, while the current shunt through the fixed biased transistors was set at 15% of the current in the main differential pair. The circuit was loaded with 1 pF capacitor. To understand the effect of the current shunting factor a, it was varied between 0.3 to 0.8 with a step of 0.1 while n and m were kept constant at 1 and 10 respectively. The following figures shows respectively how the DC gain, output resistance, transcondactance and step response behaved as the value of a changed.

As it is evident the performance of the OTA enhanced as the value of a increased since the gain reached a maximum of 53dB and the output resistance reached 2.4 MΩ. The step response of the circuit is fast and also is not affected by the changing value of a. On the other hand, it was found that the bandwidth kept on decreasing going from 333 kHz to75 kHz. Regarding the stability, which is an important factor, it was found that keeping the phase margin ≥ 60°, no more than 40% ($a = 0.4$) can be shunt through transistors M9 and M10.

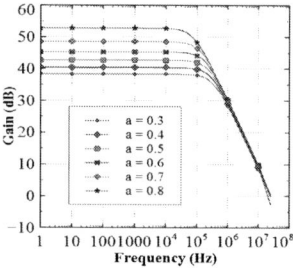

Fig. 2. The DC gain corresponding To different values of a.

Fig. 3. The output resistance corresponding to different values of a.

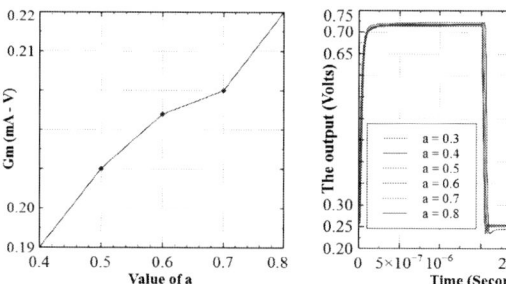

Fig. 4. The transconductance corresponding to the different values of a.

Fig. 5. The OTAs step response corresponding to the different values of a.

To understand the effect of factor n, it was varied between values 0.2 to 1.2 while $α$ and m were kept constant at 0.8 and 10 respectively. The following figures respectively shows how the DC gain, output resistance, transcondactance and step response behaved as the value of n was changed.

The results shows that the gain kept on increasing with the value of n reaching a maximum of 60dB corresponding to $n = 1.2$. The output resistance increased exponentially with the value of n starting from 7kΩ, ending at 3.1MΩ, whereas the transconductance decreased exponentially starting from 220μA

ending at 10μA. The bandwidth also decreased starting at 716 kHz and ending at 3.4 kHz. Also, the step response of all n values is nearly identical. The main difference noticed is decreasing overshoots as the value of n increases. Regarding stability, it was found that going beyond $n = 0.4$, the phase margin goes less than 60°. However, it was also noticed that as n goes higher than 1.2, the phase margin starts on increasing. Thus, through fine tuning, it was found that at $n = 1.26$ the phase margin is exactly 60° whereas the DC gain is 56dB.

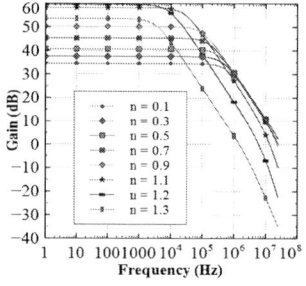 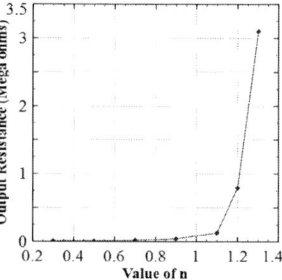

Fig. 6. The DC gain corresponding To different values of n.

Fig. 7. The output resistance corresponding to different values of n.

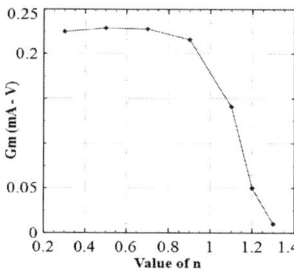 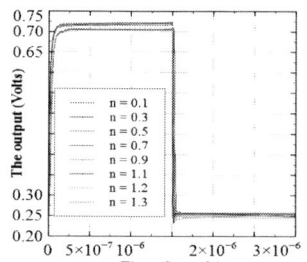

Fig. 8. The transconductance corresponding to different values of n.

Fig. 9. The OTAs step response corresponding to different values of n.

IV. CONCLUSION

The OTA examined here shows the ability to provide a high gain reaching the value of 60dB and high output resistance reaching 3.1MΩ. However, the stability of the circuit, which is a very important factor, is not always maintained as the phase margin goes less than 60° as higher gain values is achieved. Thus, other ways of enhancing the performance of the circuit should be examined that insures achieving high gain while maintaining stability.

V. REFERENCES

[1] S. A. Mahmoud and A. M. Soliman, "New cmos fully differential difference transconductors and application to fully differential filters suitable for vlsi," Microelectronics journal, vol. 30, no. 2, pp. 169–192, 1999.

[2] Libin Yao, M. S. J. Steyaert, and W. Sansen, "A 1-V 140-/spl mu/W 88-dB audio sigma-delta modulator in 90-nm CMOS", IEEE J. Solid-State Circuits, vol. 39, no. 11, pp. 1809–1818, 2004.

[3] J. Roh, "High-gain class-AB OTA with low quiescent current", Analog Integr. Circuits Signal Process., vol. 47, no. 2, pp. 225–228, 2006.

[4] J. Roh, S. Byun, Y. Choi, and S. Member, "Modulator With 83-dB Dynamic Range", Design, vol. 43, no. 2, pp. 361–370, 2008.

[5] Z. Yan, P. I. Mak, M. K. Law, R. P. Martins, and F. Maloberti, "Nested-Current-Mirror Rail-to-Rail-Output Single-Stage Amplifier with Enhancements of DC Gain, GBW and Slew Rate", IEEE J. Solid-State Circuits, vol. 50, no. 10, pp. 2353–2366, 2015.

Gap in pagination due to formatting issues.

Pages 97-98

A Low-power Low-noise Open-loop Configured Signal Folding Neural Recording Amplifier

Gauri Punekar, Venkateswarlu Gonuguntla, Palagani Yellappa, Jun Rim Choi, and Ramesh Vaddi

Abstract—**This paper proposes a design of low-power and low-noise CMOS neural recording amplifier with an open-loop configuration. The proposed design has been simulated using CMOS 0.18μm process. The proposed design with the signal folding technique, when compared to the closed-loop configured neural amplifier, has adequately minimized the total power consumption. The performance metrics such as gain, bandwidth, and input referred noise is also finely optimized.**

I. INTRODUCTION

Recent works on neural recording integrated circuits (ICs) have improved the quality of capturing the electrical activity of neurons [1]–[3] and therefore the performance of neural recording applications were also improved [4]–[6]. The general architecture of the neural recording IC is shown in Fig. 1. Design of the neural amplifier, blue highlighted box, is the focus of this study. In order to achieve a low-power and low-noise performance during the neural recording, many analog front-end pre-amplifier circuits were investigated (ex. [1], [3]). All these works improved the neural recording quality. However, the neuroscience community requires even better recordings of the neural signal to understand them. To improve the neural signal recording quality, in this paper, we propose an open-loop configuration neural amplifier with the signal folding technique. The proposed technique outperformed the existing closed-loop configured neural amplifier and has the potential to collect the improved neural signal quality.

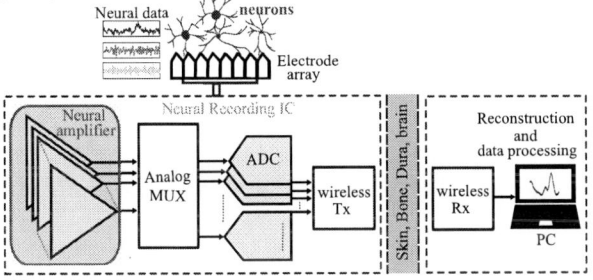

Fig. 1.   Block diagram of neural recording system.

II. DESIGN AND ANALYSIS OF THE PROPOSED NEURAL AMPLIFIER

A typical neural recording system, as shown in Fig. 1, consists of a neural amplifier to boost the signal, analog

Gauri Punekar is with Electrical Engineering Department, Shiv Nadar University, Greater Moods, UP, India;
Venkateswarlu Gonuguntla is with Medical Science Research Institute, Samsung Medical Center, Seoul, South Korea;
Palagani Yellappa and Jun Rim Choi are with School of Electronics Engineering, Kyungpook National University, Daegu, South Korea 701-701;
Ramesh Vaddi is with School of Computing, National University of Singapore, Singapore 117-417;
(E-mail: jrchoi@ee.knu.ac.kr, vaddiramesh2k9@gmail.com).

to digital converters to digitize the amplified signal, and wireless transmission to stream the data from the brain. A neural amplifier with closed-loop configuration is shown in Fig. 2(a). It is the same as in [3] except the operational transconductance amplifier (OTA) with single-stage topology. This design captures the neural signal with better quality but has less noise performance. To subdue this, we propose the design as shown in Fig. 2(b) which is similar to [3] but has an open-loop configuration and OTA with single-stage topology. In the proposed design, the output signal is reset to a common DC level whenever it is above a threshold level. OTA of the neural amplifier, as shown in Fig. 2(c), is built on a standard 0.18μm process. An input supply of 1V and a bias current (I_{bias}) of 2.37μA is supplied to OTA. Transistors in the circuit are provided with an input AC signal of 6mV_{pp} and the pairs $M3$-$M5$, $M4$-$M6$, and $M7$-$M8$ form current mirrors. The bias current supplied to the current mirror network acts as a source to the input transistors. Two comparators as in [1] are used in the design and the comparator design, shown in Fig. 2(d), has three stages: the pre-amplifier stage ($M1$ to $M7$), the latch ($M8$ to $M11$), and an output buffer ($M12$ to $M15$). A threshold voltage is given to $M1$ and the output of OTA is provided to $M2$.

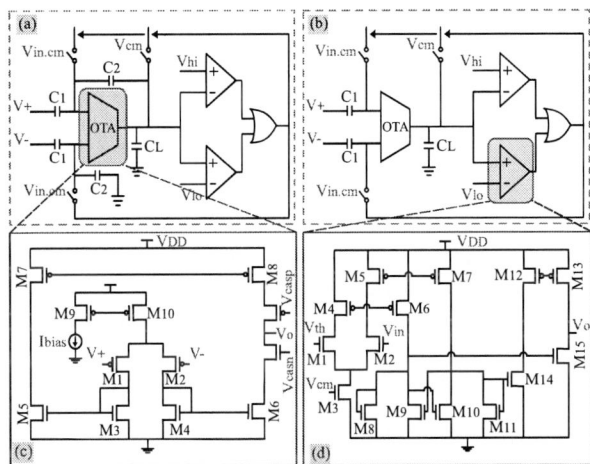

Fig. 2.   (a) Closed-loop neural amplifier with signal folding technique, (b) Open-loop neural amplifier design with signal folding technique, (c) Schematic of the OTA, and (d) Schematic of the comparator.

III. RESULTS AND DISCUSSION

The closed-loop neural amplifier design as shown in Fig. 2(a) and the proposed open-loop neural amplifier design as shown in Fig. 2(b) are simulated using 0.18μm CMOS process. The parameters: gain, power supply rejection ratio (PSRR), common mode gain, and input referred noise are

978-1-5386-7961-6/18 $31.00 © 2018 IEEE

obtained from the simulation. From the obtained parameters, common mode rejection ratio (CMRR) and the noise efficiency factor (NEF) are calculated as in [1]. The simulated results of the parameters gain, PSRR, common mode gain, and input referred noise of closed-loop design and proposed design are plotted in Fig. 3(a) and 3(b) respectively. The calculated parameters of both the designs are also summarized in Table I and it is observed that the open-loop design outperformed the closed-loop design. The comparison and

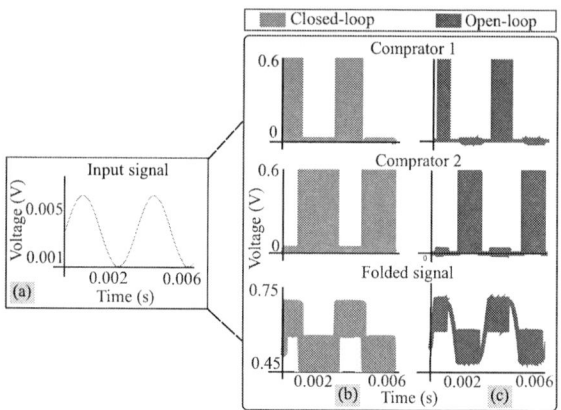

Fig. 4. (a) Transient response of AC signal 6mVpp, (b) Closed-loop configured transient response, (c) Open-loop configured transient response.

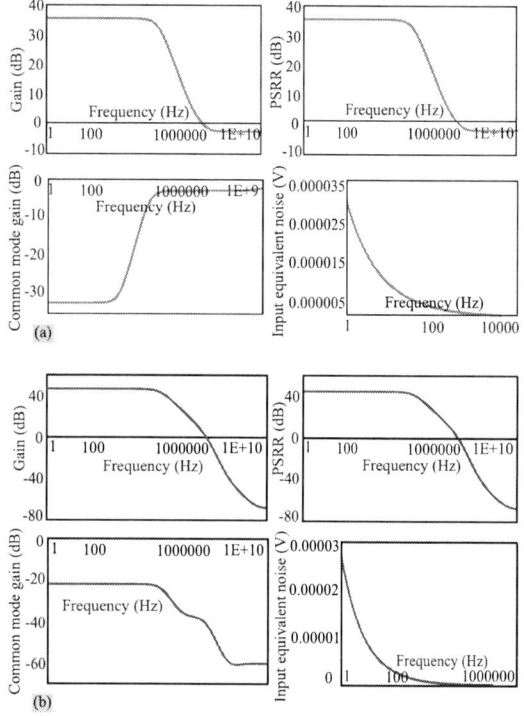

Fig. 3. (a) Simulation results of the close-loop configuration, (b) Simulation results of the open-loop configuration.

TABLE I

COMPARISON OF PERFORMANCE METRICS OF OPEN-LOOP AND CLOSED-LOOP DESIGN.

Parameters	Open-loop	Closed-loop
Process	$0.18\mu m$	$0.18\mu m$
Gain	46.4dB	35.16dB
Supply Voltage	1V	1V
Power	$6\mu W$	$6.017\mu W$
Input Referred Noise	$0.23\mu V_{rms}$	$1.081\mu V_{rms}$
PSRR	46.3dB	34.02dB
CMRR	68.2dB	38.366dB
NEF	0.05	0.28
Figure of Merit	1.425	0.81

IV. CONCLUSION

The proposed open-loop configured neural amplifier design with the signal folding technique has been proposed and found very effective in minimizing the total power consumption. The open-loop design is more suitable to the signal folding technique when compared to the closed-loop design. Optimum performance metrics are achieved with the proposed design and it can play a significant role in the neural recording enhancement.

REFERENCES

[1] R. R. Harrison, and C. Charles, "A low-power low-noise CMOS amplifier for neural recording applications," *IEEE Journal of Solid-State Circuits*, Vol. 38, June 2003.

[2] M. Yin and M. Ghovanloo, "A low-noise preamplifier with adjustable gain and bandwidth for biopotential recording applications," *IEEE*, 2007.

[3] Y. Chen, A. Basu, L. Liu, X. Zou, R. Rajkumar, G. S. Dawe, and M. Je, "A digitally assisted, signal folding neural recording amplifier," *IEEE Trans. on Biomedical Circuits and Systsems*, Vol. 8, August 2014.

[4] R. Sarpeshkar, W. Wattanapanitch, S. K. Arfin, B. I. Rapoport, S. Mandal, M. W. Baker, M. S. Fee, S. Musallam, and R. A. Andersen, "Low-power circuits for brainmachine interfaces," *IEEE Journal of Solid-State Circuits*, Vol. 46, April 2011.

[5] M. Azin, D. J. Guggenmos, S. Barbay, R. J. Nudo, and P. Mohseni, "A battery-powered activity-dependent intracortical microstimulation IC for brain-machine-brain interface," *IEEE Trans. on Biomedical Circuits and Systsems*, Vol. 2, September 2008.

[6] H. Bahrami, S. A. Mirbozorgi, A. T. Nguyen, B. Gosselin, L. A. Rusch, "System-level design of a full-duplex wireless transceiver for brainmachine interfaces," *IEEE Trans. on Microwave Theory and Techniques*, Vol. 64, October 2016.

reset circuits that are performed by the comparator and OR circuits with respect to the input signal of the closed-loop and open-loop designs are illustrated in Fig. 4. Fig. 4(a) shows the transient response to the input signal 6mVpp that is applied as input transistors. This value was chosen to approximate the amplitude which is close to that of the local field potentials. Fig. 4(b) shows the closed-loop transient response of the comparators and the signal folded output obtained by applying OR operation. Whenever Vo1 is higher than Vhi (0.7V), the comparator will give the logic high ('1'). For the low threshold level, the same comparator has been used and the inputs have been inverted. Whenever Vo2 is lower than Vlo (0.5V), the comparator will give logic high ('1'). The idea behind the comparator logic design is to ensure that the OR operation provides the logic high whenever the signal from the amplifier goes beyond the predefined range of voltages. Similar transient analysis was performed for comparators and the signal folded output was obtained for open-loop configuration as shown in Fig. 4(c).

Nano-ampere Current Sensing Technique for OLED Mobile Displays

Minhyun Jin, Hyejin Im, Minkyu Song and Soo Youn Kim

Department of Semiconductor Science
Dongguk University
Seoul, 04620, Korea
mhjin91@dongguk.edu

Abstract— **In this paper, we propose a nano-ampere current sensing technique for organic light-emitting diode (OLED) mobile displays. The nano-ampere current sensing system consists of a current integrator and an 8-bit single-slop analog to digital converter (SS-ADC). In order to reduce area and power consumption, the current integrator and a comparator of the SS-ADC share an amplifier, called differential difference amplifier (DDA). From the experimental results, we obtain 399.21 pA of accuracy with 8-bit current sensing resolution ranged from 0 to 102.2 nA. The proposed current sensing circuit was fabricated with a 90 nm CMOS technology and chip size is about 600 μm × 3000 μm.**

Keywords; Analog to digital converter; Current integrator; Differential difference amplifier: Low current sensing; Organic light-emitting diode

I. INTRODUCTION

Recently, the organic light-emitting diode (OLED) panels have been widely used in mobile applications. The OLED panels have a variety of advantages over liquid crystal display (LCD) panels: fast speed, high image quality, wide viewing angle, and prevention of visual impairment [1]. However, the operation current of the OLED (I_{OLED}) becomes low (< 10~100 nA), leading to the difficulty to compensate the variations and mismatches. In order to properly capture process variations and mismatches among other OLED pixels, the high accuracy of the current sensing technique is required. In this paper, we propose a nano-ampere current sensing circuit consisting of current-type digital to analog converter (DAC), representing 100 nA range of the pixel currents (=I_{OLED}) [2], a current integrator, and an 8-bit column-parallel single-slope analog to digital (SS-ADC). In order to obtain small area and low power consumption, a differential difference amplifier (DDA) is adopted to function both the current integrator and a comparator in SS-ADC.

II. THE PROPOSED NANO-AMPERE CURRENT SENSING CIRUCIT

Figure 1 shows the block diagram of the proposed nano-ampere current sensing circuit. In order to represent the I_{OLED} ranged from 0 to 100 nA (1 LSB= 390 pA), an 8-bit current DAC is designed. The generated I_{OLED} from DAC is converted into the voltage through a current integrator. Then, the converted voltage generates digital output codes using an 8-bit SS-ADC [3].

This work was supported in part by the MOTIE (Ministry of Trade, Industry & Energy) (project number # #100080403) and KSRC (Korea Semiconductor Research Consortium) support program for the development of future semiconductor devices and in part by This work was supported by the National Research Foundation of Korea (NRF) grant funded by the Korea government (MSIP; Ministry of Science, ICT & Future Planning) (No. S-2018-A0496 00081)..

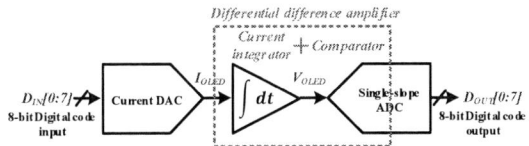

Figure 1. Block diagram of the proposed OLED current sensing.

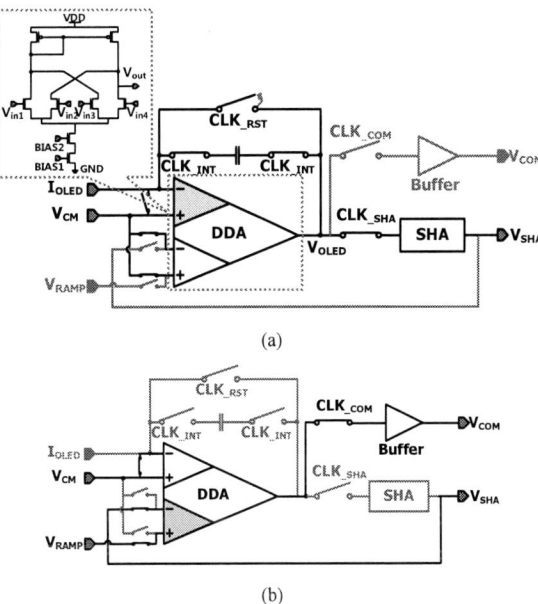

(a)

(b)

Figure 2. Operation process of (a) the current integrator and (b) the comparator of SS-ADC using DDA in analog front end block.

III. THE OPERATION OF THE PROPOSED CIRCUIT

Figure 2 shows the analog front end (AFE) block of the proposed nano-ampere current sensing circuit using a DDA which has two functions: (1) a current integrator and (2) a comparator in SS-ADC with sample and hold amplifier (SHA). Unlike conventional existing current sensing techniques [4], the area of the AFE should be small since every column consumes the area for the column-parallel SS-ADC structure. Therefore,

978-1-5386-7961-6/18 $31.00 © 2018 IEEE

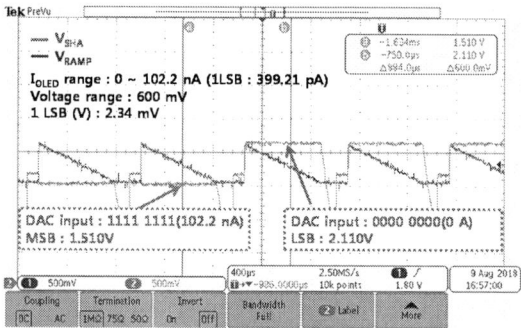

Figure 3.   Measurement results of V_{SHA} and V_{RAMP} signals.

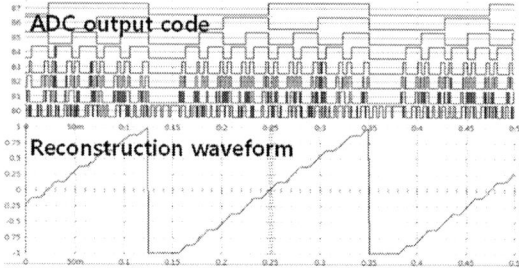

Figure 4.   Measurement results of digital codes and reconstruction signals with WinBIT.

we use a DDA comprising two operational amplifiers (op-amp) [5] to minimize the area as well as power consumption. By using the DDA rather than using separated two blocks of a current integrator and a comparator, the 16.25 % of area can be saved.

The DDA operation process is illustrated for the cases of a current integrator in Figure 2(a) and a comparator of SS-ADC in Figure 2(b). First, the current DAC generates I_{OLED} and the DDA is functioned as a current integrator, converting the I_{OLED} ranged from 0 to 100 nA into V_{OLED} ranged from 2.1 V to 1.52 V. Then V_{OLED} is sampled using SHA as shown in Figure 2(a). After finishing the process of the current integration, the DDA is operated as a comparator as shown in Figure 2(b). The sampled voltage, V_{SHA} goes to the one of inputs of the comparator with a ramping signal, V_{RAMP}. In order for the SS-ADC operation, V_{SHA} is compared to V_{RAMP}, finally flipping the output of the comparator when V_{SHA} is equal to V_{RAMP}. The comparator output voltage, V_{COM} is converted into a digital code using an 8-bit counter, SRAM cells, and latches.

IV.   EXPERIMENTAL RESULT

Figure 3 shows the measurement results of V_{SHA} and V_{RAMP} signals with an oscilloscope. Results show that V_{SHA} changes from 2.110 V to 1.510 V with about 2.34 mV of LSB, corresponding to 399.21 pA (Measured I_{OLED} current range: 102.2 nA). The 8-bit digital outputs of ADC is reconstructed with the WinBIT program as shown in Figure 4. The proposed nano-ampere current sensing circuit is designed from supply 3.3 V (analog)/1.2 V (digital) 90 nm CMOS process. Figure 5 (a) is a chip photography and Figure 5 (b) is a testing board. Table 1 is the summary of measurement results of the proposed current sensing circuit. The power consumption is 643 μW and the size of the layout chip excluding the I/O pads is 600 μm × 3,000 μm.

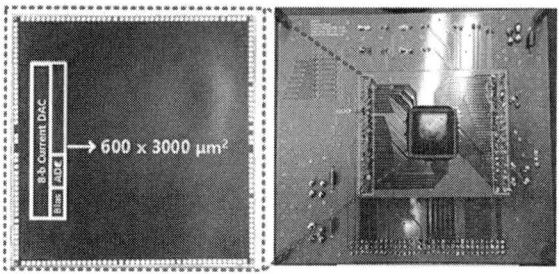

(a)   (b)

Figure 5.   (a) a chip photography and (b) a testing board.

TABLE I.   THE SUMMARY OF MEASUREMENT RESULTS.

Design Metrics	Results
Process	90 nm CMOS
Chip size	600 μm × 3000 μm
ADC structure	Single-slop
ADC resolution	8-bit
1 LSB	399.21 pA
Power consumption	643 μW
Sampling rate	15 kS/s
Supply voltages	3.3 V(Analog) / 1.2 V(Digital)

V.   CONCLUSION

In this paper, we propose a nano-ampere current sensing circuit for OLED mobile displays. In order to reduce the area and power consumption, we present a DDA sharing a current integrator and a comparator in SS-ADC. Measurement results show that the power consumption is 643 uW and 16.25 % of area are is decreased. The proposed current sensing circuit can sense I_{OLED} ranged from 0 to 102.2 nA with 399.21 pA of accuracy. Therefore, we believe that the proposed current sensing circuit helps the compensation of the process variations and mismatch of the OLED currents with high accuracy under low power and area budget.

REFERENCES

[1] Chih-Hao Chang, Szu-Wei Wu, Sung-En Lin, Chih-Wei Huang, Chung-Tsung Hsieh, Nien-Po Chen, Hsin-Hua Chang, "Production of efficient exciplex-based red, green, blue and white organic light-emitting diodes, " Active-Matrix Flatpanel displays and Devices(AM-FPD), pp. 77–80, Jul. 2015.

[2] Mostafa Chakir, Hicham Akhamal and Hassan Qjidaa. "A low power 6-bit current-steering DAC in 0.18-μm CMOS process, " Intelligent System and Computer Vision (ISCV), pp. 1-5, Jan, 2016.

[3] Byoung-Kwan Jeon ; Seong-Kwan Hong ; Oh-Kyong Kwon, "A low-power 10-bit single-slope ADC using power gating and multi-clocks for CMOS image sensors" , ISOCC, pp.157-158, Oct. 2016.

[4] Silvio Ziegler, Robert C. Woodward, "Current Sensing Techniques: A Review", IEEE Sensors Jounal, pp. 354-376, Mar. 2017.

[5] Yeonseong Hwang, Seongjoo Lee, Minkyu Song, "Design of a CMOS image sensor with a 10-bit two-step single-slope A/D converter and a hybrid correlated double sampling" , Ph.D. Research in Microelectronics and Electronics (PRIME), pp.1-4, Aug. 2014.

Elliptic OTA-C Low-Pass Filters for Analog Front-End of Biosignal Detection System

Maha S. Diab, Soliman Mahmoud

Department of Electrical and Computer Engineering
University of Sharjah
Sharjah, UAE
{U16101457, solimanm}@sharjah.ac.ae

Abstract— **This paper presents elliptic low-pass filters based on OTA-C structure for the implementation in the analog front-end (AFE) of biosignal detection systems. Second, third, fourth, and fifth-order filters are proposed, and validated using 90nm CMOS technology in LTspice. The transconductance of the OTA used is analog tuned from 0.155 nA/V to 1.4 nA/V. Equal capacitors of pF capacitance are used in implementation. Magnitude responses of the various filter orders focused on the 50 Hz notch frequency. The fourth, and fifth-order filters provide a choice of single or double notch response.**

Keywords – Biosignal; Cauer; elliptic; LPF; OTA-C

I. INTRODUCTION

The acquisition of biosignal is usually accompanied by undesired noise signal that require elimination through filters. The analog front-end (AFE) for the detection of various biosignal composes of different stages of filters (low-pass, and notch). The combination of both responses can be achieved using elliptic low-pass filters that will attenuate both the powerline interference signal at 50 Hz and higher frequency signals [1].

II. THE OTA DESIGN

The proposed filters are implemented using differential-input balanced-output (DIBO) operational transconductance amplifier (OTA). The design of the OTA is based on simple structure of input differential pair and current mirrors to provide the balanced output [2], [3]. Analog tuning is used to provide the different values of transconductance required by the filters. The biasing transistor's drain current (biasing current) controls the value the of the transconductance through V_{gate} and (W/L) ratio. Thus, providing fine tuning of transconductance values ranging from 0.155-1.4 nA/V.

III. LOW-ORDER ELLIPTIC LOW-PASS FILTERS

Referring to transfer functions of second-order and third-order elliptic filters [4], this section proposes direct implementation of the filters based on OTA-C topology. The general transfer function is re-written in terms of G_i/C_i terms, to simplify the design using integrator stages.

A. Second-Order Elliptic Low-Pass Filter

The transfer function describing the 2nd order OTA-C filter is shown in (1):

$$\frac{V_{out}}{V_{in}} = \frac{s^2 + G_2 G_{13}/C_2 C_1}{s^2 + (G_2/C_2)s + G_2 G_{11}/C_2 C_1} \quad (1)$$

The proposed circuit block design of the second-order filter is shown in Fig. 1. It consists of two lossless integrators and an output amplifier stage, and all capacitances are equal to C_1. Independent control of the notch frequency is provided by tuning of G_{13} affecting the position of the zero not the poles, or by changing the capacitance value through frequency scaling.

B. Third-Order Elliptic Low-Pass Filter

The third-order filter is also implemented based on the transfer function obtained from [4]. In previous work [5], an OTA-C third-order elliptic low-pass filter was proposed with single input balanced output topology. This work builds on the previous design to provide a fully balanced third-order filter with independent control of the notch frequency through G_{32} as shown in (2). The proposed circuit block design is shown in Fig. 2, where all the capacitors are equal to C_2.

$$\frac{V_{out}}{V_{in}} = \frac{(G_{32}/C_2) s^2 + G_{31} G_2 G_1/C_3 C_2 C_1}{s^3 + (G_{31}/C_3)s^2 + (G_{31} G_2/C_3 C_2)s + G_{31} G_2 G_{11}/C_3 C_2 C_1} \quad (2)$$

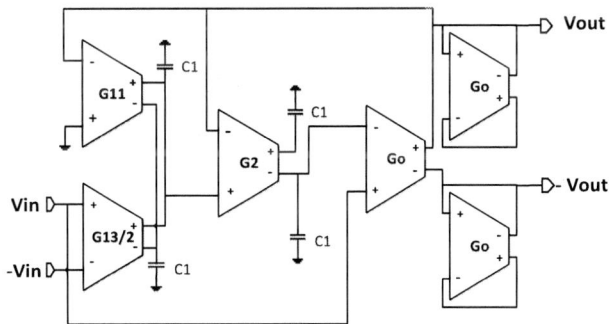

Figure 1: Second-order OTA-C elliptic filter circuit design.

978-1-5386-7961-6/18 $31.00 © 2018 IEEE   103   ISOCC 2018

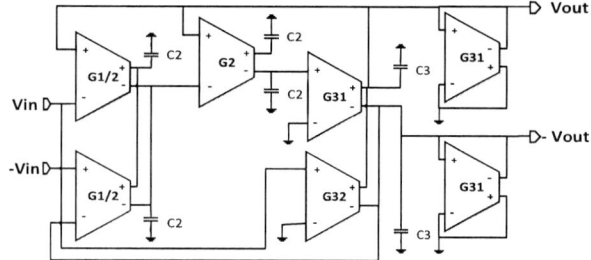

Figure 2: Third-order OTA-C elliptic filter circuit design

IV. HIGH-ORDER ELLIPTIC LOW-PASS FILTERS

The characteristics of low-pass filter used in the AFE require higher order filters with sharper characteristics that provide an improved response compared to lower-order filters.

Each of the proposed second and third-order filters provide a single notch frequency as expected from the zeros of their transfer functions. The higher order filters can provide more notches depending on the value of zeros in the transfer function. In the previous work [5], a direct design of fourth and fifth-order OTA-C filters were derived from their respective transfer functions. In this work, both filter orders are designed based on the cascading of the previously presented lower-order elliptic low-pass filters. Two second-order filters cascaded to provide fourth order filter. While second-order and third-order cascading provide the fifth-order filter. Both filters can provide single or double notches that are tunable to the desired frequency.

V. SIMLATION RESULTS

The proposed filter designs are validated through simulation results. LTspice simulator is used with 90nm CMOS model, BSIM4 (level 54) with supply voltage of ±0.6 V. The magnitude response of the second, and third-order elliptic low-pass filter are shown in Fig. 3 and Fig. 4 respectively.

The magnitude response of the fourth-order filter using two different second-order blocks provide two notches as shown in Fig. 5. A notch at 50 Hz and another at 100 Hz. Cascading two similar blocks will result in a single notch (50 Hz) with a deeper attenuation (-50 dB) as seen in Fig. 6. Similarly, the fifth order with double and single notches is presented in Fig. 7 and Fig. 8 respectively. Comparing the fourth and fifth-order responses, the fourth-order provides higher gain at passband (~9 dB) while fifth-order (~0.6 dB). However, the fifth provides attenuation in the double notch compared to the fourth's double notch. Depending on the application and during implementation, a choice between both is to be considered.

VI. CONCLUSION

An OTA-C based elliptic low-pass filters of different orders are presented. The performance of the filters is evaluated using 90 nm CMOS technology. Different responses can be achieved with a single or double notch depending on the need. The presented magnitude responses focused on a notch at 50 Hz to eliminate the powerline interference signal detected during the acquisition of biosignals, and another possible notch at 100 Hz.

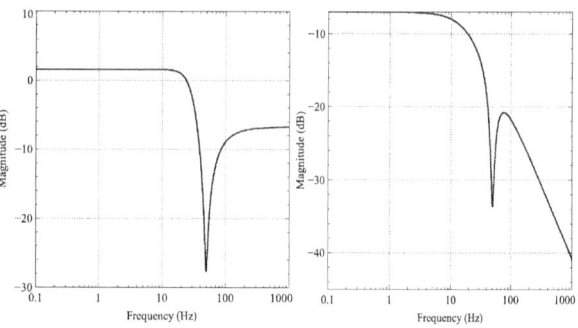

Figure 3: Second-order filter magnitude

Figure 4: Third-order filter magnitude response

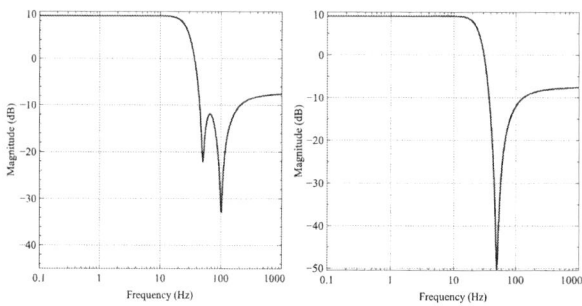

Figure 5: Fourth-order filter with double notch

Figure 6: Fourth-order filter with single notch

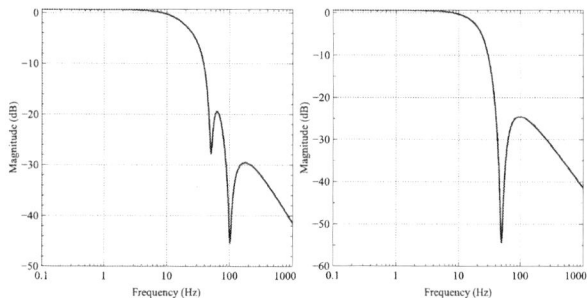

Figure 7: Fifth-order filter with double notch

Figure 8: Fifth-order filter with single notch

REFERENCES

[1] A. A. Alhammadi and S. A. Mahmoud, "Fully differential fifth-order dual-notch powerline interference filter oriented to EEG detection system with low pass feature," Microelectronics Journal, vol. 56, pp. 122–133, 2016.

[2] S. A. Mahmoud and A. M. Soliman, "New CMOS fully differential difference transconductors and application to fully differential filters for VLSI," Microelectronics J., vol. 30, pp. 169-192, 1999.

[3] S. A. Mahmoud and A. M. Soliman, "A CMOS programmable balanced output transconductor for analog signal processing," Int. J. of Electronics, vol. 82, pp. 605-620, 1997.

[4] L. Huelsman, Active and Passive Analog Filter Design: An Introduction. McGraw-Hill, 1993.

[5] M. S. Diab and S. Mahmoud, "Balanced OTA-C elliptic Cauer filters for biomedical applications," in 41st International Conference on Telecommunications and Signal Processing, (Athens, Greece), pp. 80-83, 2018.

Ka-band RF Front-End with 5dB NF and 16dB conversion gain in 45nm CMOS technology

Hyunki Jung[1], Dzuhri Radityo Utomo[1], Saebyeok Shin[1], Seok-Kyun Han[1], Sang-Gug Lee[1], and Jusung Kim[2*]

[1]Nice Laboratory, Korea Advanced Institute of Science and Technology (KAIST), Daejeon, Korea
[2]Department of Electronics and Controls Engineering, Hanbat National University, Daejeon, Korea
*jusungkim@hanbat.ac.kr

Abstract— **This paper presents the Ka-band RF front-end for the frequency channelization receiver. The main blocks of the RF front-end are the low-noise amplifier(LNA), on-chip passive Balun, down-conversion mixer and output buffer. To achieve broad bandwidth at Ka-band frequency, stagger tuned load and series peaking technique are employed. The prototype receiver front-end was designed in 45nm CMOS technology. The prototype circuit achieves >15dB conversion gain, 5dB NF and >-15dBm IIP3 with 87.6mW power consumption and 0.42mm² active area.**

Keywords: Ka-band, frequency channelization receiver, stagger tunning, series peaking inductor

I. INTRODUCTION

With the advent of 5G generation communication, the number of devices is expected to increase dramatically. The wireless transceivers are operating in the interferer-crowded scenario and thus the spectrum sensible apparatus plays an essential role.

Frequency channelization is one of the solutions to accommodate spectrum sensing in the broadband fashion. In [1], we propose the frequency channelization receiver system operating from DC to 40GHz to decompose the broadband input signal into multiple channels. To relax the requirements on the RF front-end and achieve fast frequency spectrum analysis, both parallel and series channelization are applied in the system. The aforementioned receiver [1] requires segmented broadband front-end solution. This paper is targeting the 30-40GHz band signal path in the proposed channelization receiver front-end.

II. RF FRONT-END ARCHITECTURE

Fig. 1 shows the architecture of the Ka-band RF front-end. The RF front-end consists of a Low Noise Amplifier (LNA), on-chip Balun, down conversion mixer, and output buffer.

LNA is comprised of two-stage common-source amplifier with stagger tuned load. Local oscillator (LO) signal is applied from the external 30GHz source, and 2-stage resistive feedback inverter type buffer with on-chip impedance matching network provides good LO signal to the mixer. Finally, down-converted DC-10GHz signal passes through the output buffer with 50 ohms matching at the output. Target specifications in conversion gain, noise figure (NF), and linearity (IIP3) of the front-end are 15dB, 5dB, and -15dBm, respectively.

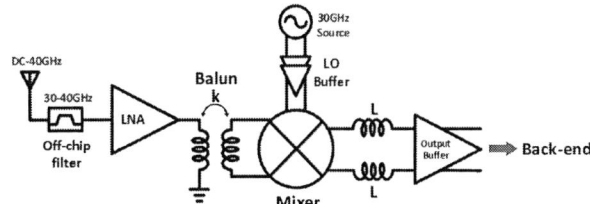

Figure 1. RF front-end architecture block diagram

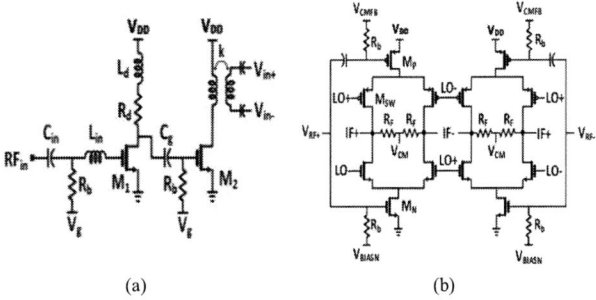

(a)   (b)

Figure 2. The schematic of (a) LNA and (b) Mixer

III. DESIGN OF BUILDING BLOCKS

A. Low Noise Amplifier(LNA)

2-stage LNA is shown in Fig. 2(a). Two stage implementation with stagger-tuned load is proposed to provide sufficient gain and wide bandwidth within the frequency range 30-40GHz [2]. 1st stage amplifier load is designed to resonate at 30GHz with series resistor R_d to lower the quality factor and increase the bandwidth. 2nd stage load, on the other hand, are resonated at 40GHz. The aggregated response due to two different peaking at the edges of the band provides wide bandwidth of 28-45GHz with the center frequency at 38GHz. Center frequency of 38GHz is to compensate for the large gain drop of the mixer at the later stage.

B. Down Conversion Mixer

The down-conversion mixer is designed in push-pull type configuration as shown in Fig. 2(b) [3]. Complementary input stage of the mixer doubles the conversion gain with the given bias current. It also improves the noise and linearity performance. There is the trade-off between conversion gain and bandwidth with the RC load. In this work, for 10GHz bandwidth and more than 0dB conversion gain, load capacitance should be less than

978-1-5386-7961-6/18 $31.00 © 2018 IEEE

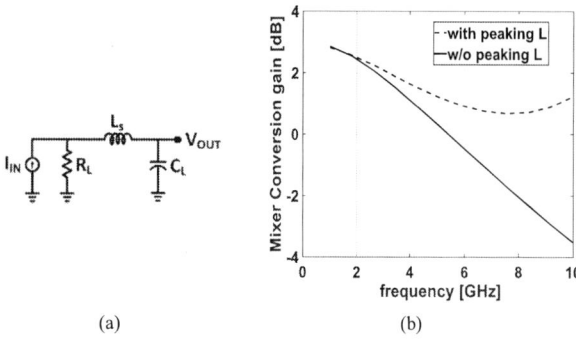

(a)   (b)

Figure 3. (a) modeling of the mixer load with series peaking inductor (b) mixer conversion gain without and with series peaking inductor

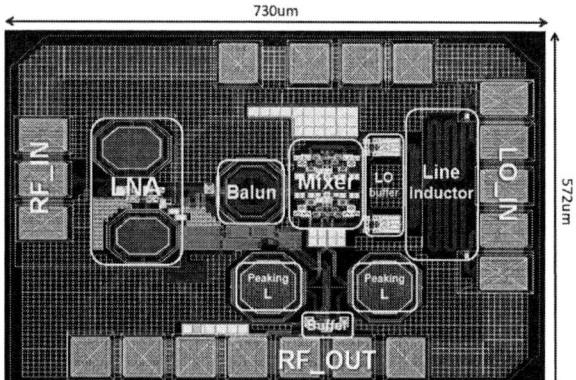

Figure 4. Chip layout of the 30-40GHz RF front-end

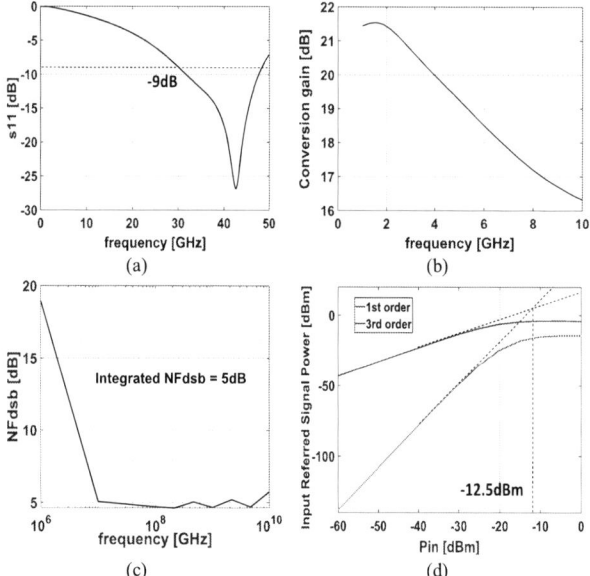

(a)   (b)

(c)   (d)

Figure 5. (a) S_{11} (b) Conversion gain (c) NF, and (d) IIP3 of the proposed RF front-end

390fF in the given bias point, 58mS of transconductance, and 41Ω of output impedance. But, with parasitic capacitance of routing and input capacitance of the output buffer at mixer load which is almost 650fF, only 6GHz bandwidth is achieved.

Therefore, the series peaking inductor at mixer load is added to minimize the gain drop. The mixer load with series peaking inductor can be modeled simply as shown in Fig. 3(a). The model gives the following equation (1).

$$\frac{V_{OUT}}{I_{IN}}(s) = \frac{R_L}{R_L L_s C_L s^2 + R_L C_L s + 1}$$

$$= \frac{R_L}{s^2 + 2\xi\omega_n s + \omega_n^2} \quad (1)$$

, where R_L is the output impedance of the mixer, C_L is the input capacitance of the output buffer and L_S is the series peaking inductor. The damping factor(ξ) and the natural frequency (ω_n) of the denominator can be expressed as (2)

$$\xi = \frac{R_L}{2}\sqrt{\frac{C_L}{L_s}}, \quad \omega_n = \frac{1}{\sqrt{L_s C_L}} \quad (2)$$

By choosing 1nH peaking inductor, we can get $\xi \cong 0.52$ and gain peaking at $\omega_n = 6.2$GHz. As a result, the gain drop of the mixer at 10GHz output is enhanced by 4.75dB as shown in Fig 3(b).

IV. SIMULATION RESULTS & CONCLUSION

The 30-40GHz RF front-end is simulated in TSMC 40nm technology. Fig. 4 shows the chip layout of the 30-40GHz RF front-end. As shown in Fig. 4, line inductors are used to transfer the good LO signal to the mixer and save the chip area. Instead of using spiral inductor, using the line inductor save about 10% of the total area. The total chip area is 0.42mm² and power consumption is 87.6mW.

S_{11} simulation is shown in Fig. 5(a) and has good input matching performance within 30-40GHz frequency range (less than -9dB). Fig 5(b) shows the simulated conversion gain. The minimum conversion gain is 16.3dB at 10GHz IF frequency which is higher than the required conversion gain of 15dB. NF and IIP3 performance are shown in Fig. 5(c) and (d), respectively. Double sideband integrated NF is 5dB and IIP3 is -12.5dBm. Simulation results meet the targeting specifications of the frequency channelization receiver.

ACKNOWLEDGMENT

This work was supported by the Civil-Military Technology Cooperation Program through the Defense Acquisition Program Administration and the Ministry of Trade, industry, and Energy, South Korea.

REFERENCES

[1] Kim, Jusung, et al. "The evolution of channelization receiver architecture: principles and design challenges." IEEE Access 5 (2017): 25385-25395.

[2] Wu, Chang-Ching, et al. "A low power CMOS low noise amplifier for ultra-wideband wireless applications." Circuits and Systems, 2005. ISCAS 2005. IEEE International Symposium on . IEEE, 2005

[3] Liu, Yao-Hong, et al. "A 1.9 nJ/b 2.4GHz multistandard (Bluetooth low energy/Zigbee/IEEE802.15.6) transceiver for personal/body-area networks." Solid-State Circuits Conference Digest of Technical Papers (ISSCC), 2013 IEEE International . IEEE, 2013.

[4] Razavi, Behzad. "RF microelectronics.", 2nd edition, 2012

A 22.8-to-32.4 GHz Injection-locked Frequency Tripler with Source Degeneration

Saebyeok Shin[1], Dzuhri Radityo Utomo[1], Hyunki Jung[1], Seok-Kyun Han[1], Sang-Gug Lee[1], and Jusung Kim[2*]

[1]Nice Laboratory, Korea Advanced Institute of Science and Technology (KAIST), Daejeon, Korea
[2]Department of Electronics and Controls Engineering, Hanbat National University, Daejeon, Korea
*jusungkim@hanbat.ac.kr

Abstract— In this paper, injection-locked frequency tripler with wide locking range and high fundamental tone rejection is presented. Undesired fundamental harmonic due to the injected signal is notched out with the source degeneration. Prototype circuit is designed in 45-nm CMOS technology and achieves 28.4 dBc of harmonic rejection ratio. The tripler exhibits 34.8% of locking range with 4 dBm of input signal power. With 1.1 V of supply voltage, the core circuit consumes 5.2 mA and occupies only 265μm x 160μm of chip-area.

Keywords; Frequency Tripler, injection-locked oscillators, notch filter, source degeneration

I. INTRODUCTION

With the advent of the 5G network, generating high frequency signal is becoming increasingly important. Stable Local Oscillator (LO) circuits at the millimeter (mm)-wave regime greatly affects the performance of the wireless transceiver targeting 5G network. However, generation of the LO circuits including the frequency synthesizer is not trivial, especially with reasonable power consumption and area efficiency [1]. Most of the electronic devices are portable, so keeping the power consumption as low as possible is highly desired.

To achieve low power consumption and frequency stability in mm-wave frequency, injection-locked frequency multiplier (ILFM) structure is the good candidate to interface frequency synthesizer at sub-harmonic of LO frequency with the up (down) converter [2,3]. However, a conventional ILFM structure does not sufficiently remove fundamental harmonic tone at the output. Efforts have been done to solve this problem by using two-stage output buffer [4] and notch filter [5]. Although the previous works solve the problem of rejecting fundamental tone, they have to sacrifice the power consumption and area efficiency due to the additional stages with many passive devices.

In this work, wide locking-range injection-locked frequency tripler (ILFT) with source degeneration is proposed. The fundamental component is suppressed utilizing the core circuit of ILFT and thus does not require additional buffers nor the filters.

II. PROPOSED DESIGN

Fig. 1(a) is a diagram of a conventional injection-locked frequency tripler. A differential square signal with fundamental frequency at f_0 is injected through M_3 and M_4. According to the Fourier series, it can be seen that the symmetric square signal

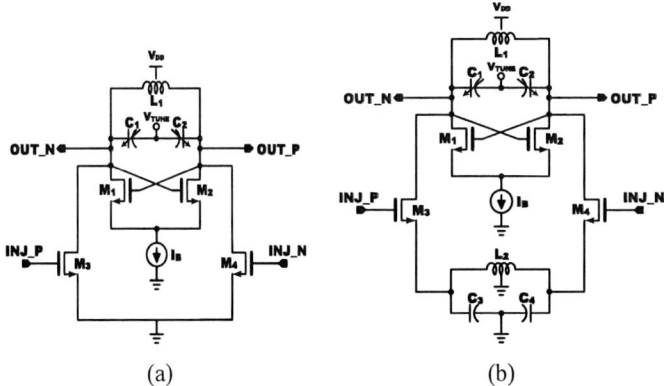

Figure 1. (a) Conventioanl ILFT and (b) Proposed ILFT

consists of frequency components that are odd multiples of the fundamental frequency. This injected signal at the drain of M_3 (M_4) is shaped by the LC tank response due to L_1, C_1, and C_2 resonating at $3f_0$. Only 3rd harmonic of the input injected signal remains with negative resistance due to M_1 and M_2.

However, in the case of the conventional ILFT, the fundamental tone at f_0 still remains and interrupts the operation of ILFT. In order to solve this problem, the presented design applies the source degeneration at the ILFT without additional power consumption. The gain of the common source amplifier can be calculated simply as (1) below.

$$A_v = \frac{V_{output}}{V_{input}} = -G_m Z_{out} \qquad (1)$$

, where G_m is the effective trans-conductance and Z_{out} is the impedance of the load. If there is sufficiently high impedance Z_S at the source of the transistor, G_m will be inversely proportional to the Z_S. This can be expressed by (2).

$$G_m = \frac{g_m}{1 + g_m \cdot Z_S} \propto \frac{1}{Z_S} \qquad (2)$$

Then, large degeneration at f_0 frequency and little or no degeneration at the desired frequency of $3f_0$ can suppress the undesired fundamental tone and maintain the desired third harmonic tone. In other words, Z_S has to be a large impedance at f_0 and small impedance at $3f_0$. This characteristic of Z_S can be achieved by creating LC tanks of L_2, C_3, and C_4 that resonates at f_0. The circuit in Fig. 1(b) is the proposed ILFT utilizing the aforementioned source degeneration technique. In

978-1-5386-7961-6/18 $31.00 © 2018 IEEE

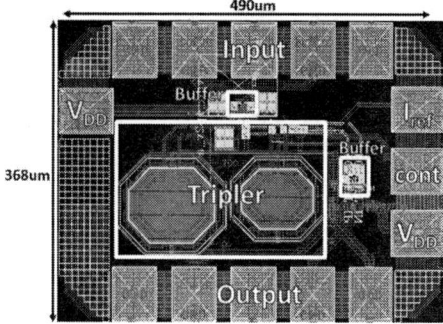

Figure 2. Chip Layout

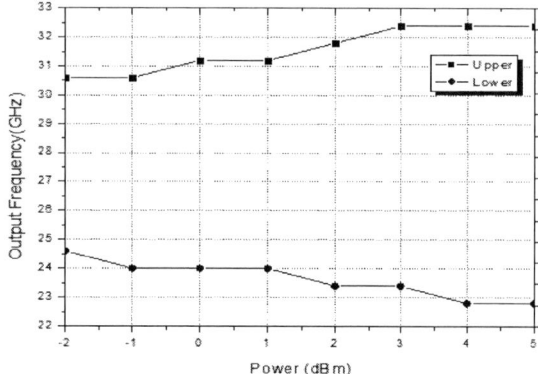

Figure 3. Simulation result of locking range

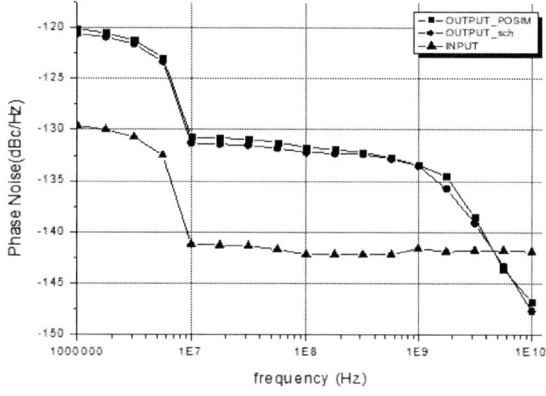

Figure 4. Simulation result of phase noise

the ideal situation, Zs exhibits the open circuit at f_0 and the short circuit at $3f_0$ and the ILFT can notch out the fundamental tone completely.

III. SIMULATION RESULTS

The tripler based on the proposed source degeneration was designed in 45-nm CMOS process in TSMC technology. A chip layout is shown in Fig. 2. The total chip area is 490μm x 368μm and the core circuit occupies only 265μm x 160μm.

The prototype circuit consumes 5.2 mA with 1.1 V voltage supply. The harmonic rejection is simulated to be 28.4 dBc and this finite rejection is related to the finite quality factor of the source degeneration inductor. Small chip area allocated for the source degeneration limited the effective quality factor to be 19 in our work.

Fig. 3 shows the locking range of the ILFT. There were no significant changes in the locking range when the injecting signal power was 4dBm and above. Then, the proposed ILFT was locked between 22.8 and 32.4 GHz output frequency, which exhibits 9.6 GHz (34.8%) locking range.

Fig. 4 shows the simulated phase noise from 1 MHz to 10 GHz. Compared to the input injecting signal, the difference between the input and output of the tripler is 9.7 dB at 100 MHz offset, which is very close to the theoretical limitation of 9.54 dB.

IV. CONCLUSION

Wide locking range injection-locked frequency tripler with source degeneration is proposed in this paper. Among many methods to obtain the desired high frequency signal, using a frequency multiplier based on injection-locked subharmonic oscillator is a promising approach. The designed circuit is modified from conventional ILFT structure by applying source degeneration to effectively remove unwanted fundamental harmonic component. As a result, the harmonic rejection ratio is up to 28.4 dBc without degrading the desired third harmonic tone power.

The ILFT designed in 45-nm technology has 34.8% of locking range when the injection power was >4 dBm, and its core part only occupies 265μm x 160μm. Phase noise of the proposed ILFT shows around 9.7 dB degradation within 1MHz to 1GHz offset frequency, which is close to the theoretical value.

ACKNOWLEDGMENT

This work was supported by the Civil-Military Technology Cooperation Program through the Defense Acquisition Program Administration and the Ministry of Trade, Industry, and Energy, South Korea.

REFERENCES

[1] J. Kim, S. Lee, and D. Choi, "Injection-locked frequency divider topology and design techniques for wide locking-range and high-order division," *IEEE Access,* vol. 5, pp. 4410-4417, 2017

[2] N. Mazor and E. Socher, "Analysis and design of an X-band-to-Wband CMOS active multiplier with improved harmonic rejection," *IEEE Trans. Microw. Theory Tech.,* vol. 61, no. 5, pp. 1924-1933, May 2012.

[3] Seong-Kyun Kim, Chanki Choi, Chenglin Cui, Byung-Sung Kim, Munkyo Seo, "A W-Band Signal Generation Using N-Push Frequency Multipliers for Low Phase Noise", *Microwave and Wireless Components Letters IEEE,* vol. 24, pp. 710-712, 2014, ISSN 1531-1309.

[4] W. L. Chan, J. R. Long, and J. J. Pekarik, "A 56-to-65GHz injection-locked frequency tripler with quadrature outputs in 90nm CMOS," *IEEE ISSCC Dig. Tech. Papers,* pp. 481-481, Feb. 2008.

[5] C. N. Kuo, and T. Z. Yan, "A 60 GHz Injection-Locked Frequency Tripler With Spur Suppression", IEEE MICROWAVE AND WIRELESS COMPONENTS LETTERS, VOL. 20, NO. 10, ,pp.560-562, Oct. 2010

Design of a Low-Power Complex Baseband Filter with Tunable Gain and Bandwidth in 65nm CMOS

Jaegyeong Choi, Jungah Kim, Yongho Lee, Seungsoo Kim, Jongsik Kim*, and Hyunchol Shin

High-Speed Integrated Circuits and Systems Lab., Kwangwoon University, Seoul, Korea
* HiDeep Inc., Seongnam, Korea

Abstract— **A low-power fourth-order complex bandpass baseband analog filter (BBA) is designed for low-IF RF receivers supporting MedRadio or Bluetooth applications. The BBA is composed of three variable gain amplifier stages and two complex biquad stages. The gain and bandwidth tuning capabilities are needed to maximize the receiver dynamic range and support the variable data rates. The center frequency and Q-factor tunings are realized for optimum filtering and amplification performances. The gain tuning range is from -15 dB to +52 dB and the bandwidth tuning range is from 250 kHz to 2.9 MHz. The image rejection ratios are found to be 39 and 42 dB at the maximum bandwidth of 2.9 MHz and the minimum bandwidth of 250 kHz conditions, respectively. The layout size is 0.42 mm^2 including the I/Q paths, while dissipating 1.5 mA from a 1-V supply.**

Keywords; Baseband filter, complex filter, image rejection, MedRadio, Bluetooth, CMOS

I. INTRODUCTION

Low-power RF receivers are highly needed for various wireless connectivity applications such as the medical radiocommunication services [1] and Bluetooth-based IoT applications. The RF receivers are usually based on either zero-IF [2] or low-IF [3-6] architecture. Compared to the zero-IF architecture, the low-IF receiver is more advantageous to alleviate flicker noise and dc offset issues. However, sufficient image rejection is critically required in low-IF receiver. The image rejection is usually carried out by employing a complex bandpass baseband analog filter at the final stage of RF receivers. In addition to the image rejection, the complex baseband filter usually requires the gain and bandwidth tunability in order to enhance the dynamic range and support scalable data rates imposed by communication standards.

In this work, a fourth-order complex bandpass baseband analog filter (BBA) is designed to provide the image rejection of more than 39 dB for a low-power low-IF RF receiver. It is also featured with the wide tuning range of bandwidth and gain.

II. CIRCIUT DESIGN AND RESULTS

Fig. 1 is the block diagram of the designed BBA. It is composed of three variable-gain amplifiers (VGA) and two 2nd-order complex bandpass filters (BPF). The total gain variation range of 67 dB is partitioned to $0 - +12$ dB for VGA1, $-3 - +21$ dB for VGA2, $-12 - +7$ dB for VGA3, $0 - +6$ dB for both BPFs. VGA1 and VGA2 have relatively higher gain than VGA3 in order to achieve better noise and linearity performances. Three VGAs are designed to have sufficiently wider bandwidth than

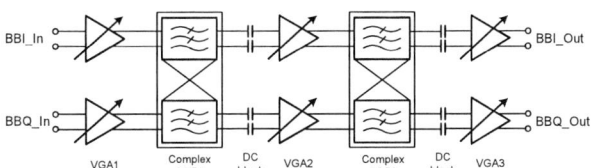

Fig. 1 Complex bandpass baseband filter block diagram

the desired bandwidth, whereas two BPFs are the main blocks to determine the total bandwidth and thus the overall channel selection characteristic of the BBA. DC offsets in the circuit can significantly degrade the overall dynamic range due to the rather high voltage gain. Thus, two dc blocking capacitors are placed between BPF1 and VGA2, and BPF2 and VGA3, so that they can efficiently suppress the dc offsets. The image rejection property is realized by inter-winding the in-phase and quadrature path biquads in the BPF.

Fig. 2 shows the VGA schematic. The gain is determined by the ratio of R_1 and R_2, and the gain tuning is realized by switched control of R_1 and R_2. Note that the R_1 of VGA1 is not variable but fixed at a constant value in order to maintain the input impedance at a constant level across the overall tuning range. The low-pass filtering capacitor C_1 is added to set the VGA bandwidth not too much wide, which helps suppress the total output noise of the BBA.

Fig. 3 is the 2nd order complex BPF. It is composed of two 2nd order biquad low-pass filters that are properly cross-coupled via 8 cross-coupling resistors R_{xa} and R_{xb}. When this BPF is fed with I/Q input signals, the overall transfer function is found to be shifted by a frequency of $(\pi C_1(R_{xa} + R_{xb}))^{-1}$ Hz compared to the original biquad low-pass response. Also, the bandpass filtering characteristic appears only at the positive frequency region, while the attenuating characteristic appears at the negative frequency region. Note that the gain and bandwidth are tuned by R_1, R_2, and C_1, the center frequency is tuned by R_{xa}, R_{xb}, and C_1, and the filter Q-factor is controlled by R_2 and R_3.

The opamp used in the BBA is based on a two-stage fully differential type, in which the first stage is a PFET-input differential pair and the second stage is a common-source stage. Common-mode feedback is applied to set the common-mode dc level at the output. Frequency compensations are carefully done by adding Miller capacitors and zero-creating resistors for the differential-mode stability as well as the common-mode stability. The phase margins for the differential and common

978-1-5386-7961-6/18 $31.00 © 2018 IEEE

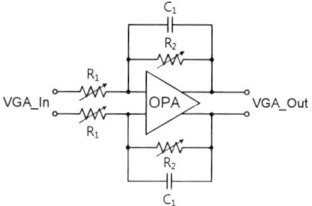

Fig. 2 Variable gain amplifier

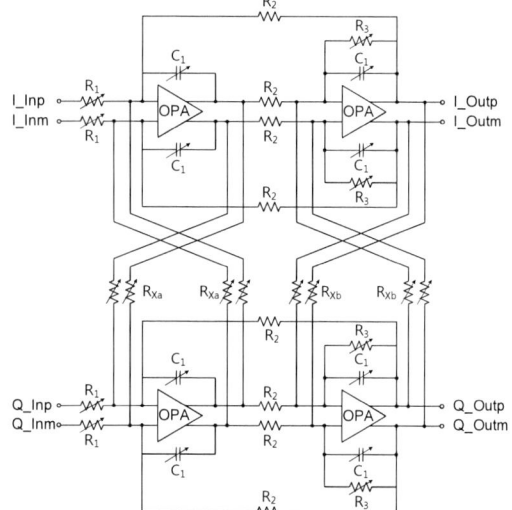

Fig. 3 Complex biquad schematic

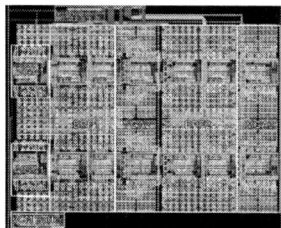

Fig. 4 Layout

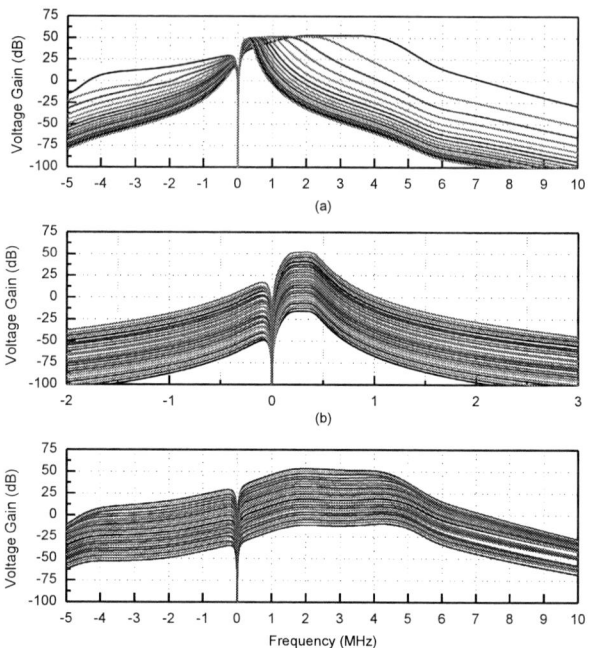

Fig. 5 (a) Bandwidth tuning at maximum gain. (b) Gain tuning at the minimum bandwidth. (c) Gain tuning at maximum bandwidth,

of Fig. 5 clearly exhibit the bandpass filtering characteristic at the positive frequency region, and at the same time, the rejection characteristic at the negative image frequency region. For instance, in Fig. 5(c) at the maximum bandwidth condition, the image rejection ratio between the positive and negative frequencies of ±2.9 MHz is found to be -39 dB. The BBA operates from a single supply voltage of 1 V, and dissipates the total current of 1.5 mA for the entire I/Q paths. The designed BBA can be successfully applicable to low-power low-IF RF receiver designs for either 400-MHz MedRadio receivers or 2.4-GHz Bluetooth receivers.

ACKNOWLEDGMENT

This work was supported by the IITP under the grants 2016-0-00421(Harvest Energy Based IoT Devices) and 2016-0-00111 (Dual/Wide-Band MICS/BAN Transceiver for Capsule Endoscopy).

REFERENCES

[1] FCC Rules and Regulations Part 95, Medical Device Radiocommunication Service (MedRadio), 2009

[2] M. Vidojkovic *et al.* "A 0.33nJ/b IEEE 802.15.6/Proprietary-MICS/ISM-Band Transceiver with Scalable Data Rate from 11kb/s to 4.5 Mb/s for Medical Applications," ISSCC 2014

[3] P. D. Bradley, "An Ultra Low Power, High Performance Medical Implant Communication System (MICS) Transceiver for Implantable Devices", IEEE BioCAS, 2006

[4] J. Liu *et al.*, "An Ultra-Low Power 400 MHz OOK Transceiver for Medical Implanted Applications," ESSCIRC 2011

[5] N. Cho *et al.*, "A 10.8 mW Body Channel Communication/MICS Dual-Band Transceiver for a Unified Body Sensor Network Controller", IEEE JSSC, vol. 44, no. 12, Dec. 2009.

[6] D. Lee *et al.*, "Low Power FSK Transceiver using ADPLL with Direct Modulation and Integrated SPDT for BLE Application," A-SSCC 2016

modes are found to be more than 60 and 37 degrees, respectively, across the whole PVT corner variations.

The BBA is designed and laid out in 65nm CMOS process. Fig. 4 shows the top layout view of the total circuit. The active area is 760 x 560 μm². In order to minimize the unwanted degradation of the image-rejection ratio, which is typically created by the amplitude and phase mismatches between the I/Q-path signals, the I/Q signal paths are carefully laid out to keep the symmetry and balance as best as possible. Many passive elements and routing lines are laid below the MIM capacitors to minimize the layout area.

Fig. 5 shows the post-layout simulation results. Fig. 5(a) exhibits the filter bandwidth can be tuned from 250 kHz to 2.9 MHz, for which the center frequency is also properly tuned to support the low-IF receiver's frequency planning. Fig. 5(b) and (c) are the gain tuning characteristics when the bandwidths are set to the minimum value of 250 kHz and the maximum value of 2.9 MHz, respectively. The gain variation range is between -15 and +52 dB with 3 dB step. The frequency response curves

A 24-28GHz Reconfigurable CMOS Power Amplifier in 22nm FD-SOI for Intelligent SoC Applications

Jill C. Mayeda and Donald Y.C. Lie
Dept. of Electrical and Computer Engineering
Texas Tech University (TTU)
Lubbock, TX 79409-3102
Donald.Lie@ttu.edu

Jerry Lopez
NoiseFigure Research (NFR) Inc. & Texas Tech University
3911 4th Street
Lubbock, Texas 79415

Abstract— **A 2-stage 24 to 28 GHz reconfigurable CMOS power amplifier (PA) is designed in a 22nm fully-depleted silicon-on-insulator (FD-SOI) technology. The differential PA uses stacked FETs and digitally-controlled neutralization capacitors and output matching capacitors with adjustable body-bias and on-chip baluns to increase output power P_{OUT}, reconfigurability, and wideband performance. Post-layout SPICE simulations show S_{21} of 27 dB, peak power-added efficiency (PAE) of 29.1% at P_{OUT} =17.5 dBm and P_{1dB} =13 dBm can be achieved at 28 GHz. The digitally-controlled neutralization caps can tune S_{21} by 6 dB at 28 GHz, while varying body bias of the FD SOI NMOS can enable ~2 dB fine gain tuning, and the output matching caps can further optimize PA's performance for intelligent SoC applications.**

Keywords: body biasing, CMOS FD-SOI, CMOS PA, millimeter-wave (mmWave), reconfigurable PA

I. INTRODUCTION

It is well-known that RF and millimeter-wave (mm-Wave) power amplifiers (PA) can often dominate the overall transmitter performance, as its performance such as power-added efficiency (PAE), P_{OUT} and bandwidth dictate the performance for the entire wireless system. Especially for mm-Wave phased-array system where considerably more PAs are needed to be integrated in the RF front-end modules (FEMs) to achieve beamsteering to compensate for the path loss and blocking effects, highly efficient and reconfigurable mm-Wave PAs are critical and desirable. For commercial mm-Wave to be successful, the PAE, P_{OUT}, linearity, bandwidth, reliability, cost and form factors of mm-Wave PAs need to be very good. It is thus very attractive to have reconfigurable mm-Wave PAs that can cover 24 to 28 GHz bands to reduce the cost and size of the RF FEM. We report here a highly-efficient reconfigurable 24-28 GHz CMOS PA in a 22nm FD-SOI process, where body bias, neutralization caps, and output matching caps can all be controlled to optimize and reconfigure the PA performance.

II. RECONFIGURABE CMOS PA DESIGN

A. Stacked PA Schematics

Figure 1 shows the schematics of the reconfigurable 2-stage CMOS PA, which includes the switches for digitally controlled neutralization caps [1], and the output matching caps that affect the I-V waveforms of the power devices. It consists of a 1-stage differential common-source driver with neutralization caps that drives the 2nd stage stacked-FET PA with its neutralization and output matching caps [2]. On-chip baluns from the design kit are used for I/O and interstage broadband matching. All SPICE simulation in this paper are with post-layout parasitics extraction. Body bias nodes for each device are tied together for tuning (0V as simulation baseline).

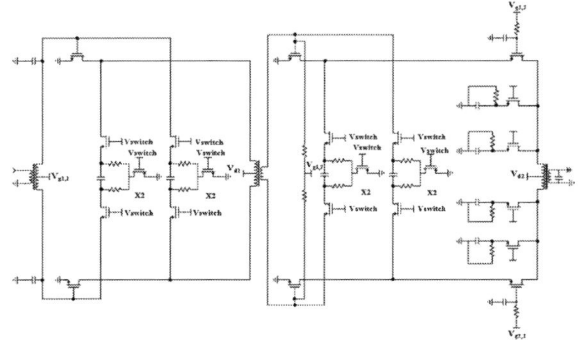

Fig. 1 A simplified schematics of the reconfigurable CMOS PA

B. PA Small Signal Performance

Fig. 2 plots the post-layout simulated small signal power gain S_{21} of the reconfigurable PA under 16 different ON/OFF switching conditions on neutralization caps. The simulations indicate that the frequency of the peak S_{21} can be controlled from around 27 GHz (case 0000; all OFF) to 21 GHz (case 1111; all ON). The digitally-controlled neutralization caps can adjust S_{21} by 6 dB at 28 GHz, while body bias of FD SOI NMOS can enable ~2 dB S_{21} tuning (stage 1 and 2, 0/0V, 0/1V, and 1/1V). Further reconfigurability can be achieved by tuning the output caps, dynamic gate and supply biasing [3].

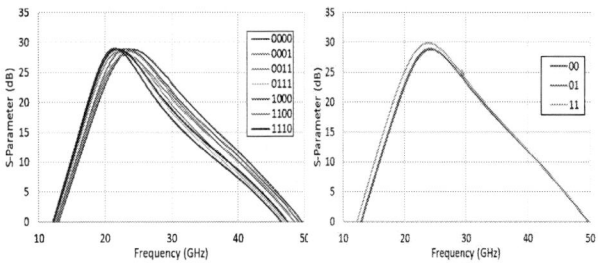

Fig. 2 Post-layout simulated S_{21} of the PA under different ON/OFF conditions on neutralization caps (LEFT) and body biasing (RIGHT)

We wish to acknowledge the funding sources from TTU Keh-Shew Lu Regents Chair Endowment and the US DoD (Dept. of Defense). Thanks to the GlobalFoundries University Program, in particular Dr. D. Harame, Ms. D. Wang, Dr. N. Cahoon, Dr. A. Joseph at GlobalFoundries for IC fabrication.

978-1-5386-7961-6/18 $31.00 © 2018 IEEE

C. PA Large Signal Performance

Fig. 3 presents the large signal simulated performance of the reconfigurable PA at 24 GHz and 28 GHz, respectively. At P_{IN} = 0 dBm, this PA reaches a PAE of 28.8% and P_{OUT} =17.5 dBm at 24 GHz, and a PAE 26.2% and P_{OUT} =17.1 dBm at 28 GHz. It appears the switching of neutralization cap values can effectively adjust PA's frequency response, enabling it to cover wider bandwidth with good performance.

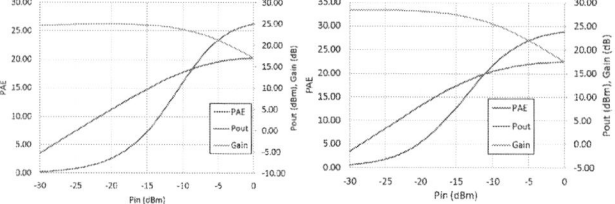

Fig. 3(a) Post-layout simulation of the reconfigurable CMOS PA at 28 GHz for case 0000 (LEFT); (b) at 24 GHz for case 1111 (RIGHT)

Fig. 4 shows simulated PA performance as we vary the body bias only or the output cap bank only at 28 GHz (both from 0 to 1V). Additional fine tuning in gain/P_{OUT} can be provided, but considerably less than switching the neutralization caps.

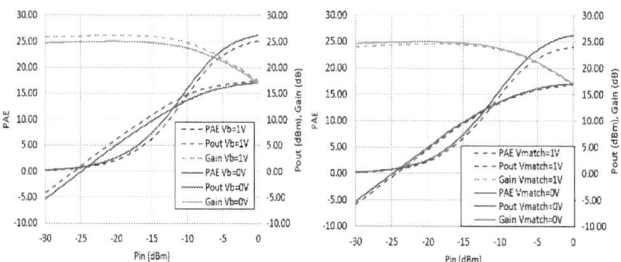

Fig. 4 Simulation results of the PA performance at 28 GHz from tuning of body bias (LEFT) or the output capacitors bank (RIGHT)

The PA was laid out and being fabricated, where it is 785x560 μm with pads (core 550x200 μm). More PA linearity simulation will be presented in the conference. Table I shows its performance summary table vs. state-of-the-arts CMOS

PAs in the literature near 28 GHz. One can see our PA has excellent reconfigurability and power gain across 24-28 GHz, and also with the smallest die size, making it potentially attractive for various mm-Wave SoC and 5G-like applications.

VI. CONCLUSION

We have reported a novel reconfigurable CMOS PA operating within the 24 to 28 GHz band in 22nm FD-SOI. The proposed 2-stage differential PA uses stacked FETs and digitally-controlled neutralization caps and output matching caps, in addition to dynamic body biasing to realize PA reconfigurability for 24-28 GHz intelligent SoC applications. The post-layout SPICE simulated PA performance has been compared with the state-of-the-art CMOS PA data in the literature, and our PA has exhibited excellent reconfigurability, large power gain, and the smallest die size.

REFERENCES

[1] Z. Deng and A. Niknejad, "A Layout Based Optimal Neutralization Technique for mm-Wave Differential Amplifiers", *IEEE RFIC Symp.*, pp. 355-358 (2010)

[2] A. Ezzeddine and H. Huang, in IEEE Radio Freq. Integr. Circuits Symp., Jun. 2003, pp. 215–218

[3] J.-H. Chen, S.R. Helmi, R. Azadegan, F. Aryanfar and S. Mohammadi "A Broadband Stacked Power Amplifier in 45-nm CMOS SOI Technology," *IEEE J. Solid-State Circuits*, 48, 11, pp. 2775-2784, 2013

[4] S. Shakib, H.-C. Park, J. Dunworth, V. Aparin, and K. Entesari, "A Highly Efficient and Linear Power Amplifier for 28-GHz 5G Phased Array Radios in 28-nm CMOS", *IEEE J. Soild-State Circuits*, 51,12, pp. 3020-3036, 2016

[5] J.A. Jayamon, J.F. Buckwalter, and P.M. Asbeck, "Multigate-Cell Stacked FET Design for Millimeter-Wave CMOS Power Amplifiers," *IEEE J. Soild-State Circuits*, 51, 9, pp. 2027-2039, Sept. 2016.

[6] B. Moret, V. Knopik and E. Kerherve, "A 28GHz Self-Contained Power Amplifier for 5G applications in 28nm FD-SOI CMOS", IEEE Latin American Symp. on Circuits & Systems (LASCAS), pp. 1-4, 2017

[7] T. Hanna, N. Deltimple and S. Frégonèse, "A Class-J Power Amplifier for 5G Applications in 28nm CMOS FD-SOI Technology", IEEE SBCCI '17, pp. 110-113, Brazil, 2017

[8] S. Shakib, *et al.*, "A Wideband 28GHz Power Amplifier Supporting 8×100MHz Carrier Aggregation for 5G in 40nm CMOS," *IEEE ISSCC*, pp. 44-45, Feb. 2017

TABLE I: SUMMARY TABLE OF OUR CMOS PA DESIGN NEAR 28 GHz

Reference	Technology	Design	Freq. (GHz)	Psat (dBm)	P1dB (dBm)	Peak PAE (%)	PAE @P1dB (%)	Gain S21 (dB)	Reconfigurable ?	Chip area (mm2; no pads)
This work*	22nm FD CMOS SOI	2-stage stacked	28	17.5	13	29.1	14.1	27	No	0.06
This work*	22nm FD CMOS SOI	2-stage stacked; reconfigurable	24	17.5	13.1	28.8	14.6	28.4	Yes (backbias & digital cap banks)	0.11
This work*	22nm FD CMOS SOI	2-stage stacked; reconfigurable	26	17.1	13	26.2	14.1	25	Yes (backbias & digital cap banks)	0.11
IEEE JSSC'16 [4]	45nm CMOS SOI	Multigate-cell (4 gates)	29	24.8	20	29	16	13	No	0.3
IEEE JSSC'16 [5]	28nm CMOS	2-stage CS with neutr. cap & Ldeg	30	14	13.2	35.5	34.3	15.7	No	0.16
IEEE LASCAS'17 [6]	28nm FD CMOS SOI	segmented biased push-pull	28	18.7	15.4	12.4	9	17.5	No	0.66
IEEE SBCCI'17* [7]	28nm FD CMOS SOI	2-stage CS; class F	28	16.2	13.2	39	31.5	15	No	0.96**
IEEE ISSCC'17 [8]	40nm CMOS	3-stage; 1st stage cascode VGA; 2nd & 3rd CS neutr. cap	27	15.1	13.7	33.7	31.1	22.4	Yes (VGA 9x1dB)	0.23

* Post-layout simulated data; ** With pads

Gap in pagination due to formatting issues.

Pages 113-114

PHY Layer Design of OFDM-VLC System based on SoC using Reuse Methodology

Erwin Setiawan[1], Trio Adiono[2], Syifaul Fuada[3]

University Center of Excellence on Microelectronics, Institut Teknologi Bandung
IC Design Laboratory, PAU Building, ITB Campus, Jl. Tamansari No. 126, Bandung 40132, Indonesia
Tel. +62-22-2506280/Fax. +62-22-2508763
Email: erwin.ouyang@gmail.com[1], tadiono@stei.itb.ac.id[2], syifaulfuada@pme.itb.ac.id[3]

Abstract— A System-on-Chip (SoC) architecture for OFDM-based Visible Light Communication (VLC) system is presented in this paper. It is developed with AXI4 system bus standard, which is generally used in the SoC design. The PHY layer can be accessed by ARM Cortex-A9 processor using memory-mapped I/O method. Several peripherals such as USB, Ethernet, and SDIO are also integrated to the processor. The system can run Linux OS based on Ubuntu 16.04. The system has been implemented on Xilinx Zynq-7000 SoC platform.

Keywords— AXI4 system bus; OFDM; PHY layer; SoC; visible light communication

I. INTRODUCTION

System-on-Chip (SoC) is a system that is realized on a single chip. The system consists of many components such as processor, co-processor, memory, peripherals, and system bus. In a new perspective of the SoC design, the digital blocks do not need to be designed from scratch. It generally employs the digital components that already available as IP cores in which they have been designed and verified by the third party. This methodology is called reuse [1]. With this methodology, we can integrate many IP cores to build a complex system, so the designer's productivity will be increased.

In previous work [2], a network-enabled visible light communication (VLC) system is rapidly developed on SoC. The PHY layer is built based on UART protocol and achieves 900 Kb/s throughput. In previous work [3], a hardware/software model of DCO-OFDM based VLC is developed on SoC. The DMA is used in this system for data transfer between processing element and memory buffer. The system can achieve throughput of ≈6 Mb/s. This paper has correlation with other our as presented in [4]. In this work, we design a complete system (OFDM techniques as application case) for VLC system. The OFDM digital blocks consist of several complex components such as Fast Fourier Transform (FFT) and Forward Error Correction (FEC).

By using reuse methodology, we can reuse the IP cores that are already available. We only design the IP cores from scratch when that IP cores are not available. Later, we integrate those IP cores to build a complete OFDM system.

This paper discusses the implementation of PHY layer design for OFDM-VLC system. In the design, we use our own designed IP cores and reuse the customized Xilinx IP cores. The IP cores are integrated with AXI4 system bus standard, which is generally used in industry. The ARM Cortex-A9 processor that runs Linux OS can access the PHY layer that is implemented on FPGA. The design is implemented on Xilinx Zynq-7000 SoC platform.

II. SoC ARCHITECTURE AND PHY LAYER DESIGN

A. SoC Block Diagram

The proposed architecture of our SoC is shown in Fig 1. The Processing System (PS) of Zynq-7000 consists of dual core ARM Cortex-A9 processor and hard IP core peripherals, such as USB, Ethernet, SDIO, UART, and GPIO. The Programmable Logic (PL) is an FPGA that can be used to implement custom logics. The PHY layer is implemented in the PL. The PS and PL can exchange data by using AXI4 system bus.

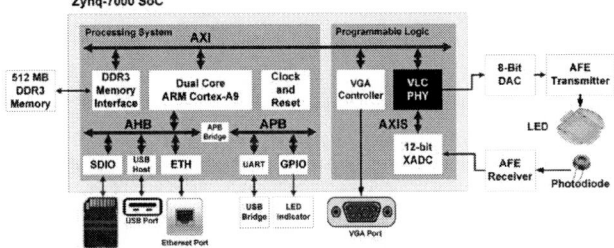

Fig. 1. SoC architecture of OFDM-VLC system

B. Memory-Mapped Controller

A memory-mapped controller is designed inside the VLC PHY block. The memory-mapped controller is used for interfacing with the processor when writing or reading data to the PHY layer. There are two memory-mapped controllers, for TX and RX. Inside the TX controller, there is a configuration register and data registers. Inside the configuration register, the modulation type (BPSK, QPSK, or QAM-16) and guard interval length can be configured. There is also a busy flag that indicates status of the TX. Inside the RX controller, there is also a configuration register and data registers. Inside the configuration register, the demodulation type (BPSK, QPSK, or QAM-16) and

978-1-5386-7961-6/18 $31.00 © 2018 IEEE   115   ISOCC 2018

synchronizer threshold value can be configured. There is also a ready flag that indicates status of the RX.

C. Software Layer

The software layer for this system is shown in Fig 2. There are kernel space and user space. Inside the kernel space, there are architecture-dependent code, Linux kernel, and System Call Interface (SCI). While inside the user space, there are *GNU C library* and user's TX and RX applications. The architecture-dependent code are drivers for peripherals. The SCI provides interface between user space and kernel, so the user's applications can request a service from the kernel. The user's TX and RX applications are applications that can access the PHY through the SCI. The TX and RX applications access the SCI to map the physical address of the PHY to the virtual address of the applications process.

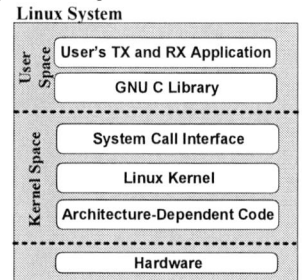

Fig. 2.   Software layer for the system

D. PHY Layer Design

The detailed of this chapter has been discussed in [4], the design of PHY layer including an explanation of PHY blocks diagram, i.e. Reed-Solomon (RS) encoder and decoder, modulator and demodulator, IFFT and FFT, preamble generator and synchronizer, OFDM framing and deframing.

III. RESULTS AND DISCUSSION

In previous work [4-5], we elaborated the result of PHY layer timing diagram. While in this paper, we only show the result of latency, throughput, and resource utilization. The detailed as follows.

A. Resource Utilization

As presented in [4], the PHY layer for transmitter and receiver are implemented in PL. The hardware resources utilization of this PHY layer is shown in Table I. The result is obtained from Xilinx Vivado® implementation report.

TABLE I.   PHY LAYER RESOURCES UTILIZATION

Resource	Utilization	Available	Utilization (%)
LUT	13167	17600	74.81
LUTRAM	901	6000	15.02
FF	17167	35200	48.77
BRAM	15	60	25.00
DSP	33	80	41.25
IO	76	100	76.00

B. Throughput

The PHY layer is synthesized by using Xilinx Vivado® and the synthesis report shows maximum FPGA clock that can be used is 106.74 MHz, so we used 100 MHz of FPGA clok. The throughput is calculated by using (1). The total data bits are 31-bits, 62-bits, and 124-bits for BPSK, QPSK, and QAM-16, respectively. The throughput for transmitter and receiver are shown in Table II.

$$Throughput = \frac{1}{Processing\ time} \times Total\ bit \qquad (1)$$

TABLE II.   THROUGHPUT OF THE TRANSMITTER AND RECEIVER

Modulation	Transmitter	Receiver
BPSK	6.67 Mb/s	8.82 Mb/s
QPSK	13.31 Mb/s	17.59 Mb/s
QAM-16	26.51 Mb/s	34.98 Mb/s

C. Latency

The latency for each PHY block (as elaborated in [4]) is shown in Fig 3. The largest latency is occurred on RS decoder block at the receiver. The latency of RS decoder is related to the number of bit error that can be corrected.

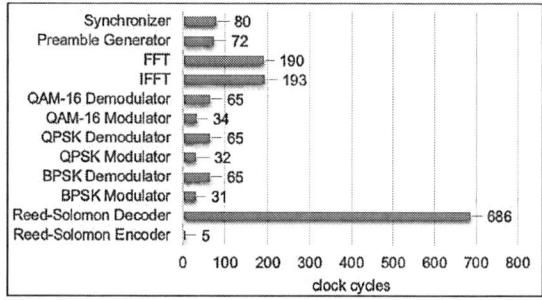

Fig. 3.   Latency for each PHY block

IV. CONCLUSION

The proposed PHY layer design of OFDM-VLC system is implemented on SoC. By using reuse methodology, a complete OFDM-VLC system is designed rapidly with low effort. The maximum data throughput of 26.51 Mb/s is achieved by using 100 MHz FPGA clock.

ACKNOWLEDGMENT

This work is supported by Newton Fund 2017 Programme with Industry Academia-Partnership (IAPP) scheme (No. Ref. IAPP1\100074).

REFERENCES

[1] M. Keating and P. Bricaud, Reuse Methodology Manual for System-on-Chips Designs, 3rd Ed, Dordrecht: Kluwer Academic Pub., 2002, pp. 1-6.

[2] T. Adiono, S. Fuada, and R. A. Saputro, "Rapid development of system-on-chip (SoC) for network-enabled visible light communications," *Int. J. of Recent Contributions from Engineering, Science & IT*, Vol. 6(1), pp. 107-119, 2018.

[3] T. Adiono and A. P. Putra, "Hardware/software model of DCO-OFDM based visible light communication SoC using FPGA," *Proc. of Int. SoC Design Conference* (ISOCC), Seoul, South Korea, November 5-8, 2017, pp. 92-93.

[4] E. Setiawan, T. Adiono, and S. Fuada, "Modelling the OFDM-based PHY layer in SoC for visible light communication," *Unpublished.*

[5] E. Setiawan, T. Adiono, and S. Fuada, "Experimental Demonstration of OFDM Visible Light Communications based on System-on-Chip," *Proc. of Int. Symposium on Electronics and Smart Devices* (ISESD), October 2018.

Model-Based Parallelizer for Embedded Control Systems on Single-ISA Heterogeneous Multicore Processors

Zhaoqian Zhong
PDSL (Parallel and Distributed Systems Lab)
Graduate School of Informatics, Nagoya University
Nagoya, Japan
zhaoqian@ertl.jp

Masato Edahiro
PDSL (Parallel and Distributed Systems Lab)
Graduate School of Informatics, Nagoya University
Nagoya, Japan

Abstract— **This paper presents a model-based parallelization approach for embedded systems on single instruction set architecture (ISA) heterogeneous multicore processors, wherein the core assignment of Simulink blocks is determined based on the control design constraints and characteristics of single-ISA heterogeneous multicore processors. The proposed method first groups blocks hierarchically and forms tens of top-level clusters that contain blocks of the same attribute. For these clusters, a mixed-integer linear programming (MILP) formulation determines the core assignment solution considering load balancing and minimization of inter-core communication across cores with different performance. Finally, each top-level cluster is assigned to cores on multicore processors based on the core assignment solution from the MILP formulation and expanded to the block level by the block dependency. We evaluate the proposed approach by generating parallel code on a single-ISA heterogeneous multicore processor to determine its effectiveness.**

Keywords; model-based design; MATLAB Simulink; single-ISA heterogeneous multicore processors; parallelization

I. INTRODUCTION

Recently, as embedded control systems such as automotive control systems are becoming larger and more complex, model-based design (MBD) with platforms such as MATLAB/Simulink [1] is becoming increasingly common. A Simulink model is a brief and descriptive block diagram that can be automatically translated to sequential source code for embedded implementation on single-core processors. On the other hand, heterogeneous multicore processors of single instruction set architecture (ISA) [2] are widely used because they have potential benefits over homogeneous multicore processors. On a single-ISA heterogeneous multicore processor, such as ARM big.LITTLE architectures, cores may execute the same instruction set; however, they offer different capabilities and performance such as different clock frequencies and power consumption. To implement the control models described in Simulink on a single-ISA heterogeneous multicore processor, it is important to partition the generated control software based on the control design constraints and characteristics of single-ISA heterogeneous multicore processors for parallel execution.

In this paper, a model-based parallelization approach is proposed to parallelize embedded systems built in the Simulink MBD environment on single-ISA heterogeneous multicore processors. In this approach, a hierarchical clustering method is proposed to group blocks of the same attribute to top-level clusters and use a mixed-integer linear programming (MILP) formulation to discover the optimal core assignment solution.

II. PROPOSED APPROACH

Fig. 1 shows an overview of the proposed approach for model-level parallelization. It is used to find the parallelization solution at the model level for given Simulink models and target single-ISA heterogeneous multicore processors, and generates parallel codes for execution on the target processor.

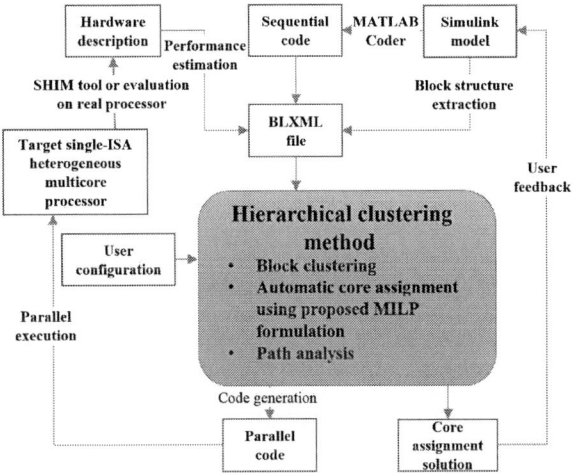

Figure 1. Overview of proposed approach.

Our proposed approach takes the control model designed with Simulink, a description of target single-ISA heterogeneous multicore processor, and a user configuration file as the input. The user configuration file contains information, such as whether the users prefer some special blocks to be assigned to a specified core, and which cores on the processor should be utilized to parallelize the input model.

This work was supported by JSPS KAKENHI Grant Number 16H02800.

978-1-5386-7961-6/18 $31.00 © 2018 IEEE

First, we generate the sequential C code of the input model with MATLAB Coder, and utilize tools from SHIM [3] to evaluate the target heterogeneous multicore processors. SHIM is a hardware abstraction description standardized by Multicore Association that provides tools to roughly estimate software performance at the instruction level. On a single-ISA heterogeneous multicore processor, we set the core of the lowest frequency as the base core and use the SHIM tool to generate a SHIM data file, which obtains some of the architectural features such as the number of clock cycles an instruction may require to execute on this core. Furthermore we require performance information, such as the inter-core communication overhead, core processing speed, and core utilization, which are used in our proposed parallelization method. Such information should be provided by the user or evaluated on real processors. We then combine the sequential C code and SHIM data with the block-level structure of the model and generate a block-level structure XML (BLXML) file. BLXML is a description standardized by our laboratory that contains block-level structure information, the code of each block, and the number of cycles required to execute the code of each block on the base core based on the SHIM data file. We call such generated block groups clusters and assign these clusters to the cores on the target multicore processors. Because we use MILP to solve core assignment problem, parallelization on too many blocks may lead to a large number of MILP variables and constraints, and the solver may run for a very long time to determine the optimal solution. In contrast, the utilization of clusters instead of blocks greatly reduces the scale of the MILP problem to avoid an excessive solver time.

We assign clusters to the target multicores with our MILP formulation. In our formulation, the cores are grouped by the inter-core overheads between them. We assume that inter-group overheads between cores of different core groups are the same, and in each core group, the inter-core overhead is also the same. We evaluate some of the single-ISA heterogeneous multicore products such as ODROID-XU4, in which the values of the overheads between two big cores or two LITTLE cores are always at a very close range, and the overheads between a big core and a LITTLE core are also similar. The objective function in our MILP formulation minimizes both the communication cost between different core groups and the communication cost between the cores in the core groups to reduce the execution time of the entire application. In addition, we assign different workload limits on the different cores to distribute the workloads of the clusters. For example, if a core is two times faster than the base core, the maximum workload assigned to this core should be two times larger than that of the base core. By doing so, the workload can be distributed among each given core, and cores of different performance can be used efficiently.

Finally, we generate the parallel code of the model based on the core assignment solution and BLXML file. The block diagram of the input model is translated into a graph based on the communicating sequential process (CSP) theory, and a path analysis is performed on the block dependency to determine the execution sequence of the blocks on each core. The generated code is implemented in POSIX Threads.

III. EXPERIMENT

To study the scalability and efficiency of our approach, we parallelize a motor control model, which is abstracted from a real automotive control evaluation model, using our approach, and execute the parallel code on an ODROID-XU4 single-board computer. ODROID-XU4 is one of the latest single-board computing devices and involves a Samsung Exynos 5422 processor that includes four Cortex-A15 and four Cortex-A7 cores in a big.LITTLE configuration. We set the CPUFreq Governor policy to the performance mode to ensure that the cores operate at the highest clock frequency and use the Cortex-A7 LITTLE core as the base core. In our experiment, we use up to four cores because of the block number in the motor control model.

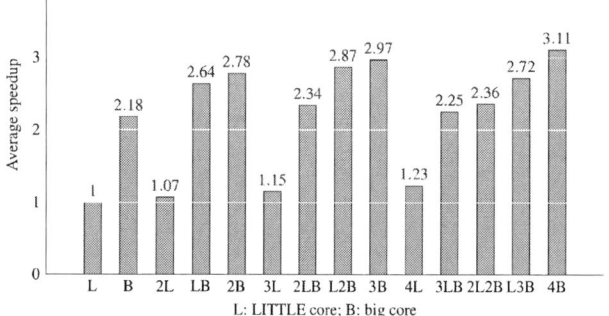

Figure 2.   Acceleration performance.

Fig. 2 shows the acceleration performance of the generated codes executed on ODROID-XU4 compared to the execution time when using only one LITTLE core. As shown by the results, using the proposed approach leads to a reasonable speedup on real processors. Although using more cores may lead to more inter-core communication and result in a lower performance increase for a real scenario model, our approach demonstrates the potential to generate an effective parallelization solution when multiple models are implemented on a single-ISA heterogeneous multicore processor.

IV. CONCLUSIONS

In this paper, we addressed a model-based parallelization approach, specifically, a hierarchical clustering method based on MILP, to parallelize embedded control systems designed on the MATLAB/Simulink platform for single-ISA heterogeneous multicore processors. We paralleled a real model to an ODROID-XU4 board with our approach and observed a reasonable speedup performance.

REFERENCES

[1]   SIMULINK, Mathworks. Simulation and Model-Based Design. 2015.

[2]   R. Kumar, K.I. Farkas, N.P. Jouppi, P. Ranganathan, and D.M. Tullsen. "Single-ISA heterogeneous multi-core architectures: The potential for processor power reduction," *MICRO*, pp. 81–92, 2003.

[3]   G. Masaki, F. Arakawa, and M. Edahiro. "Establishing a standard interface between multi-manycore and software tools-SHIM," IEEE COOL Chips XVII, pp. 1–3, 2014.

978-1-5386-7961-6/18 $31.00 © 2018 IEEE

Efficient Implementation of Multiple Interleavers in IDMA for 5G

Byeong Yong Kong and In-Cheol Park

School of Electrical Engineering

KAIST

Daejeon, Republic of Korea

bykong@kaist.ac.kr; icpark@kaist.edu

Abstract—**This paper presents an efficient implementation of multiple interleavers in interleave division multiple access (IDMA) for the 5G telecommunication. Rather than focusing on designing a single interleaver efficiently, in this paper, all the interleavers in an IDMA system are designed as a whole. More specifically, the interleaving formulae of multiple interleavers are mapped to a multiple constant multiplication problem, and common subexpressions are eliminated from the adder tree. As a result, the proposed architecture replaces costly multipliers with a similar number of adders, effectively reducing the hardware resources required to implement an IDMA system.**

Keywords—5th generation (5G); interleave division multiple access (IDMA); interleaver; multiuser detection; nonorthogonal multiple access (NOMA)

I. INTRODUCTION

Nonorthogonal multiple access (NOMA) is an essential technology that enables the massive connectivity of the 5th generation (5G) telecommunication standard [1]. Interleave division multiple access (IDMA) is one of promising NOMA candidates for the 5G that differentiates multiple users by their unique interleaving patterns [2], [3]. Due to such a principle, an IDMA system inherently contains numerous interleavers, making efficient implementation of those highly important. There have been several architectures reported in the literature that aim at efficient realization of an interleaver [4], [5].

Unlike the previous works [4], [5] that focus on designing a single interleaver, in this paper, we consider all the multiple interleavers in a system as a whole. More specifically, we first formalize the design of multiple interleavers as a multiple constant multiplication (MCM) problem. Then we apply common subexpression elimination (CSE) technique to reduce the number of adders. As a result, all the interleavers in a system can be implemented by using only a feasible amount of adders instead of costly multipliers.

II. BACKGROUND

Fig. 1 illustrates an IDMA system for U users that includes U respective transmitters and a single receiver accommodating all the U users by its own. A multiple access channel connects the transmitters and the receiver. Each user in IDMA employs a unique interleaver associated with an interleaving pattern

---

This work is supported by the National Research Foundation of Korea under Grant NRF-2017R1E1A1A01076992, the Center for Integrated Smart Sensors under Grant CISS-2012M3A6A6054192, and BK21+.

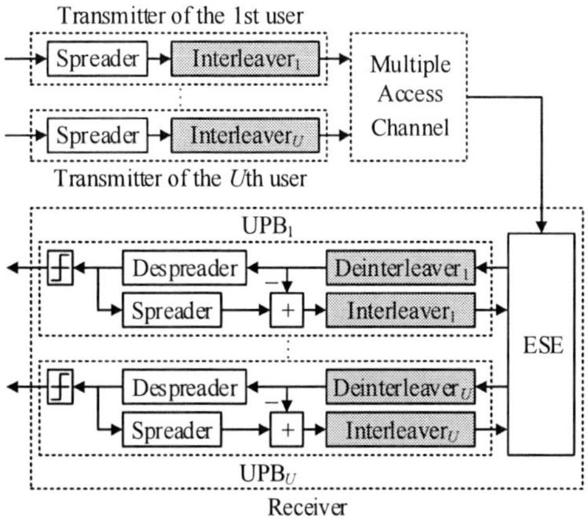

Figure 1. IDMA system for U users.

different from the others'. The receiver is composed of an elementary signal estimator (ESE) and U user-specific processing blocks (UPBs), each of which is dedicated to a user.

To achieve an optimal multiple-access capacity, U perfectly random interleaving patterns are demanded. However, such a randomness induces a prohibitive complexity in an interleaver, being avoided in practice. Instead, a multistage algebraic interleaver [6] has been widely employed in the literature [7]–[10] to replace a random interleaver without degrading the error rate notably. A 3-stage algebraic interleaver known to be near-optimal calculates $\pi_u(j) = \pi_{u3}(\pi_{u2}(\pi_{u1}(j)))$, where

$$\pi_{uk}(j) = \left(\frac{K_{uk} j(j+1)}{2}\right) \bmod J = \left(\frac{K_{uk} i}{2}\right) \bmod J \qquad (1)$$

for all $u = 1, 2, \ldots, U$, $k = 1, 2, 3$, and $j = 0, 1, \ldots, J-1$. J is the interleaving length, K_{uk} is a randomly chosen odd number called *key*, and $i = j(j + 1)$ by definition. Note that $\{j\}$ is a set of ordered addresses and $\{\pi_u(j)\}$ is that of interleaved addresses. For example, $\{j\} = \{0, 1, 2, 3\}$, and $\{\pi_u(j)\} = \{2, 1, 0, 3\}$.

III. PROPOSED ARCHITECTURE

The evaluation of (1) necessitates three multiplications, one addition, one division, and one modular operation. The division by two is straightforward as it is equivalent to a right shift by

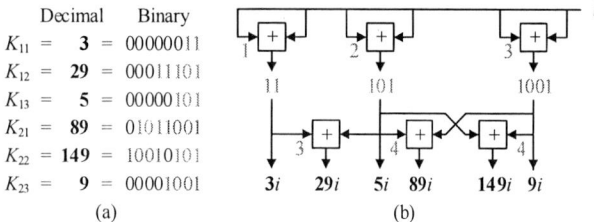

Decimal	Binary
$K_{11} = $ **3** $= 00000011$	
$K_{12} = $ **29** $= 00011101$	
$K_{13} = $ **5** $= 00000101$	
$K_{21} = $ **89** $= 01011001$	
$K_{22} = $ **149** $= 10010101$	
$K_{23} = $ **9** $= 00001001$	

(a)　　　　　　(b)

Figure 2. (a) Decimal and binary representations of keys. (b) Adder tree.

TABLE I. NUMBER OF OPERATORS

U	Naïve		SAA		Proposed	
	$\times$	$+$	$\times$	$+$	$\times$	$+$
2	6	0	0	12 (1.00)	0	6 (0.50)
4	12	0	0	32 (1.00)	0	14 (0.44)
8	24	0	0	101 (1.00)	0	27 (0.27)
16	48	0	0	191 (1.00)	0	43 (0.23)

one bit. If J is a power of two as usual, the modular operation is also trivial, since it is equal to taking $\log_2 J$ least significant bits. On the contrary, the multiplications are costly and need to be performed in a sophisticated manner. If we implement the multiplications in a naïve manner, we require $3U$ multipliers to serve U users, which is highly unfavorable in the 5G associated with a very large U.

We now explain how the multiplier-based architecture can be replaced by a multiplierless one. In the preceding section, we defined $i = j(j + 1)$ in (1). Grounded on the definition, the calculation of $\{\pi_u(j)\}$ for all $u = 1, 2, \ldots, U$ can be thought of as multiplying i with various keys, followed by the trivial division and the insignificant modular operation. In other words, the set of multiplications can be mapped to the MCM problem. In solving the MCM problem, we can decompose a multiplication into a set of shifts and additions, and apply the CSE techniques to reduce the number of additions.

Let us exemplify the application of the CSE to a simple 2-user IDMA system. A set of keys in the system is listed in Fig. 2(a). By observing the keys represented in a binary format, we can identify three common subexpressions, i.e., 11, 101, and 1001 colored red, green, and blue, respectively. By sharing the common subexpressions, we can synthesize an adder tree depicted in Fig. 2 (b), where gray numbers at the edge represent left-shift amounts. The adder tree takes $i = j(j + 1)$ as input, and each of the outputs are destined to be divided by two and undergo the modular operation. The naïve multiplier-based implementation demands 6 multipliers. By employing the shift-and-add (SAA) architecture without any optimization, we can substitute the 6 multipliers with 12 adders. Applying the CSE technique, we can further diminish the operators from 12 to 6 adders, which is remarkable reduction in hardware complexity.

Note that the adder tree in Fig. 2(b) is not unique. The adder cost can be further lowered by identifying common terms from canonical or minimal signed digit representations [11] and by applying a high-performance multiplier-block synthesis algorithm, e.g., [12].

IV. EVALUATION

For randomly generated sets of keys, Table I summarizes the numbers of multipliers and adders required to implement U interleavers. The numbers in parentheses are normalized with respect to the SAA architecture. As analyzed in Section III, a naïve architecture demands $3U$ multipliers. On the other hand, the proposed architecture needs about $3U$ adders. Compared to the SAA architecture without any optimization, on average, the

proposed one requires about 64% less adders. It is worth noting that the effectiveness of the proposed scheme may vary to a certain extent along with actual values of keys and the CSE algorithm applied.

V. CONCLUSION

A multiplierless architecture for multiple interleavers in an IDMA system has been presented. The algebraic formulae of the interleavers has been mapped to the MCM problem, and the CSE technique has been applied to an intact SAA architecture to remove adders from the tree. Consequently, computationally expensive multipliers have been successfully substituted with a similar amount of adders.

REFERENCES

[1] Y. Liu, Z. Qin, M. Elkashlan, Z. Ding, A. Nallanathan, and L. Hanzo, "Nonorthogonal multiple access for 5G and beyond," *Proc. IEEE*, vol. 105, no. 12, pp. 2347–2381, Dec. 2017.

[2] L. Ping L. Liu, K. Wu, and W. K. Leung, "Interleave-division multiple-access," *IEEE Trans. Wireless Commun.*, vol. 5, no. 4, pp. 938–947, Apr. 2006.

[3] C. Zhang, Y.-H. Huang, F. Sheikh, and Z. Wang, "Advanced baseband processing algorithms, circuits, and implementations for 5G communication," *IEEE J. Emerg. Sel. Topics Circuits Syst.*, vol. 7, no. 4, pp. 477–490, Dec. 2017.

[4] W. Belaoura, M. Djeddou, and K. Ghanem, "GRP-based interleaver for IDMA systems over frequency selective channel," *Electron. Lett.*, vol. 51, no. 18, pp. 1462–1464, Aug. 2015.

[5] S. Wu, X. Chen, and S. Zhou, "A parallel interleaver design for IDMA systems," in *Proc. Int. Conf. Wireless Commun. Signal Process. (WCSP)*, Nanjing. China, 2009, pp. 1–5.

[6] O. Y. Takeshita and D. J. Costello, "New classes of algebraic interleaver for turbo-codes," in *Proc. IEEE Int. Symp. Inf. Theory (ISIT)*, Cambridge, MA, USA, 1998, p. 419.

[7] R. Dodd, C. Schlegel, and V. Gaudet, "DS-CDMA implementation with iterative multiple access interference cancellation," *IEEE Trans. Circuits Syst. I, Reg. Papers*, vol. 60, no. 1, pp. 222–231, Jan. 2013.

[8] S. Yoshizawa, M. Nozaki, and H. Tanimoto, "VLSI implementation of an interference canceller using dual-frame processing for OFDM-IDMA systems," *IEICE Trans. Fundam. Electron. Commun. Comput. Sci.*, vol. E98-A, no. 3, pp. 811–819, Mar. 2015.

[9] T. T. T. Nguyen, L. Lanate, S. Yoshizawa, and H. Ochi, "Low latency IDMA with interleaved domain architecture for 5G communications," *IEEE J. Emerg. Sel. Topics Circuits Syst.*, vol. 7, no. 4, pp. 582–593, Dec. 2017.

[10] B. Y. Kong and I.-C. Park, "A memory-efficient IDMA architecture based on on-the-fly despreading," *IEEE J. Solid-State Circuits*, to be published. DOI: 10.1109/JSSC.2018.2863950.

[11] I.-C. Park and H.-J. Kang, "Digital filter synthesis based on minimal signed digit representation," in *Proc. Des. Autom. Conf. (DAC)*, Las Vegas, NV, USA, 2001, pp. 468–473.

[12] B. Y. Kong and I.-C. Park, "FIR filter synthesis based on interleaved processing of coefficient generation and multiplier-block synthesis," *IEEE Trans. Comput.-Aided Des. Circuits Syst.*, vol. 31, no. 8, pp. 1169–1179, Aug. 2012.

Hybrid Decoding for Polar Codes

Soyeon Choi
Dept. of Electronics Engineering
Chungnam National University
Daejeon, South Korea
sychoi.cas@gmail.com

Hoyoung Yoo
Dept. of Electronics Engineering
Chungnam National University
Daejeon, South Korea
hyyoo.cnu@gmail.com

Abstract—**Among various polar decoding algorithms, SC-List [2] and SC-Flip [3] suffer from high hardware complexity and long decoding latency, respectively. In this paper, a novel hybrid decoding algorithm is proposed to achieve affordable hardware complexity with a suitable decoding latency. According to the experimental results, the proposed method affords a comparable error-correcting performance to that of SC-List [2] and SC-Flip [3] counter parts.**

Keywords—polar codes; SC-List decoding; SC-List decoding; hybrid decoding;

I. Introduction

The polar code [1] is the first error-correcting code that can achieve channel-capacity provably. Although Successive-Cancellation (SC) decoding [1] has been traditionally used to implement a polar decoder, the error-correcting performance is not suitable for next-generation communication and storage systems. To improve the error-correcting performance, SC-List decoding [2] and SC-Flip decoding [3]-[4] with flip are recently proposed by taking more chances to find a valid codeword. SC-List decoding [2] provides a good error-correcting performance, but it suffers from high hardware complexity. On the other hands, SC-Flip [3] requires a small hardware circuity, but it deteriorates a decoding latency. In this paper, we propose a hybrid decoding algorithm that combines SC-List [2] and SC-Flip decoding [3] resulting in affordable hardware complexity with a suitable decoding latency. The proposed method suggests a good candidate for a polar decoder with a stringent hardware requirement.

II. Review of Polar Decoding

A. Polar codes

Let us consider the polar (N, K) code, where N and K denote a code length and a message length, respectively. Among the N bit-channel, K most reliable bit-channels are used for information bits and $(N-K)$ remaining bit-channels are used for frozen bits. The index sets of information and frozen bits are denoted by A and A^c, respectively. The codeword x is generated by matrix multiplication of the generator matrix G and message vector u.

Given received vector y, Successive-Cancellation (SC) [1] decoding has been traditionally used to bring a polar decoder in practice. Due to its serial nature, SC decoding alternatively computes soft information corresponding to (1) and (2) in log-likelihood ratio (LLR) domain. Although SC decoding [1] is the first decoding algorithm, its error-correcting performance is not comparable to other capacity-achieving codes such as LDPC and turbo codes since it makes a hard-decision when the message bit is estimated as (3).

$$f(LLR_i, LLR_{i+1}) \approx \text{sign}(LLR_i)\,\text{sign}(LLR_{i+1})\,\min(|LLR_i|, |LLR_{i+1}|)\,, \quad (1)$$

$$g(LLR_i, LLR_{i+1}, \hat{u}) = (-1)^{\hat{u}}\,LLR_i + LLR_{i+1}\,, \quad (2)$$

$$\hat{u}_i = \begin{cases} 0 \,, & LLR_i \geq 0, i \in A \\ 1 \,, & LLR_i < 0, i \in A \\ u \,, & i \in A^c \end{cases} \quad (3)$$

B. SC-List and SC-Flip decoding

To improve error-correcting performance, SC-List [2] decoding algorithm with list size of L inspects L most likely codewords simultaneously, whereas SC decoding [1] inspects only one most likely codeword. Since SC-List [2] searches more possible codewords, it always provides a stronger error-correcting capability compared to SC decoding [1]. As the list size of L increases, the error correcting capability is more improved. However, severe hardware complexity is inevitable in SC-List [2] decoding since the hardware to search valid codewords at the same time is proportionally increased as the larger L is adopted.

Similar to SC-List [2] decoding, SC-Flip [3] decoding with flipping bits of F provides a more chance to inspects possible codewords. Whereas SC-List [2] decoding searches L codewords simultaneously, SC-Flip [3] decoding searches 2^F codewords in a sequence. More precisely, the standard SC [1] decoding is firstly performed to estimate a message vector $\hat{u}_i^N$. If the CRC is success, the decoding is terminated. Otherwise, the SC decoding is performed for T additional attempts to identify which bit is an error by flipping the most unreliable bit. Although SC-Flip [3] decoding succeeded in reducing hardware complexity, it necessitates a long decoding latency. In general, the error-correcting performance of SC-Flip [3] with flipping bit of F is similar to that of SC-List [2] with list size of $L = 2^F$.

III. Proposed Method

From an implemental points of view, SC-List [2] and SC-Flip [3] decoding provides extreme candidates in terms of hardware complexity and decoding latency. In other words,

978-1-5386-7961-6/18 $31.00 © 2018 IEEE

The proposed hybrid algorithm

Initialize: $\hat{u}_0^{N-1} \leftarrow \text{SC}(y_0^{N-1}, \varnothing)$.

    if CRC($\hat{u}_0^{N-1}$) = fail **then**

        $U \leftarrow i \in A$ of smallest $|LLR_i|$
        for $t = 0$ **step** 1 **until** T **begin**
            $k \leftarrow U(t)$.
            $\hat{u}_0^{N-1} \leftarrow \text{SCL}(y_0^{N-1}, k)$.

            if CRC($\hat{u}_0^{N-1}$) = success **then**

                break.
            end if
        end for
    end if
Output: $\hat{u}_0^{N-1}$.

Figure 1. The proposed hybrid decoding algorithm

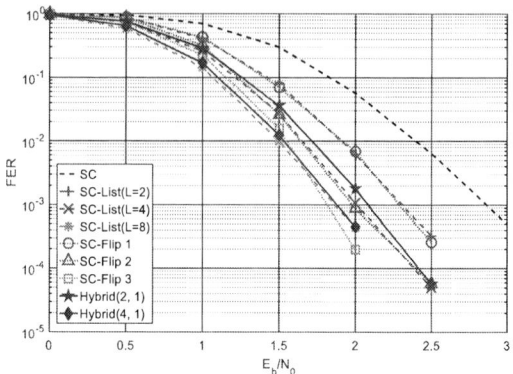

Figure 2. FER performance of SC-List, SC-Flip, and proposed decoding

TABLE I.   Comparison of hardware complexity

	Area	Time	Area×Time
SC-List Decoding	L	1	L
SC-Flip Decoding	1	$F \cdot T$	$F \cdot T$
Proposed Decoding	l	$f \cdot t$	$l \cdot f \cdot t$

SC-List [2] is the best candidate in terms of decoding latency but the worst candidate in terms of hardware complexity. Similarly, SC-Flip [3] is the best candidate in terms of hardware complexity but the worst candidate in terms of decoding latency. In this paper, we propose a hybrid decoding algorithm for affordable hardware complexity with a suitable decoding latency by combining SC-List [2] and SC-Flip [3] decoding algorithms. Figure 1 describes the proposed hybrid decoding algorithm with list size l, flipping bit f, and additional attempt t. At first, standard SC [1] decoding is performed to find the set of flipping index U, which is a set of the least reliable bits. When CRC is successful, the proposed algorithm is terminated as SC-Flip [3]. Otherwise, additional t attempts are tried to identify which bit is an error by flipping the least reliable bit in U. Unlike SC-Flip [3] decoding which employs SC [1] decoding for additional T attempts, the proposed hybrid decoding algorithm employs SC-List [2] decoding to accelerate decoding latency. Note that SC and SCL(y_0^{N-1}, k) in Fig. 1 denote the SC and SCL decoding process with a flipped bit at the index of k. Since the proposed hybrid decoding provides an intermediate decoding algorithm between extreme SC-List [2] and SC-Flip [3] decoding algorithms, it can be a good design candidate for a polar decoder with a stringent hardware requirement in both area and time.

IV. EXPERIMENTAL RESULTS

We compare SC-List [2], SC-Flip [3], and the proposed decoding algorithms for Polar (1024, 512) codes with 16-bit CRC codes. AWGN channel is used for a transmission channel, and a codeword is modulated by BPSK. Figure 2 shows the frame error rate (FER) of various decoding algorithms for different channel environment. FER of the proposed decoding algorithm with $l = 2$ and $f = 1$ is similar to that of SC-List decoding with $L = 4$ and that of SC-Flip decoding with $F = 2$. In addition, FER of the proposed decoding algorithm with $l = 4$ and $f = 1$ is similar to that of SC-List decoding with $L = 8$ and that of SC-Flip decoding with $F = 3$. As a result, the proposed method affords a comparable error-correcting performance to that of SC-List and SC-Flip counter parts. Table I summarize a complexity comparison in terms of area and time. Generally speaking, $l \times f$ in the proposed decoding algorithm is the same as L in SC-List [2] and F is SC-Flip [3]. Note that l and f equipped in the proposed decoding algorithm are always smaller than L in SC-List [2] and F and T in SC-Flip [3].

V. CONCLUSION

In this paper, proposed decoding algorithm for polar codes has been newly proposed by combining SC-List [2] and SC-Flip [3] decoding algorithms. The proposed decoding algorithm requires a lower hardware complexity compared to SC-List [2] decoding and provides a shorter decoding latency compared to SC-Flip [3] decoding. Thus, the proposed algorithm can be used for a polar decoder with a stringent requirement.

ACKNOWLEDGMENT

This work was supported by the National Research Foundation (NRF) grant funded by the Korea government (MSIP) (2017R1C1B5015962) and by the IC Design Education Center (IDEC).

REFERENCES

[1] E. Arıkan, "Channel polarization: A method for constructing capacity achieving codes for symmetric binary-input memoryless channels," *IEEE Trans. Inform. Theory*, vol. 55, pp. 3051–3073, 2009

[2] I. Tal and A. Vardy, "List decoding of polar codes," *in Proc. IEEE Symp. Inform. Theory*, Saint Petersburg, Russia, 2011, pp. 1–5.

[3] O. Afisiadis, A. Balatsoukas-Stimming, and A. Burg, "A low-complexity improved successive cancellation decoder for polar codes," *in Proc. 48th Asilomar Conf. Signals Syst. Comput.*, Nov. 2014, pp. 2116–2120.

[4] L. Chandesris, V. Savin, and D. Declercq, "An improved SCFlip decoder for polar codes," *in IEEE Global Communications Conference (GLOBECOM)*, Dec 2016, pp. 1–6.

The Hardware Acceleration of SC Decoder for Polar Code towards HLS Optimization

Yujie Huang, Yujie Cai, Minge Jing, Jun Han, Yibo Fan, Xiaoyang Zeng

State Key Laboratory of ASIC & System
Fudan University
Shanghai, 201203, China
Email: mejing@fudan.edu.cn

Abstract— With the extensive use of hardware acceleration for complex algorithms, fast hardware implementations of those algorithms with HLS become necessary. However, high performance hardware generated with HLS needs numerous skills and the successive-cancellation (SC) decoding algorithm for polar code, which is widely used in 5G era, can't be converted from *c* code to *verilog* code with HLS directly because of the recursion in it. In this paper, we firstly find out the operation laws of SC decoding algorithm; secondly, we expand the recursion in the SC decoding algorithm according to the laws; thirdly, we convert float point data to fixed point which reduces LUT resources by 63.5% and simplify the function consisting of exponential and logarithm operation through Taylor Expansion to reduce hardware resource and accelerate the hardware; finally, we apply pipeline in *for* loop which reduces cycles by 29% and call on-chip RAM in HLS to quickly generate high performance SC decoder which consumes low hardware resources.

Keywords; polar code; HLS; SC decoder

I. INTRODUCTION

With the development of new technologies, the complexity of the algorithm is getting higher and higher. Besides, as the wide application of FPGA in algorithm acceleration, fast hardware implementation of the algorithm becomes necessary. As a result, the Xilinx tool called High-Level Synthesis (HLS), is wildly used by programmers to quickly implement the algorithms on hardware. The programmers just need to input *c* code into HLS and HLS will output the *verilog* code corresponding to the *c* ode [1][2]. But, the performance of hardware generated from different styles of *c* code vary greatly. Furthermore, numbers of skills, such as converting floating point to appropriate fixed point, should be appropriately applied in HLS to achieve high performance hardware with low hardware resources consumption. And the 5G era is coming soon where polar code is used which makes high performance decoder for polar code necessary [3]. So in this paper, we use HLS to generate successive-cancellation (SC) decoder for polar code.

The rest part of this paper is as follows. Related works on decoder for polar code are introduced in Section 2. In section 3, the implementation process of the SC decoder for polar code through HLS is described in detail. The implementation result is shown in Section 4 and this paper is concluded in Section 5.

II. RELATED WORK ON POLAR CODE DECODER

At 2012, Chuan Zhang use pipelining and parallel processing to reduce the latency of SC Decoder when code length is long [3]. Meanwhile, at 2014, B Yuan and KK Parhi [4] use 2-bit Decoding (INT2) to simplify the decoder, which means converting the data format using in SC Decoder from float point to fixed point is possible. They also made some research on LLR-based SC List Decoder at 2015 [5]. In all their work, the decoders are finely designed which will consume much time.

III. IMPLEMENTATION OF DECODER WITH HLS

In this section, the implementation process of SC decoder for polar code is described in detail. Firstly, we expand the recursion in SC algorithm to avoid failure in HLS caused by recursion. Secondly, *c* code of SC algorithm with arbitrary-bit input is quickly generated automatically with our python script. Thirdly, we convert the data format from float to INT4 to reduce the cost of hardware resources. Fourthly, we use Taylor Expansion to simplify *F* function consisting of exponential and logarithm operation. Finally, we apply pipeline command in *for* loop to improve the performance and on-chip RAM call command in the arrays to save the LUT resources.

A. Expansion of the recurison

Here, we need to expand the recursion by finding out the operation laws of SC algorithm. The operation laws are described below where N means the number of layers.

Intra-layer operation law. There are 2^N data per layer and the data in layer K is divided into 2^K groups according to the parity. And, the data in the same group can be calculated in parallel by $2^{(N-K)}$ G functions or F functions.

5234->5234235->523423523423

Figure 1. Example of the inter-layer operation law of 32-bit SC decoding algorithm (5 layers).The underlined parts are symmetrical about the corresponding middle numbers.

Inter-layer operation law. As shown in Fig.1, each number T means the calculation is from layer N-T+1 to the last layer. Besides, a calculation of a group of data in the last layer will be performed after each number which means there is a number *1* after each number in Fig.1 where the number *1* is omitted to facilitate the display of the law. We can find that the numbers are generated symmetrically in Fig.1.

U parameter operation law. When G function is applied, it needs the U parameter which is generated from all the calculated U values. There are two operations on an array of all the calculated U values. One operation represented by *0* is grouping the array according to the parity and then *XOR* within each group. For example, the array is (u_0, u_1, u_2, u_3), then it will be $((u_0, u_2), (u_1, u_3))$ after being grouped according to the parity, and $(u_0\ XOR\ u_2, u_1\ XOR\ u_3)$ will be obtained by XOR within each group. The other operation represented by *1* is extracting the even items in the array. For example, (u_0, u_1, u_2, u_3) will become (u_1, u_3) after the operation. When one group of data in layer K is going be calculated by G functions in parallel, the U parameters used by those G functions are respectively generated by the combinations of the two operations which can be represented by the $(N-K)$-bit binary form of numbers from 0 to $2^{(N-K)}-1$. For example, if N is 6 and K is 4, then the operations are *00, 01, 10,* and *11*. And the U parameters are separately the last items of the arrays obtained by the array of all the calculated U values after the above operations.

G and F function usage laws. In each layer, G and F function are used interchangeably between each group.

B. Generation of c code by python script

After achieving the above laws, another problem happens that the c language implementation of the SC decoder for 1024-bit polar code in the form of expanding the recursion has a huge amount. As a result, we write a python script to automatically generate the c language for arbitrary-bit SC decoder according to those operation laws.

C. Convert float point data to fixed point

Converting the data format from float point to fixed point is a good idea for the implementation of many algorithms, so as the SC decoder. We evaluate the influence of data format changing by monitoring the error rate, and find that changing the data format into INT4 have no inference on the error rate. So we implement our design with INT4.

D. Simplify F function

The exponential and logarithm operations in each F function will consume much logical resources. We simplify the function $ln(\ (1+exp(a+b))/(exp(a)+exp(b))\ )$ into $a*b/2$ by Taylor Expansion, with no influence on the error rate, which will not only save hardware resources but also accelerate the hardware.

E. Application of commands in HLS

Through experiment we find that unrolling all the *for* loop in 1024-bit decoder will cause failure in HLS, so unrolling command is not used. As a result, we apply RAM call command in all the arrays to use the on-chip RAMs which makes us save the register resources. Then, we apply pipeline command in each *for* loop to improve the performance.

IV. EXPERIMENT RESULT

In this section, we use Xilinx tool HLS to convert c code to *verilog* code on the VCU119 FPGA platform and then use Vivado to synthesize the *verilog* code. We compare the four kinds of hardware: fixed-point pipeline, float-point pipeline, fixed-point without pipeline, float-point without pipeline. As shown in Table 1, it is identified that fixed-point pipeline SC decoder takes 22525 cycles to finish 1024-bit decoding while only consumes 30411 LUT and fixed-point without pipeline only consumes 14860 LUT. Furthermore, pipeline reduces cycles by 29% in the case of fixed point, and fixed point reduces the LUT resources by 63.5% in the case of pipeline.

TABLE I. 1024-BIT SC POLAR CODE DECODER

Kind of hardware	Cycle	RAM	LUT
fixed-point pipeline	22525	33	30411
float-point pipeline	31738	130	31840
fixed-point without pipeline	31743	33	14860
float-point without pipeline	32253	130	40682

V. CONCLUSION

In this paper, we generate a SC decoder for polar code with HLS. Firstly, we find out four operation laws of the SC decoding algorithm to expand the recursion in it to avoid failure in HLS. Secondly, a python script is applied to generate c code of arbitrary-bit decoder according to those operation laws. Thirdly, we convert the data format from float to INT4 to reduce the cost of logical resources. Fourthly, we use Taylor Expansion to simplify F function consisting of exponential and logarithm operation. Finally, pipeline command is applied in *for* loop to improve the performance and on-chip RAM call command is applied in the arrays to save the resources. And the results show that a SC decoder with high performance and low LUT consumption is generated by HLS. Furthermore, pipeline reduces cycles by 29% and fixed point reduces the LUT resources by 63.5%.

REFERENCES

[1] Cong, Jason, et al. "High-Level Synthesis for FPGAs: From Prototyping to Deployment." IEEE Transactions on Computer-Aided Design of Integrated Circuits and Systems 30.4(2011):473-491.

[2] Canis, Andrew, et al. "LegUp:high-level synthesis for FPGA-based processor/accelerator systems." Acm/sigda International Symposium on Field Programmable Gate Arrays, FPGA ACM, 2011:33-36.

[3] Zhang, Chuan, B. Yuan, and K. K. Parhi. "Reduced-latency SC polar decoder architectures." IEEE International Conference on Communications IEEE, 2012:3471-3475.

[4] Yuan, Bo, and K. K. Parhi. "Low-Latency Successive-Cancellation Polar Decoder Architectures Using 2-Bit Decoding." IEEE Transactions on Circuits & Systems I Regular Papers 61.4(2014):1241-1254.

[5] Yuan, Bo, and K. K. Parhi. "Reduced-latency LLR-based SC List Decoder for Polar Codes." Edition on Great Lakes Symposium on Vlsi ACM, 2015:107-110.

FPGA-based Optical Character Recognition for Handwritten Mathematical Expressions

Bobbi Winema Yogatama[a,1], Jhonson Lee[a,2], Suksmandhira Harimurti[a,b], Trio Adiono[a,b,3]

[a] *School of Electrical Engineering and Informatics, Institut Teknologi Bandung, Bandung, Indonesia*
[b] *University Center of Excellence on Microelectronics Institut Teknologi Bandung, Bandung, Indonesia*
[1]bwyogatama@gmail.com, [2]jhonsonlee10@gmail.com, [3]tadiono@stei.itb.ac.id

Abstract—This paper presents the hardware design of optical character recognition for handwritten mathematical expressions using field programmable gate array (FPGA). The OCR is based on feedforward neural networks. The purpose of this research is to increase the speed and reduce the energy consumption of the learning process of neural networks by performing forward and backpropagation in the hardware, instead of software. To optimize the speed and the hardware area of the design, we proposed a parallel architecture and hardware sharing design. The simulation result of our system indicated that our proposed approach achieved an accuracy of 72.5% on a 25 MHz FPGA. Based on simulation measurement, it significantly consumes a very low power, while being as fast as the baseline method in terms of wall clock time.

Keywords—Optical Character Recognition, Neural Networks Accelerator, AXI4-Lite protocol, FPGA

I. Introduction

Optical Character Recognition (OCR) has been an interesting research topic since such system was first patented by Tauschek in 1935 [1]. Many researches focus on optimizing the software in terms of computation and method. Therefore, most of them propose an OCR implemented in software. However, in this paper, we propose a novel approach to implement OCR in hardware. Moreover, we also focus on enhancing OCR processing speed and decreasing its power consumption significantly.

As OCR is a complex system that requires lots of processing schemes, we limit our discussion to only OCR implementation as hardware accelerator. Preprocessing schemes to segment the mathematical expressions into individual symbols will not be discussed in this paper. In real application, the proposed system will act as a neural networks accelerator which cooperates with its main processor in conducting real character recognition from obtained mathematical expressions.

The OCR is based on feedforward neural networks and designed to recognize handwritten mathematical symbols. As there are many mathematical symbols, we limit our experiment to only mathematical expressions that include numbers (0-9) and basic operators (+, -, x, :) that are obtained from Kaggle datasets. The system is built using feedforward neural networks with 2 hidden layers (16 neurons in each layer) and supports hardware for both training and inference. It is important to note that our neural networks accelerator design can also be adapted for other applications that use different layer sizes and number of neurons while still using feedforward neural networks.

II. Hardware Design

A. Hardware Architecture

Our design supports both inference and training for neural networks. Therefore, our hardware design consists of two main components: the forward propagation and the backpropagation module. There are several methods in implementing the feedforward neural networks, e.g. bit-slice, single instruction multiple data, and systolic array [2]. In this paper, we alternatively utilize McLaurin series to approximate the sigmoid function (tansig) that is segmented into 10 parts. Every processing element in the forward propagation will be responsible for the calculation of input *activation* (z) and output *activation* (a) in each neuron. Meanwhile, the backpropagation module works in similar way to the forward propagation module, except the processing elements in backpropagation will be responsible for the calculation of *delta* and the updates of *weight* and *bias*. The hardware architecture of the proposed accelerator design is shown in Figure 1.

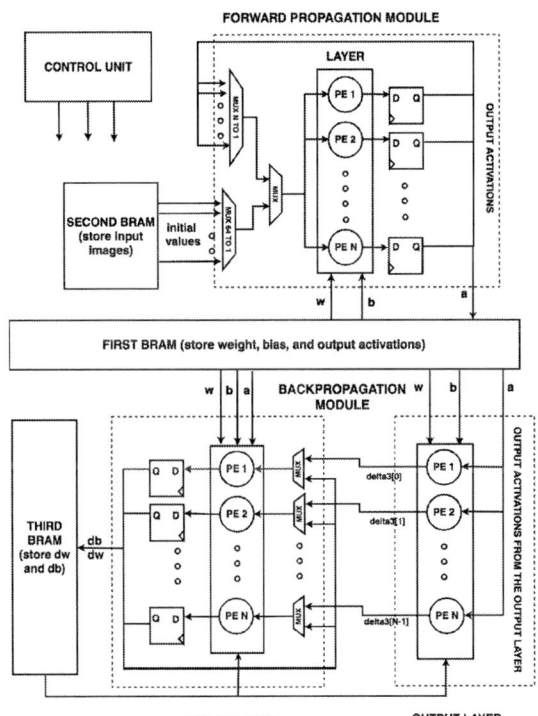

Fig. 1. Hardware Architecture

We also apply several optimizations into our design to further enhance its performance. Parallel processing, hardware sharing, and block RAM utilization are incorporated in the design. In our design, we 'share' or 'reuse' the digital circuits used to calculate the parameters in each layer to save the area of our design. By utilizing hardware sharing, our design will be able to adapt with any number of hidden layers without increasing the total area of the design. On the other hand, since each neuron is independent, the computation in each neuron is also parallel-processed to reduce the latency of the design.

To optimize the registers usage, we also utilize Block RAM in our design. We use 3 BRAM in our design to store the data. The First BRAM stores the weight, bias, and output activations (a) of each neuron, the Second BRAM stores the dataset for the training mode, while the Third BRAM stores the Δw and Δb of each layer.

B. System on a Chip Design

Another challenge in implementing digital design is to integrate the digital circuit with a main processor. In this case, we need to transfer 64-pixel binary images from main processor to digital circuit. On the opposite site, processing result from digital circuit is transferred back to the main processor. Those processes are done using AXI4-Lite protocol. This mechanism applies only to 64-pixel binary image which can be stored in 64-bit register. However, AXI4 standard protocol is required for cases with complex input images (not binary) or identification process with many images at once in which more registers are necessary.

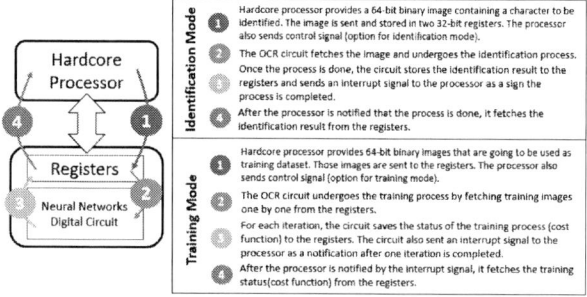

Fig. 2. Visualization for Data Transfer Mechanism

III. SIMULATION

To evaluate the performance of our design, we will compare the overall accuracy of our design with the overall accuracy of the software (MATLAB) running on 2.4 GHz Intel i7 CPU as our baseline method. We used a total of 7000 datasets (500 dataset for each class) for our training. The result will be compared with the result from the timing simulation of the proposed design. With the same number of dataset and the same number of iterations, the baseline method achieved an accuracy of 80.462% while the proposed approach achieved a lower accuracy of 72.5% simulated in Xilinx Vivado using XC7Z035 (25 MHz clock speed). The decrease in accuracy occurs due to approximation of sigmoid

function and bit precision limitation. The cost function of the proposed approach resulting from MATLAB simulation can be seen in Figure 3 (learning rate: 2×10^{-5}; 35000 iterations).

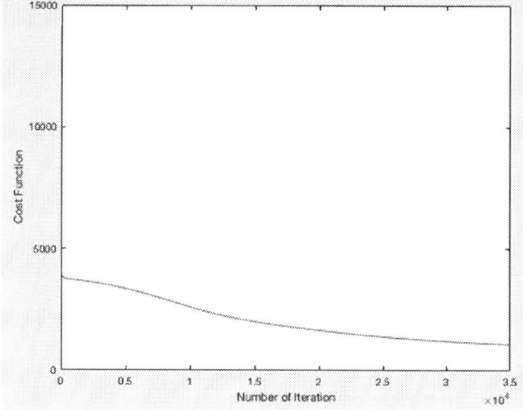

Fig. 3. The Cost Function of The Proposed Design

We also calculated the time required to finish the training using the same dataset. Our calculation showed that with the same clock speed, our accelerator is about 30 times faster compare to the baseline method (2.4 GHz Intel i7 CPU). To evaluate the power consumption of our design, we compared the power consumption and clock speed of our accelerator with typical CPU or GPU in Table I.

Table I. The Comparison of Power Consumption and Clock Speed

	Clock Speed	Power
FPGA (XC7Z035)	25 MHz	0.52 – 2.53 W
CPU (Intel i7)[3]	2.8 – 4.2 GHz	35 – 140 W
GPU (NVIDIA GeForce)[3]	1 – 1.7 GHz	75 – 250 W

IV. CONCLUSION

The proposed OCR design as a hardware accelerator provides fast computation and low-power consumption. The OCR system is based on feedforward neural network which provides both training and inference mode. Design optimization is done using hardware sharing and parallel architecture principles. AXI4-Lite protocol provides data transfer schemes between the logic design and the main processor. The simulation result indicated the system could finish the training at about the same time as 2.4 GHz CPU using 80 MHz clock speed. The simulation shows the system consumes significantly lower power than common CPU and GPU. Operation with higher clock frequency will surely exceed common CPU performance.

V. REFERENCE

[1] D. Berchmans, S. S. Kumar, "Optical Character Recognition: An Overview and an Insight", 2014.

[2] T. Adiono, G. Melliola, S. Harimurti, "Design of Neural Network Architecture using Systolic Array Implemented in Verilog Code", in ISESD. Bandung. October 2018.

[3] Brown, Nicholas, "How Much Power Do Computer Consume?", in Kompulsa, https://www.kompulsa.com/much-power-computers-consume/ , 2018.

Transfer Learning-based Vehicle Classification

So Yeon Jo, Namhyun Ahn, Yunsoo Lee, and Suk-Ju Kang
Department of Electronic Engineering
Sogang University
Seoul, Republic of Korea
josoyun123@gmail.com, neition503@gmail.com, profitshore@gmail.com, sjkang@sogang.ac.kr

Abstract— In this paper, we propose a transfer learning-based vehicle classification from the convolutional neural network (CNN) pre-trained on a large scale dataset. It is possible to construct deep neural networks effectively for new problems with a limited scale vehicle dataset. The proposed system is divided into two stages. First, the vehicle area is detected on the roadway video by Haar-like features. Second, the transfer learning-based vehicle classification using GoogLeNet classifies vehicle models. Experimental results show that the proposed system has a high accuracy of 0.983, which is 0.326 higher than that of the conventional method without transfer learning.

Keywords; Vehicle Classification; Transfer Learning; Deep Learning

I. INTRODUCTION

Vehicle classification is a key element of the intelligent transportation system and has various applications such as traffic flow statistics, intelligent parking systems, and driver assistance systems [1]. In the past, several researches have been done on the vision-based vehicle classification using support vector machine (SVM) [2] to train classification models. The conventional method is not robust due to unstable feature extraction from illumination change. Because convolutional neural network (CNN) learns itself without the need for people to extract features, it complements drawbacks of the conventional method and has contributed to the rapid development of the image classification [3].

CNNs have a very large number of trained parameters, and hence the use of small dataset limits their performance including overfitting. Therefore, this paper proposes a transfer learning-based vehicle classification which can be used in a limited scale dataset. The proposed system has two stages. First, the vehicle area is detected on the roadway video by Haar-like features [4]. Second, the transfer-learning based vehicle classification model classifies vehicle models. In the transfer learning, the earlier layers of GoogLeNet pre-trained on ILSVRC-2012 are fixed and gradients are backpropagated only through the higher-level portion with the vehicle dataset.

II. PROPOSED SYSTEM

The overall architecture of the proposed system is shown in Fig. 1. After detecting the vehicle area on the roadway video, the top 10 vehicles with high frequency of appearance are selected and a total of 10,000 dataset, which has 1,000 for each vehicle, are constructed. Using this dataset, a vehicle

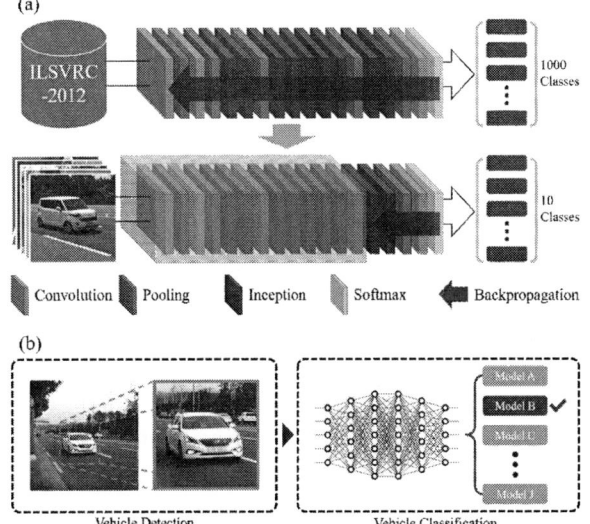

Figure 1. Overall architecture of a transfer learning-based vehicle classification: (a) training process and (b) inference process.

classification model is built by the transfer learning with pre-trained GoogLeNet as shown in Fig. 1 (a). The inference process using the vehicle classification model is divided into vehicle detection and classification as shown in Fig. 2 (b).

Using a little dataset for training a deep neural network (DNN) can easily lead to overfitting. In this situation, transfer learning can be an appropriate solution. Transfer learning is a technique to solve new problems faster and with better performance by applying previously learned knowledge to new datasets [5]. Therefore, transfer learning can achieve better performance with a relatively small dataset.

A. Training Process

In the proposed system, we use GoogLeNet consisted of nine inception modules as a deep neural network. As shown in Fig. 1 (a), the earlier layers of GoogLeNet pre-trained on ILSVRC-2012 are fixed and gradients are backpropagated only through the higher-level portion with vehicle dataset to produce the final vehicle classification model.

The training process in transfer learning is calculated as follows:

978-1-5386-7961-6/18 $31.00 © 2018 IEEE

TABLE I. ACCURACY OF VEHICLE CLASSIFICATION WITHOUT TRANSFER LEARNING AND WITH TRANSFER LEARNING

		The model of vehicle										
		Model A	Model B	Model C	Model D	Model E	Model F	Model G	Model H	Model I	Model J	Total
Without transfer learning		1.00	0.57	0.76	0.58	0.76	0.29	0.65	0.95	0.35	0.66	**0.657**
With transfer learning	Method 1	0.98	0.51	0.89	0.77	0.86	0.60	0.64	0.93	0.91	0.90	**0.799**
	Method 2	0.95	0.96	0.98	0.98	0.98	1.00	0.96	1.00	0.99	1.00	**0.980**
	Method 3	0.96	0.92	1.00	0.98	1.00	0.98	1.00	1.00	1.00	0.99	**0.983**
	Method 4	0.99	0.71	0.97	0.98	0.98	1.00	0.97	1.00	1.00	1.00	**0.960**
	Method 5	1.00	0.95	0.88	0.99	0.93	0.98	0.95	1.00	1.00	0.97	**0.965**

$$w_k^l = w_{k-1}^l + \alpha\, p_k^l$$
$$b_k^l = b_{k-1}^l + \alpha\, p_k^l \qquad (1)$$

where w_k^l and b_k^l denote the weight and the bias, α and p_k^l denote the learning rate and the search direction. w_k^l and b_k^l are updated in the direction of p_k^l. p_k^l is calculated as follows:

$$p_k^l = \begin{cases} \dfrac{\partial F}{\partial w_k^l} & (l \geq n) \\ 0 & otherwise \end{cases} \qquad (2)$$

where F denotes the loss function, l and n denote the order of the layer to be updated and the order of the earliest layer to be trained on the new dataset. If l is greater than or equal to n, the derivative of F is selected, otherwise 0 is selected as p_k^l.

B. Inference Process

As shown in Fig. 1 (b), the inference process is divided into vehicle detection and classification. In vehicle detection, the vehicle area is detected on the roadway video by Haar-like features. In vehicle classification, the transfer learning-based vehicle classification model distinguishes 10 vehicles among Hyundai Starex to Kia K5 represented by Model A to Model J.

III. EXPERIMENTAL RESULTS

Table 1 shows the results of comparing the accuracy of the vehicle classification with and without transfer learning. The proposed system was tested with five methods. In all methods, the earlier layers of GoogLeNet pre-trained on ILSVRC-2012 were fixed and not updated. In method 1, only the linear layer, the last convolutional layer, was trained on the vehicle dataset. In method 2, the last inception module as well as the linear layer was trained. Similarly, experiments were conducted by increasing the number of the inception modules to train one by one in method 3 to method 5.

Without transfer learning, the total accuracy was 0.657 which is the lowest accuracy among the results. It shows that proper learning has not been done due to the small dataset. All of the five methods of the proposed system have higher accuracy than that of the previous method. In particular, model 3 has the highest accuracy of 0.983, 0.326 higher than that of the conventional method. Comparing accuracy from method 2 to method 5, the total accuracy of them is higher than that of method 1 by 0.161 or more. Front layers of CNNs detect low-level features like edges, less dependent on the final application, while back layers detect more specific features of the original dataset. Thus, it can be seen the part of pre-trained GoogLeNet detecting the specific features of ILSVRC-2012 is not only the linear layer but also the inception modules.

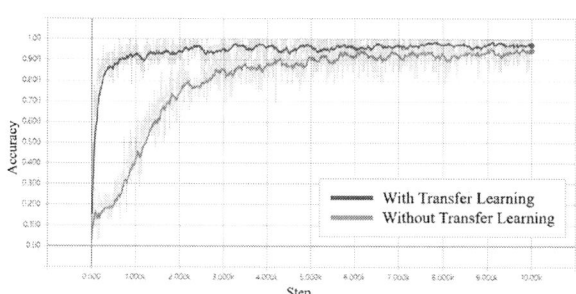

Figure 2. Graph for comparing the accuracy of vehicle classification with transfer learning and without transfer learning by step.

Fig. 2 shows the validation accuracy each step of the vehicle classification. It is shown that the proposed system has higher initial accuracy at the very beginning of learning, shorter amount of time it takes to fully learn, and higher final accuracy than that of the conventional method without transfer learning. This aspect can be seen as the reason transfer learning improves the performance of the vehicle classification.

IV. CONCLUSION

In this paper, we proposed a transfer learning-based vehicle classification. The proposed system showed higher performance with a limited scale dataset than conventional method. Additionally, we found the range of layers, not fixed but trained, affects the performance of transfer learning.

ACKNOWLEDGMENT

This work is supported by the Korea Institute of Energy Technology Evaluation and Planning (KETEP) and the Ministry of Trade, Industry & Energy (MOTIE) of the Republic of Korea (No. 20161210200560), and the MSIT(Ministry of Science and ICT), Korea, under the ITRC(Information Technology Research Center) support program(IITP-2018-0-01421) supervised by the IITP(Institute for Information & communications Technology Promotion.

REFERENCES

[1] Wen, Xuezhi, et al, "A Rapid Learning Algorithm for Vehicle Classification," Information Sciences, vol. 295, pp. 395-406, 2015.

[2] C. Cortes, V. Vapnik, "Support-vector Networks," Machine Learning, vol, 20, no. 3, pp. 273-297, 1995.

[3] A. Krizhevsky, I. Sutskever, and G. E. Hinton, "Imagenet Classification with Deep Convolutional Neural Networks," Advances in Neural Information Processing Systems, 2012.

[4] P. Viola, and M. Jones, "Rapid Object Detection Using a Boosted Cascade of Simple Features," IEEE Computer Society Conference on Computer Vision and Pattern Recognition, 2001.

[5] S. J. Pan, and Q. Yang, "A Survey on Transfer Learning," IEEE Transcation on Knowledge and Data Engineering, vol. 22, no. 10, 2010.

Fixed-Point Quantization of 3D Convolutional Neural Networks for Energy-Efficient Action Recognition

Hyunhoon Lee, Younghoon Byun, Seokha Hwang, Sunggu Lee, and Youngjoo Lee

Department of Electrical Engineering, POSTECH, Pohang, Korea,
youngjoo.lee@postech.ac.kr

Abstract— **In this paper, 3D convolutional neural networks (CNNs) are simplified to reduce the energy consumption of the action recognition process. Instead of using floating-point weights and input values, which results in a huge amount of processing energy, we introduce a systematic way to quantize all the values of 3D CNNs without degrading the recognition accuracy. Simulation results show that, compared to the baseline CNN architecture, the proposed method significantly reduces the computational complexity as well as the memory requirements.**

Keywords; deep learning; quantization; 3D convolutional neural network (3D CNN); action recognition;

I. INTRODUCTION

In order to recognize actions in video streams, 3D convolutional neural networks (CNNs) have been recently proposed due to their powerful recognition accuracy compared to the traditional 2D CNNs [1]–[4]. By expanding the direction of convolutions to the time domain, as shown in Fig. 1, 3D CNNs can find additional features related to the moving actions. However, in general, 3D CNNs require more computational complexity as well as larger numbers of filter parameters. As the algorithm-level studies have been focusing on increasing the recognition accuracy, moreover, the recent works tend to include more layers, consequently requiring more computing resources.

For example, as shown in Fig. 2, the depth of 3D ResNet has been continuously increased, and the pre-trained 3D ResNet-34 includes 33 3D convolutional layers to achieve the accuracy of 84.10% for UCF-101 datasets [6]. In terms of the required resources, however, the 3D ResNet-34 uses 12.5G multiply-accumulate (MAC) operations with 63M filters, which cannot

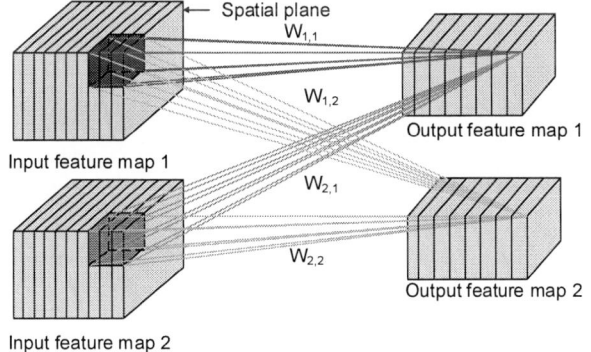

Fig. 1. Operation of 3D convolution filter.

This research was supported by the Ministry of Science and ICT, Korea, under the ITRC support program (IITP-2018-2016-0-00309), and by the Sports Promotion Fund of Seoul Olympic Sports Promotion Foundation from Ministry of Culture, Sports and Tourism (s072016022016).

ResNet-10

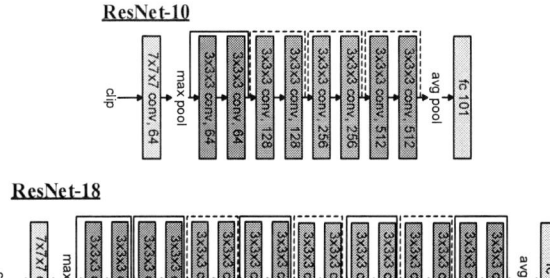

ResNet-18

ResNet-34

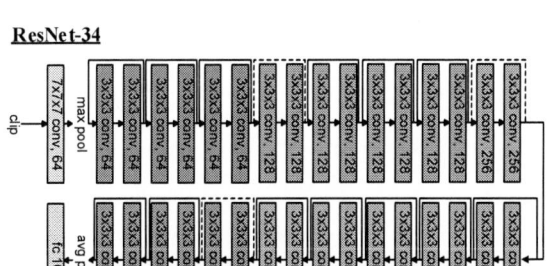

Fig. 2. Different 3D ResNet architectures.

be directly applied to the resource-limited embedded systems. As the action recognition will play a critical role in the next-generation mobile system, it is urgent to develop the hardware-friendly optimizations for reducing the complexity of 3D CNNs. For the entry-level optimization step, in this work, we find the optimal fixed-point quantization of 3D CNNs, reducing the required storage size as well as the recognition energy.

II. FIXED-POINT QUATIZATION OF 3D CNN

It is well known that fixed-point quantization is one of the most effective techniques for reducing the complexity of CNNs [5]. In order to minimize the accuracy degradation caused by the quantization effects, however, it is strongly required to develop a systematic way to select the optimal precision for the given networks. Targeting the three pre-trained 3D ResNets in Fig. 2, we first find the optimal precision of filter weights by changing their quantization levels as summarized in Table I. Compared to the baseline networks, which are based on the 32b floating-point

978-1-5386-7961-6/18 $31.00 © 2018 IEEE   129   ISOCC 2018

TABLE I.   ACCURACY CHANGES WITH QUANTIZED FILTERS

	ResNet-10	ResNet-18	ResNet-34
Baseline	76.95	82.42	84.11
12bit	76.90	82.21	83.93
11bit	76.87	82.45	84.22
10bit	76.90	**82.08**	84.19
9bit	77.03	81.07	**84.22**
8bit	**76.50**	76.82	75.63
7bit	68.23	4.23	1.35

TABLE II.   ACCURACY CHANGES WITH QUANTIZED FEATURES

	ResNet-10	ResNet-18	ResNet-34
Baseline*	76.50	82.08	84.22
18bit	76.26	82.05	84.32
17bit	76.32	81.89	84.48
16bit	76.32	81.87	84.30
15bit	**75.47**	**81.73**	**83.80**
14bit	74.78	79.46	82.05
13bit	71.42	74.81	76.90

* Baseline accuracies are results of the optimally-quantized filters.

numbers, we allow an accuracy drop of no more than 1%. Hence, we select 8b, 10b, and 9b for representing the filter weights of ResNet-10, ResNet-18, and ResNet-34, respectively. After fixing the fixed-point filter weights, we perform a similar process to find the optimal precisions for feature-map values as depicted in Table II. This process is simple but quite effective to reduce the overall complexity while guaranteeing the accuracy.

III.   EXPERIMENTAL RESULTS

For the quantitative analyses on the network complexity as well as the energy consumption, we design the major processing units of each 3D CNNs and synthesize them in a 65nm CMOS process. As shown in Table III, based on the optimization steps discussed in the previous section, we select the optimal fixed-point quantization of filter weights and feature maps, which can relax the complexity of the given network while providing the acceptable recognition accuracy. For each network, Fig. 3(a) shows the required memory size of each 3D CNN. Compared to the 32b floating-point number system, the optimal quantization accordingly reduces the memory size. For example, the required memory size for the ResNet-34 network is reduced by 71%, leading to the cost-effective action recognition system.

In terms of energy consumption, the effects of the proposed quantization are also attractive in practice as the previous 3D networks usually necessitates the power-starving floating-point processing units [3]. By introducing the energy-efficient fixed-point operators, in this work, the required energy per single action recognition is remarkably reduced as depicted in Fig. 3(b). For the ResNet-34 case, for example, the proposed work saves the processing energy by up to 69% with the 9b and 15b fixed-point representations for filter weights and the feature maps, respectively. Therefore, the proposed scheme effectively relaxes the hardware costs of 3D CNNs, allowing the energy-efficient embedded action recognition.

TABLE III.   OPTIMIAL QUANTIZATION OF EACH 3D CNN

	ResNet-10	ResNet-18	ResNet-34
Weight	8b (1, 7)	10b (1, 9)	9b (1, 8)
Feature map	15b (10, 5)	15b (9, 6)	15b (9, 6)

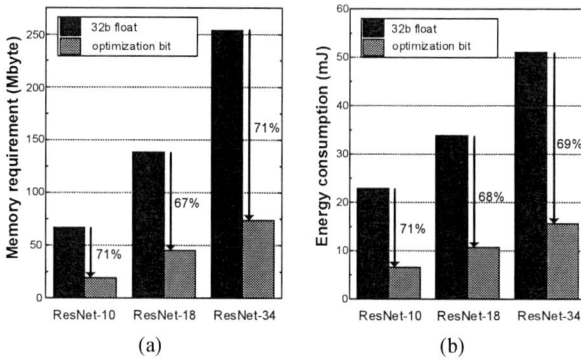

Fig. 3. (a) The memory size for storing the data of 3D CNNs, and (b) the required energy for each action recognition.

IV.   CONCLUSION

In this paper, we have presented a systematic quantization scheme for achieving energy-efficient 3D CNN processing for embedded system action recognition. By selecting the optimal fixed-point precisions for filter weights as well as feature maps, which would be the first step for simplifying the given network, the 3D CNN adopting the proposed fixed-point quantization consumes significantly reduced energy without degrading the recognition accuracy. Several case studies on 3D ResNets show that the proposed quantization method can save the recognition energy by almost 70% compared to the straightforward 3D CNNs with a 32b floating-point number system.

REFERENCES

[1] S. Ji, W. Xu, M. Yang, and K. Yu. "3D convolutional neural networks for human action recognition," *IEEE Trans. on Pattern Anal. and Mach. Intell.*, vol. 35, no.1, pp. 221–231, Jan. 2013.

[2] D. Tran, L. Bourdev, R. Fergus, L. Torresani, and M. Paluri. "Learning spatiotemporal features with 3D convolutional networks," in *Proc. IEEE Int. Conf. Comput. Vis. (ICCV)*, 2015, pp. 4489–4497

[3] K. Hara, H. Kataoka, and Y. Satoh, "Can spatiotemporal 3D CNNs retrace the history of 2D CNNs and ImageNet?," in *Proc. IEEE Conf. Comput. Vis. Pattern Recognit. (CVPR)*, 2018, pp. 6546–6555

[4] G. Varol, I. Laptev, and C. Schmid, "Long-term Temporal Convolutions for Action Recognition," " *IEEE Trans. on Pattern Anal. and Mach. Intell.*, vol. 40, no.6, pp. 1510-1517, jun. 2017.

[5] S. Han, H. Mao, and W. J. Dally, "Deep compression: Compressing deep neural network with pruning, trained quantization and huffman coding," in *Proc. Int. Conf. Learning Repr. (ICLR)*, 2016.

[6] K. Soomro, A. R. Zamir, and M. Shah. "UCF101: A dataset of 101 human actions calsses from videos in the wild." Technical Report CRCV-TR-12-01, UCF Center for Research in Computer Vision, 2012.

Implementation of 3D Hand Gesture Recognition System using FPGA

Tsung-Han Tsai
Dept. of Electrical Engineering
National Central University
Taoyuan City, Taiwan

Yuan-Chen Ho
Dept. of Electrical Engineering
National Central University
Taoyuan City, Taiwan

Yih-Ru Tsai
Dept. of Electrical Engineering
National Central University
Taoyuan City, Taiwan

Abstract— **In this paper, a VLSI design with dual-camera to construct the depth map and recognize hand gestures was proposed. The proposed system adopts adaptive depth filter which can separate foreground to segment the interesting object under complicated environment. We also proposed dynamic gesture recognition by using depth and coordinate information. The system contains one static gesture and two dynamic gestures and has been implemented in FPGA and the average accuracy with each gesture is 83.98%.**

Keywords: VLSI, FPGA, Hand gesture recognition, SAD matching, Object labeling

I. INTRODUCTION

As the technology advances, more and more researches focus on gesture recognition. It helps us contact to people with disabilities by using sign language finger alphabet easily and control some robots and appliances without controller. The old control devices like remote controller, keyboard and mouse have been replaced to remote control by some hand gestures. In Human computer interface (HCI), gesture recognition is an important topic which needs to have breakthrough [1]. Traditional gesture recognition devices are not convenient and will be constrained by environment for example, data gloves [2]. In order to let users feel unrestrained and comfortable during using the product, a hardware architecture system is developed with dual-camera to deal with it.

The algorithm of dual-camera system usually require high computation time especially implemented in embedded system. The algorithm is not easily run in real-time computing when using CPU in embedded system to develop a real product. In this paper, a hardware architecture system is proposed to speed up the overall run time with high efficiency. It can recognize one static gesture as "fist", two dynamic gestures as "up, down, left, right" and 3D gesture as "push" respectively. The two dynamic gestures can interact with each other, so the system can recognize gestures like "up push", "down push", etc. On the other hand, the design is implemented in SMIMS development board with Xilinx Artix-7 to demonstrate the HCI system.

II. THE PROPOSED SYSTEM

There are several modules in the system to implement the overall hand gesture recognition system. The design flow and hardware block diagram is illustrated in Fig.1 and Fig.2

respectively. The input of the our system is YUV data which has been transform from RGB by Left Gray and Right Gray module. There are three kinds of memory used in the system. Two 3840x40 SRAM are used to save the Y value of left and right image pixels. Two 960x40 SRAM are used to save the U and V data of the right image. Two 256x16 register-file memory are used to save temporal data. All the module share these six memory to save the different results.

The proposed system adopts dual-camera with stereo matching algorithm to construct the depth map [3]. In order to keep accuracy of depth map, we use histogram equalization to reduce the light effect in different cameras, and fit epipolar geometry with image rectification. After acquiring the depth information, skin detection is adopted to remove the non-skin-like color by YUV data. With the interaction of skin information and depth map, the system can segment hand easily, because of the hand in camera is always exist nearer than the face. Next, we use object labeling to remove the little noise and get the coordinate and the centroid of hand. Finally, our system utilizes the information of hand and object labeling modules to develop hand gesture recognition applications, which can recognize gestures of fist or palm, push or pull and the 4-directions movement.

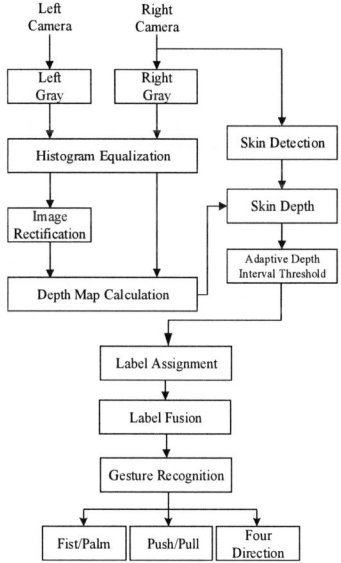

Fig. 1. Design flow of the overall system

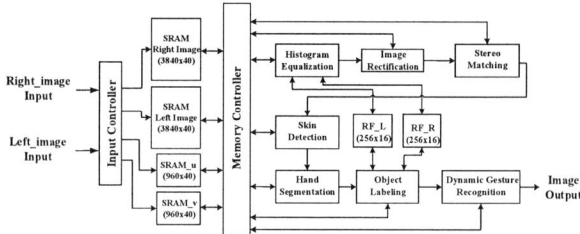

Fig. 2. Block diagram of proposed system

III. EXPERIMENTAL RESULTS

To verify the overall system, we implement the hardware architecture to SMIMS VEXA7-200 development board. The development board has two FMC daughter boards to control the input and output information and a FPGA with Xilinx Artix-7. The experiment results are illustrated in Fig. 3 by PC software simulation. Fig. 3(a) and (b) show the detection of continuous hand gestures to recognize behavior which can be applied for the movement of virtual mouse. Also this technique can be implemented on car's center console to control by hand gestures rather than tangible touch by hand. Consequently it can decline the risk of traffic accidents, such as making some adjustment to volume and air-condition temperature, accepting or rejecting a phone call etc. Furthermore, our dual-camera system can acquire the depth map information, thus we can provide some requested by movement actions in depth-axis direction. This cannot be realized in traditional single-camera system. The result is illustrated in Fig 3(c) and (d).

The real experiment environment is constructed in Fig. 4. The resource usage of FPGA is listed in Table I. The accuracy of the proposed system is shown in Table II. It shows that the average accuracy of the proposed system is 83.98%.

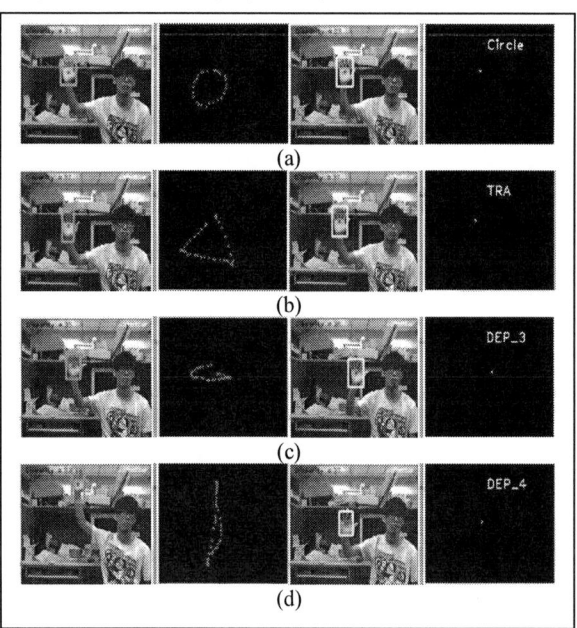

Fig. 3. Result of behavior recognition: (a) circle, (b) triangle, (c) horizontal deep circle and (d) vertical deep circle

Fig. 4. FPGA experiment environment.

TABLE I.   The resource usage of FPGA

Element	Total available	Used	Utilization
Slice Registers	29200	6051	2%
Slice LUTs	134600	8204	6%
Used as logic	134600	7969	5%
Bonded IOBs	400	178	44%

TABLE II.   The accuracy of proposed system

	Item			
	Fist	*Push*	*Four Direction*	*Push + Four Direction*
Accuracy	91.2%	84.8%	81.3%	78.6%

IV. CONCLUSIONS

In this paper, a hardware architecture system with dual-camera is proposed to construct the depth map and recognize hand gestures to reach the goal of 3D hand gesture recognition. Preprocessing module is chosen to enhance the accuracy of stereo matching. SAD algorithm is adopted to match the pixel and construct the depth map. Adaptive hand gesture threshold removes the most noise. Then object labeling gets the information to remove the rest noise and do gesture recognition. This whole system is implemented in SMIMS development board to demonstrate the HCI system and verify to overall architecture. The dynamic gestures can interact with each other to gain more applications. The average accuracy of all gestures is 83.98%. With the aid for the proposed system, it can recognize gestures even under the complicated environment and applied in home appliances to control them.

REFERENCES

[1] S. Mitra and T. Acharya, "Gesture recognition: A survey," IEEE Transactions on Systems, Man, and Cybernetics, Part C (Applications and Reviews), vol. 37, no. 3, pp. 311–324, Apr. 2007.

[2] P. Kumar, S. S. Rautaray, and A. Agrawal, "Hand data glove: A new generation real-time mouse for human-computer interaction," International Conference on Recent Advances in Information Technology (RAIT), pp. 750-755, May 2012.

[3] P. Stefania, C. Daniela, Z. Paolo, C. Pasquale, "SAD-Based Stereo Matching Circuit for FPGAs", IEEE International Conference on Electronics, Circuits and Systems, pp. 846–849, Dec. 2006. 2.

Multi-Mode LSTM Network for Energy-Efficient Speech Recognition

Junseo Jo, Seokha Hwang, Sunggu Lee, and Youngjoo Lee

Department of Electrical Engineering, POSTECH, Pohang, Korea,
youngjoo.lee@postech.ac.kr

Abstract— We newly introduce a novel processing scenario of long short-term memory (LSTM) network for the energy-efficient speech recognition. Compared to the conventional single-mode processing based on the fixed computing scheme, the proposed LSTM processing contains multiple operating cells providing attractive tradeoff between the recognition accuracy and the energy consumption. For the case study, the state-of-the-art LSTM network is modified to have two types of processing cells, strong and weak cells, which are dedicated to the accuracy-aware and energy-aware LSTM sequences, respectively. By allocating as many weak cells with low energy as possible, experimental results show that the proposed work saves the energy consumption for speech recognition by 75% compared to the original network.

Keywords; LSTM; Low-power architecture; Speech recognition;

I. INTRODUCTION

Recently, the LSTM based recurrent neural network (RNN), has received great interests due to its excellent performance in speech recognition systems by preserving temporal memories for long periods using forget, input and output gates [1]–[3]. Applying the bidirectional LSTM networks associated with the mel-frequency centrum coefficients (MFCC), DeepSpeech in [4] is one of the most accurate recognition systems, achieving a character-level error rate of less than 10%. After calculating the MFCC values, as depicted in Fig. 1, three fully-connected (FC) layers are firstly applied to find different features as many as possible, and then two LSTM networks having the different directions followed by additional FC layers are operated to extract the received character. Based on the word-level libraries, finally, the connectionist temporal classification (CTC) process eliminates redundant characters, making the output sentence.

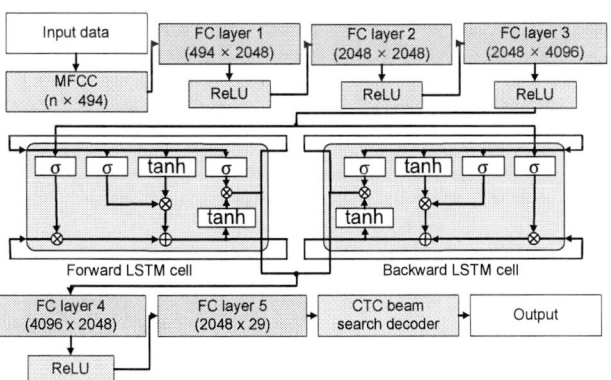

Fig. 1. DeepSpeech architecture

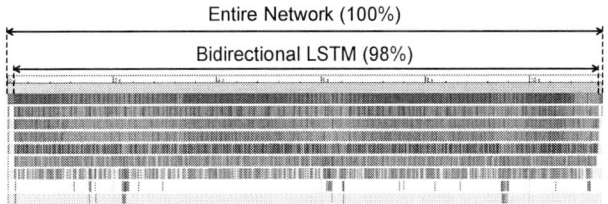

Fig. 2. Timeline of DeepSpeech processing

Although DeepSpeech network shows an attractive accuracy, however, it requires a huge amount of computational complexity caused by two n-length LSTM processing sequences, where n is determined by the number of windows issued to the MFCC calculator. Note that the value of n varies according to the length of the captured speech, and typically ranges from 500 to 1000 [5]. Therefore, the bidirectional LSTM in DeepSpeech requires more than 1000 same LSTM operations, each of which consists of numerous matrix multiplications as well as several look-up tables (LUTs) for nonlinear functions. As shown in Fig. 2, therefore, the bidirectional LSTM covers more than 98% of total processing costs, leading to the energy-starving recognition system. Therefore, it is urgent to develop the low-power LSTM processing scenario with the acceptable accuracy.

II. MULTI-MODE LSTM PROCESSING

In the bidirectional LSTM sequences of DeepSpeech, as shown in Fig. 3(a), total n cells are repeatedly processed in both directions, consuming a huge amount of energy. Even though a single LSTM cell can be simplified with any technique [6], in general, all the LSTM sequences in the improved architecture are still based on the identical cells. Hence, in this work, the previous architecture having the same simplified LSTM cell is referred to as the single-mode LSTM architecture.

To further reduce the energy consumption of bidirectional LSTM, on the other hand, we newly introduce the multi-mode LSTM processing. We first define several LSTM cells having the different processing accuracies on their operations, i.e., the multiple modes. It is natural that we consume more processing energy for the stronger cells associated with the more precise computations. Therefore, we may place weaker LSTM cells to non-critical positions for removing the excessive computations. This energy-accuracy tradeoffs can be achieved by adjusting various factors including quantization level, filter pruning, data compression, and even approximate computation.

This work was supported by Samsung Research Funding & Incubation Center of Samsung Electronics under Project Number SRFC-TB1703-07.

978-1-5386-7961-6/18 $31.00 © 2018 IEEE 133 ISOCC 2018

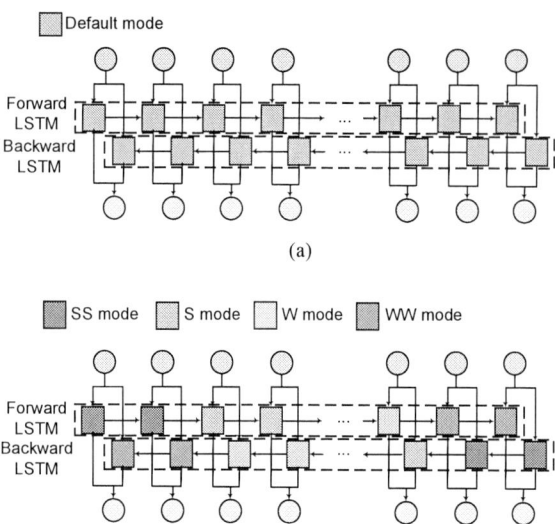

Fig. 3. The conceptual diagram of (a) the conventional single-mode LSTM processing and (b) the proposed multi-mode operation.

After designing the different LSTM cells, we then arrange the new energy-efficient processing scenario based on the important observation that that weak LSTM cells suffers from the erroneous results, and the errors tend to be accumulated at the subsequent cells. Therefore, it is necessary to process the earlier LSTM cells with stronger modes. On the other hand, the later LSTM steps can tolerate more aggressive approximations, allowing the usage of weak modes. Starting from the first LSTM cell of each direction, therefore, the proposed energy-efficient DeepSpeech network gradually changes its processing mode from the strongest to the weakest, minimizing the overall energy consumption without degrading the recognition accuracy.

III. CASE STUDY: DUAL-MODE QUANTIZATION

To show the effects of the multi-mode LSTM processing, as a case study, we divide the n-length bidirectional LSTM network into two regions, i.e., the accuracy-ware and the energy-aware regions. Then two LSTM cells, denoted as strong and weak cells, are defined to have the different quantization levels by using 19 and 7 bits, respectively. According to the numerous simulations, the ratio of two regions is carefully determined to 1:9, which means that the accuracy-aware region only covers the first 10% of LSTM steps to minimize the energy consumption under the 1% accuracy drop as summarized in Table I. Note that the straightforward quantization based on the single-mode 19-bit operations results the similar level of recognition accuracy, but it obviously has the limitation on the amount of energy reduction. For the quantitative comparison, we also design the multi-mode LSTM operator in a 65nm CMOS process. By allocating the dual-mode operations dynamically, as shown in Fig. 4, the proposed scheme reduces the energy consumption by 75% and 52% compared to the baseline network using 32-bit floating-point numbers and the straightforward single-mode architecture with the 19-bit quantized values, respectively.

TABLE I. PERFORMANCE OF LSTM NETWORKS

LSTM architecture	Word error rate (%)
Baseline (32-bit floating-point)	8.4733
Single-mode (19-bit fixed-point)	9.3589
Dual-mode (19-bit / 7-bit fixed-point)	9.4409

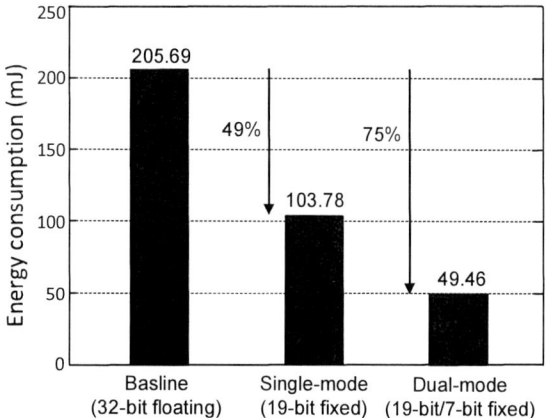

Fig. 4. The energy reduction from the proposed dual-mode operation.

IV. CONCLUSION

In this paper, we have introduced the multi-mode LSTM network for the energy-efficient speech recognition. Allocating the different processing modes properly, the proposed scheme provides more attractive operating point reducing the processing energy of speech recognition. Compared to the straightforward single-mode approach, the case study based on the dual-mode quantization shows that our architecture saves more than 50% of recognition energy while maintaining the recognition accuracy.

REFERENCES

[1] S. Hochreiter and J. Schmidhuber, "Long short-term memory," *Neural Comput.*, vol. 9, no. 8, pp. 1735–1780, 1997.

[2] H. Sak, A. Senior, and F. Beaufays, "Long short-term memory based recurrent neural network architectures for large vocabulary speech recognition," arXiv prepring arXiv:1402.11281, 2014.

[3] T. N. Sainath, O. Vinyals, A. Senior, and H. Sak, "Convolutional, Long Short-Term Memory, Fully Connected Deep Neural Networks," in *Proc. IEEE Int. Conf. Acoust., Speech, Signal Process. (ICASSP)*, 2015, pp. 4580–4584.

[4] A. Hannun *et al.*, "Deepspeech: Scaling up end-to-end speech recognition," arXiv preprint arXiv:1412.5567, 2014.

[5] A. Graves, A. Mohamed, and G. Hinton, "Speech recognition with deep recurrent neural networks," in *Proc. IEEE Int. Conf, Acoust., Speech, Signal Process. (ICASSP)*, 2013, pp. 6645-6649.

[6] S. Han *et al.*, "ESE: Efficient speech recognition engine with sparse LSTM on FPGA," in *Proc. ACM/SIGDA Int. Symp. Field-Programmable Gate Arrays*, 2017, pp. 75–84.

A Wideband Differential VCO Based on Multiple-path Loop Architecture

Tomotaka Tanaka, Fumiya Naito,
Makoto Nakamura and Daisuke Ito
Gifu University
1-1 Gifu City, Gifu, 501-1193, Japan
Email: t_tanaka@ieee.org

Keiji Kishine
University of Shiga Prefecture
2500, Hikone Hassaka, Shiga, 522-8533, Japan

Abstract—A new delay cell design for VCOs based on a ring oscillator with a multiple-path loop architecture is presented. The proposed delay cell uses the Gilbert cell as an interpolating inverter to widen the oscillation frequency tuning range. We designed and fabricated a four-stage multiple-path ring VCO that applies the proposed delay cells in 0.18-μm CMOS technology. It has about twice the tuning range of the conventional one and can oscillate over the wide tuning range of 0.82–1.72 GHz.

I. Introduction

A phase locked loop (PLL) is an essential component that provides a timing clock for high-speed systems. In the PLL, a voltage controlled oscillator (VCO) which directly provides the output of the PLL is a principal building block. VCOs can be mainly composed of a ring oscillator or LC oscillator. Compared to LC VCOs, ring VCOs tend to have poor phase noise and slow-speed operation however they usually have a wider tuning range and a small die area [1], [2]. Sun et al. have proposed the ring VCO using a multiple-path loop architecture (hereinafter called "multiple-path ring VCO") that can achieve high-speed operation and a wide tuning range, as shown in Fig. 1 [3]. The wide tuning range makes it possible to increase the margin for frequency variations caused by PVT (process, voltage, and temperature) variations [4] and to apply the VCO to a multirate communication system.

In this paper, we propose a new delay cell design for a multiple-path ring VCO to achieve a wider tuning range. By applying the Gilbert cell to the delay cell, a control current range is expanded, so that the tuning range can be widened. The simulated result shows that the tuning range of the proposed VCO is about twice as wide as the conventional one. In addition, we fabricated the proposed VCO with the new delay cells using 0.18-μm CMOS technology. From experimental results, the IC has the wide tuning range of 0.82–1.72 GHz.

II. Delay Cell Design for a Multiple-path Ring VCO

A. Topology of a multiple-path ring VCO

The fundamental topology of a multiple-path ring VCO is composed of N major loop inverters A_k and N interpolating inverters B_k, as shown in Fig. 1. The interpolating inverters form additional signal paths to introduce sub-feedback (or sub-feedforward) loops into the major loop. Since the phase relationship between each node (X_{k-1}, X_k) remains constant due to the symmetrical structure, the sub-feedback loops can

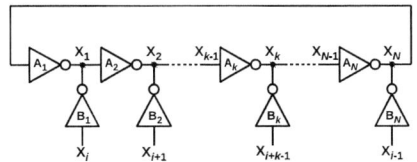

Fig. 1. Fundamental topology of a multiple-path ring VCO.

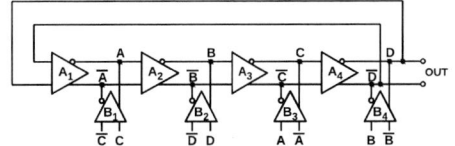

Fig. 2. A differential multiple-path ring VCO (when $N = 4$, $i = 3$).

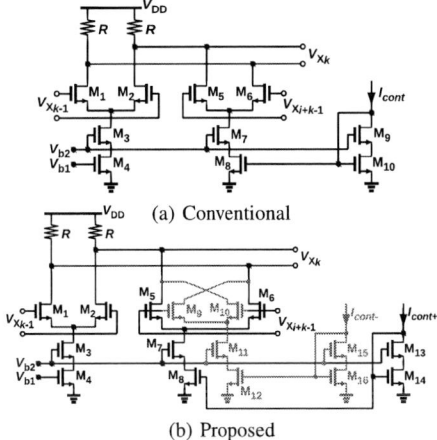

(a) Conventional

(b) Proposed

Fig. 3. Delay cell for a multiple-path ring VCO.

reduce the tap-to-tap delay. Thus, the multiple-path ring VCO can operate at higher frequencies than typical ring oscillators.

The minimum required gain of each major loop inverter $g_m R$ and the oscillation frequency ω_{osc} of the N-stage multiple-path ring VCO have been derived in [3] as follows:

$$g_m R = (g_m R)'(1 + K_a g_{m,\mathrm{sub}} R), \tag{1}$$

$$\omega_{\mathrm{osc}} = \omega'_{\mathrm{osc}}(1 + K_f g_{m,\mathrm{sub}} R), \tag{2}$$

where $K_a = \cos(\phi(i-1))$, $K_f = K_a + \sin(\phi(i-1))/\tan\phi$. Here, g_m and $g_{m,\mathrm{sub}}$ are the transconductances of the major and interpolating inverters, R is the loading resistance on each node, and $(g_m R)'$ and ω'_{osc} are the minimum required gain and the oscillation frequency for a N-stage typical ring oscillator, respectively. ϕ is the amount of phase shift between each node ($\phi := \pi/N + \pi$), and i is a feedback index number that

978-1-5386-7961-6/18 $31.00 © 2018 IEEE

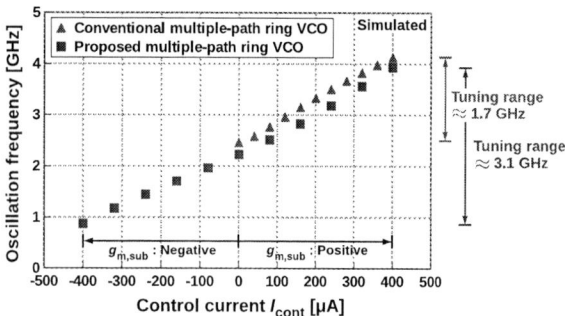

Fig. 4. Simulated frequency tuning characteristics.

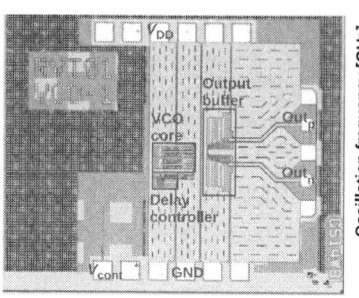

(a) Chip photograph

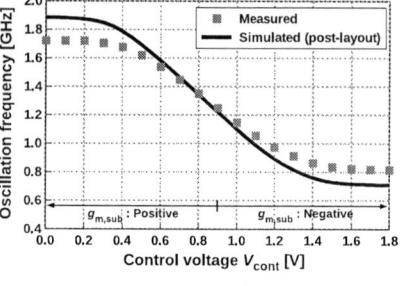

(b) Frequency tuning characteristics

Fig. 5. Fabricated and experimental results of proposed VCO.

represents the number of inverters in each sub-feedback loop. Let us consider the case of four-stage and three inverters in each sub-feedback loop ($N = 4$ and $i = 3$) as shown in Fig. 2. Note that $K_a = 0$ and $K_f = 1$. In this case, the minimum required gain and the oscillation frequency are expressed as

$$g_m R = (g_m R)', \qquad (3)$$
$$\omega_{osc} = \omega'_{osc}(1 + g_{m,sub}R). \qquad (4)$$

From Eqs. (3) and (4), the multiple-path ring VCO can operate at higher frequencies than the typical ring oscillator with the same gain condition.

B. Conventional delay cell for a multiple-path ring VCO

As shown in Fig. 3(a), the conventional delay cell for a multiple-path ring VCO based on differential CML inverters is composed of two differential pairs (M_1–M_4 and M_5–M_8) corresponding to a major loop inverter A_k and an interpolating inverter B_k, respectively. From Eq. (2), the oscillation frequency of the multiple-path ring VCO linearly depends on the transconductance $g_{m,sub}$. This $g_{m,sub}$ can be controlled by the bias current, which is a replica of the control current I_{cont}. In the conventional VCO, the frequency control was performed by changing $g_{m,sub}$ in a monotonic direction. For example, as I_{cont} increases, $g_{m,sub}$ also increases, thereby (from Eq. (2)) raising the oscillation frequency.

C. Proposed delay cell for a multiple-path ring VCO

We proposed the delay cell using the Gilbert cell architecture to achieve a wide tuning range as shown in Fig. 3. In this design, the interpolating inverter has the Gilbert cell architecture (M_5–M_{12}) composed of two differential pairs whose outputs are connected with opposite phases. The bias currents in the interpolating inverter are replicas of I_{cont+} and I_{cont-}, and the transconductance $g_{m,sub}$ can be controlled by the differential control current $I_{cont} = I_{cont+} - I_{cont-}$. When I_{cont} is positive ($I_{cont+} > I_{cont-}$), $g_{m,sub}$ is a positive value. As I_{cont} increases, $g_{m,sub}$ becomes more positive, raising the oscillation frequency in the same manner as the conventional operation. Conversely, when I_{cont} is negative ($I_{cont+} < I_{cont-}$), $g_{m,sub}$ is a negative value and, as I_{cont} decreases, $g_{m,sub}$ becomes more negative, lowering the oscillation frequency. Therefore, $g_{m,sub}$ is extended to the negative value region by applying the Gilbert cell architecture, leading to about twice the wider tuning range than the conventional delay cell. Figure 4 shows the simulated

tuning characteristics for the designed VCOs with 0.18-μm CMOS technology. While the conventional VCO has a tuning range of 1.7 GHz, the proposed VCO has a tuning range of 3.1 GHz. These results mean that the proposed VCO can obtain about twice the tuning range of the conventional one.

III. EXPERIMENTAL RESULTS

The proposed multiple-path ring VCO was fabricated in 0.18-μm CMOS technology. Figure 5(a) shows the chip photograph of the proposed VCO, and the VCO core size is 0.1×0.13 mm^2. The measured frequency tuning characteristic is shown in Fig. 5(b). A center frequency is 1.25 GHz at the supply voltage $V_{DD} = 1.8$ V, and the wide operating frequency range of 0.82–1.72 GHz is confirmed. The linear operating range of the control voltage accords with the post-layout simulation. This result indicates that using the proposed delay cell enables to extend the controllable range of the transconductance of the interpolating inverter $g_{m,sub}$.

IV. CONCLUSION

We have proposed a new delay cell design for a multiple-path ring VCO. The proposed delay cell is based on a differential CML inverter and applies the Gilbert cell as an interpolating inverter. Utilizing the proposed delay cell, the VCO can oscillate with the tuning range about double wider than the conventional one. The proposed VCO was fabricated using 0.18-μm CMOS technology, and a wide tuning range of 0.82–1.72 GHz was achieved.

ACKNOWLEDGMENTS

Part of this research was supported by JSPS KAKENHI Grant Number JP17K06381. This work was supported by VDEC, the University of Tokyo in collaboration with Rohm Corp. and Toppan Printing Corp.

REFERENCES

[1] Son Sanquan, Kim Byungsub and Xiong Wei, "The oscillation frequency of CML-based multipath ring oscillators," J. Semiconductor Technology and Science, vol. 15, no. 6, pp. 671–677, Dec. 2015.

[2] Nanhe Lal, Ashish Raman and Balraj K., "Wide Tuning Range CMOS VCO for Radio Frequency Application," International J. Computer Applications, vol. 67, no. 13, pp. 8–13, Apr. 2013.

[3] Lizhong Sun and T. A. Kwasniewski, "A 1.25-GHz 0.35-μm monolithic CMOS PLL based on a multiphase ring oscillator," IEEE J. Solid-State Circuits, vol. 36, no. 6, pp. 910–916, Jun. 2001.

[4] J. W. Moon, K. C. Choi and W. Y. Choi, "A 0.4-V, 90 $\sim$ 350-MHz PLL With an Active Loop-Filter Charge Pump," IEEE Transactions on Circuits and Systems II, vol. 61, no. 5, pp. 319–323, May. 2014.

Gap in pagination due to formatting issues.

Pages 137-139

Digital PHY Design Methodologies for High-Speed and Low-Power Memory Interface

Kwanyeob Chae, Billy Koo, Jihun Oh, Sanghune Park, Jongshin Shin, and Jaehong Park
Foundry/ASIC&IP
Samsung Electronics Co. Ltd.
Hwaseong-City, Korea
corean@samsung.com

Abstract—**This paper presents practical high-speed and low-power design methodologies for digital PHY in deep sub-micron technologies. The standard-cell-based design approaches with automated place and route shorten the design time dramatically. In addition, robust digital design flow can be applied for wide range of operation considering model-hardware-correlation in deep sub-micron technologies. Eventually, all-digital PHY improves power efficiency of the system with wide-voltage-range DVFS. Simplified architecture with calibration logic helps improve logic speed with minimized area and power. The designed PHY with proposed design methodologies shows 1.6Gbps at 520mV and 6.6Gbps at 780mV, which allows extreme power efficiency and performance. In addition, the wide range of voltage scaling is allowed depending on the target frequency.**

Keywords; all-digital PHY; MHC; DVFS; calibration

I. INTRODUCTION

Modern mobile processors demand high bandwidth for higher performance with stringent power budget [1]. As a result, improving power efficiency of PHY operating in high frequency is quite challenging [2]. In this requirement, there are two major technical challenges in memory interface circuits. First, wide-range operation is mandatory for maximized power efficiency [3]. In recent application processors, the battery life has become the most important factor. To extend the battery life, the power efficiency of the system should improve. As a result, dynamic-voltage-frequency-scaling (DVFS) like in Fig. 1 for maximized power efficiency of the system is mandatory technique in mobile system. To support this DVFS feature, the PHY should operate under wide voltage range. If the PHY support voltage scalability, it helps maximize power efficiency of the memory system. When the system is running at the minimum operating frequency, the supply voltage can be lowed to minimize power consumption.

To achieve the wide operating range of the PHY, standard-cell-based all-digital PHY is developed to support the above-mentioned DVFS feature. As a result, it is implemented with digital sign-off methodologies considering model-hardware-correlation (MHC). If the process variation associated with MHC is not considered, it is hard to achieve the minimum operating voltage due to unexpected delay variation. In recent technologies, there is a gap between the model and the hardware due to the complex local layout effects. Thus, it is important to consider MHC in PHY design stage.

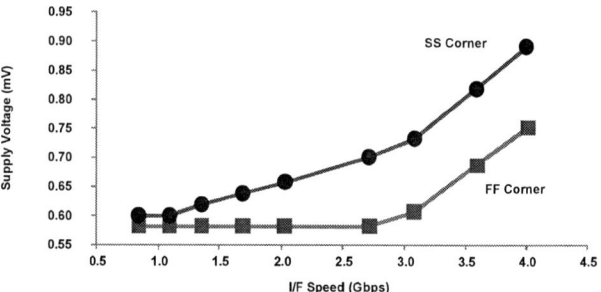

Fig. 1. Wide operating range for high speed I/F.

This work also considers minimizing design overhead by utilizing calibration circuit. The digital calibration logic helps improve valid-window-margin (VWM) with minimized hardware complexity and overhead.

II. ALL-DIGITAL PHY FOR WIDE-VOLTAGE-RANGE

A. Standard-Cell-Based Design

All-digital PHY shown in Fig. 2 is proposed to implement the PHY with conventional digital design flow. It indicates that auto layout tool can be used in implementing the proposed PHY. In addition, digital sign-off flow can be applied and various verification methods such as coupling noise, timing considering process variation, wide voltage range sign-off, IR-drop, and jitter analysis, can be utilized. Thus, the confidence level of the design even under new process technologies can be increased, and the probability of failure can be reduced dramatically. To enable all digital design of the PHY, all digital delay-locked-loop (DLL), which can be implemented with standard cells [4] like in Fig. 3, is used for DQ-DQS centering under process-voltage-temperature (PVT) variations.

B. MHC-Aware Design

To achieve low voltage operation in digital design, MHC should be considered during implementation process. Robust cells, which are tolerant to PVT variations, should be selectively used. In this work, delay sensitivities to voltage variation of the standard cells are analyzed like in Fig. 4, and sensitive cells to voltage variations are excluded in the design. In addition, the adaptive design margining is used for low voltage sign-off. At under-drive (UD) or super-under-drive (SUD) voltage conditions, device models have reduced

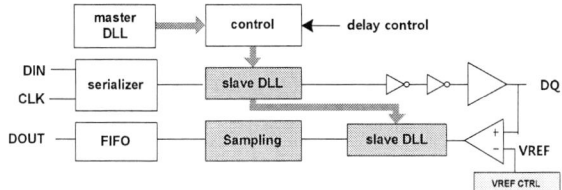

Fig. 2. Digital PHY architecture.

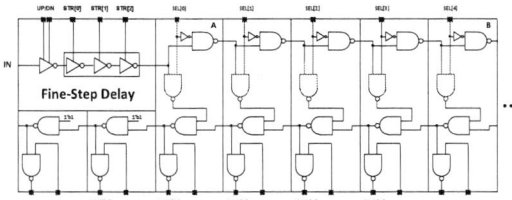

Fig. 3. Digital delay line.

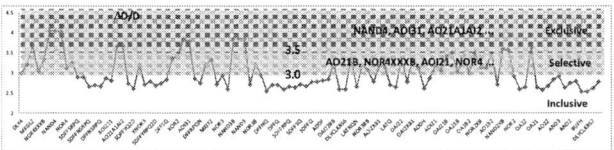

Fig. 4. Delay-sensitivity to voltage variation.

correlation with the real hardware compared with nominal-drive (ND). Thus, additional design margin is included considering MHC at low voltage.

C. Calibration Logic for Minimized Design Overhead

To achieve higher speed, clock skew and the duty performance is very critical. As a result, the critical paths tend to be over-designed considering unknown factors such as external board and interconnection environments. If skew calibration utilizing training sequence is allowed, the H-tree clock structure shown in Fig. 5 can be replaced to the chain structure. The fly-by architecture can minimize the number of clock cells, and thus, the power and the area can be saved [2]. Even if, the H-tree shows the good performance in the clock

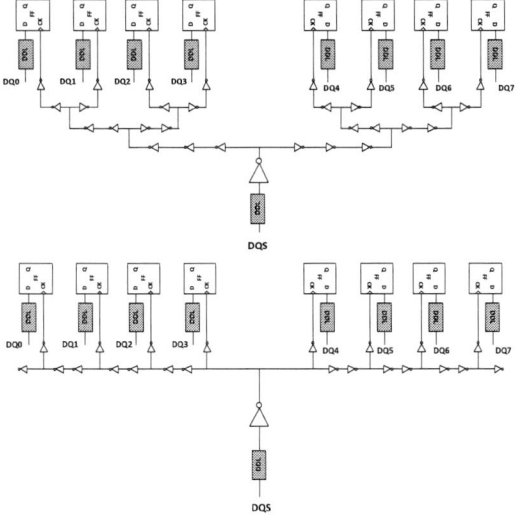

Fig. 5. Chain structure for power optimization instead of H-tree structure.

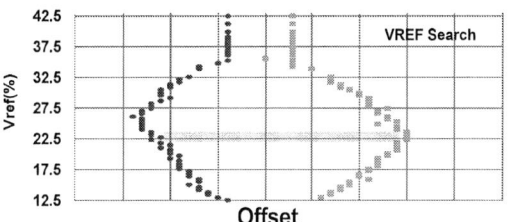

Fig. 6. Measured VREF search results.

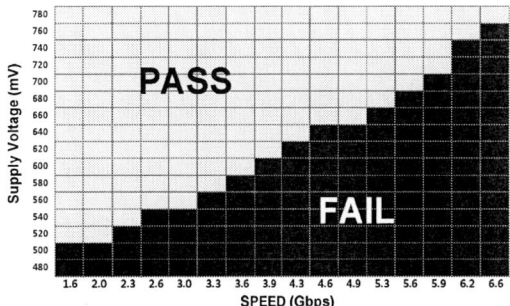

Fig. 7. Measured operating range.

skew management, the design burden is quite excessive. Since the calibration logic operates at the initial boot-up, the power overhead coming from the calibration logic can be neglected. In addition, the optimal reference voltage search with VREF training shown in Fig. 6 for maximizing duty performance can reduce the I/O design overhead [5].

III. MEASUREMENT RESULTS

The digital PHY was implemented in Samsung 10nm technology. The combined design methodologies shown in this paper contributed to low voltage operation down to 520mV at 1.6Gbps operation and high speed up to 6.6Gbps at 780mV. The measured VWM with VREF training is shown in Fig. 6. The shmoo plot of the designed PHY is shown in Fig. 7.

IV. CONCLUSION

The simplified all-digital architecture with calibration logic helps improve logic speed with minimized area and power. The all-digital PHY implemented with the proposed digital design flow considering MHC showed wide operating range considering DVFS. In addition, low voltage operation even at high speed was achieved.

REFERENCES

[1] T.Y. Oh et al., "A 3.2Gb/s/pin 1.0V LPDDR4 SDRAM with Integrated ECC Engine for Sub-1V DRAM Core Operation," IEEE ISSCC, pp. 430-431, 2014.

[2] K.Y. Chae et al., " A 690mV 4.4 Gbps/pin all-digital LPDDR4 PHY in 10nm FinFET technology," IEEE ESSCIRC, pp. 461-464, 2016.

[3] K.Y. Chae et al., " Resilient pipeline under supply noise with programmable time borrowing and delayed clock gating," IEEE TCAS II, pp. 173-177, 2014.

[4] K.Y. Chae, "Delay Cells and Delay Line Circuits Having The Same,"U.S. Patent 7,486,125.

[5] S.M. Lee, et al., "A 0.6V 4.266Gb/s/pin LPDDR4X Interface with Auto-DQS Cleaning and Write-VWM Training for Memory Controller," ISSCC, pp.398-399, 2017.

978-1-5386-7961-6/18 $31.00 © 2018 IEEE

A Two-Step Time-to-Digital Converter using Ring Oscillator Time Amplifier

Min Kim, Kyung-Sub Son, Namhoon Kim, Chang Hang Rho, and Jin-Ku Kang

Dept. of Electronics Engineering, Inha University
100, Inha-ro, Nam-Gu, Incheon, 402-751, Korea
jkang@inha.ac.kr

Abstract— A two-step time-to-digital converter using a ring oscillator time amplifier is presented. The time amplifier structure does not accumulates the error in the iterative process of time. There are 8 bits in total, of which 4 bits are obtained in the coarse conversion and 4 bits are obtained in the fine conversion by amplifying the remaining time. The TDC circuit occupied an area of 0.34 mm² using 180 nm CMOS process. The effective number of bits is 7.42bits. The TDC circuit has shown 10.5 ps resolution for a 50 MHz. The DNL and INL are 0.7(LSB) and 0.5(LSB), respectively. The power consumption is 1.34 mW with a 1.8 V supply.

keywords— *DPLL (Digital-Phase locked loop), TDC (Time-to-Digital Converter), Two-Step TDC*

I. INTRODUCTION

Various high-resolution time-to-digital converters (TDC) have been proposed, such as a two-step TDC. This TDC has advantages such as fast conversion speed and low jitter accumulation. In the first step, quantization is performed with a low-resolution quantizer, and the remaining error is amplified by a time amplifier. In the second step, time information is converted into digital information through a high-resolution quantizer.

In this case, the error must be accurately transmitted to the next stage to ensure linearity. The time amplifier of the conventional latch structure has a disadvantage in that the operation region in which the linearity is guaranteed is limited [1]-[2]. Therefore, the pulse-train time amplifier was developed, which is an iterative time amplifier [3]. Although the conventional iterative time amplifier has the advantage of high linearity, but it has a disadvantage of error caused by the delay cell in the iterative process.

We propose a time amplifier using a ring oscillator structure. The proposed structure minimizes the delay mismatch and improves the stability in the remaining error transmission part. The proposed two-step TDC has high linearity because there is no accumulation of error in the iterative process. It is advantageous in terms of power consumption even when the number of iterations is large, and it reduces the quantization noise by increasing the number of iterations.

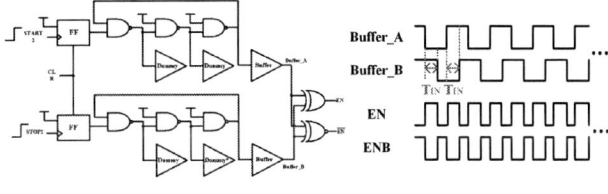

Fig. 1. Block diagram of the proposed two-step TDC circuit.

II. SYSTEM ARCHITECTURE

The block diagram of the proposed TDC is shown in Fig. 1. The coarse TDC converts the time to a digital code and then transfers the remaining error after the coarse conversion to the input of the time amplifier. In the time amplifier, the iterative process is repeated for the desired number of iterations. The time information is converted into a digital code, and the repeated time is supplied to the fine TDC. Each TDC converts thermometer code to binary code through an encoder and outputs the final output. The output from the coarse conversion is the upper bit, and the output from the fine conversion is the lower bit.

Fig. 2. (a) Block diagram of ring oscillater time amplifier, and (b) timing diagram.

The proposed time amplifier is shown in Fig. 2. The inputs are the START2 signal and STOP2 signal output through the coarse TDC and the residue generator, and the output is the EN signal and ENB signal. The proposed time amplifier is composed of a ring oscillator. The remaining NAND gates except for the first one are connected to VDD so that the input is always '1'. The first NAND gate of the input stage is connected to the output of a flip-flop.

The D flip-flop at the input stage supplies power when a signal is received. Since the VDD voltage is applied to D of the flip-flop, the flip-flop keeps supplying the output '1' when the rising signal is input to the clock input. When the output of the flip-flop is '0', the ring oscillator does not oscillate since

978-1-5386-7961-6/18 $31.00 © 2018 IEEE   143   ISOCC 2018

the first NAND gate is off, but it operates as an oscillator when the output of the flip-flop becomes '1'. Therefore, it starts to oscillate in accordance with the input signal.

When the START2 signal arrives, it oscillates at a constant period from the time when the START2 signal arrives. When the STOP2 signal is received, it oscillates at a constant period from the time when the STOP2 signal is received. As a result, both oscillators oscillate with the difference between the START2 signal and the STOP2 signal. *Dummy** is connected to the counter, and if the oscillator repeats for a desired number of times, the counter outputs the CLR signal and stops the oscillation. Therefore, the proposed architecture has an advantage of not needing additional buffers or intentional delay cells, even if the number of iterations increases, and time information error does not accumulate. It is also efficient in power consumption.

The difference between the two rising and falling signals can be output as pulses when the two oscillating signals are input and the XOR gate and XNOR gate are passed. At this time, the pulse must be made them the complementary EN and ENB signals so that the fine TDC can convert the time information into digital code. This is because the time amplifier is an iterative time amplifier, so the delay cell of the fine TDC must use a gated-delay cell or gated-ring oscillator. Therefore, we used the structure of a fine TDC [3] for linearity and uniform load capacitance. The gated-delay cell receives the pulse information iteratively, operates the delay cell when the pulse is '1', and stores the remaining time information in the form of a voltage in the gate capacitor when the pulse is '0'.

III. SIMULATION RESULTS

The proposed TDC has been verified with 180 nm CMOS process. Fig. 3 shows the output code with respect to input time difference from 0 to 400 ps. Since the coarse conversion resolution is $Q_C = 168$ ps and the amplification is 16 times in the simulation, the effective resolution of the fine conversion is $Q_F = Q_C/16 = 10.5$ ps. As shown in the graph, the output is constant and linear with increasing time. Fig. 4 shows a graph of the DNL and INL obtained through the output code from 0 to 400ps.

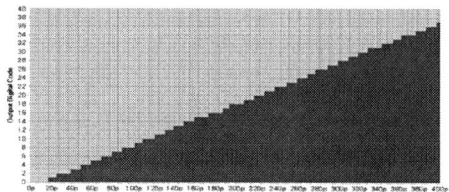

Fig. 3. Output digital code with respect to time difference

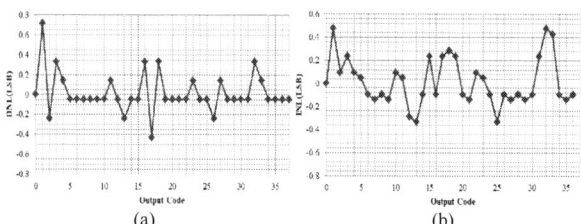

Fig. 4. Simulation result of (a) DNL, and (b) INL.

Table I is performance summery. The resolution has a critical effect on the speed of the delay cell, depending on the process (the smaller the process, the better the resolution). Theoretically, when the minimum channel length of a transistor is reduced by half, the speed is doubled. Therefore, considering a resolution of 10.5 ps at 180 nm, the resolution will be about 1.4 ps for a 65 nm process and 2.63 ps for a 90 nm process. However, since the time amplification factor is 16 in this design, the number of bits is greater by one. Considering the simulation, it is expected that the resolution of the actual measurement will be similar or better. Also, since the delay-line TDC structure is used, there is an advantage of fast conversion speed. However, because of the time it takes to generate repeated pulses in the time amplifier, the conversion rate is 50 MHz.

Table I. Performance Summery

	This work
Technology	180nm CMOS
Scheme	Two-Step TDC
TA Scheme	NAND-Gate ring-oscillator
Number of bits	8
Conversion rate	50 MS/s
Resolution	10.5 ps
Max DNL (LSB)	0.7
Max INL (LSB)	0.5
ENOB	7.42
Total Circuit Area	0.34 mm²
Power	1.34 mW

*ENOB(Effective Number Of Bits) = Bits − $\log_2$(INL +1)

IV. COCLUSION

We have proposed two-step TDC using ring oscillator time amplifier. The proposed time amplifier eliminates the accumulated error in the iterative process. This is achieved because the signal passes through the same delay cell at regular intervals from the time when the input signal information is input to the ring oscillator time amplifier. Since the error does not accumulate in the iterative process. In addition, there is an advantage in that an intentional delay cell is not included in the iterative process, which beneficial for power consumption.

ACKNOWLEDGMENT

This research was supported by the MSIT(Ministry of Science and ICT), Korea, under the ITRC(Information Technology Research Center) support program(IITP-2017-1711050923) supervised by the IITP(Institute for Information & communications Technology Promotion). The authors also thank IDEC for CAD tool support.

REFERENCES

[1] M. J. Lee ,and Asad A. Abidi, "A 9b, 1.25ps Resolution Coarse-fine Time-to-digital converter in 90nm CMOS that amplifies a time residue", IEEE Symposium on VLSI Circuits, pp. 168-169, Oct. 2007.

[2] S.- K. Lee, Y.- H .Seo, H.-J. Park, and J.- Y. Sim, "A 1GHz ADPLL with a 1.25ps minimum-resolution Sub-Exponent TDC in 0.18μm CMOS", IEEE J. Solid-State Circuits, vol. 45, no. 12, pp. 2874-2881, Dec. 2010.

[3] K. S. Kim, Y.- H. Kim, W. S. Yu, and S. H. Cho, "A 7bit, 3.75ps resolution Two-step time-to-digital converter in 65nm CMOS using Pulse-Train Time amplifier", IEEE J. Solid-State Circuits, vol. 48, no. 4, pp. 1009-1017, Apr. 2013.

Low Power - High Speed Magnitude Comparator Circuit Using 12 CNFETs

Jitendra Kumar Saini[1,*], Avireni Srinivasulu[2,†], *SM-IEEE*

[1, 3]Dept. of Electronics & Communication Engineering
Manipal University Jaipur
Jaipur-303007, Rajasthan, INDIA
*jitendraksaini@bitmesra.ac.in, †avireni_s@yahoo.com

Renu Kumawat[3,‡], *SM-IEEE*

[2]Dept. of Electronics & Communication Engineering
JECRC University
Jaipur-303905, Rajasthan, INDIA
‡renu.kumawat.2015@ieee.org

Abstract—**In VLSI domain there is an unending demand for low power, high speed and smaller chip area based circuits. Under these constraints, 1-bit magnitude comparator circuit (1B-MCC) is proposed using 12 Carbon Nanotube Field Effect Transistor (CNFET). However, the proposed analysis processes have been conducted in terms of power, delay and power-delay product (PDP). In order to arrive at a fair comparison between earlier designs and the proposed design, 2-bit magnitude comparator circuits (2B-MCC) are favored as they are designed in 45 nm CMOS technology at supply voltage (V$_{DD}$) of 0.7 V. Later on the proposed design was extended to CNFET technology at 32 nm with V$_{DD}$ of 0.7 V to gain the benefits of CNFET technology.**

Keywords; CNFET; magnitude comparator; high speed;

I. INTRODUCTION

Ubiquitous demand to shrink the IC form-factor (high-density) and to scale it as per demand (integrity) is the challenge every company needs to deal with on daily basis. Other essential considerations are low power consumption and higher operational speeds. CMOS came in with a promise to deal with these issues but with recent technological advancements we see a new player in the market i.e. CNFET [1-2].

The aim of this paper is to demonstrate the improvement in circuit performance of 2-bit magnitude comparator by reducing the number of transistors and length of the critical path. 1-bit magnitude comparator circuit designed using 12 CNFETs (denoted as 1B-MCC) has been proposed in this paper, which is further extended to 2-bit magnitude comparator circuit (2B-MCC). We have also tried to enhance the overall circuit performance, when compared with existing 2B-MCC designs.

The remaining part of the paper is organized as follows: Introduction of CNFET in Section 2. Section 3, discusses the structural implementations of 1B-MCC, while simulation results of 1B-MCC and comparisons of 2B-MCC with earlier designs are summarized in Section 4. Section 5 covers the conclusion drawn.

II. CARBON NANO TUBE FIELD EFFECT TRANSISTOR

Carbon Nano Tubes (CNT) are structurally nano scale (10^{-9}) uniaxial honeycomb lattice with benefits of graphite i.e. carbon. They can have structural variations characterized as Armchair, Zigzag and Chiral. Further they can either be single layered or multi-layered, the single layered structure is referred to as single-walled and multi-layered structure is referred to as multi-walled CNT. This ability to form single-walled or multi-walled structure enables CNT to possess characteristics of metals and semiconductors both [3-4].

The single-walled or multi-walled CNT when roll-up to form a cylindrical or concentric cylindrical structure, it further enhances the properties of CNT sheets. The connections between these sheets binding them together are measured by Chirality Vector (C) shown in Figure 1 and is represented as

$$C = na1 + ma2 \qquad (1)$$

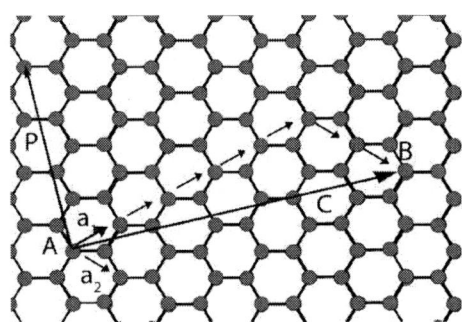

Figure 1. Representation of Chiral vector [3]

The type of CNT will depend on the n and m factor, i.e. If n = m, then type of CNT will be Armchair. If m = 0, then type of CNT will be Zigzag. If m < n or m > n, then type of CNT will be Chiral. Here a_1, a_2 are unit vectors in equation (1) [5-6].

The characteristics like mobility, drive current and current voltage of CNFET are same as of MOSFET. The CNFET is a 4 terminal device with heavily doped CNT acting as the drain / source while lightly doped CNT acts as the gate. Considering the performance as a vector, the CNFET is segregated into Schottky Barrier or MOSFET. The threshold can be measured by dividing the band-gap of CNFET by a factor of 2e, represented as:

$$Vth = E_{bg}/2e = 0.436/D_{CNT} \text{ (nm)} \qquad (2)$$

Here V_{th} is the threshold Voltage, E_{bg} is the energy band-gap of CNFET, e is the electron charge and D_{CNT} is the diameter of the rolled-up CNT [7].

$$D_{CNT} = d = 0.0783\sqrt{m^2 + n^2 + mn} \qquad (3)$$

978-1-5386-7961-6/18 $31.00 © 2018 IEEE

III. PROPOSED 1-BIT MAGNITUDE COMPARATOR CIRCUIT

The proposed 1B-MCC consists of 2 inputs (A and B) and three outputs (A < B, A = B, A > B) as shown in Figure 2. The proposed circuit uses only 12 transistors for the implementation of 1B-MCC. The relation between inputs and outputs of 1B-MCC is given in the form of equations (4), (5), (6) respectively

$$(A < B) = \overline{A}B \tag{4}$$

$$(A = B) = \overline{A + B} \tag{5}$$

$$(A > B) = A\overline{B} \tag{6}$$

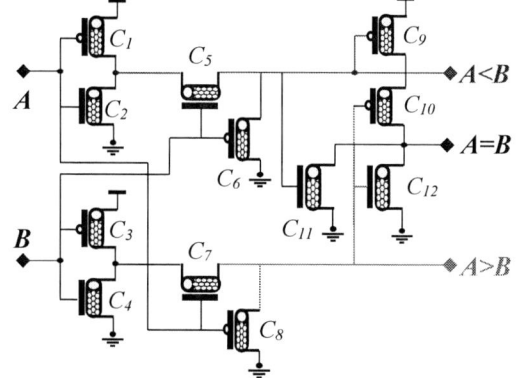

Figure 2. Proposed 1-Bit magnitude comparator circuit

IV. SIMULATION RESULTS

The proposed 1B-MCC are simulated on 32 nm CNFET technology using Cadence Virtuoso CAD Tool at supply voltage of +0.7 V, threshold $V_{th} = 0.289$ V with Chirality vector (19, 0). Figure 3 depict the simulation waveforms of 1B-MCC for all the test cases. Simulation results shows that the proposed design provides full output swing.

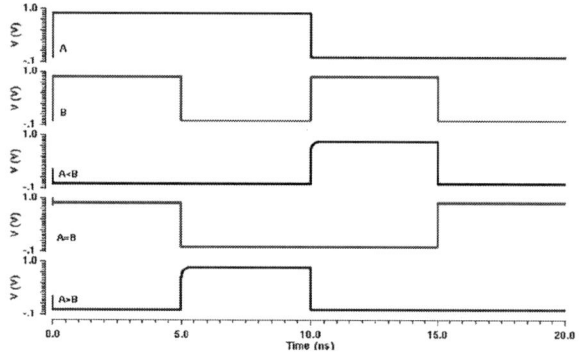

Figure 3. Simulated input and output waveforms of Fig. 2.

For the purpose of comparison, the proposed 1B-MCC is extended to a 2B-MCC. The simulation results of all the 2-bit magnitude comparators [5-7] including the proposed 2B-MCC are summarized in Table 1. Magnitude comparators in this table are arranged in the order of their transistor count. Performance analysis of all these designs along with the proposed designs is done in terms of power, delay and power-delay product. It can be seen from Table I that the power-

delay product is significantly lower for the proposed 2B-MCC as compared to other 2B-MCC designs.

TABLE I. COMPARATIVE ANALYSIS OF CANDIDATE DESIGNS

2 Bit MC, V_{DD}=0.7V, Technology (nm)	Transistors Count	Power (µW)	Delay (ns)	PDP (fJ)
Conventional CMOS [5] CMOS 45 nm	54	0.62	8.4	5.208
TGL Based [6] CMOS 45 nm	74	0.47	8.49	3.990
P. TGL Based [7] CMOS 45 nm	30	0.28	7.63	2.136
X-E Logic Based [7] CMOS 45 nm	25	0.33	7.96	2.626
Proposed with CMOS CMOS 45 nm	24	0.09	4.73	0.425
Proposed with CNFET CNFET 32	24	0.02	2.08	0.041

V. CONCLUSION

We have proposed low power, high speed magnitude comparator circuit using 12 CNTFETs. The performance of the proposed 2-bit magnitude comparator and earlier 2-bit magnitude comparator circuits are compared based on transistor count, power, delay and power-delay product. Among all these design structures, the proposed magnitude comparator has yielded the best performance and hence it can be used to design N-Bit magnitude comparator. It can thus be concluded that by using proposed magnitude comparator for implementing high performance complex structures, such as, multipliers, arithmetic logic unit, and mobile communication devices, etc. the intrinsic benefits of proposed magnitude comparator could be fully exploited.

REFERENCES

[1] J. Deng and H. S. P. Wong, "A compact SPICE model for carbon-nanotube field-effect transistors including nonidealities and its application-Part I: Model of the intrinsic channel region," *IEEE Trans. Electron Devices,* vol. 54, no. 12, pp. 3186–3194, Dec. 2007. DOI: 10.1109/TED.2007.909030

[2] Jitendra Kumar Saini, A. Srinivasulu and B. P. Singh, "A new low-power full-adder cell for low voltage using CNFETs", *in proc. of IEEE International Conference on Electronics, Computers and Artificial Intelligence,* pp. 1-5, July 2017. DOI: 10.1109/ECAI.2017.8166425.

[3] P. Kavitha, M. Sarada, K.Vijayavardhan, Y. Sudhavani and Avireni Srinivasulu, "Carbon nano tube field effect transistors based ternary Ex-OR and Ex-NOR gates", *Current Nanoscience,* vol. 12, no. 4, pp. 520-526, Aug 2016. DOI: 10.2174/1573413712666151216221629

[4] P. L. Kartik, K. Balakrishna, M. Sarada and A. Srinivasulu, "A new low voltage high performance Dual Port 7- CNT SRAM Cell with improved differential reference based sense amplifier", *International Journal of Sensors, Wireless Communications and Control ,* vol. 7, no. 3, pp. 246 - 254, 2017. DOI: 10.2174 /2210327908666180110154612

[5] D. Sharma. Logic Design, EE Department IIT Bombay:Microelectronics group. pp. 1–34, 2006.

[6] Anjuli, S. Anand "Two-bit Magnitude Comparator Design Using Different Logic Styles", *International Journal of Engineering Science Invention,*vol. 2 no. 1 pp.13-24, January. 2013

[7] D. N. Mukherjee, S. Panda and B. Maji "Performance Evaluation of Digital Comparator Using Different Logic Styles", *IETE Journal of Research,* vol. 64 no. 3, pp. 422-429, Mar. 2018. DOI: 10.1080/ 03772063.2017.1323564.

978-1-5386-7961-6/18 $31.00 © 2018 IEEE

Gap in pagination due to formatting issues.

Pages 147-148

Estimation of Leakage Distribution Utilizing Gaussian Mixture Model

Hyunjeong Kwon
Department of Electrical Engineering
Pohang University of Science and Technology
Pohang 790-784, Korea
kwonhj@postech.ac.kr

Young Hwan Kim
Department of Electrical Engineering
Pohang University of Science and Technology
Pohang 790-784, Korea
youngk@postech.ac.kr

Seokhyeong Kang
Department of Electrical Engineering
Pohang University of Science and Technology
Pohang 790-784, Korea
shkang@postech.ac.kr

Abstract— **In this paper, we propose a novel method which utilizes the Gaussian Mixture Model (GMM) to estimate the leakage distribution of a circuit. Our proposed method assumes that the leakage distribution can be represented using the GMM which can cover any continuous function. After the GMM clustering using the leakage simulation data, the leakage distribution of the input circuit can be obtained. The experimental results with the K-S test showed that the proposed method exhibited 1.82e+05 times larger p-value and 7.74e-01 times smaller K-S statistics compared to the state-of-the-art benchmark method on average.**

Keywords; Statistical leakage analysis; Gaussian mixture model;

I. INTRODUCTION

With the advent of nanotechnology in semiconductor design, the process variations cause the uncertainty in leakage currents. The methods considering the process variations in leakage analysis are called statistical leakage analysis (SLA).

The SLA is classified into analytic model-based methods and sampling-based methods. One of the recent studies related to the analytic model-based methods assumes that the leakage distribution follows the generalized extreme value (GEV) distribution which is closer to the actual leakage distribution than lognormal distribution which had been widely used function [1]. However, the shape of leakage distribution changes as the technology advances. Therefore, the analytic model-based methods which assumes that the leakage distribution follows a particular continuous function are not proper because the semiconductor technology will continuously advance.

Sampling-based methods have tried to address the complexity problem of the Monte-Carlo (MC) simulation. State-of-the-art study in this area is the SLA using quasi Monte-Carlo simulation (QMC) [2]. Even if the QMC greatly reduced the number of simulation for the convergence, the terminating condition of the simulation is hard to be defined because the error bound is too complicated to calculate [3]. Also, the histogram of the QMC data is lack of information because of its wide bins resulted from the reduced number of data [4].

To address the disadvantages of the previous work, we propose a new SLA which uses the GMM as the leakage distribution. Our method efficiently address the complexity problem of MC simulation by using rapidly converged samples in leakage

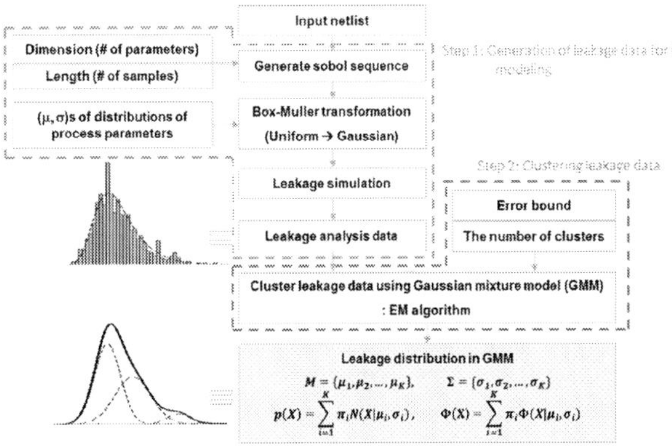

Fig. 1. Overall flow of the proposed method which consists of two steps.

simulation. In addition, the accuracy of estimated leakage distribution using the GMM is significantly improved compared to the analytic model-based methods by using the GMM.

II. PROPOSED METHOD

The overall flow of our method is shown in Fig. 1. Our proposed method consists of two steps. First, the leakage data are obtained for the GMM modeling. Next, the obtained leakage data are clustered to form the GMM which represents the leakage distribution of the input circuit. The detailed process will be described in the following sections.

A. Step 1: Generation of leakage data for GMM modeling

First, after reading the input netlist, the parameter samples for leakage simulation are extracted. Our proposed method aims to reduce the number of required leakage data for the convergence. To achieve this, our proposed method uses the sobol sequence [5] which converges faster than the pseudo random numbers. The sobol sequence is the set of samples from the uniform distribution. Because the parameter distributions are usually assumed to be normal distributions, the uniformly distributed sobol sequence is transformed to be normally distributed samples. Using the rapidly converged parameter samples from normal distributions, the leakage simulation is performed and the leakage data are obtained.

This research was supported by the MSIT(Ministry of Science and ICT), Korea, under the "ICT Consilience Creative program" (IITP-2018-2011-1-00783) supervised by the IITP(Institute for Information & communications Technology Promotion)"

978-1-5386-7961-6/18 $31.00 © 2018 IEEE
149
ISOCC 2018

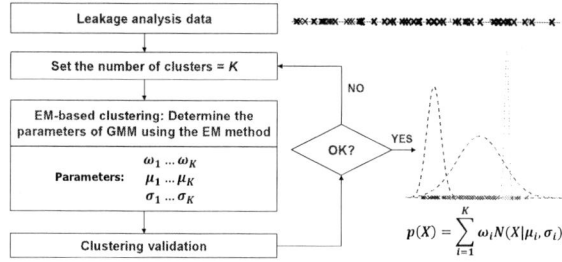

Fig. 2. Flow of the Step 2 of the proposed method.

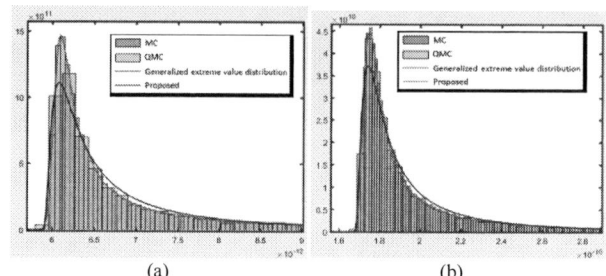

Fig. 3. Estimated leakage distributions using QMC, GEV, and the proposed method compared to MC simulation.

B. Step 2: Clustering leakage data

Using the leakage data collected in *Step 1*, the proposed method clusters the data into the several components of Gaussian distributed samples to form the GMM using the traditional expected-maximization algorithm (Fig. 2). To determine the number of clusters in this step, the cluster validation measure should be determined first. In the proposed method, following measures are used for cluster validation. First, the sum of squares within a cluster (SSW) measures how close the data in a cluster. The smaller SSW is the better. SSW is defined as follows.

$$SSW = \sum_i \sum_{x \in C_i} (x - m_i)^2 \qquad (1)$$

where x is the data included in i^{th} cluster, C_i and m_i is the mean of C_i. Second, the sum of squares between clusters (SSB) measures how well-separated is a cluster from the other clusters. The larger SSB is the better. SSB is defined as follows.

$$SSB = \sum_i |C_i| (m - m_i)^2 \qquad (4)$$

where $|C_i|$ is the number of data in C_i and m is the mean of all observed data.

Finally, $score = \frac{SSW}{SSB}$ is used for the clustering validation. The smaller $score$ represents the better clustering. As the number of clusters increases, the $score$ usually decreases. As shown in Fig. 2, if the $score$ is not saturated, the clustering is performed with larger number of clusters. On the other hand, if the $score$ is saturated where the $score$ difference is less than the user defined bound, the clustering is finished.

III. EXPERIMENTAL RESULTS

In experiments, we used five ISCAS 85 circuits. The leakage data are obtained through HSPICE simulation. The numbers of simulations used for the proposed method and MC simulation were 10,000 and 200,000, respectively. The accuracy comparison was performed by Kolmogorov-Smirnov (K-S) test for histogram using QMC data, fitting the QMC data to GEV distribution, and the proposed method.

Fig. 3 shows the estimated leakage distributions of QMC, GEV, and the proposed method compared to MC simulation for c17 and c432 circuits. As shown in Fig. 3, the proposed method estimates the leakage distribution closest to the MC simulation result. For the numerical comparison, K-S test is used. In this test, the larger p-value and the smaller K-S statistic are better. As shown in Table I, average p-value and the K-S statistic of the proposed method were 1.82e+05 and 7.74e-01 times compared

TABLE I. K-S TEST FOR GEV, QMC AND THE PROPOSED METHOD

		GEV	QMC	Proposed
c17	P-value	3.38E-290	5.48E-13	4.49E-11
	K-S statistic D	3.87E-02	8.05E-03	7.41E-03
c432	P-value	1.26E-203	9.40E-11	5.45E-06
	K-S statistic D	3.34E-02	7.52E-03	5.52E-03
c499	P-value	2.85E-37	1.28E-08	2.81E-06
	K-S statistic D	1.26E-02	5.95E-03	5.03E-03
c880	P-value	3.83E-216	6.30E-13	5.35E-07
	K-S statistic D	3.40E-02	8.17E-03	5.93E-03
c1355	P-value	1.89E-52	3.53E-04	5.59E-02
	K-S statistic D	2.18E-02	5.87E-03	3.78E-03

to the QMC. This result means that the proposed method exhibited the most similar distribution with the MC simulation results.

IV. CONCLUSION

We proposed a novel method to estimate the leakage distribution using the GMM for the accuracy. The proposed method assumes that the leakage distribution follows the GMM which can cover any arbitrary continuous function. Therefore, the proposed method is robust to the change of the shape of the leakage distribution as the technology advances. The K-S test showed that the average p-value and the K-S statistic of the proposed method were 1.82e+05 and 7.74e-01 times compared to the QMC.

REFERENCES

[1] H. Aghababa, A. Khosropour, A. Afzali-Kusha, B. Forouzandeh, and M. Pedram, "Statistical estimation of leakage power dissipation in nano-scale complementary metal oxide semiconductor digital circuits using generalized extreme value distribution," *IET Circuits, Devices & Systems*, vol. 6, no. 5, pp. 273-278, Nov. 2012.

[2] V. Veetil, D. Sylvester, D. Blaauw, S. Shah, and S. Rochel, "Efficient smart sampling-based full-chip leakage analysis for intra-die variation considering state dependence," in *Proc. Des. Autom. Conf.*, 2009, pp. 154–159.

[3] B. Tuffin, "Randomization of Quasi-Monte Carlo Methods for Error Estimation: Survey and Normal Approximation," *Monte Carlo Methods and Applications* (MCMA), vol. 10, no. 3-4, pp 617-628.

[4] D. Freedman and P. Diaconis, "On the histogram as a density estimator: L_2 theory," *Probability Theory and Related Fields*, vol. 57, no. 4, pp. 453-476, Dec. 1981.

[5] I. M. Sobol, "The distribution of points in a cube and the approximate evaluation of integrals," *Zh. Vychisl. Mat. i Mat. Fiz.*, vol. 7, pp. 784–802, 1967.

System Level Power Reduction for YOLO2 Sub-modules for Object Detection of Future Autonomous Vehicles

YoungBae Kim, Qiang Tong and Ken Choi
DA-lab, Electrical and Computer Engineering Department
Illinois Institute of Technology
Chicago, IL, USA
{ykim102, qtong}@hawk.iit.edu, kchoi12@iit.edu

Eunchong Lee, Sung-Joon Jang and Byeong-Ho Choi
Intelligent Image Processing Research Center
Korea Electronics Technology Institute(KETI)
Seongnam-si, KOREA
{sjjang0626, bhchoi}@keti.re.kr

Abstract— **AI-based object detection is one of the critical design components for the success of level 3 or level 4 autonomous vehicles. Currently, GPU based system design is a popular way to implement Convolutional Neural Networks(CNN) or Recurrent Neural Networks(RNN) for the system. The thing is that the feasibility of the GPU-based implementation for real cars is quite low because of the huge power consumption of GPU. In this paper, we claimed that a holistic approach from system level to circuit level is necessary for ultra-low power design. As an initial step for the approach, we proposed system-level power reduction techniques that can be applied to advanced CNN algorithms such as YOLO2. By applying proposed system level techniques such as loop unrolling, declare small function as inline, arguments passing, branching, and common expressions elimination, we demonstrated that the proposed low-power schemes can be reduced up to 86.95% for YOLO2 sub-modules compared with the original one using system-level power-estimation CAD tools (Simple Scalar and WATTCH [1]).**

Keywords; System level, power reduction, YOLO2, Loop optimization

I. INTRODUCTION

It is very convincing to assume that only hardware, not software, consumes most of the power when using many types of microprocessors, but in fact, as has been shown in many recent researches, hardware as well as software has a significant impact on power consumption.

One of the major causes of power consumption is computer systems that are affected by memory systems. The memory system occupies a significant portion of the portable computer's power budget and can be a major source of memory-intensive DSP applications such as video processing. Therefore, we must take this point seriously when implementing system level design.

Another problem we should keep in mind is that in modern computer system, the OS maintain a stack for each program at runtime. The stack is used to keep historical data in the registers. It is mostly used for function call. When the program calls a function, the OS push the states of registers into stack and pop them out after the function returns. Maintaining the stack is an extra cost, so it is not a good idea

to call functions when it is not necessary. Especially, recursive function is not a good idea, since it calls a function recursively, in which case a large needed and frequent access
of it is expected.

II. TESTING PLATFORM WATTCH

For this research, we used a simulator called "WATTCH". WATTCH is an architectural simulator that estimates CPU power consumption.

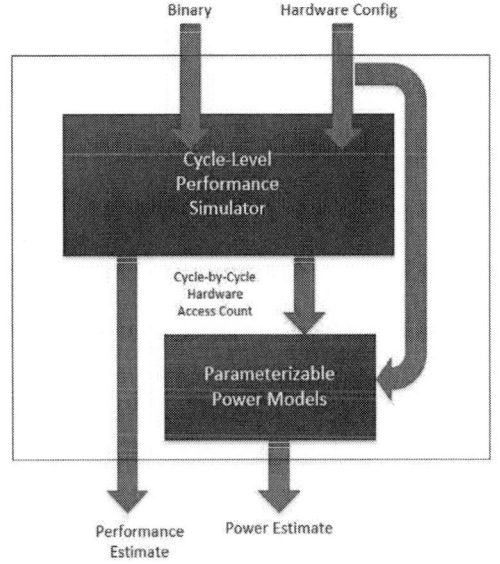

Figure 1. Overall Structure of the WATTCH

The power estimation is based on a suite of parametrizable power models for different hardware structures and on per-cycle resource usage counts generated through cycle level simulation. The power models have been integrated into the "Simple Scalar" architectural simulator.

The Simple Scalar tool set performs fast, flexible, and accurate simulation of modern processors that implement the Simple Scalar architecture (a close derivative of the MIPS

This work is supported by the Industrial Core Technology Development Program of MOTIE/KEIT, KOREA. [#10083639, Development of Camera-based Real-time Artificial Intelligence System for Detecting Driving Environment & Recognizing Objects on Road Simultaneously]

978-1-5386-7961-6/18 $31.00 © 2018 IEEE

architecture). The most complicated and detailed simulator in the tool set is *sim-outorder*. It is used to collect results by WATTCH. Simple Scalar provides a simulation environment for modern out-of-order processors with 5-stage pipelines: fetch, decode, issue, writeback and commit. The power oriented modifications provided by WATTCH (whose modules are integrated within Simple Scalar) track which units are accessed on each cycle and compute the power values associated with those units accordingly. These power modules have been verified against industrial circuits and have been found to be within 10% for low level power estimation.

III. LOOP OPTIMIZATION IN CNN

The convolution layer is the most efficient convolutional neural networks(CNN) algorithm because it performs the most computation. CNN's convolution requires large design space because it requires three-dimensional multiply and accumulate(MAC) operations with four levels of loops. However, previous studies have used limited loop optimization techniques to produce results that are lower than what we expected in terms of power dissipation or performance. So, if we do not thoroughly study circuit loop optimization before the hardware design phase, the resulting accelerator can make little use of data reuse and efficiently manage data movement.

A. DIRECTIVES LOOP UNROLLING

Loop unrolling is effective technique to exploit parallelism between loop iterations. It creates multiple copies of the loop body and adjust the loop iteration counter accordingly. a Small loops can be unrolled for higher performance, with the disadvantage of increased code size. When a loop is unrolled, a loop counter needs to be updated less often and fewer branches are executed. If the loop iterates only a few times, it can be fully unrolled, so that the loop overhead completely disappears.

```
void binarize_cpu(float *input, int n, float *binary)
{
    int i;
    for(i = 0; i < n; i+=2){
        binary[i] = (input[i] > 0) ? 1 : -1;
        binary[i+1] = (input[i+1] > 0) ? 1 : -1;
    }
}
```

Figure 2.   Directives loop unrolling

we can see that by unrolling a small loop, the instruction counts and the total power consumption both decrease a lot. This means that we can have both speed and power reduction if we use loop unrolling properly. Below loop optimization method shows loop parallelization, two processors can split up the number of loop iterations by unrolling once and then distributing. By using this method, the first processor takes the odd iterations and the second processor takes the even ones.

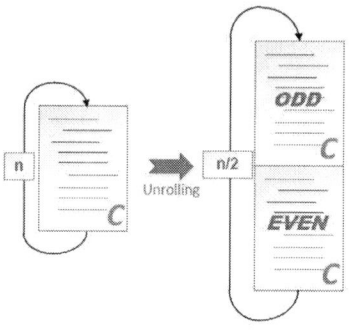

Figure 3.   Loop parallelization by unrolling

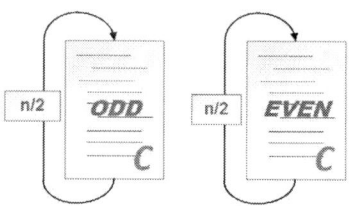

Figure 4.   Loop parallelization by distribution

```
for(i = 0; i < net.n; i+=2)// unrolling
    {
    if(i%2==1)
    {
        layer l = net.layers[i];
    if(l.type == CONVOLUTIONAL)
        {
                rgbgr_weights(l);
                break;}
    }
    if(i%2==0)
    {
        layer l = net.layers[i+1];
        if(l.type == CONVOLUTIONAL)
        ...
        ...
    }
```

Figure 5.   Loop parallelization

It is to limit the amount of loop overhead as a proportion of the commands, to decrease the number of control hazards in the total run of the loop.

TABLE I.   APPLIED LOOP UNROLLING TO YOLO

	Before	After
Avg. power per instruction	38.1397	36.1954
Total instruction executed	364783	302786
Total Power	13912702.6739	10959453.9537
Power saving	N/A	21.23%

As can be seen in Table I, you can see that the total number of instruction and total power consumption are reduced by unrolling even if it is a small loop. In other words, using the same loop unrolling in the same way can reduce both speed and power a lot.

B. DECREASING INDEX

Most for loop, including CNN and YOLO, require index variables as the counter for loop iterations, and most people write for loop. But we propose the following form to differentiate from the existing method.

$$for (i=100; i>0;i--) \{...\}$$

This approach is very effective. Because the CPU provides independent instructions to provide whether the variable is 0 (for example, JNZ in the MIPS instruction set) and the 8051 microcontroller provides DJNZ(decrement jump of not zero). That is, instead of comparing a variable to another number, it is more convenient to determine if the variable is 0.

```
void normalize_net(char *cfgfile, char *weightfile, char *outfile)
{
    gpu_index = -1;
    network net = parse_network_cfg(cfgfile);
    if(weightfile){
        load_weights(&net, weightfile);
    }
    int i;
    for(i = net.n; i > 0; i--)   //decreasing index
    {
        layer l = net.layers[i];
                ...
                ...
```

Figure 6.   Decreasing index

TABLE II.   APPLIED LOOP DECREASING INDEX TO YOLO

	Before	After
Avg. power per instruction	36.7198	38.1397
Total instruction executed	390841	364783
Total Power	14351600.5796	13912702.6739
Power saving	N/A	3.058%

As can be seen in Table II, you can see that the total number of instruction and total power consumption are reduced by loop decreasing index.

C. LOOP ORDERS FOR ARRAY INDEXING

Various array indexing is used in YOLO. However, there is a restriction on our test platform for checking the power dissipation after applying our method. We used the following example to apply loop orders for array indexing. The following shows a two-dimensional array with row-first and column-first styles.

Before	After
int arr[100][10]; void example(void){ int i=0, j=0; for(i=0;i<100;i++){ for(j=0;j<10;j++){ arr[i][j]=1;	int arr[100][10]; void example(void){ int i=0, j=0; for(i=0;i<10;i++){ for(j=0;j<100;j++){ arr[i][j]=1;

Figure 7.   Loop orders for array indexing

As can be seen in the results, we can see the overall improvement such as total power and instruction execute after applying our method.

TABLE III.   LOOP ORDERS FOR ARRAY INDEXING

	Before	After
Avg. power per instruction	35.0946	33.5407
Total instruction executed	148527	115680
Total Power	5212490.1271	3879987.4230
Power saving	N/A	25.57%

IV.   OTHER DESIGN METHOD FOR CONSIDERING THE POWER DISSIPATION

Each function call has to operate on the stack, but the stack operation incurs a lot of additional cost. So, for good function design considering power consumption we have to reduce these additional costs as much as possible in the following way. 1) make the arguments as less as possible 2) declare the frequently called small function as a MACRO or inline function 3) don't call a function too frequently, especially avoid using recursive implementation of algorithm.

A. DECLARE SMALL FUNCTION AS INLINE

An 'inline' indication for the function. It shows that the compiler does not compile the function as a normal function. It is only deployed at compile time. Additional operations on the stack make function calls considerably more expensive in terms of power dissipation.

```
Inline void operations(char *cfgfile)
{
    gpu_index = -1;
    network net = parse_network_cfg(cfgfile);
    int i;
    long ops = 0;
    for(i = 0; i < net.n; i+=3)
            ...
            ...
```

Figure 8.   Declare small function as inline

TABLE IV. DECLARE SMALL FUNCTION AS INLINE

	Before	After
Avg. power per instruction	38.5724	38.0983
Total instruction executed	259777	247827
Total Power	10020226.8730	9441791.1957
Power saving	N/A	5.77%

B. ARGUMENTS PASSING

If the parameter is very large during the function argument passing, the pass-by-address is more efficient than the pass-by-value.

Especially suitable for complex structures. Passing by value creates a copy of each arguments that needs access to memory. This copying process takes time and consumes power.

```
int main(struct l* CONVOLUTIONAL, struct l*batch_normalize)
{
if(CONVOLUTIONAL->x == batch_normalize->x &&
CONVOLUTIONAL->y == batch_normalize->y)
    {
        CONVOLUTIONAL->value = batch_normalize->value
        CONVOLUTIONAL->key = batch_normalize->key;
    }
}
```

Figure 9. Arguments passing

Passing by value is more efficient because it only passes address to the parameter. The only risk is that we can change the raw data from what we did not expect. However, this problem can be handled by declaring the 'const' property on the parameter.

TABLE V. ARGUMENT PASSING

	Before	After
Total instruction executed	6809	254
Total Power	687221.7207	89728.6343
Power saving	N/A	86.95%

C. BRANCHING

Most modern CPUs all have branch prediction capabilities to speed up execution. With the branch prediction method, the modern CPU will suffer a prediction error when the CPU incorrectly predicts the branch. Incorrect prediction penalties vary from design to design, but it usually takes much longer to recover from incorrect prediction. As shown below, CNN or YOLO has a large number of branches as shown below. We can predict and apply these various branches in the following way. We should put the most possible branch in condition 1, second most branch in condition 2, and vice versa. When a machine runs this code section, it stops the branching whenever it finds a match. Putting the most probable branch in front makes the machine execute the code section more efficient without redundant branching determination.

```
if(condition1) {
operation1         //put most probable branch here

}else if(condition2) {
operation2         //put second most probable branch here

}else if(condition3) {
operation3         //put third most probable branch here

}else{
operation4         //put least probable branch here
}
    ...
```

Figure 10. Branching

D. COMMON EXPRESSIONS ELIMINATION

Common expression elimination helps reduce the number of operations by storing a variable that is computed multiple times. In the example, instead of computing a floating-point division with each operation, we can compute it once and store the result in another variable. This way, the value is computed once at the beginning of the for loop and used in the expressions below it.

Before	After
```int main() { int i=1; for(i=1;i<500;i++){ float v1 =1./i + 0.15; float v2 =1./i + 0.62; float v3 = 1./i + 0.45; float v4 =1./i + 0.244; } }```	```int main() { int i=1; for(i=1;i<500;i++){ float tmp = 1./i; float v1 = tmp + 0.15; float v2 = tmp + 0.62; float v3 = tmp + 0.45; float v4 = tmp + 0.244; } }```

Figure 11. Common expressions elemination

TABLE VI. COMMON EXPRESSIONS ELIMINATION

	Before	After
Avg. power per instruction	93.4508	71.9462
Total instruction executed	26738	25764
Total Power	2498686.2018	1853621.7799
Power saving	N/A	25.82%

V. CONCLUSION

In this paper, we aim to present a way to achieve ultra-low power by suggesting a change in coding style in system level, rather than a complex hardware design change for low power. Especially, we focused on reducing power consumption through loop optimization at system level. We reduced power consumption by 21.23% by loop unrolling using loop parallelization, and reduced power consumption by 25.57% through loop orders for array indexing. In addition, we have reduced total power consumption by utilizing various system level techniques. In particular, we achieved a power saving of up to 86.95% through the arguments passing and common expressions elimination reduced power by 25.82%. Through the paper, we present and briefly summarize the techniques to overcome the power dissipation for the Convolutional Neural Networks(CNN) in the system level and provide general coding tips for designs based on ultra low power.

ACKNOWLEDGMENT

We thank our colleagues from KETI and KEIT who provided insight and expertise that greatly assisted the research and greatly improved the manuscript.

REFERENCES

[1] David Brooks, Vivek Tiwari, Margaret Martonosi. "Wattch: A Framework for Architectural-Level Power Analysis and Optimizations". Proceedings of the 27th International Symposium on Computer Architecture, June 2000.

[2] Kaushik Roy, Mark C. Johnson, Software design for low power, low power design in deep submicron electronics, Kluwer Academic Publishers, Norwell, MA,1997

[3] Abraham Silberschatz, Peter B. Galvin, Greg Gagne. " Operating System Concepts". John Wiley & Sons Inc.2012

[4] Vivek Tiwari, Sharad Malik, Andrew Wolfe, Mike Tien-Chien Lee. "Instruction Level Power Analysis and Optimization of Software" Proceedings of International Conference on VLSI Design. P326-328, January 1996.

Gap in pagination due to formatting issues.

Pages 156-157

A Simple Circuit Model for PWM-1-Controlled DC-DC Converter and Its Analysis

Kojiro Tamura[†], Yuki Kametaka[†], Takuji Kousaka[‡], Hirokazu Ohtagaki[†], Hiroyuki Asahara[†]

[†] *Department of Electrical and Electronic Engineering, Okayama University of Science,*
1-1 Ridai-cho, Kita-ku Okayama-shi, Okayama, 700-0005 JAPAN.

[‡] *Department of Electrical and Electronic Engineering, Chukyo University,*
101-2 Yagoto Honmachi, Showa-ku, Nagoya-shi, 466-8666, JAPAN.
tamura@nonlineargroup.net

Abstract—**In this report, we study a simple circuit model for PWM-1-controlled DC-DC converter. The original circuit model is based on the PWM-1 voltage-mode-controlled DC-DC converter. However, in the original circuit, the number of dimensions of the circuit is more than two, which means that it is not easy to analyze the circuit dynamics rigorously. Therefore, we propose a simple circuit model, which is PWM-1 current-mode-controlled DC-DC converter. In this report, we investigate nonlinear phenomena in the proposed model.**

I. Introduction

In the field of power electronics, DC-DC converter is a typical example of the power conversion circuit. Since nonlinearlity of the circuit due to the switching action, rich nonlinear phenomena including chaotic attractor are observed in the DC-DC converters. Moreover, there are many kinds of the controller for the DC-DC converters. The analysis of the nonlinear phenomena in the circuit has been done well since decades ago, and it is still continuing presently [1]. In this report, we develop a simple circuit model for PWM-1-controlled DC-DC converter which is proposed in Ref. [3]. The circuit dynamics is based on the PWM-1 voltage-mode-controlled DC-DC converter studied in Ref. [2]. However, the number of dimensions of the original circuit is more than two, which means that it is not easy to analyze the nonlinear phenomena rigorously. Therefore, we have proposed a simple circuit model, which is PWM-1 current-mode-controlled DC-DC converter in Ref. [3]. However, our proposed circuit was not suitable for studying nonlinear phenomena because bifurcation structure was not the same with that of observed in the original circuit. Therefore, we improve the circuit model and investigate the circuit dynamics in this report.

II. Circuit Dynamics and Analytical Result

The circuit model is shown Fig. 1. The circuit is a DC-DC boost converter. It boosts the output voltage to an arbitrary value. This circuit has a main circuit and a control circuit, the main circuit is shown above, and the control circuit is shown below. Current control is a method to control the switching action by comparing a given reference current value with the actual current value. In this research, we used PWM-1 control. One characteristic of PWM-1 control is including a sample

hold circuit, which is sampling the current value at every period of the reference signal and is holding it during the clock interval. The reference signal is defined as follows.

$$S(t) = I_{\mathrm{L}} + \frac{1}{T}(I_{\mathrm{H}} - I_{\mathrm{L}})(t \bmod T), \tag{1}$$

The circuit parameters are $R = 20[\Omega]$, $L = 330[\mathrm{uH}]$, $C = 100[\mathrm{uF}]$, $E = 10[\mathrm{V}]$, and $f = 30[\mathrm{kHz}]$. Assuming that the inductor current is i, and a time when the clock pulse is applied

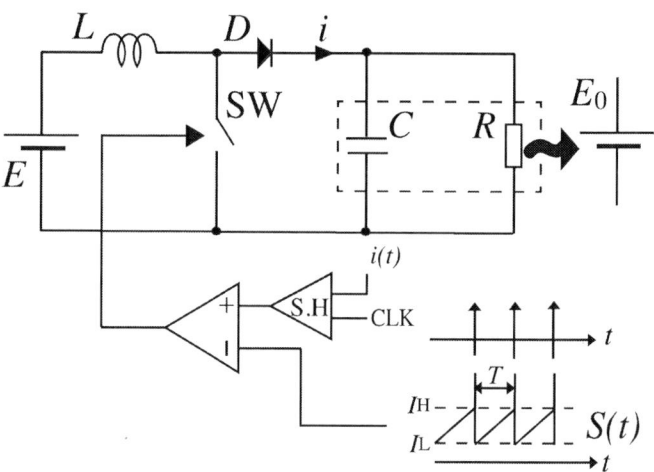

Fig. 1. Circuit model.

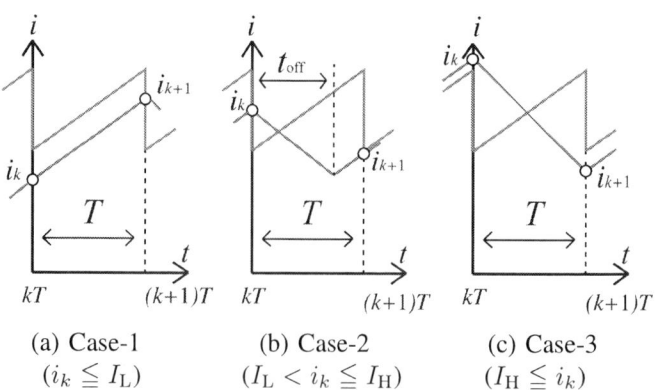

(a) Case-1
$(i_k \leqq I_{\mathrm{L}})$

(b) Case-2
$(I_{\mathrm{L}} < i_k \leqq I_{\mathrm{H}})$

(c) Case-3
$(I_{\mathrm{H}} \leqq i_k)$

Fig. 2. Waveform behavior.

is $t = kT$, where $k = 0, 1, 2, 3, \cdots$. Thus, solution of the circuit equation is given by

$$i(t) = \begin{cases} i_k + \dfrac{E}{L}(t - kT), & \text{SW : ON} \\ i_k + \dfrac{E - E_0}{L}(t - kT). & \text{SW : OFF} \end{cases}, \quad (2)$$

where i_k is an initial value at $t = kT$. Capacitor voltage E_0 in Eq. (2) is defined as follows.

$$E_0 = \sqrt{\hat{i}RE}. \quad (3)$$

In Eq. (3), the inductor current was defined as follows in Ref. [3].

$$\hat{i} = \frac{I_H + I_L}{2}, \quad (4)$$

where I_H and I_L are the maximum and the minimum values of the reference signal. On the other hand, we redefine Eq. (4) with

$$\hat{i} = \frac{I_{max} + I_{min}}{2}, \quad (5)$$

where I_{max} and I_{min} are the maximum and the minimum values of the inductor current during the clock interval. Figure 2 shows waveform behavior of the inductor current. The waveform behaviors during a clock interval is classified into three cases. The switch keeps ON during clock interval if $i_k \leq I_L$ is satisfied, whereas it keeps OFF if $i_k > I_H$ is satisfied. On the other hand, the switch, which is initially connected to ON position, turns OFF at a time when $i_k = i_{ref}(t)$ is satisfied. Therefore, the return map can be defined as follows.

$$i_{k+1} = \begin{cases} i_k + \dfrac{E}{L}T & (i_k \leqq I_L), \\ \hat{i}_k + \dfrac{E}{L}\left(1 - \dfrac{i_k - I_L}{I_H - I_L}\right)T & (I_L < i_k \leqq I_H), \\ i_k + \dfrac{E - E_0}{L}T & (I_H \leqq i_k), \end{cases} \quad (6)$$

where $\hat{i}_k$ can be defined as follows.

$$\hat{i}_k = i_k + (E - E_0)\frac{i_k - I_L}{L(I_H - I_L)}T. \quad (7)$$

Figure 3 shows a 1-parameter bifurcation diagram in the previous model, whereas Fig. 4 shows that in the improved model. The bifurcation parameter I_L is varied from $I_L = 0.2$ to $I_L = 1$. Note that we plot I_L and I_H in the figures because these parameters are the borders that divides waveform behaviors during clock interval, i.e. three cases of return maps. It is clear that the bifurcation phenomena occur around $I_L = 0.4$ and $I_L = 0.8$, respectively. Note that the period doubling bifurcation may occur around $I_L = 0.4$ and immediately after that border-collision bifurcation occurs, and the period-one solution bifurcates to nonperiodic solution. In addition, a bifurcation phenomenon which occurs around $I_L = 0.8$ may be a kind of the border-collision bifurcation because a part of the solution collides with I_L. On the other hand, there are rich bifurcation phenomena in the improved model. For example,

the period doubling bifurcation may occur around $I_L = 0.3$, and after that the period doubling bifurcation including the border-collision bifurcation, which is the same mechanism with previous model, may occur around $I_L = 0.4$. Moreover, we can observe that the period doubling and border-collision bifurcations occurring by increasing the bifurcation parameter. Actually, rich nonlinear phenomena, such as period doubling and border-collision bifurcations, are observed in the PWM-1 voltage-mode-controlled DC-DC converter [2]. Therefore, we conclude that improved model is better to investigate the qualitative property of it.

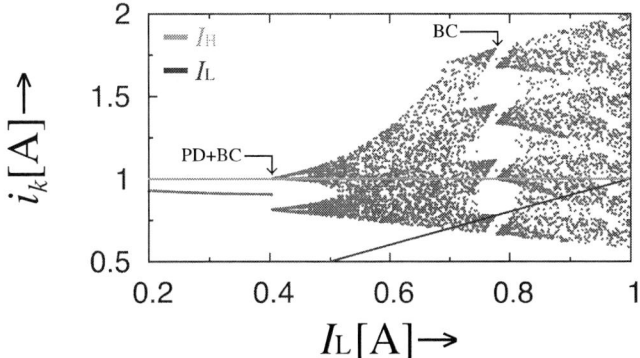

Fig. 3. 1-parameter bifurcation diagram (previous model).

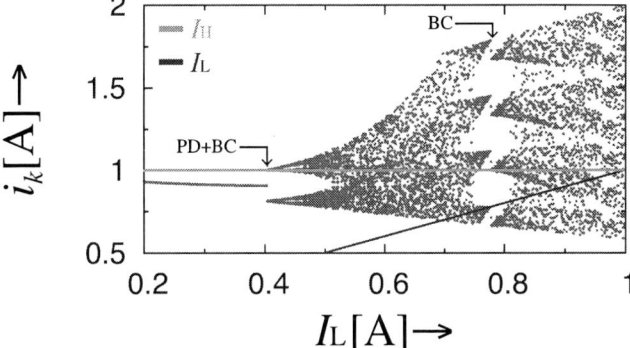

Fig. 4. 1-parameter bifurcation diagram (improved model).

III. Conclusion

In this report, we studied a simple circuit model for PWM-1-controlled DC-DC converter. First, the circuit model was shown, and then the circuit dynamics was explained. Finally, we calculated 1-parameter bifurcation diagrams. We clarified that the proposed model is better to investigate fundamental characteristic of the practical circuit. The experimental study of this circuit model is our future work.

References

[1] S. Banerjee G. C. Verghese, "Nonlinear phenomena in power electronics: attractors, bifurcations," Chaos, and Nonlinear Control, *IEEE Press*, 2001.

[2] S. Maity, D. Tripathy, T. K. Bhattacharya, S. Banerjee, "Bifurcation analysis of PWM-1 voltage-mode-controlled buck converter using the exact discrete model," *IEEE Transaction on Circuits and Systems I*, vol. 54, no. 5, pp. 1120–1130, 2007.

[3] R. Miki, H. Ohtagaki, T. Kousaka, K. Shinohara, H. Asahara, "PWM-1 current-controlled DC-DC converter," Proc. of 2017 Taiwan and Japan Conference on Circuits and Systems, p. 17, 2017.

Analysis and Design of Phase-Controlled Class-D ZVS Inverter

Tatsuki Ohsato[*], Yuta Yamada,[*], Xiuqin Wei,[†], and Hiroo Sekiya[*]

[*]Graduate School of Science and Engineering, Chiba University, Chiba, 263-8522 Japan

Email: t.osato@chiba-u.jp

Email: sekiya@faculty.chiba-u.jp

[†]Department of Electrical Engineering, Chiba Institue of Technology. Chiba, 275-0016 Japan

Email: xiuqin.wei@p.chibakoudai.jp

Abstract—This paper presents a design of phase shift-controlled class-D ZVS inverter by the steady-state analytical expression. With the expressions of the MOSFET body diode effects and harmonic components of resonant currents, the design values, which let the inverters achieve zero-voltage switching continuously in the controlled range are derived. The validity of the design was confirmed by experiment results.

Keywords; Soft-switching, MOSFET; Class-D inverter; body diode; Phase-shift controlled

I. INTRODUCTION

The class-D inverter is one of the typical resonant inverters[1], which can be widely used for applications such as induction heating, transmitter part of the wireless power transfer, and plasma power supply. By applying the zero-voltage switching (ZVS) technique on the class-D inverter, it is possible to achieve high power-conversion efficiency at high frequencies.

For practical application, it is necessary to adjust the output power. Phase-shift controlled is one of the effective method for controlling the output power. By the phase shift between the two inverters adjusted to the load variation, the output power can be controlled correctly. However, phase-shift controlled will cause the non-ZVS condition. the ZVS condition is continuously maintained against the variation of the phase shift[2]. Distribution map such as zero-voltage switching region, contours of power-conversion efficiency and output power are useful for design. Numerical algorithm for deriving zero-voltage switching region of a phase-shift controlled class-D inverter has been proposed [3].

As the resonant inverter usually takes long transient time for converging to the steady state, the numerical calculation of a differential equation becomes time consuming. So, for the steady state analysis, we can assume the output voltage and the output current as periodic functions.

In the conventional analysis, analytical expression including anti-parallel diode of MOSFET effect for class-E inverter is presented[4]. By using this analysis, each of the switching conditions is fixed by corresponding group of the circuit parameters, respectively. Most analysis assumed the resonant current as sinusoidal. However, the waveforms of the resonant current with varying phase shift are often non-sinusoidal. By considering harmonic component of resonant current, the

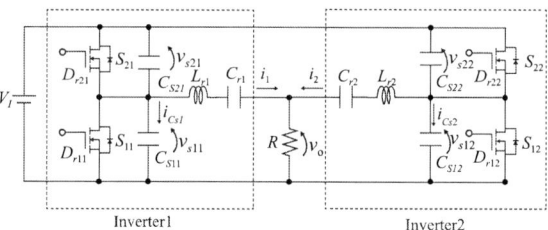

Fig. 1. Circuit schematic of out of phase controlled class D inverter

accurate waveforms of the resonant current can be obtained[5].

This paper presents a steady-state analysis for the phase-shift controlled class-D ZVS inverter. By considering harmonic components, high accurate waveform at any phase shift can be achieved. In the analysis, the phase-shift controlled class-D ZVS inverter, which achieves the ZVS condition at entire control regions can be designed. The validity and effectiveness of the design were confirmed by carrying out circuit experiments.

II. THE PHASE-CONTROLLED CLASS-D ZVS INVERTER

The output power voltage is controlled by changing the phase shift differences between the drive signals of the two inverters connected in parallel or series. Figure 1 shows the circuit topology of the phase-controlled class-D ZVS inverter. The phase-controlled class-D ZVS inverter consists of two class-D inverters connected in parallel. Each class-D inverter is composed of input voltage source V_I, two switching devices S_{1j} and S_{2j}, shunt capacitor C_{sj}, series-resonant circuit $L_{rj} - C_{rj}$, and load resistor R, where the subscription j is a label of the inverters.

III. ANALYSIS OF STEADY WAVEFORM

A. Assumptions

In this paper, the analysis is based on the following assumptions.

    a)   The resonant current i_j is taken the harmonic components into account.

    b)   All the switching devices, which are MOSFET and body-diodes of the MOSFET, work as ideal switching devices. Namely, zero switching time and infinite off-resistance are assumed.

c) The switches S_{11}, S_{21} turn on at $\theta = 0$ and $\theta = \pi$, respectively, with duty ratio D. Additionally, the class-D inverters 1 and 2 are driven with phase shift ϕ.

d) The equivalent series resistances of all passive components are ignored. This is because they are sufficiently small and not to affect the waveforms.

B. ZVS region map

In this analysis, ϕ_{ini} is defined as the phase shift between two inverters in the nominal state. The nominal state means that both inverters are at the zero-voltage switching and zero-derivative switching conditions simultaneously. The switching-state boundary conditions are considered. At the boundary of the ZVS and non-ZVS, the switch voltage reaches zero at turn-on instant of the driving signal. In other words, the ZVS operation can be achieved without anti-parallel diode operation at the boundary. Namely, we have the switching-state boundary conditions as

$$v_{sj}(2\pi) = 0. \tag{1}$$

A zero-voltage switching map of the switching states is drawn on ϕ_{ini} and ϕ space, where ϕ_{ini} is the phase that both inverters 1 and 2 at the switching-state boundary condition. Additionally, the slope of the voltage at turn on instant are specified as

$$\left. \frac{dv_{sj}}{d\theta} \right|_{\theta = 2\pi} = \alpha. \tag{2}$$

From (1)-(2), algebraic equations are obtained. The selection of the unknown parameters in this equation depends on renewed parameter. Figure 2 shows the ZVS region map on ϕ_{ini} and ϕ space for fixed $\alpha = 0$. In Fig. 2, the solid lines are obtained from the conditions $\phi_{ini} = \phi$. On the other hand, the dashed line is obtained by applying the condition of (1). From Fig. 2, the widest ZVS region can be obtained at $\phi_{ini} = 14$ degree.

IV. VERIFICATION OF SWITCH STATE

For confirming the validity of the design by analytical expression, experimental measurements were carried out. The design specifications were given as follows: operating frequency f = 400 kHz, switch on-duty ratio D = 0.4, DC-input voltage V_I = 80 V and load resistance R = 25 Ω. Resonant capacitance C_{r1} = 3.54 nF and C_{r2} = 3.36 nF, Resonant inductance L_{r1} = 55.58 μH and L_{r2} = 53.16 μH. Four IRF510 MOSFETs were used as switching devices. Therefore, shunt capacitance C_{s1} = C_{s2} = 90 pF were given as parasitic capacitance. Figure. 3 shows the waveforms for ϕ = 0, 14 and 30 degrees from analytical and experimental measurements. It is seen from Fig. 3 (a) that only Inverter 2 achieved the ZVS condition. However, it is seen from Fig. 3 (b) and (c) that both Inverters 1 and 2 achieved the ZVS condition. All the experimental switching states described above agreed with the ZVS region map well. It can be stated that the measured values agreed with the analytical expression quantitatively, which showed the validity of the obtained ZVS region map.

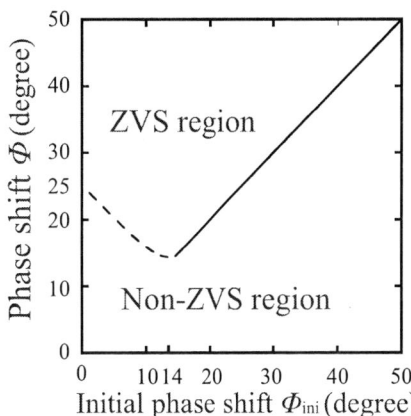

Fig. 2. ZVS region of the phase-controlled class-D inverter on ϕ_{ini} space.

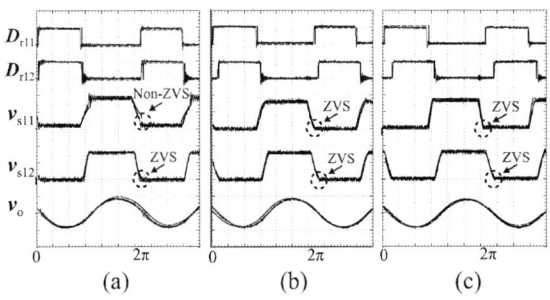

Fig. 3. Analytical (dashed line) and experimental (solid line) waveforms for ϕ_{ini} = 14 degrees and α = 0.(a)For ϕ = 0 degrees. (b)For ϕ = 14 degrees. (c)For ϕ = 30 degrees.

V. CONCLUSION

This paper has presented the design of the phase shift controlled class-D ZVS inverter by steady-state analytical expressions. By considering the harmonic components of the resonant current and the anti-parallel diode of the MOSFET, the simple design equations that the relationships between the circuit parameters and responses are obtained. We can design the phase shift controlled class-D inverter which achieving ZVS at entire control regions from the obtained ZVS region map. The analytical expressions showed good agreement with experimental results.

REFERENCES

[1] M. K. Kazimierczuk, "Class-D voltage-switching MOSFET power amplifier," IEE Proc. B, vol. 138, no. 6, pp. 285-296, Nov. 1991.

[2] D. Kawamoto, H. Sekiya, H. Koizumi, I. Sasase, and S. Mori, "Design of phase-controlled class E inverter with asymmetric circuit configuration," IEEE Trans. Circuits Systs. II: express Briefs, vol. 51, no. 10, pp. 523-528, Oct. 2004, .

[3] Y. Yamada, T. Nagashima, Y. Ibuki, Y. Fukumoto, and H. Sekiya, "Numerical algorithm for distribution-map derivations of switching converters," in proc. IEEE IECON, Nov. 2014, pp. 4338-4344.

[4] T. Nagashima, X.Wei, T. Suetsugu, M. K. Kazimierczuk and H. Sekiya, "Waveform equations, output power, and power conversion efficiency for class-E inverter outside nominal operation," IEEE Trans. Ind. Electron., vol. 61, no. 4, pp. 1799-1810, Apr. 2014.

[5] X. Wei, T. Nagashima, M. K. Kazimierczuk, H. Sekiya and T. Suetsugu, "Analysis and design of class-E_M power amplifier," IEEE Trans. Circuits Syst. I, vol. 61, no. 4, pp. 976-986, Apr. 2014.

Design of Two Template Cellular Neural Networks for Color Image Processing

Yasuteru Hosokawa

Dept. of Media and Information Systems,
Shikoku University
Furukawa, Ohjin, Tokushima, Japan
hosokawa@keiei.shikoku-u.ac.jp

Yoko Uwate and Yoshifumi Nishio

Dept. Electrical and Electronic Engineering,
Tokushima University
2-1 Minami-Josanjima, Tokushima, Japan
{uwate, nishio}@ee.tokushima-u.ac.jp

Abstract— **In this study, TTCNNs for color image processing were designed. In this design, Bayer arrangement is applied to TTCNNs. Some simulation results were shown for a possibility of one-layer color image processing using TTCNNs.**

Keywords; Cellular Neural Networks; Color Image Processing; Bayer Filter

I. INTRODUCTION

One of important advantages of Cellular Neural Networks (CNNs) [1][2] is a simple and uniform structure. The simple and uniform structure makes an IC implementation of CNNs easily. However, almost all modified CNNs proposed by many researchers became complicated systems by the modify.

Two Template CNNs (TTCNN) [3], which are one kind of Modified CNNs, keeps the simple and uniform structure of a conventional CNN. The difference from a conventional CNN is the number of signal lines for cloning template values only. The structure of one Cell is the same. Therefore, the simple and uniform structure is kept.

In this study, TTCNNs for color image processing are designed. In order to process a color image, Bayer arrangement, which is a well-known arrangement of a color filter array of an image sensor, is applied to TTCNNs. By this design, TTCNNs can process color image by one layer. Thus, this design has a possibility of hardware coupling an image sensor and TTTCNN.

II. TWO TEMPLATE CELLULAR NEURAL NETWORKS

Figure 1 shows a system model of Two Template CNN (TTCNN). Different cloning template values can be set between Cell alpha and Cell beta. Arrangement of two kinds of cells is a checkered pattern. In case of applied same cloning template values, this system becomes same as a conventional CNN. The difference from a conventional CNN is only this point. Therefore, TTCNN has an almost same structure as a conventional CNN. This feature means that the simple and uniform structure of a conventional CNN is kept. State equations are described as follows.

State equation of Cell alpha:

$$\frac{dx_{ij}}{dt} = -x_{ij} + I_\alpha + \sum_{c(k,l)} A_\alpha(i,j;k,l)y_{kl} + \sum_{c(k,l)} B_\alpha(i,j;k,l)u_{kl} \tag{1}$$

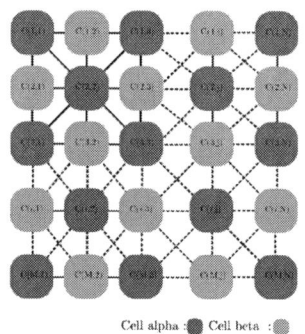

Cell alpha : ● Cell beta : ●

Figure 1. System Model of TTCNN.

State equation of Cell beta:

$$\frac{dx_{ij}}{dt} = -x_{ij} + I_\beta + \sum_{c(k,l)} A_\beta(i,j;k,l)y_{kl} + \sum_{c(k,l)} B_\beta(i,j;k,l)u_{kl} \tag{2}$$

Output function:

$$y_{ij} = 0.5(|x_{ij} + 1| - |x_{ij} - 1|). \tag{3}$$

Only some subscripts of state equations are different from a conventional CNN. Thus, the characteristic and structure of cells is same as a conventional CNN. Some template sets for this system were proposed in [6].

III. COLOR IMAGE PROCESSING BY TTCNN

In this section, Bayer arrangement and its application to TTCNN are described. Bayer arrangement is very simple as one-layer color image processing. This arrangement is used in almost all image sensors of one-layer. By applying this arrangement to TTCNN, one-layer color image processing of TTCNN can be realized.

A. Bayer Arrangement

Bayer arrangement patented by Bryce E. Bayer in 1975 is a kind of color image filter. Photo diodes used in an image sensor can obtain strength of light only. Therefore, in order to processing a color image, processing three primary colors is needed. Bayer arrangement can process three primary colors on one photo diodes array. This arrangement is used in almost all one-layer color image sensors because it is very simple and it can be realized with low costs.

978-1-5386-7961-6/18 $31.00 © 2018 IEEE

ISOCC 2018

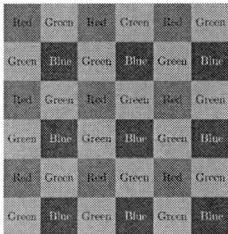

Figure 2. Bayer arrangement.

Processing of Bayer filter is as follows. First, an input image is filtered by Bayer arrangement as shown in Fig. 2. Next, each color pixel data is generated using four filtered pixels data which are a red, a blue, and two green. This process is known as demosaicing. The number of obtained color image data is equal to the number of photo diodes minus boundary pixels because filtered pixels data excepted boundary pixels is used four times. We can obtain color image data from two steps processing only.

B. Applying Bayer arrangement to TTCNN

Bayer arrangement is similar to cell arrangement of TTCNN. Cell alpha of TTCNN is corresponding to Red or Blue of Bayer arrangement. Cell beta is corresponding to Green. Hence, Bayer arrangement is applied to TTCNN for color image processing. Some of simulation results are shown in Figs. 3 and 4.

Figures 3 show some simulation result of applying CIE 1931 xy color space diagram [5] as an input image. Fig. 3 (a) is the input image. Figure 3 (b)-(d) are simulation results in case of applying following cloning template values, respectively.

$$\mathbf{A}_\alpha = \begin{pmatrix} 1 & 1 & 1 \\ 1 & 1 & 1 \\ 1 & 1 & 1 \end{pmatrix}, \quad \mathbf{A}_\beta = \begin{pmatrix} -1 & 1 & -1 \\ 1 & 0.5 & 1 \\ -1 & 1 & -1 \end{pmatrix}, \quad \mathbf{B}_\alpha = \begin{pmatrix} 0 & 0 & 0 \\ 0 & 1 & 0 \\ 0 & 0 & 0 \end{pmatrix}, \quad \mathbf{B}_\beta = \begin{pmatrix} 1 & 0 & 1 \\ 0 & 1 & 0 \\ 1 & 0 & 1 \end{pmatrix}, \quad (4)$$

$$I_\alpha = 0, \qquad I_\beta = -1$$

$$\mathbf{A}_\alpha = \begin{pmatrix} 0 & 1 & 0 \\ 1 & -2 & 1 \\ 0 & 1 & 0 \end{pmatrix}, \quad \mathbf{A}_\beta = \begin{pmatrix} 2 & 1 & 2 \\ 1 & 1 & 1 \\ 2 & 1 & 2 \end{pmatrix}, \quad \mathbf{B}_\alpha = \begin{pmatrix} -1 & 0 & -1 \\ 0 & 4 & 0 \\ -1 & 0 & -1 \end{pmatrix}, \quad \mathbf{B}_\beta = \begin{pmatrix} 1 & 0 & 1 \\ 0 & 1 & 0 \\ 1 & 0 & 1 \end{pmatrix}, \quad (5)$$

$$I_\alpha = -2, \qquad I_\beta = 1$$

$$\mathbf{A}_\alpha = \begin{pmatrix} 1.6 & 1.4 & 1.6 \\ 1.4 & 0.8 & 1.4 \\ 1.6 & 1.4 & 1.6 \end{pmatrix}, \quad \mathbf{A}_\beta = \begin{pmatrix} -1 & 1.8 & -1 \\ 1.8 & 0.4 & 1.8 \\ -1 & 1.8 & -1 \end{pmatrix}, \quad \mathbf{B}_\alpha = \begin{pmatrix} 0 & 0.6 & 0 \\ 0.6 & -2 & 0.6 \\ 0 & 0.6 & 0 \end{pmatrix}, \quad \mathbf{B}_\beta = \begin{pmatrix} -1 & -1 & -1 \\ -1 & 1.6 & -1 \\ -1 & -1 & -1 \end{pmatrix},$$

$$I_\alpha = 0.6, \qquad I_\beta = 0.1 \tag{6}$$

In Fig. 3 (b), purple is extracted. In Fig. 3 (c), cyan and yellow are extracted. In Fig. 3 (d), white is extracted as purple.

Another case is shown in Figs. 4. Figure 4 (a) shows an input image which shows a sky, a forest, and a river. Figure 4 (b) is a simulation result in case of applying following cloning template values.

$$\mathbf{A}_\alpha = \begin{pmatrix} -3 & 1 & 2 \\ -3 & -5 & -3 \\ -2 & 4 & -5 \end{pmatrix}, \quad \mathbf{A}_\beta = \begin{pmatrix} 3 & 2 & -2 \\ 3 & -4 & -3 \\ 3 & 2 & 0 \end{pmatrix}, \quad \mathbf{B}_\alpha = \begin{pmatrix} 2 & 1 & -1 \\ 3 & 4 & 3 \\ 3 & 4 & -5 \end{pmatrix}, \quad \mathbf{B}_\beta = \begin{pmatrix} -2 & 3 & 2 \\ -3 & -4 & 2 \\ -5 & 4 & 3 \end{pmatrix},$$

$$I_\alpha = 3, \qquad I_\beta = -1 \tag{7}$$

The forest part is recognized clearly as orange color. Thus, processing some simple color image can be confirmed.

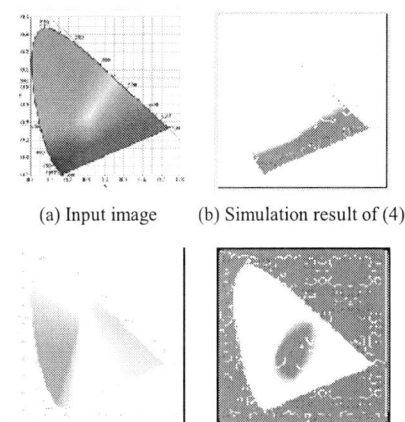

(a) Input image   (b) Simulation result of (4)

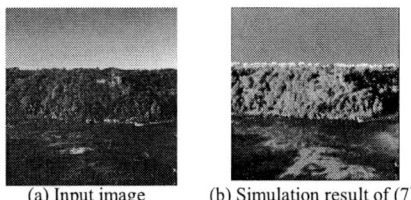

(c) Simulation result of (5)   (d) Simulation result of (6)

Figure 3. Simulation results of applying CIE 1931 xy color space diagram [5].

(a) Input image   (b) Simulation result of (7)

Figure 4. Simulation results of a picture of a sky, a forest, and a river.

IV. CONCLUSION

In this study, TTCNNs for color image processing are designed. In order to process a color image, Bayer arrangement, is applied to TTCNNs. By this design, it was confirmed that TTCNNs can process color image by one layer. This result shows a possibility of hardware coupling an image sensor and TTTCNN.

ACKNOWLEDGMENT

This work was partly supported by JSPS Grant-in-Aid for Scientific Research 16K06357.

REFERENCES

[1] L. O. Chua and L. Yang, "Cellular neural networks: Theory," *IEEE Trans. Circuits and Systems*, vol. 35, no. 10, pp. 1257– 1272, 1988.

[2] L. O. Chua and L. Yang, "Cellular neural networks: Applications," *IEEE Trans. Circuits and Systems*, vol. 35, no. 10, pp. 1273–1290, 1988.

[3] J. Fujii, Y. Hosokawa and Y. Nishio, "Wave Phenomena in Cellular Neural Networks Using Two Kinds of Template Sets," *Proc. of International Symposium on Nonlinear Theory and its Applications*, pp. 23-26, 2007.

[4] Bryce E. Bayer, "Color Imaging Array, " *U.S. Patent 3971065*, Mar. 5, 1975.

[5] BenRG, File:CIE1931xy_blank.svg, Wikipedia Commons, https://comm ons.wikimedia.org/wiki/File:CIE1931xy_blank.svg

[6] R. Yamane, Y. Hosokawa and Y. Nishio, "Image Processing Using Two-Template CNN with Conventional Cloning Templates" *Proceedings of IEEE Workshop on Nonlinear Circuit Networks (NCN'10)*, pp. 5-8, Dec. 2010.

Synchronization Phenomena of Coupled Chaotic Circuits Network with Coupling Strength Depending on Number of Degree

Kyohei Fujii, Shuhei Hashimoto, Yoko Uwate and Yoshifumi Nishio
Dept. Electrical and Electronic Engineering,
Tokushima University
2-1 Minami-Josanjima, Tokushima 7708506, Japan
Email: fujii, s-hashimoto, uwate, nishio@ee.tokushima-u.ac.jp

Abstract— **We investigate the influence of direct coupling and indirect coupling for synchronization state in the network. The node of the network is expressed by chaotic circuit. This complex network is designed that a node has two or more links except for one node with scale free property. The coupling strength is determined depending on the number of links between the nodes. In this investigation, we find that not only the direct coupling but also the indirect coupling by the path is strongly related for synchronization state.**

Keywords; Synchronization, Complex network, Chaotic Circuits

I. INTRODUCTION

In our living life, complex network can be seen in various fields such as airport network, computer network and neurons in the brain. Furthermore, synchronization phenomena observed in complex network have been investigated by many researchers [1], [2]. Especially synchronization phenomena of coupled oscillatory systems are very interesting. There are various investigations of synchronization using chaotic circuits. However, few studies focus on the difference in coupling strength [3], [4].

In this study, we focus on the path and coupling strength in coupled chaotic circuit network. We design a network that each node has different coupling strength depending the number of degree and the network has scale free property. In order to make the network which the total coupling strength of node in the network has nearly uniform, we apply the following rule to set the coupling strength. When the difference of number of degree between two nodes is large, the coupling strength between two nodes is set to small value.

By using the computer simulations, we investigate synchronization between hubs when one link is opened. Furthermore, we also investigate synchronization between hubs when another link is opened.

II. CIRCUIT MODEL

Figure 1 shows the chaotic circuit which is three dimensional autonomous circuit proposed by Shinriki *et al*. This circuit consists of one negative resister, two capacitors, one inductor and dual-directional three diodes [5]. And this circuit equation is shown in Eq. (1).

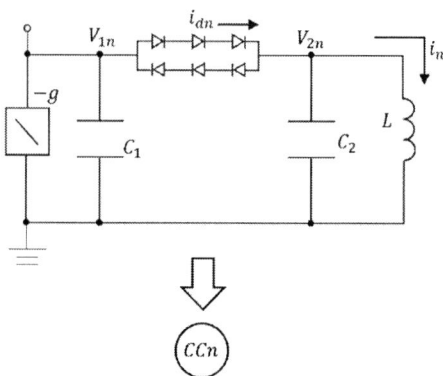

Figure 1: Circuits model.

$$
\begin{cases}
L\dfrac{di_1}{dt} = V_{2n}, \\[2mm]
C_1\dfrac{dV_{1n}}{dt} = gV_{1n} - i_{dn}, \\[2mm]
C_2\dfrac{dV_{2n}}{dt} = i_{dn} - i_n.
\end{cases}
\tag{1}
$$

The normalized equation of this circuit is given as follows:

$$
\begin{cases}
\dfrac{dx}{d\tau} = z_n, \\[2mm]
\dfrac{dy}{d\tau} = \alpha\gamma y_n - \alpha f(y_n - z_n), \\[2mm]
\dfrac{dz}{d\tau} = f(y_n - z_n) - x_n.
\end{cases}
\tag{2}
$$

The nonlinear function f() corresponds to the i-v characteristics of the nonlinear resistor consisting of the diodes and are described as follows:

$$
f(y_n - z_n) =
\begin{cases}
\beta(y_n - z_n - 1) & (y_n - z_n > 1), \\[2mm]
0, & (|y_n - z_n| < 1), \\[2mm]
\beta(y_n - z_n + 1), & (y_n - z_n < -1).
\end{cases}
\tag{3}
$$

978-1-5386-7961-6/18 $31.00 © 2018 IEEE

III. SYSTEM MODEL

This complex network is designed that each node has two or more links except for one node with scale free property. Only three nodes are set to the hubs (CC18, CC19, CC20). All coupling strength value are fixed with a certain rule. This rule is based on the different degrees between nodes. The value of coupling strength is determined by the number of degree of the node. Namely, the coupling strength is set to the certain value by calculating the difference of number of degree between the nodes and add 1.0 for every links.

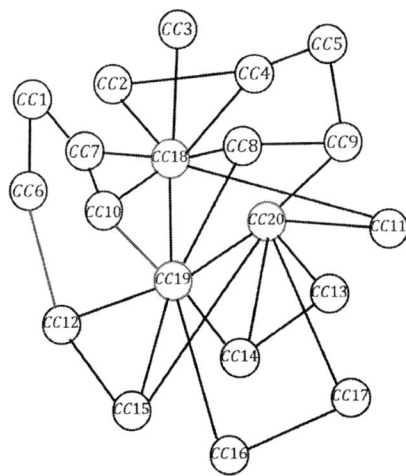

Figure 2: Network model.

IV. SIMULATION RESULT

In order to analyze synchronization states, we define the synchronization by the following equation.

$$|y_j - y_i| < 0.03 \qquad (i, j = 1, 2, \cdots, 10) \qquad (4)$$

First, we investigate the synchronization by changing the coupling strength. Figure 3 shows the synchronization rate with the coupling strength.

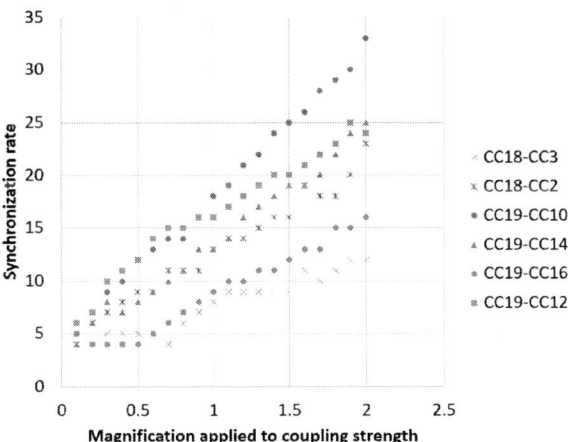

Figure 3: Synchronization rate of same strength link.

From this result, we confirm that the synchronization state is affected not only direct coupling but also indirect coupling. Next, we focus on the link of CC18 and CC19. The synchronization between CC18 and CC19 is measured when only CC10 - CC19 is opened or when only CC6 – CC12 is opened.

Table 1: Synchronization rate with open one link

Location of open link	CC10-CC19	CC6-CC12
Synchronization rate of CC18-CC19 [%]	46	36

The number of circuits between the two nodes is different for the both cases of open link (CC10-CC19 and CC6-CC12). Table 1 shows the result of synchronization rate when one link is opened. We confirm that the synchronization rate has higher value when the link between CC10-CC19 is opened.

V. CONCLUSION

In this study, we have investigated influence of path by using scale-free coupled chaotic network by changing the coupling strength. In this result, we found an effect of the path and coupling strength. Next, we have investigated synchronization between hubs when one indirect link is opened. We confirm that the effect of the coupling strength on synchronization is larger than the effect of the open link.

ACKNOWLEDGMENT

This work was partly supported by JSPS Grant-in-Aid for Scientific Research 16K06357.

REFERENCES

[1] M. Uchida, S. Shirayama, "Analysis of Network Structure and Model Estimation for SNS" IPSJ journal, Vol. 47, No. 9, Sep. 2006.

[2] H. Kori, N. Masuda, "Synchronization of Coupled Oscillators on Complex Networks" Journal of the Robotics Cociety of Japan, Vol.26, No.1, pp.6-9, 2008.

[3] K. Ago, Y. Uwate, Y. Nishio, "Investigation of Synchronization in Coupled Chaotic Circuit Network with Local Bridge", IEEE Workshop on Nonlinear Networks December 12-13, 2014.

[4] K. Oi, Y. Uwate, Y. Nishio, "Influence of Regional Change in Synchronization of Complex Networks in Coupled Parametrically Excited Oscillators", IEEE Workshop on Nonlinear Circuit Networks. December 9-10, 2016.

[5] K. Ago, Y. Uwate and Y. Nishio, "Synchronization of Coupled Chaotic Circuits with Parameter Dispersion in Small World Network", Proceedings of International Symposium on Nonlinear Theory and its Applications, pp. 431-434, Dec.2015.

Design of Convolutional Neural Network for Classifying Depth Prediction Images from Overhead

Shu Sumimoto, Yuichi Miyata, Ryuta Yoshimura, Yoko Uwate and Yoshifumi Nishio

dept. of Electrical and Electronic Engineering
Tokushima University
2-1 Minami-Josanjima, Tokushima 7708506, Japan
E-mail: sumimoto, y.miyata, yoshimura, uwate, nishio@ee.tokushima-u.ac.jp

Abstract— **We predict depth of some objects, such a person, chairs and a soccer ball and so on, in overhead images with Fully Convolutional Residual Networks (FCRN) [1]. This networks can predict depth of RGB images taken by monocular cameras. And we classify images predicted depth. Thus we aim at differentiating person or other objects.**

Keywords; Neural Network; Image classification; Depth prediction

I. INTRODUCTION

Image recognition by deep learning is used in various field. An automatic drive vehicle uses image recognition by deep learning when it avoids dangers. In this way, it is important for a machine to use image recognition by deep learning. Drones are also recently infiltrating various fields, for example delivery, rescue, guard and so on. It is necessary that drones fly safely. Then image recognition by deep learning is becoming important for that drones fly safely. Therefore, I used YOLOv3 for drones to fly safely. It is a popular object detection algorithm. Also it is suited to avoid any dangers, because it can recognize objects quickly. However, standing people in overhead images from the view of drones are not able to be recognized by using YOLOv3. It learns shapes of humans, so it is difficult for YOLOv3 to recognize humans that hide a part of a body.

In this study, we investigate the prediction of the depth of some objects, such humans, chairs and cars in overhead images with FCRN. This system can predict depth of images taken by a monocular camera, so it costs lower than any systems. Depth prediction images from FCRN have 3D data, so Convolutional Neural Network(CNN) can get more data from them than 2D images. We classify the RGB images and the depth prediction images taken by a monocular camera and we aim at differentiating human or other objects by using CNN. Also, the camera has two position, one is lower position and another is higher position. We compare learning and test accuracies of image classification with two camera position.

II. PROPOSED SYSTEM

We propose to classify depth prediction images with FCRN. First, we prepare four types images from overhead, images of a human and a chair taken by a camera which is close to objects, and images of a human and a car taken by a camera which is far from objects.

Second, we predict depth of overhead RGB images of a human, a chair and car taken by monocular cameras in Figs. 1 and 2.

Third, we classify the RGB images and the predicted depth images of a human and a chair or a car with a CNN which has 2 convolutional layers, 2 pooling layers and 2 fully connected layers. When CNN learns, images are compressed. Therefore, training and test images are 28×28 pixels. Learning rate of this CNN is 0.00005.

We compare the learning and test accuracies when a camera is close to an object and when a camera is far from an object. When they are close to a camera, we classify images of a human and a chair. On the other hand, when they are far from a camera, we classify images of a human and a car.

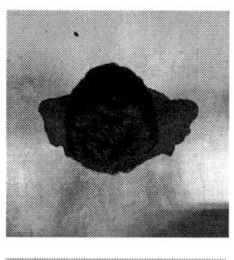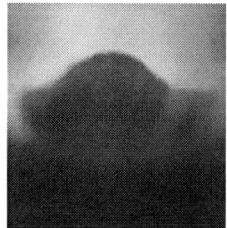
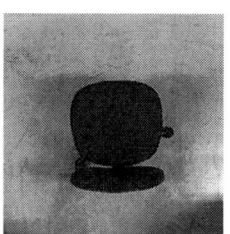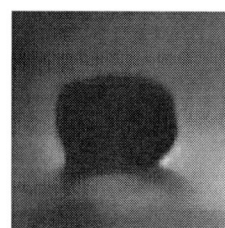

Figure 1. RGB and depth prediction images when a camera is close to objects. (training data)

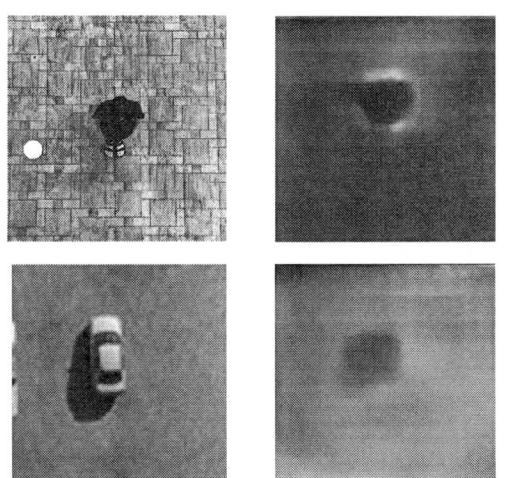

Figure 2.   RGB and depth prediction images when a camera is far from objects. (training data)

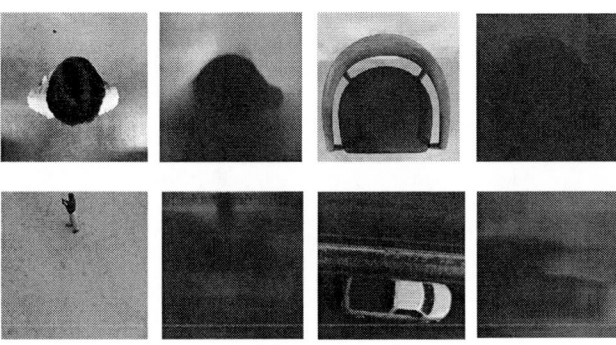

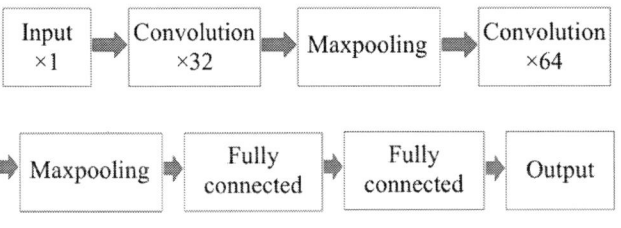

Figure 3.   RGB and depth prediction images. (trest data)

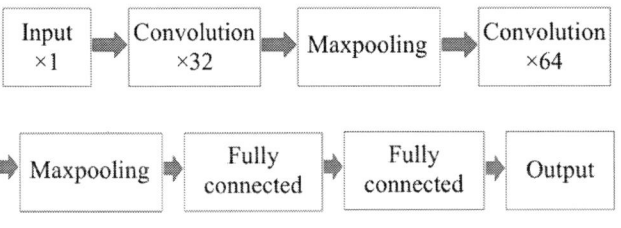

Figure 4.   Structure of CNN.

III. SIMULATION RESULT

We define as the learning steps 300, the number of the training data sets 80 which are 40 overhead images of a human and 40 overhead images of a chair and the number of the test data sets 10 which are 5 overhead images of a human and 5 overhead images of a chair. They are taken by a camera which is close to an object. Table 1 shows average of learning and test accuracies when we classify only RGB images 10 times and we classify only predicted depth images 10 times.

We define as the learning steps 100, the number of the training data sets 80 which are 40 overhead images of a human and 40 overhead images of a car and the number of the test data sets 10 which are 5 overhead images of a human and 5 overhead images of a car. They are taken by a camera which is far from an object. Table 2 shows average of learning and test accuracies when we classify only RGB images 10 times and we classify only predicted depth images 10 times.

From Tables 1 and 2, when a camera is close to an object, the test accuracy of classifying RGB images is higher than predicted depth images. On the other hand, when a camera is far from an object, the test accuracy of classifying predicted depth images is higher than RGB images.

TABLE I. AVERAGE OF LEARNING AND TEST ACCURACIES HEN A CAMERA IS CLOSE TO AN OBJECT

	RGB images	*Predicted depth images*
Learning accuracy	0.97	0.90
Test accuracy	0.87	0.72

TABLE II. AVERAGE OF LEARNING AND TEST ACCURACIES HEN A CAMERA IS FAR FROM AN OBJECT

	RGB images	*Predicted depth images*
Learning accuracy	0.83	0.76
Test accuracy	0.38	0.58

IV. CONCLUSION

From these simulation results, we consider it is effective for classifying images taken by a camera which is far from an object to predict the depth of images.

However, the learning accuracies of classifying predicted depth images are lower than RGB images. Then we will try to raise accuracies of classifying predicted depth images.

REFERENCES

[1] Iro Laina, Christian Rupprecht, Vasileios Belagiannis, Federico Tombari, Nassir Navab, "Deeper Depth Prediction with Fully Convolutional Residual Networks"

[2] Joseph Redmon, Santosh Divvala, Ross Girshick, Ali Farhadi, "You Only Look Once: Unified, Real-Time Object Detection"

[3] Joseph Redmon, Ali Farhadi, "YOLO9000: Better, Faster, Stronger"

[4] Joseph Redmon, Ali Farhadi,"YOLOv3: An Incremental Improvement"

Analysis of Chaotic Circuit Networks with One-Way Coupling

Akari Oura*, Kyohei Fujii*, Yoko Uwate*, and Yoshifumi Nishio*

dept. of Electrical and Electronic Engineering,
Tokushima University

2-1 Minami-Japan, Tokushima 770-8506, Japan
Email: {oura, fujii, uwate, nishio}@ee.tokushima-u.ac.jp

Abstract— **In this study, we investigate the synchronization phenomena of the complex network in which the chaotic circuits are coupled with one direction. We observe the synchronization by using the voltage difference between the nodes expressed by the chaos circuit. Moreover, we also observe the synchronization phenomena when the coupling strength is changed.**

Keywords; Complex network; Synchronization Phenomena; Chaotic Circuit.

I. INTRODUCTION

Synchronization phenomena are observed everywhere. For example, flashing firefly, crowing tree frogs, beating rhythm of the heart. Chaos is applied to biology, medical science and engineering. Recently, many scientists study synchronization phenomena of chaotic circuits. Especially, synchronization of chaos is very interesting phenomena. Additionally, coupled systems of chaotic elements produce many kinds of complexity phenomena such as clustering, chaos propagation and so on. Studying synchronization of chaos is expected useful in a variety of fields.

In this study, we investigate synchronization phenomena of coupled chaotic circuits with one-way coupling. Each node is applied chaotic circuit and connected with resistance via buffer. By using computer simulations, we observe interesting synchronization phenomena in the proposed chaotic circuits network.

II. NETORK MODEL

Figure 1 shows basic chaotic circuit. In this study, chaotic circuits are applied to nodes of the network. A proposed network model is shown in Fig. 2. The network consists of a chaotic circuit connected in one direction.

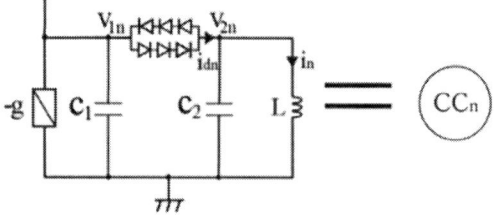

Figure 1: Circuit model.

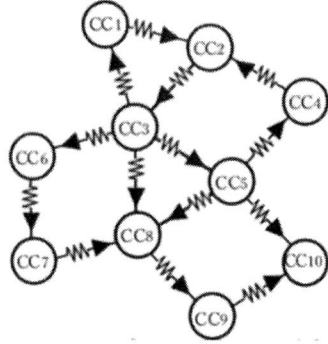

Figure 2: Network model (n=10).

First, the circuit equations are given as follows:

$$\begin{cases} L \dfrac{di_n}{dt} = v_{2n} \\ C_1 \dfrac{dv_{1n}}{dt} = gv_{1n} - i_{dn} - \dfrac{1}{R}\sum_{CC_n}(v_{1n}-v_{1k}) \\ C_2 \dfrac{dv_{2n}}{dt} = -i_n + i_{dn} \end{cases}$$

$$(n = 1,2,3,\dots,10),$$

we approximate the $i - v$ characteristics of the nonlinear resistor consisting of the diodes,

$$i_{dn} = \begin{cases} G_d(v_{1n}-v_{2n}-a) & (v_{1n}-v_{2n} > a) \\ 0 & (|v_{1n}-v_{2n}| \le a) \\ G_d(v_{1n}-v_{2n}+a) & (v_{1n}-v_{2n} < -a). \end{cases}$$

By using the parameters and variables as follows:

$$i_n = \sqrt{\frac{C_2}{L}}ax_n, v_{1n} = ay_n, v_{2n} = az_n$$

$$t = \sqrt{LC_2}\tau, "\cdot" = \frac{d}{d\tau}, \alpha = \frac{C_2}{C_1}$$

$$\beta = \sqrt{\frac{L}{C_2}}G_d, \gamma = \sqrt{\frac{L}{C_2}}g, \delta = \frac{1}{R}\sqrt{\frac{L}{C_2}}.$$

The normalized circuit equations are given as follows:

$$\begin{cases} \dot{x} = z_n \\ \dot{y} = \alpha\gamma y_n - \alpha\beta f(y_n - z_n) - \alpha\delta f(y_k - y_n) \\ \dot{z} = \beta f(y_n - z_n) - x_n, \end{cases}$$

where the nonlinear function corresponding to the characteristics of the nonlinear resistor of the diodes and are described as follows:

$$f(y_n - z_n) = \begin{cases} y_n - z_n - 1 & (y_n - z_n > 1) \\ 0 & (|y_n - z_n| \le 1) \\ y_n - z_n + 1 & (y_n - z_n < -1). \end{cases}$$

By computer simulations, we observe the synchronization by calculating the voltage and observing the voltage difference.

III. SIMULATION RESULT

In this study, we fix parameters as $\alpha = 0.5$, $\beta = 20$, $\gamma = 0.5$ and $\delta = 0.22$ on all circuits. The simulation results are shown in Figs. 3 to 5. Figures 3 and 4 show the phase difference.

We observe completely in-phase state, in CC1-CC2 to CC5 (Fig. 3). However, we did not observe in-phase state in CC1-CC6 to CC10 (Fig. 4). We can say that there are two groups in the chaotic network depending on the synchronization state.

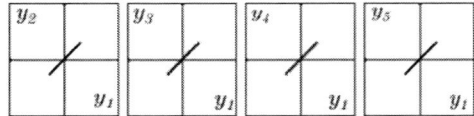

Figure 3: Phase difference (CC1-CC2, CC1-CC3, CC1-CC4, and CC1-CC5). $\delta = 0.22$.

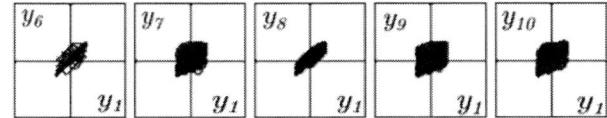

Figure 4: Phase difference (CC1-CC6, CC1-CC7, CC1-CC8, CC1-CC9, and CC1-CC10). $\delta = 0.22$.

Figure 5 shows the voltage difference. We did not observe difference of voltage in CC1-CC2 to CC5. However, we observed difference of voltage in CC1-CC6 to CC10. We observed that the voltage difference is smaller in CC1-CC6 than in CC1-CC9 to CC10.

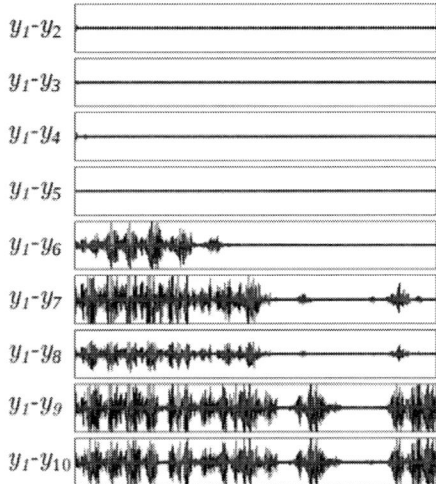

Figure 5: Voltage difference. $\delta = 0.22$.

IV. CONCLUSION

We studied synchronization phenomena of coupled chaotic circuit with one-way coupling. Synchronization phenomena was observed certain part of the circuit with completely in-phase state. In this study, we used ten chaotic circuits.

For the future works, we will use more number of chaotic circuits. By using more chaotic circuits, we expect to observe complicated phenomena such as clustering. Also we expect to observe chaos synchronization. Also, it is expected that differences in characteristics will be observed by changing network connection.

ACKNOWLEDGMENT

This work was partly supported by JSPS Grant-in-Aid for Scientific Research 16K06357.

REFERENCES

[1] K. Kaneko, "Clustering, Coding, Switching, Hierarchical Ordering, and Control in a Network of Chaotic Elements," Physica D, vol. 41, pp. 137-172, 1990.

[2] K. Ago, "Synchronization Phenomena of Coupled Chaotic Circuit Network with Bridge" IEEE Workshop on Nonlinear Circuit Networks (NCN'13), pp. 27-29, Dec. 2013

[3] K. Ago, "Clustering of Coupled Logistic Map with Bridge" Heisei 20th Conference on Electrical Relations Association Shikoku Branch Conference Presentation paper, no. 1-23, p. 23, Sep. 2013.

[4] M. Inoue and H. Hata, "Foundations and development of chaos science", 1999, pp.171–199.

Area-Delay Product Efficient Design for Convolutional Neural Network Circuits Using Logarithmic Number Systems

Tso-Bing Juang

Department of Computer Science and Information
Engineering
National Pingtung University (NPTU)
Pingtung City, Taiwan, R. O. C.
tsobing@mail.nptu.edu.tw

Cong-Yi Lin[1] and Guan-Zhong Lin[2]

Department of Computer Science and Information
Engineering
National Pingtung University (NPTU)
Pingtung City, Taiwan, R. O. C.
[1]d40214104@gmail.com
[2]c27350627@gmail.com

Abstract— **In this paper, we have proposed area-delay product (ADP) efficient design for convolutional neural network (CNN) circuits using logarithmic number systems (LNS). By employing LNS-based schemes, the area overhead for large amount of conventional multipliers required in CNN circuits can be tremendously reduced. Simulation results show that our proposed design can achieve lower-error with almost 60% ADP savings compared with the conventional multipliers-based design, which is suitable for deep learning applications.**

Keywords: Convolutional neural network (CNN), deep learning, logarithmic number system (LNS), computer arithmetic, VLSI design.

I. INTRODUCTION

As for the progress of deep learning technology in recent years, people can design more comprehensive systems to make computer can think as experts. Convolutional neural network (CNN) is one of commonly used among many methods in deep learning [1-2]. Fig. 1 is the execution example of CNN, where it consists of three layers to produce the output, and each layer consists of some multiplications and additions performed by input numbers and the fixed coefficients stored in Filter array.

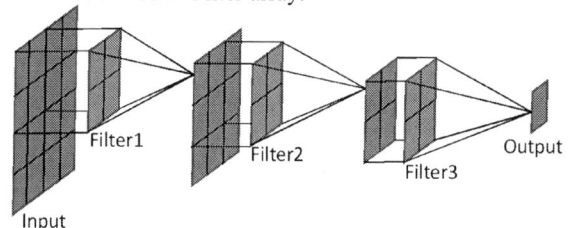

Fig. 1: The architecture of CNN circuits used in our work.

As we can observe in Fig. 1 and Fig. 2, the area costs required are very huge; therefore some work are necessary to take to reduce the computation overhead. In previous years, we have been successfully designed many works about logarithmic number system (LNS)-based schemes to speed up the executions for many computation-intensive applications in [3-4], we expect to make use of LNS-based schemes to reduce the computation overhead especially for the multiplications since LNS-based method could simplify the conventional multiplications into simple additions.

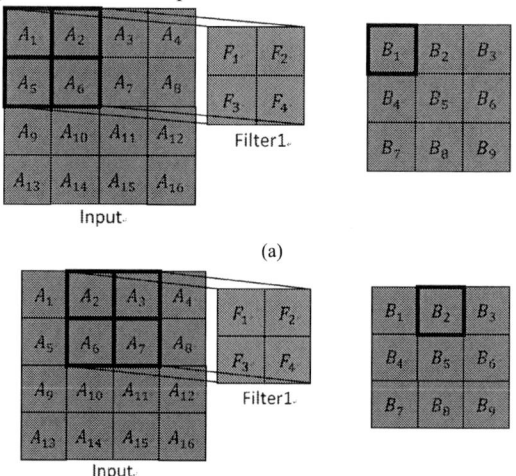

Fig. 2: Execution example for each layer of CNN for obtaining the values of (a) B_1 ($B_1 = A_1 \times F_1 + A_2 \times F_2 + A_3 \times F_3 + A_4 \times F_4$) and (b) B_2 ($B_2 = A_2 \times F_1 + A_3 \times F_2 + A_6 \times F_3 + A_7 \times F_4$), respectively.

Our purpose is to use LNS schemes applied in CNN circuits shown in Fig. 1 to set up a design example and to show that LNS can help to reduce the computation overhead with tolerable errors.

This paper is organized as follows. In Section II we will introduce the architecture of LNS schemes, and then LNS-based CNN circuits will be given in Section III. Simulation results and comparisons of our designed CNN circuits are shown in Section IV, and Section V concludes the paper.

II. ARCHITECTURE OF LNS-BASED SCHEMES

Fig. 3 is the architecture of LNS schemes, where three steps are required. For example, to perform $x \times y$, x and y should be converted to $\log_2 x$ and $\log_2 y$ in the first stage, and to

978-1-5386-7961-6/18 $31.00 © 2018 IEEE

be added in the second stage to produce the value of $\log_2 x + \log_2 y$, finally to be converted to the anti-logarithmic value of $2^{\log_2 x + \log_2 y} = x \times y$.

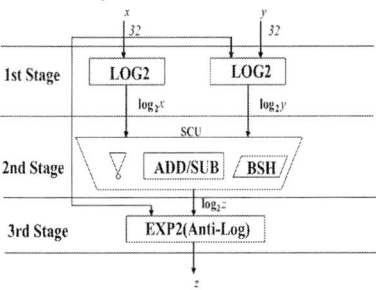

Fig.3: The architecture for LNS-based schemes [3].

III. PROPOSED LNS-BASED CNN CIRCUITS

According to Fig. 1, there are 56 multiplications and 42 additions required for performing CNN operations. If tremendous multiplications can be replaced by LNS-based additions, we can reduce many computation efforts. Referred to Section II, taking the value of B_1 for example, since $B_1 = A_1 \times F_1 + A_2 \times F_2 + A_3 \times F_3 + A_4 \times F_4$, using LNS schemes, which can be converted to $B_1 = 2^{\log_2 A_1 + \log_2 F_1} + 2^{\log_2 A_2 + \log_2 F_2} + 2^{\log_2 A_3 + \log_2 F_3} + 2^{\log_2 A_4 + \log_2 F_4}$
Besides, since all the coefficients in Filter1~Filter3 are fixed, the values of $\log_2 F_1$ to $\log_2 F_4$ can be computed first without the overhead using the first stage.

TABLE I: The stored contents of all coefficients in CNN.

Coefficients	x	$\log_2 x$
Filter1-F_1	1.0001	0.0001011001100011111011011
Filter1-F_2	1.0101	0.0110010001101110111101010
Filter1-F_3	1.0011	0.00111111011110000010110101
Filter1-F_4	1.001	0.000101011100000000011010001
Filter2-F_1	1.01	0.0101001001101001111000000100
Filter2-F_2	1.101	0.101110011010100000000010001
Filter2-F_3	1	0.000000000000000000000000000
Filter2-F_4	1.011	0.0111010110011101010010011110
Filter3-F_1	1.0001	0.0001011001100011111011011
Filter3-F_2	1.0011	0.00111111011110000010110101
Filter3-F_3	1.0101	0.0110010001101110111101010
Filter3-F_4	1.001	0.000101011100000000011010001

We use Eq. (1) and (2) to perform the lower-error logarithmic and anti-logarithmic conversions required in simplifying the computation overhead of multiplications, respectively, that were been proposed in our previous work [3-4].

$$\log_2(1+x)' = \begin{cases} x + \dfrac{x_{-2} \vee x_{-3}}{32} + \dfrac{x_{-2} \vee x_{-4}}{64} + \dfrac{x_{-2} \vee x_{-4}}{128}, & 0 \le x < 0.5 \\ x + \dfrac{\overline{x_{-2} \wedge x_{-3}}}{32} + \dfrac{\bar{x}_{-2}}{64} + \dfrac{\bar{x}_{-2}}{128}, & 0.5 \le x < 1 \end{cases}$$

$$\vee : \text{logic-OR} \quad \wedge : \text{logic-AND}$$

(1)

$$(2^m)' = \begin{cases} 1 + x - \left[\dfrac{x_{-2}}{16} + \dfrac{\overline{x_{-3} \vee x_{-2}}}{32} + \dfrac{x_{-4}}{64} + \dfrac{x_{-3}}{128} \right] & 0 \le x < 0.5 \\ 1 + x - \left[\dfrac{x_{-2}}{16} + \dfrac{x_{-3} \vee x_{-2}}{32} + \dfrac{x_{-4}}{64} + \dfrac{x_{-3}}{128} \right] & 0.5 \le x < 1 \end{cases}$$

(2)

IV. SIMULATION RESULTS AND COMPARISONS

We implement our LNS-based CNN circuits using Verilog HDL and all the circuits are synthesized by Synopsys. Table I is the comparisons of our proposed LNS-based with conventional implementations (i.e., multipliers are generated by Design ware), our proposed LNS-based CNN circuits could achieve almost 60% ADP (Area-Delay Product) savings. We also use Python programming language to test the accuracy of our LNS-based CNN circuits using 10,000,000 randomly inputs, Table II demonstrates that our implementations could lead highly accurate CNN computations with ADP efficiency.

Table I: Comparisons of our proposed LNS-based CNN circuits with conventional implementations. (TSMC 0.18 μm technology)

Methods	Area(μm²)	Delay(ns)	ADP Ratio
Conventional	23697540.81	70.68	100%
Proposed	5596735.83	124.26	41.50%

Table II: Error information for our proposed LNS-based CNN circuits.

Maximum Errors	10.22
Minimum Errors	1.11
Maximum Error Percent	5.07%
Minimum Error Percent	0.75%
Average Errors	5.08

V. CONCLUSIONS

In this paper we have proposed ADP-efficient implementations of LNS-based CNN circuits used in deep learning, simulation results show that our work can achieve almost 60% ADP savings over conventional implementations. We are confident that using LNS-based schemes can efficiently reduce the computation overhead compared with conventional implementations in CNN circuits.

ACKNOWLEDGMENT

This work was supported by the Ministry of Science and Technology (MOST) in Taiwan under contract number MOST MOST 106-2221-E-153-001-.

REFERENCES

[1] Abadi, M., Agarwal, A., Barham, P., Brevdo, E., Chen, Z., Citro, C., Corrado, G. S., Davis, A., Dean, J.,Devin, M., et al., "Tensorflow: Largescale machine learning on heterogeneous systems," Software available from tensorflow.org. 2015.

[2] Goodfellow, I., Bengio, Y., and Courville, A., "Deep learning," MIT Press, 2016.

[3] Tso-Bing Juang, Pramod Kumar Meher and Kai-Shiang Jan, "High-Performance Logarithmic Converters Using Novel Two-Region Bit-Level Manipulation Schemes," Proc. of VLSI-DAT (VLSI Symposium on Design, Automation, and Testing), pp. 390-393, April 2011.

[4] Tso-Bing Juang, Han-Lung Kuo and Kai-Shiang Jan, "Lower-Error and Area-Efficient Antilogarithmic Converters with Bit-Correction Schemes," Journal of the Chinese Institute of Engineers, Vol. 39, No. 1, pp. 57-63, Jan. 2016.

Neuromorphic Properties of Memristor towards Artificial Intelligence

Chun Zhao, Zong Jie Shen, Guang You Zhou, Ce Zhou Zhao, Li Yang, Ka Lok Man, Eng Gee Lim

AI University Research Centre (AI-URC), Xi'an Jiatong-Liverpool University, Suzhou, China

Chun.Zhao@xjtlu.edu.cn; Cezhou.Zhao@xjtlu.edu.cn

Abstract—**Recent implementations of memristors have opened up the possibility of making brain-like artificial intelligence neuromorphic computing systems, including highly scalable and low-power neural networks. In fact, it has been demonstrated that a memristors can be implemented as an artificial synapse or as a protruding core of an artificial neuron. This paper reviews the neuromorphic properties of memristors, as well as the similarities of neural computation, synapses, and neurons.**

Keyword: Memristor; Neuromorphic; Artificial Neural Network

I. INTRODUCTION

Designing brain-like computing devices that can learn and adapt to the new environment is a key step in implementing artificial intelligence (AI) [1-3]. Based on the deterministic von Neumann architecture, the memory and processing units are physically separate and cannot be effectively learned and adapted [4-5]. However, brain-like strategies rely on a non-deterministic approach that interacts with a large number of simple processing units, neurons, which play an important building block role in learning and decision making [6].

Communication between neurons is obtained through synapses, and memory is stored in the brain through synaptic strength [7]. Learning is done by a mechanism called synaptic plasticity that increases or decreases synaptic strength. A memristor (memory resistor) is a non-volatile, two-terminal device that is a metal-insulator-metal structure with a non-linear relationship presented between current and voltage history. Variable resistance and nano-sized memristors make them potential candidates for electronic synaptic applications and have spurred the potential to develop computing low-power intelligent brain-like hardware arrays [5]. It is worth noting that synaptic learning rules, such as spike timing-dependent plasticity (STDP), have been replicated and examined in hardware devices like memristor for artificial neural networks (ANN) [8].

A briefing to the fundamental neurobiological mechanisms that support information processing/storage is introduced. This paper discusses the neuromorphic properties of memory resistors and possession of the qualities for computing fundamental properties from individual neurons and synapses. The recent implementation of memory resistors opens up the potential capability of creating new neuromorphic computing systems, including highly possible to scale and low-power ANN.

II. RESULTS AND DISCUSSION

STDP were firstly demonstrated in nanoscale memristors in 2010 [9]. The designed memristor crossbar structure consists of a bottom-tungsten and top chrome/platinum nanowire electrode and a sputtered Ag and a Si layer acitive with Ag/Si ratio along the depth change, as shown in Figure 1 (a). The conductance value of the sample is constantly following the voltage sweep, while the slope of the current voltage (current-voltage) is picked up for each subsequent sweep scan and finally scanned (demonstration (b) in Figure 1). Applying bias results in analog switching, with a continuous conduction leading edge motion (Ag ions moving from the enriched Ag region to the other,), enabling the first demonstration of STDP in the memory resistor. To a further extent, the change in conductance (synaptic weight of synaptic synapses) as a function of time (Δt) peaks due to presynaptic neurons and postsynaptic CMOS integration in Figure 1 (c), smaller time difference between the pulses, the variation in the conductance of the memristor is greater. Both of these features are very consistent with the STDP function of biological synapses.

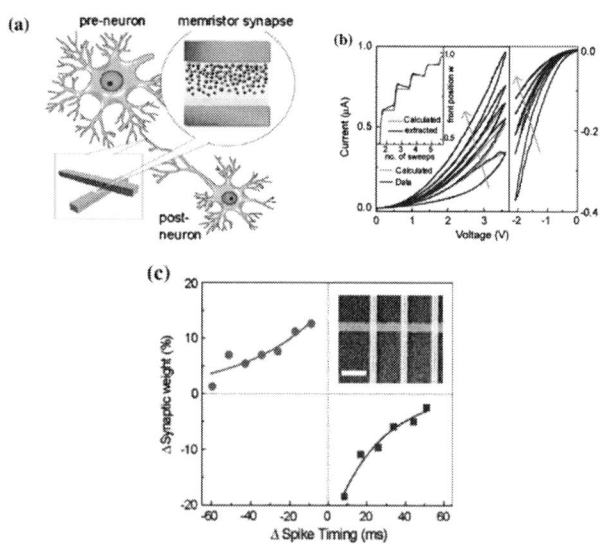

Figure 1. (a) Design of memristor. (b) I-V property and (c) Synaptic weight as a function of neuron spike relative time [9].

978-1-5386-7961-6/18 $31.00 © 2018 IEEE

Thereafter, some of the memory structures observed in synaptic behavior, including the titanium dioxide$_{2-x}$ / TiO$_y$ bilayer system, presenting multilevel conductance due to the movement of oxygen between the titanium dioxide$_{2-x}$ and TiO$_y$ layers [10], volatility and non-volatile rectification of WO$_{3-x}$-based nanoionic devices [11], Ni-doped graphene oxide [12], nano-organic memory field effect transistors [13] and Ag / conductive polymers / Ta crossed the memory system [14]. However, a practical problem before large-scale industrial applications of neural networks memristor memory is re-synaptic weights vary dependent on initial conductance. Noteworthy, there is significant breakthrough happened in 2016 [15]. Experimental with a Pt/Alumina/Titanium Dioxide$_{2-x}$/Ti/Pt memory resistor in a 12 x 12 structure array, succeed demonstrating the STDP behavior with the average conductance of the adaptive memory resistor, enabling plasticity on the state of the initial conductance device.

In 2015 [16] the implementation of STDP was represented in a secondary memristor. The Pd dynamics / Ta$_2$O$_{5-x}$ / TaO$_y$ / Pd structure was interpreted to adapt state variables. One variable is dependent on the area of the conduction filament, which originated from the diffusion of oxygen vacancy, determining the conductance of memristor. Another variable is the dynamics of the diffusion of sample temperature change (as shown in Figure 2 (a)). It is related to the biological situation than the first-order memory resistor, because in biological synapses, the weight is not regulated by spikes, but by secondary-order variables (such as postsynaptic calcium ions diffusion density). By giving pulses to the prepared structure, the condition of the gradual change in conductance is observed (the duration of the pulse and the time interval between the pulses), thereby making plastic variation in Figure 2 (b). However, for a large enough interval and a small enough set, the heat from separative pulse dissolved before consequent pulse arrives, and a gradual switching is obtained. It enables weight varied in frequency-dependent and STDP via parallel input peaks and applications in biorealistic manner (as shown in Figure 2(c)). Similarly, in the same year, the realization of secondary memory resistor was reported to be a state variable, that is, the conductive filament area between the electrodes and the mobility of oxygen holes. When the stimulation pulse [17-21] was used, oxygen vacancies was found to increase. Also, the mobility of holes increases as well.

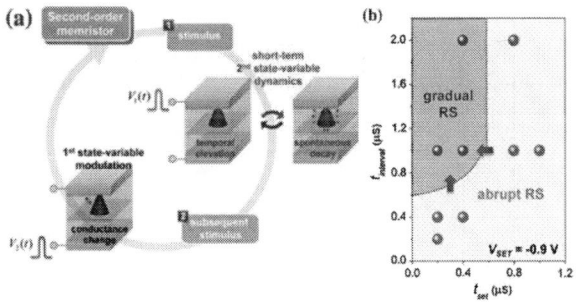

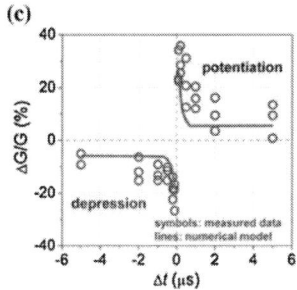

Figure 2. (a) Second-order memristor mechanism (b) Obtained t$_{Set}$ versus t$_{interval}$ operating window. (c) STDP in memristor [16]

III. SUMMARY

Memory resistors (Memristors) show remarkable characteristics for an potential building block for ANN, due to its low-power, nano-sized, highly scalability. And most importantly memristor imitate the core aspects of ion channel dynamics on synaptic and neuronal membranes. Memory resistors obtain the potential to enable computational advances by building blocks for bio-inspired computational systems, an alternative to the von Neumann architecture, in which storage and processing are integrated and fabricated in one single device.

ACKNOWLEDGMENT

This research was funded in part by the National Natural Science Foundation of China (21503169, 2175011441, and 61704111), Key Program Special Fund in XJTLU (KSF-P-02 and KSF-A-07).

REFERENCES

[1] Yuasa, S., et al. Nature Materials, 3, 868, 2004
[2] Strukov, D. B., et al. Nature, 453, 80, 2008
[3] Sandberg, et al. Neurocomputing, 32, 987, 2000
[4] Kiselev, S. I., et al. Nature, 425, 380, 2003
[5] Kandel, et al. J of Neurophysiology, 88, 507, 2003
[6] Bo, L., et al. Chinese J of Electronics, 1, 1, 2006
[7] Hopfield, J. J. PNAS, 79, 2554, 1982
[8] McCulloch, et al. Bull of Math Bio, 5, 115, 1943
[9] Jo, S., et al. Nano Letters, 10, 1297, 2010
[10] Seo, et al. Nanotechnology, 22, 254023, 2011
[11] Yang, R., et al. ACS Nano, 6, 9515, 2012
[12] Pinto, S., et al. App Phy Lett, 101, 063104, 2012
[13] Alibart, F., et al. Adv Func Mat, 22, 609, 2012
[14] Li, S., et al. J of Mat Chem C, 1, 5292, 2013
[15] Prezioso, M., et al. Sci Rep, 6, 21331, 2016
[16] Kim, S., et al. Nano Letters, 15, 2203, 2015
[17] Du, C., et al. Adv Func Mat, 25, 4290, 2015
[18] Q. Lu, et al. Materials, 8 (8), 4829-4842, 2015.
[19] Y. Mu, et al. IEEE T on Nuclear Science, 64 (1), 673-682, 2017.
[20] C. Zhao, et al. International SoC Design Conference, 306-309, 2011.
[21] Y. Mu, et al. Nuclear Instruments and Methods in Physics Research Section B: Beam Interactions with Materials and Atoms, 372, 14-28, 2016.

Memristor-based Neuromorphic Implementations for Artificial Neural Networks

Chun Zhao, Guang You Zhou, Ce Zhou Zhao, Li Yang, Ka Lok Man, Eng Gee Lim

AI University Research Centre (AI-URC), Xi'an Jiatong-Liverpool University, Suzhou, China

Chun.Zhao@xjtlu.edu.cn; Cezhou.Zhao@xjtlu.edu.cn

Abstract— **Remarkable computing complexity and power is obtained by combining multiple neurons and synapses computational units into high integrated brain-like network system. In order to simulate the biological learning rules in artificial synapses, an artificial neural network (ANN) capable of performing complicated functions is constructed. A significant challenge is to design a circuit that maintains a simple neuron structure, occupies a small silicon area, and implement only one electronic device as an artificial synapse. However, in traditional electronics, the area of silicon occupied by synaptic circuits can vary significantly. As observed and demonstrated, memory resistors exhibit such characteristics, making them the most promising candidates in scalable neural networks. Findings of various memristor network structures that support artificial neural network classification or information storage functions are presented.**

Keyword: Memristor; Neuromorphic; Artificial Neural Network

I. INTRODUCTION

Within traditional electronics, the human-brain-like system induces significant processing power and complexity by integrating computational components (ie, synapses and neurons) into ultra-complex network systems [1]. Consequently, intensive attempts to simulate the law of biological learning in artificial synapses and to institute/design artificial neural networks (ANN) with complex functions [2]. In the construction of artificial neural networks, a fundamental method is to implement the computational model required by the nervous system directly on semiconductor technology using very large scale integrated circuits (VLSI) [3]. However, the synapse circuit area can vary greatly because the layout design required a considerable number of transistors [4]. Executing the expansive network of the brain with transistors on a single chip may be a major challenge, since an expansive number of transistors are required [5]. Therefore, traditional silicon implementations are impractical and to simulate synaptic functions into a single scalable device is needed. Memristors (memory resistor) have been proved and found of these characteristics to be the most promising devices for neural networks. This article discusses examples of various memory resistor network structures for classification or data storage in neural networks [6].

II. NETWORK ARCHITECTURES OF ANN

The network (the architecture shown in Figure 1) is signal transmitted from source node (neuron) to multiple nodes based on events through edges (synapses) [7]. In most neural network models, synapses are dynamic connection between source (presynaptic) and sink (postsynaptic) neurons [4]. The synaptic signal from the source is modified by the transfer function and passed to the sink receiver. In order to promote neurons communication, the action potential transmits in the form of digital signal pulses. The output of an ANN node is regarded to be dependent on the sum of all input information. A state variable of the sink receiver is depending in part on the history of the incoming information from from the synapse. The variable, together with the source information, determine the synaptic state variable evolution.

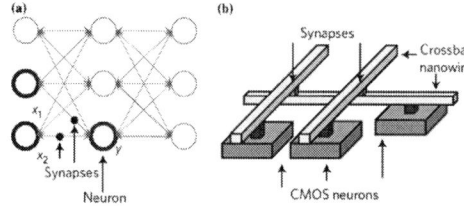

Figure 1. (a) Symbol graph and (b) architectures of crossbar-network [7].

III. RESULTS AND DISCUSSIONS

Recent advances in the implementation of memristive devices have revived attentions in ANN. Variable edge weights are the essential feature of neural networks and serves as the basic element of their extensive adaptive functions [2]. The conductance of an edge varies with the voltage drop, which is caused by the positive source spike and the negative sink spike. Edges are realized using memory nanodevices and bipolar pulse pairs enable communication, which develop a complete electronic neural network possible [6]. The actual implementation of an artificial network via memristors enable the weight to be renewed parallelly because of sufficient interconnectivity towards substantial synapses [7]. The data or information stored in memristor are multiplied by the input logical information.

A typical applications based on memristor is location detection. A random system of polymer-coated silver and oxidized passivated nickel nanowires whose position is not determinant, performance differences are averaged, and IV memory behavior is also demonstrated [8]. The morphology of the silver nanowire random system is shown in Figure 2 (a). Specified location or position in the network is locally activated by a voltage pulse via an atomic force microscope

978-1-5386-7961-6/18 $31.00 © 2018 IEEE   174   ISOCC 2018

metal processed tip. Figures 2 (b) - (f) shows the current diagrams from a given voltage pulse to the selected area.

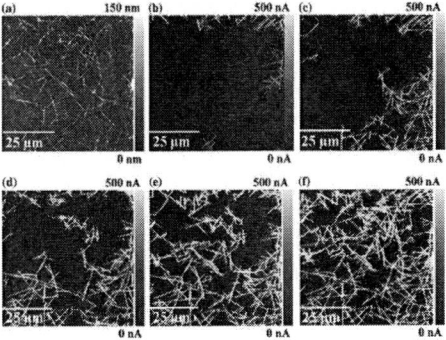

Figure 2. (a) The morphology of the Ag nanowire random network. (b) – (f) The graph shown is the result of applying a voltage pulse in the region [8].

The atom exchange system of $Ag/Ag_2S/Ag$ in Turing's Type unorganized framework was prepared by SU-8 photoresist and Ag_2S nanowires, and a filament structure was formed within the atomic scale (shown in Figure 3 (a) to (f)) [9]. Completing the electroforming, the entire network of high-connected atomic switches is evaluated by macroscopic electrode repeatability, presenting an isolated memristor hysteresis loop. In addition, infrared mapping identifies the distributed energy consumption of the system, resulting in the system functional. Finally, sudden behavior variation, such as the assembly behavior of neuron, can also be deduced by observing massive metastable conductance states produced by the separate system configuration.

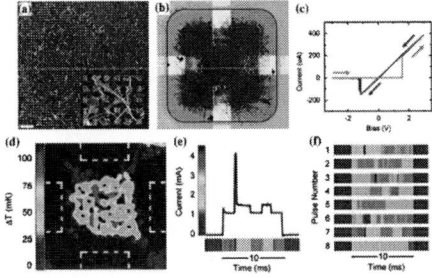

Figure 3. (a) Amplify and (b) Amplify a random nanowire network. (c) A compression hysteresis loop associated with the electrical behavior of the entire system. (d) Application of voltage to display network infrared mapping of distributed power consumption. (e), (f) Metastable switching in conductance states [9].

Cross-arrays are considered to be possible structure for brain-based nanoelectronic processing. In the cross-linked structure, a double-ended memory resistance synapse is built on each intersection of the presynaptic and postsynaptic neurons.

This two-dimensional ANN is considered as a building block, where input pulse to a specific output pulse fully linked by adaptive synapses. In addition, larger unit density than the traditional semiconductor technology architecture can be facilitied in the crossbar structure, and the number of unit density can be further improved by 3D overlay technology introduced. Alumina/TiO_{2-x} was fabricated to compact density 12×12 (200×200 nm) memory system without traditional circuit component in Figure 4 (a) [10]. A key progress is to achieve maximum efficiency in the I-V nonlinearity through the thickness of the alumina (as shown in Figure 4(b)) to succeed negligible variability [11-14].

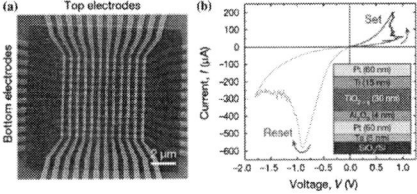

Figure 4. (a) Memory resistor overview; (b) Current-voltage property of a single memory resistor [10].

IV. CONCLUSIONS

Memristors show notable attributes that make them an excellent basis for building neuromorphic systems to imitate the core feature of ion channel dynamics on synaptic and neuronal cell. Therefore, many mechanisms are related to neural calculation and processing.

ACKNOWLEDGMENT

This research was funded in part by the National Natural Science Foundation of China (21503169, 21750110441, and 61704111), Key Program Special Fund in XJTLU (KSF-P-02 and KSF-A-07).

REFERENCES

[1] U. Karmarkar, et al. J of Neurophysiology, 88(1), 507–513, 2002.

[2] I. Ebong, et al. Proceedings of the IEEE, 100(6), 2050–2060, 2012.

[3] H. Kim, et al. IEEE T on Circuits and Systems, 59(1), 148–158, 2012.

[4] A. Vincent, et al. IEEE T on Bio Cir and Sys, 9(2), 166–174, 2015.

[5] Y. Zhou, et al. Proceedings of the IEEE, 103(8), 1289–1310, 2015.

[6] M. Suri, et al. IEEE T on Electron Devices, 60(7), 2402–2409, 2013.

[7] J. Yang, et al. Nature Nanotechnology, 8(1), 13–24, 2013.

[8] P. Nirmalraj, et al. Nano Letters, 12(11), 5966–5971, 2012.

[9] A. Stieg, et al. Advanced Materials, 24(2), 286–293, 2012.

[10] M. Prezioso, et al. Scientific Reports, 6(February), 21331, 2016.

[11] Q. Lu, et al. Materials, 8 (8), 4829-4842, 2015.

[12] Y. Mu, et al. IEEE T on Nuclear Science, 64 (1), 673-682, 2017.

[13] C. Zhao, et al. International SoC Design Conference, 306-309, 2011.

[14] Y. Mu, et al. Nuclear Instruments and Methods in Physics Research Section B: Beam Interactions with Materials and Atoms, 372, 14-28, 2016.

Automatic Shading Detection System for Photovoltaic Strings

Jieming Ma
Dept. of Computer Science and
Software Engineering, Xi'an Jiaotong-
Liverpool University, Suzhou, China
Sch. of Elec and info. Engineering,
Suzhou University of Science and
Technology, Suzhou, China
jieming.ma@xjtlu.edu.cn

Ziqiang Bi, Ka Lok Man, Yong Yue
Dept. of Computer Science and
Software Engineering,
Xi'an Jiaotong-Liverpool University,
Suzhou, China
{ziqiang.bi, ka.man,
yong.yue}@xjtlu.edu.cn

Jeremy S. Smith
Dept. of Electrical Engineering and
Electronics
University of Liverpool
Liverpool, UK
J.S.Smith@liverpool.ac.uk

Abstract— **Partial shading is one of the main factors that affect the output power of series-parallel photovoltaic (PV) strings. However, physical irradiance measurement instruments (e.g. pyranometers and pyrheliometers) are seldom used in commercial PV systems due to their high cost. This paper proposes an automatic shading detection system for estimating the shading rate by using voltage sensors. The main features of the proposed method are the utilization of a reduced number of sensors, a simple switching control strategy and the high detection rate. The feasibility and effectiveness of the proposed shading detecting system is validated through experiments.**

Keywords--Partial shading scenarios; shading detection; photovoltaic systems.

I. INTRODUCTION

Even in near-ideal solar locations, partial shade from passing clouds, neighboring buildings, trees or dust is inevitable in photovoltaic (PV) strings. Early studies have shown that partial shade can dramatically cut an entire solar string's output [1]. Bypass diodes serve as a protection mechanism that allows the PV module to continue producing power in partial shading scenarios (PSS) [2]. Experiments demonstrate that the power-voltage (*P-V*) characteristic curves of PV strings obtain multiple peaks in PSS. Therefore, knowing the PSS information is of significance for maximum power point tracking (MPPT) control and system maintenance.

Zheng et al. [3] demonstrated that the number of local power points depends on the bypass diode configurations and string topology in PSS. The occurrence of partial shade is usually identified by a sudden significant change in output power. However, it is hard to distinguish the PSS from the rapidly changing atmospheric conditions. Jieming et al. [4] proposed a partial shading detection method in accordance with the relations between module voltage and string voltage, but the system cost is high. Silvestre et al. [5] proposed reference thresholds for detecting PSS based on the errors between simulated and measured capture losses. Although many shading detection methods have been proposed in recent years, few of them can estimate the shading rate [6] of the partially shaded PV systems. In this paper, an automatic shading

detection system has been proposed to estimate the shading rate of PV strings. The main features of the detection system are not only the high detection rate, but also the utilization of the reduced number of sensors and the simple control strategy.

II. SHADING DETECTION SYSTEM

A. Voltage measurement system

The sub-string voltage – string voltage (V_{ss}-V_s) curves of a PV string with a shaded module are shown in Fig. 1. The bypass diode allows current to pass around M2. The number of unshaded modules would be equal to V_s/V_{ss1}, where V_{ss1} is the module voltage of the unshaded M1. In other words, the shading rate χ (the proportion of shaded modules in the PV string), can be estimated as soon as one unshaded module voltage is measured. Thus, the proposed shading detection system estimates the shading rate by measuring the module voltages.

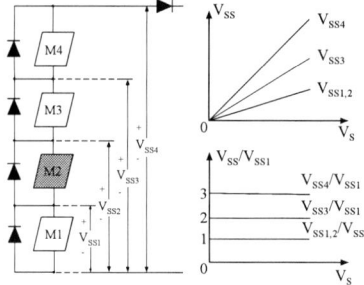

Fig. 1. The V_{ss}-V_s characteristics of a PV string in PSS.

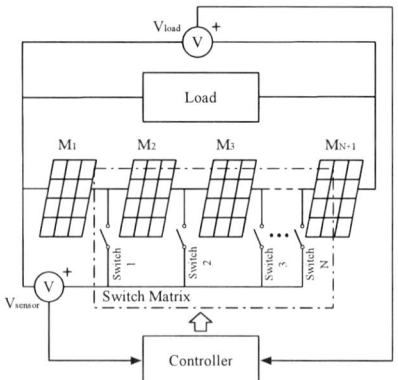

This research is supported by the National Natural Science Foundation of China (Grant No. 61702353), the Natural Science Foundation of Jiangsu Province (Grant No. BK20160355), the Science and Technology Project of Ministry of Housing and Urban-Rural Development (Grant No. 2016-K1-019) the Suzhou Science and Technology Project (SYG201603), XJTLU Research Development Funds (RDF-14-02-32 & -17-02-04) and XJTLU Key Program Special Fund (Grant No. KSF-P-02).

978-1-5386-7961-6/18 $31.00 © 2018 IEEE

Fig. 2. Block diagram for the proposed shading detection system.

Fig. 2 shows the block diagram of the proposed shading rate detection system for a PV string with (N+1) modules. Only two voltage sensors are utilized in the voltage measurement system. One is used to measure the load voltage V_{load} and the other one is used to measure the module voltage V_{sensor}. A switch matrix is applied to select the measured sub-string.

B. Switching control strategy

A module is considered as the shaded module if the measured sub voltage is less than threshold voltage $V_{threshold}$ ($V_{threshold} = 1V$ in this paper). A switching control strategy is designed to measure the voltage of the first insolated module as shown in Algorithm 1. The situation that all the modules in the string are shaded is not included as it is trivial for this research. Let i be the switch index ($1 \leq i \leq N$). If a shaded module is detected, switch i will be opened and switch $i+1$ will be closed. As soon as the system gets the voltage of an insolated module V_{sensor}, the number of the insolated PV modules $N_{insolated}$ can be expressed as:

$$N_{insolated} = round\left(\frac{V_{load}}{V_{sensor}}\right). \tag{1}$$

The shading rate χ can be calculated by the following equation:

$$\chi = 1 - \frac{V_{load}}{(N+1)V_{sensor}}. \tag{2}$$

Algorithm 1 Switching control strategy.

1:   $i \leftarrow 1$
2:   **while** True **do**
3:     Read sensor values
4:     **if** $V_{sensor} < V_{threshold}$ **then**
5:       **if** $i < N$ **then**
6:         $i \leftarrow i + 1$
7:       **else**
8:         $N_{insolated} \leftarrow 1$
9:         **break**
10:      **end if**
11:     **else**
12:       $N_{insolated} \leftarrow round(V_{load}/V_{sensor})$
13:       **break**
14:     **end if**
15:     Close switch i after opening other switches
16: **end while**

III. EXPERIMENTAL RESULTS AND DISCUSSIONS

The overview of the experimental setup is shown in Fig. 3. The experimental detection system consists of four 10W PV modules, a DC electronic load (ITECH IT8512A+), an oscilloscope (GWINSTEK GDS-2202A), a voltage sensor module, a relay module and a central controller (UDOO NEO FULL). Experiments were conducted to evaluate the feasibility of the proposed method. The insolated PV modules received about 600 W/m² and shaded modules received 55 W/m².

Fig. 4 shows the time-sequence results in a rapid changing PSS. The values of V_{sensor} and V_{load} are respectively shown by yellow and cyan lines on the oscilloscope. The bottom red wave is the calculated $N_{insolated}$. Four different PSS were manually set in the experiment. In T1, all modules were insolated and $N_{insolated} = 4$. In T2, T4 and T6, it is observed that $N_{insolated}$ was correctly estimated. Due to the sampling interval, the right switches cannot be closed in T3 and T5, which leads to wrong estimates. As soon as controller finishes sampling, the $N_{insolated}$ can be predicted. Besides the above 4 PSS, 10 more

possible PSS were set to verify the correctness of the proposed method. No error was found in the experiments.

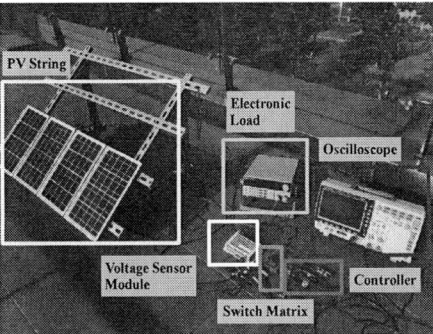

Fig. 3. Experimental setup of the validation system.

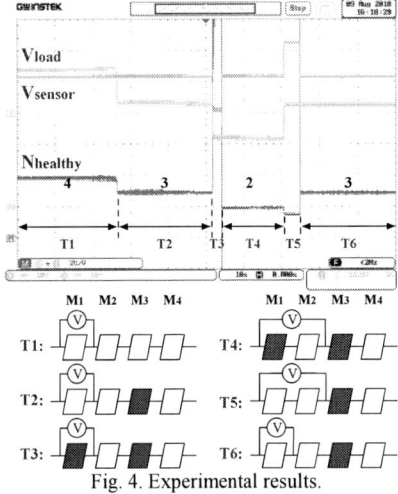

Fig. 4. Experimental results.

IV. CONCLUSIONS

This paper has presented a novel automatic shading detection system for PV strings. The shading is detected by using a voltage measurement system with the proposed switching control strategy. Experimental results have shown that the system has the ability to estimate the shading rate effectively.

REFERENCES

[1] H. Heydari-doostabad, R. Keypour, M.R. Khalghani and M.H. Khooban. "A new approach in MPPT for photovoltaic array based on extremum seeking control under uniform and non-uniform irradiances." Solar Energy, 94, pp. 28-36, 2013.

[2] J. Storey, P. R. Wilson and D. Bagnall, "The Optimized-String Dynamic Photovoltaic Array," in IEEE Transactions on Power Electronics, vol. 29, no. 4, pp. 1768-1776, 2014.

[3] H. Zheng, S. Li, R. Challoo, and J. Proano, "Shading and bypass diode impacts to energy extraction of pv arrays under different converter configurations," Renewable Energy, vol. 68, no. 7, pp. 58–66, 2014.

[4] J. Ma, T. Zhang, Y. Shi, X. Li and H. Wen, "Shading pattern detection using electrical characteristics of photovoltaic strings," IEEE International Conference on Power Electronics, Drives and Energy Systems (PEDES) , Conference Proceedings, 2016, pp. 1-4.

[5] S. Silvestre, A. Chouder, and E. Karatepe, "Automatic fault detection in grid connected pv systems," Solar Energy, vol. 94, no. 4, pp. 119–127, 2013

[6] J. Ma, X. Pan, K. L. Man, X. Li, H. Wen and T. O. Ting, "Detection and Assessment of Partial Shading Scenarios on Photovoltaic Strings," in IEEE Transactions on Industry Applications. Eearly Acces, doi: 10.1109/TIA.2018.284

A Multiband Rectenna for Self-sustainable Devices

Zhao Wang[1], Heng Zhang[1], Zhenzhen Jiang[1], Mark Leach[1], Jingchen Wang[1], Kalok Man[2], Eng Gee Lim[1]

[1]Dept. of Electrical and Electronic Engineering,
[2]Dept. of Computer Science and Software Engineering,
Xi'an Jiaotong-Liverpool University. Suzhou, P.R.China
Enggee.lim@xjtlu.edu.cn

Abstract—**This paper presents the design of a multi-band cross-polarized dipole antenna for use in energy harvesting. The antenna is designed to work from 1.53 GHz to 2.47 GHz in its lower lower frequency band and from 4.90 GHz to 5.63 GHz in its upper frequency band, covering the GSM1800/1900, UMTS2100, 2.4G, and 5G Wi-Fi bands. The design and simulation are covered which show that the antenna is cross-polarised with omni-directional radiation patterns.**

Keywords: Rectenna; cross-polarisation; multiband.

I. INTRODUCTION

With the rapid development of wireless communication systems, wireless RF energy has become widely used during the past few decades, in systems such as radio, cellular mobile, Bluetooth and WLAN to name a few [1]-[3]. In tandem with the explosive increment in ambient RF signals as a result of this, the ambient wireless power density is growing and researchers have realised the feasibility of harvesting this ambient RF energy for use in various remote systems.

For some low-power electronic devices, such as remote electronic sensors, limited battery life becomes the bottle-neck for their extensive application. In addition, the power sources of these devices, typically batteries, add to the size of the device and also cause environmental pollution [4]. Harvesting ambient wireless RF energy is a solution for this situation. Powering low-power electronic devices by using an energy harvesting antenna is not only environmentally friendly but also self-sustaining [5]. It has been shown previously that a network of low-power wireless electronic sensors can be powered by an energy harvesting system, which harvests energy from ambient RF signals [6]-[7].

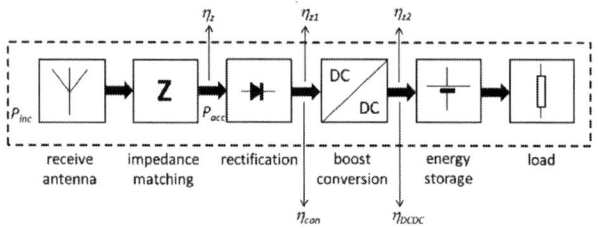

Figure 1: A block diagram of a typical rectifying-antenna system.

A "Rectenna", i.e. the rectifying-antenna system, is a device combining an antenna and a rectifier circuit, which is popular in harvesting ambient wireless RF energy [1]-[3]. The antenna can collect the ambient wireless signals, and the rectifier circuit can convert these energies into useable DC power [3]. In general, this DC power will be stored in an energy storage device, and then it can be delivered to a load. A

block diagram of a typical rectifying-antenna (rectenna) system is shown in Figure 1.

The primary purpose of this paper is to present the design and simulation of a new low-profile antenna which is able to operate over commonly used frequency bands, including GSM1800/1900, UMTS2100, 2.4G and 5G Wi-Fi bands, for use in a rectenna circuit.

II. ANTENNA DESIGN

In this paper, CST microwave studio has been used to perform simulations of the antenna designed to meet the band requirements specified.

The proposed multiband antenna is designed to cover the GSM1800/1900, UMTS2100, 2.4G and 5G Wi-Fi bands. The basic structure of the antenna is divided into three main sections; the square substrate made of FR4, the irregular diamond-shaped copper radiators on the front side of the substrate (as shown in Figure 2), and the H-slotted microstrip line perpendicular to the substrate to feed the antenna (as shown in Figure 3).

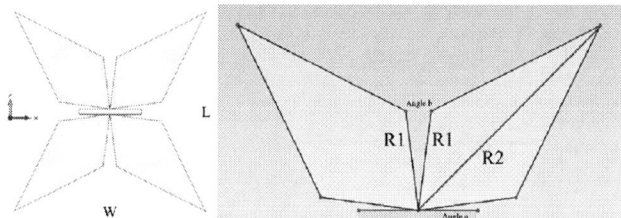

Figure 2: Front view of the cross-polarised dipole antenna with feed line.

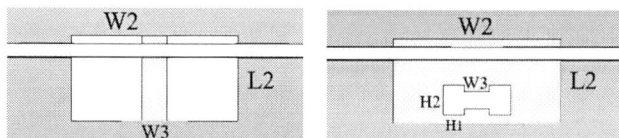

Figure 3: Top and Bottom view of the antenna feed line with H-slot.

The four irregular diamond-shaped quadrangles formed four arms of the planar dipole antenna, making the antenna cross-polarised. The size of the irregular diamond-shaped quadrangle (*R1* and *R2*) and two angles between this quadrangle with two axes (angle *a* and *b*) are crucial to the antenna performance. The H-slotted microstrip line is selected as a feed to suppress any higher order harmonics that may result from connection to a rectifier [3].

These key parameters highlighted in Figure 2 and Figure 3 have all examined and optimised by simulation. The optimised parameters of the antenna are listed in Table 1.

Table 1: Parameters of the cross-polarized dipole antenna with feed line.

Name	Value (mm)	Description
h	1.5	The thickness of substrate FR4.
t	0.035	The thickness of copper.
W	70	The width of substrate for antenna.
L	70	The length of substrate for antenna.
l	26.5	The length of two sides of sector.
R1	17.19	The length of the adjacent side.
R2	29.27	The length of the diagonal line.
a	6.5	Angle between conductor and x axis.
b	6.5	Angle between conductor and y axis.
W2	20	The width of the substrate for the microstrip feed line.
L2	10	The length of the substrate for the microstrip feed line.
W3	2.9	The width of the microstrip feed line.
H1	2.5	The width of the H-shaped slot.
H2	3.5	The length of the H-shaped slot.

III. RESULTS

A. Return Loss

From the simulation results shown in Figure 4, it can be seen that the -6 dB bandwidth of the antenna is from 1.44 GHz to 3.21 GHz in the lower frequency band and from 4.86 GHz to 5.34 GHz in the higher frequency band, covering the frequency desired bands as required.

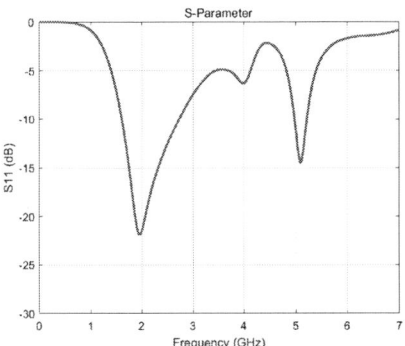

Figure 4: Return loss of the cross-polarised dipole antenna

B. Radiation Pattern

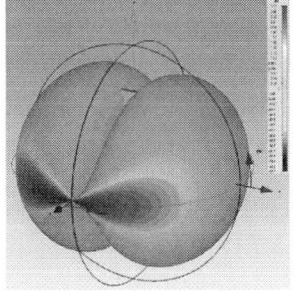 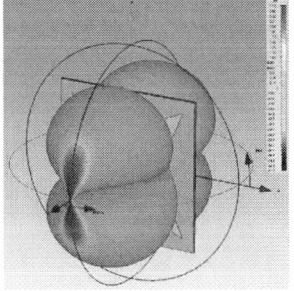

Figure 5: Simulated 3D Cross-polarization and Co-polarization Radiation Pattern at 2.2 GHz.

The simulated results of the co-polar and cross-polar radiation patterns of the antenna at 2.2 GHz are plotted in Figure 5. It can be seen that this antenna can receive RF waves with either vertical polarisation or horizontal polarisation at this frequency, which means this antenna is indeed dual-polarised due to the cross-polarised dipole structure used.

The 2-D radiation patterns of the antenna are shown in Figure 6. They show that the antenna is omnidirectional at this frequency, which is desirable as the ambient RF energy it is designed to receive will be incident from all angles.

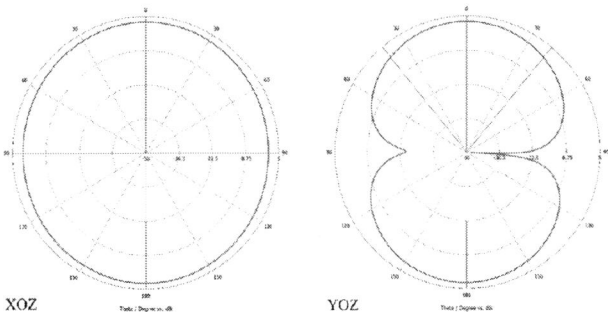

Figure 6: Simulated 2D radiation patterns at 2.2 GHz.

The radiation patterns at 5 GHz are slightly distorted, but show similar features of cross-polarisation and are also omni-directional.

IV. CONCLUSION

A new multi-band cross-polarised dipole antenna has been designed and simulated able to cover the GSM1800/1900, UMTS2100, 2.4 and 5G Wi-Fi bands. This antenna is cross-polarised and has omni-directional radiation patterns in both design bands. In the future, the high frequency performance will be further improved to cover the ISM 5800 band and the antenna will be manufactured and tested for verification purposes.

ACKNOWLEDGEMENT

The authors would like to express their sincere gratitude to CST AG for providing the CST STUDIO SUITE® electromagnetic simulation software package under the China Key University Promotion Program, and the comprehensive support on it. This work is partially supported by the XJTLU Research Development Fund (PGRS-13-03-06, RDF-14-03-24 and RDF-14-02-48) and XJTLU Key Programme Special Fund (KSF-P-02). The authors would also like to acknowledge the support of the Suzhou Municipal Key Laboratory for Broadband Wireless Access Technologies.

REFERENCES

[1] C. Song et al., "A broadband efficient rectenna array for wireless energy harvesting," 2015 9th European Conference on Antennas and Propagation (EuCAP), Lisbon, 2015, pp. 1-5.

[2] S. Ladan et al., "Highly Efficient Compact Rectenna for Wireless Energy Harvesting Application," in IEEE Microwave Magazine, vol. 14, no. 1, pp. 117-122, Jan.-Feb. 2013.

[3] C. Song et al., "A High-Efficiency Broadband Rectenna for Ambient Wireless Energy Harvesting," in IEEE Transactions on Antennas and Propagation, vol. 63, no. 8, pp. 3486-3495, Aug. 2015.

[4] H. Jabbar, Y. S. Song and T. T. Jeong, "RF energy harvesting system and circuits for charging of mobile devices," in IEEE Transactions on Consumer Electronics, vol. 56, no. 1, pp. 247-253, February 2010.

[5] S. Kim et al., "Ambient RF Energy-Harvesting Technologies for Self-Sustainable Standalone Wireless Sensor Platforms," in Proceedings of the IEEE, vol. 102, no. 11, pp. 1649-1666, Nov. 2014.

[6] T. Le, K. Mayaram and T. Fiez, "Efficient Far-Field Radio Frequency Energy Harvesting for Passively Powered Sensor Networks," in IEEE Journal of Solid-State Circuits, vol. 43, no. 5, pp. 1287-1302, May 2008.

[7] D. Bouchouicha et al., "Ambient RF energy harvesting," 2010 Int. Conf. Renew. Energies and Power Qual.

Oscillation Quenching in Coupled van der Pol Oscillators with Different Frequencies

Yoko Uwate

Dept. of Electrical and Electronic Engineering,
Tokushima University
2-1 Minami-Josanjima, Tokushima 770-8506, Japan
uwate@ee.tokushima-u.ac.jp

Yoshifumi Nishio

Dept. of Electrical and Electronic Engineering,
Tokushima University
2-1 Minami-Josanjima, Tokushima 770-8506, Japan
nishio@ee.tokushima-u.ac.jp

Abstract—**In this study, we investigate oscillation quenching in coupled van der Pol oscillators with different frequencies. The network topology is inspired from real brain network. Here, we consider the network which consists of two modules including connector and provincial hubs. We confirm oscillation quenching from the proposed system.**

Keywords—oscillation quenching; coupled oscillatory network

I. INTRODUCTION

In our research group, we have focused on synchronization phenomena observed from nonlinear oscillatory networks [1], [2]. This is because the results of synchronization in complex networks are useful deeper understanding of control methods in power networks, communication systems and so on. Also, they can be used as an alternative approach, apart from existing ones, for describing mode-locking phenomena in biological networks. Therefore, we consider the schematic brain networks as one of complex biological networks. Because, we would like to propose modeling of synchronization in brain by using coupled electrical oscillatory circuits, in order to make clear the mechanism of functional operation in brain.

Recently, the relationship between structural and functional network in biological neural network has attracted their attention from many researchers. Hartelt et al. has discovered the network topology in the pre-BotC which consists of densely connected clusters with rare inter-cluster links. And, the hubs of the network behave quiescent output [3].

We apply this phenomenon for the coupled oscillatory systems using electrical oscillator. Namely, one node is set to lower frequency than the others. We also investigate the influence of location of oscillator with different frequencies.

II. PROPOSED SYSTEM

A. Network Model

A network model composed of 13 nodes is shown in Fig. 1 [4]. There are two important hubs in this network, "Connector hub" and "Provincial hub". The both hubs are high-degree nodes. "Connector hub" shows a diverse connectivity by connecting two sub-networks. "Provincial hub" primarily connects nodes in the same sub-network. There are two modules and the node is expressed by van der Pol oscillator as

shown in Fig. 2. The van der Pol oscillators are coupled by a resistor.

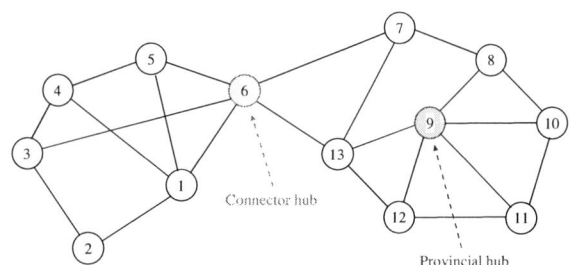

Fig. 1. Network model.

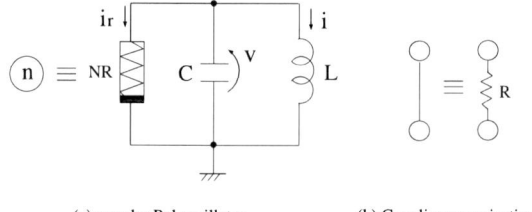

(a) van der Pol oscillator. (b) Coupling organization.

Fig. 2. Ocillator and coupling organization.

Next, we develop the expression for the circuit equation of the network model. The normalized circuit equations governing the circuit are expressed as

$$
\begin{cases}
\dfrac{dx_k}{d\tau} = \varepsilon\left(1 - \dfrac{1}{3}x_k{}^2\right)x_k - y_k - \gamma \displaystyle\sum_{n \in S_k}\left(y_k - y_n\right) \\
\dfrac{dy_k}{d\tau} = x_k \\
\hspace{5cm} (k = 1, 2, ...13).
\end{cases}
\tag{1}
$$

In these equations, γ is the coupling strength, ε denotes the nonlinearity of the oscillators. For the computer simulations, we calculate circuit equations using the fourth-order Runge-Kutta method with the step size h =0.005. The parameter of the nonlinearity of oscillator is fexed as ε =0.1.

978-1-5386-7961-6/18 $31.00 © 2018 IEEE

B. Adding Different Frecuency

In this simulation, one node of the network has lower frequency than the others. The standard frequency is $\omega = 1.0$ and the lower frequency is set to $\omega = 0.47$. In order to investigate influence of location of oscillator with different frequencies, we focus on several nodes as follows.

Case-1) Connector hub (oscillator no. 6)
Case-2) Provincial hub (oscillator no. 9)
Case-3) Other node (oscillator no. 2)
Case-4) Other node (oscillator no. 10)

III.   OSCILLATION QUENCHING

Figure 4 shows the simulation result of amplitude change with the coupling strength. From these results, we confirm amplitude death at certain range. Furthermore, the range of amplitude death of connector hub is smaller than the provincial hub.

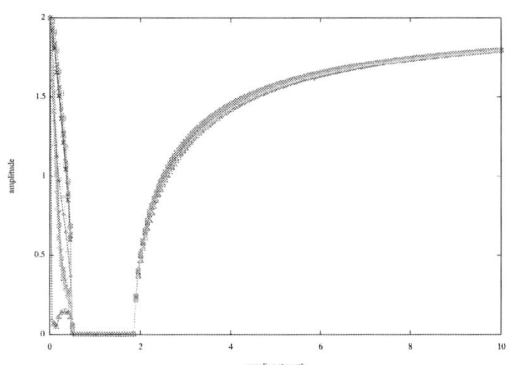

(a) Case-1: Connector hub (oscillator no. 6).

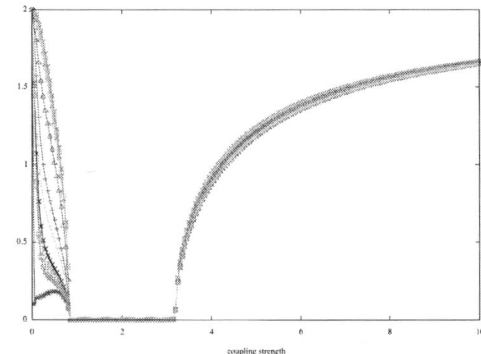

(b) Case-2: Provincial hub (oscillator no. 9).

Fig. 4. Amplitude of connector and provincial hubs.

Next, the simulation results of the amplitude change in Cases 3 and 4 are shown in Fig. 5. By increasing the coupling strength, we also observe the amplitude death in the both cases. We can see that the range of amplitude death of Cases 3 and 4 is longer than Cases 1 and 2.

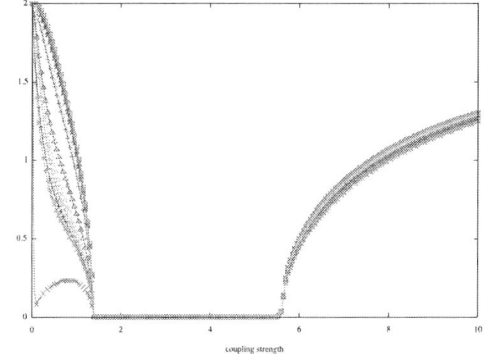

(a) Case-3: Oscillator no. 2.

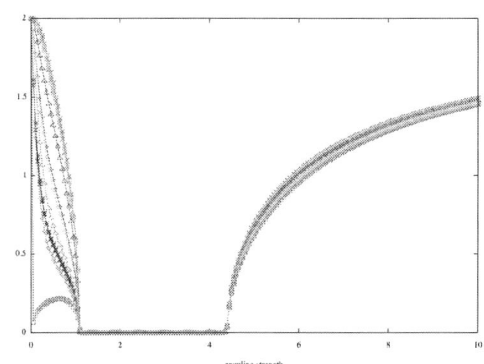

(b) Case-4: Oscillator no. 10.

Fig. 5. Amplitude of other nodes.

IV.   CONCLUSIONS

In this study, we have investigated oscillation quenching observed in coupled oscillatory system with different frequencies. We have confirmed amplitude death at certain range with the coupling strength from the proposed system. The range of amplitude death is smallest when the lower frequency is set to the connector hub. In the future work, we would like to investigate the relationship between the amplitude death and importance of the nodes.

REFERENCES

[1] Y. Setou, Y. Nishio and A. Ushida, "Synchronization Phenomena in Resistively Coupled Oscillators with Different Frequencies", IEICE Transactions on Fundamentals, vol. E79-A, no. 10, pp. 1575-1580, Oct. 1996.

[2] Y. Uwate and Y. Nishio, "Synchronization in Several Types of Coupled Polygonal Oscillatory Networks", IEEE Transactions on Circuits and Systems I, vol. 59, no. 5, pp. 1042-1050, May 2012.

[3] N. Hartelt, E. Skorova, T. Manzke, M. Suhr, L. Mironova, S. Kugler and S. L. Mironov, "Imaging of Respiratory Network Topology in Living Brainstem Slices," Mol. Cell. Neurosci. 37, pp. 425–431, 2008.

[4] E. Bullmore and O. Sporns, "Complex Brain Networks: Graph Theoretical Analysis of Structural and Functional Systems," Nature, Reviews, Neuro, vol. 10, pp. 186-198, Mar. 2009.

Gap in pagination due to formatting issues.

Pages 182-193

A design of rectifier for 13.56MHz wireless power transfer receiver with all digital delay-locked loop

Joonho Park
Soongsil University
School of Electronic Engineering
Seoul, Korea
drpark@soongsil.ac.kr

Yong Moon
Soongsil University
School of Electronic Engineering
Seoul, Korea
moony@ssu.ac.kr

Abstract— This paper proposes a rectifier for the wireless power transfer receiver for medical implant devices. The proposed bridge rectifier uses MOSFET instead of diode. If the output voltage is higher than the input voltage, the reverse current flows and this affects the efficiency. Therefore, we added the reverse current detector that detects the current flowing through the MOSFET and blocks the reverse current. The reverse current detector is designed as all digital delay-locked loop. The proposed rectifier is designed using 0.35um CMOS process. The input voltage is possible up to 3.3V and the rectifier operates at the frequency of 13.56MHz. When the load current is 24mA, the power efficiency is 82.69% and the maximum transferred power transfer is 82.8mW.

Keywords; Rectifier, DLL, wireless power transfer, 13.56MHz

I. INTRODUCTION

One of the major problems with medical implant devices is the limited lifetime of the battery. The replacement of the battery may require surgery. One way to avoid this problem is to recharge the battery using a wireless power transmission system. Using this kind of technique, power can be transmitted from the transmit coil to the receive coil via self-resonant coupling. However, depending on the frequency, the human body could be affected. Low transmission frequencies improve the permeability of the magnetic field inside the human body, but the power conversion efficiency(PCE) is reduced. On the other hand, high transmission frequencies can cause current loss, tissue heating, and electromagnetic compatibility (EMC) problems.[1]

Another important aspect of the design is that the recharge time of the battery should be limited to a maximum of few hours. Because the heat generated by the power loss of the coil conductor can destroy biological tissue. And the choice of maximum current should be taken cautious, as the electromagnetic fields affecting the human body are limited by international guidelines. [2]

In this paper, a rectifier using 13.56MHz frequency band is designed and the current is limited to 24mA. To improve the efficiency, the designed rectifier adopts a body-biasing circuit. The reverse current detector is designed as all digital type in order to minimize the influence of noise caused by the surrounding environment.

II. THE PROPOSED ARCHITECTURE

A. Block diagram

Fig. 1 is the TOP block diagram of the designed rectifier with all digital delay-locked loop. The wireless power signal is coming through the antenna coil is applied to the active rectifier across the matched filter. Since the initial condition of the rectifier output voltage(V_{rec}) is 0V, so each sub-block can not be operated. Therefore, the rectifier starts operation in the passive mode. When the V_{rec} rises above the certain voltage, the LDO regulator operates and supplies the power to each sub-block. When the rectifier voltage rises to the desired level, the passive / active selector block switches the rectifier to active mode. At the same time, the POR(Power On Reset) initializes the logic gates in each sub-block.

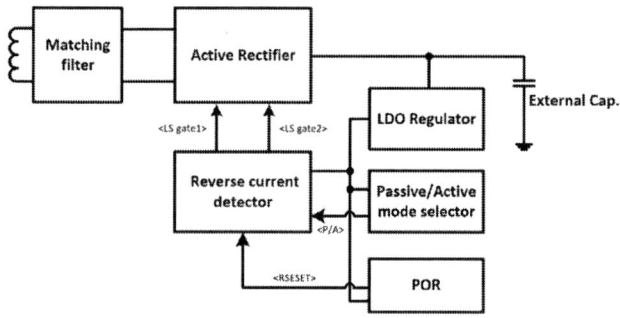

Fig. 1 Top block diagram of the designed rectifier

B. Active rectifier

Fig. 2 is the circuit diagram of the active rectifier. In the basic full-bridge rectifier circuit structure, MOSFET is used instead of diode. The disadvantage of the active rectifier is that when the input AC is lower than the rectifier output voltage(V_{rec}), the reverse current flows and adversely affects the efficiency. Therefore, we need to turn off the MOSFET by adding the reverse current detector. In addition, the body biasing circuit is designed. This circuit avoids the latch-up of the PMOS and prevents breakdown, and improves the efficiency by decreasing the threshold voltage.[3]

978-1-5386-7961-6/18 $31.00 © 2018 IEEE

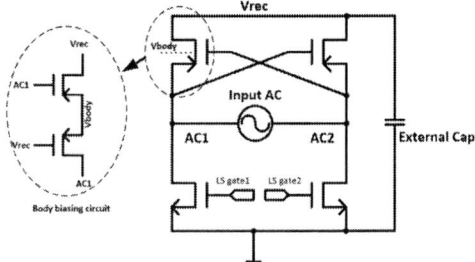

Fig. 2 Schematic of the active rectifier

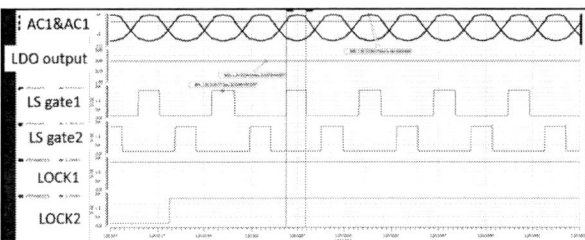

Fig. 4 Top-level simulation result (@RL=155Ω)

C. Reverse current detector

Fig. 3 is the block diagram of the reverse current detector. The reverse current detector is designed as all digital delay-locked loop to minimize the noise influence of the surrounding environment. [4] When the input AC falls below the rectifier output voltage, the comparator outputs high state ('1'). This value passes through the delay line. And the digital phase detector receives the feedback of the delay line output and comparator output. It starts to detect the phase. When the comparator output is fast, it outputs the UP signal and when the feedback signal is fast, it outputs the DN signal. If the phases are the same, the LOCK signal is generated. 6-bit counter operates with 3 signals. When the Up signal come in, the counting value is increased. When the DN signal come in, the counting value is decreasing. When the LOCK signal is input, the counting stops. A 6-bit signal is received by the delay line block to control the speed of the signal. The Delay Line consists of two paths. One is fast path, and the other is slow path. 6 bit signal selects the fast path or the slow path in real time.

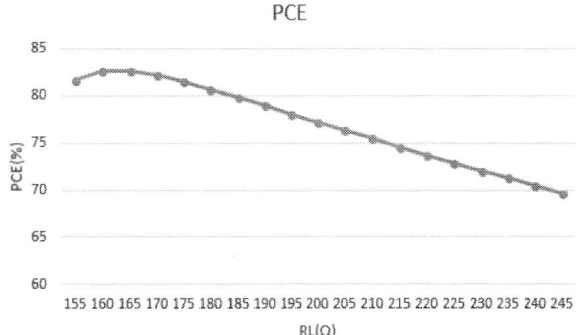

Fig. 5 PCE of the rectifier vs load resistance

IV. CONCLUSION

This work shows the rectifier designed by using all digital delay-locked loop to minimize the influence of ambient noise. The rectifier operates in the frequency band of 13.56MHz and is designed by 0.35 um CMOS process. It can current supply up to 24mA and the maximum efficiency is 82.69%.

The proposed receiver could be used in wireless power transfer for implant devices.

ACKNOWLEDGMENT

This work has been studied with support from the Ministry of Trade, Industry and Energy in 2018, 'Creative industry convergence characterization human resource development business' (N0000717)

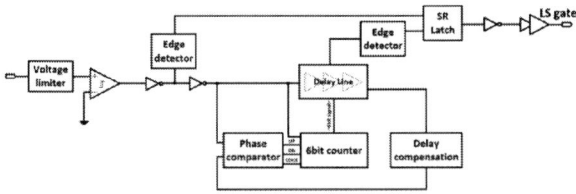

Fig. 3 Block diagram of the reverse current detector

III. SIMULATION RESULTS

Fig. 4 shows the total simulation results of the designed rectifier. As result of the simulation after stabilizing to a certain extent, it can be confirmed that the LOCK signal becomes high state. If the input AC is higher than the output of the rectifier, the LS gate changes high state, and if the input AC becomes lower than the output of the rectifier, the LS gate changes to low sate.

Fig. 5 shows the PCE(Power Conversion Efficiency) of the designed rectifier. When the load is 160Ω, the power conversion efficiency is 82.69% and the transferred maximum power is 82.8mW.

REFERENCES

[1] Tommaso Campi, Silvano Cruciani, Federica Palandrani, Valerio De Santis, Akimasa Hirata, Mauro Feliziani, "Wireless Power Transfer Charging System for AIMDs and Pacemakers", IEEE transactions on microwave theory and techniques, vol. 64, no. 2, February 2016

[2] 2016Guidelines for limiting exposure to time varying electric, magnetic, and electomagnetic fields(Up to 300GHz), International Commission on Non-Ionizing Radiation Protection (ICNIRP) Guidelines

[3] Byeong Wan Ha, Choon Sik Cho, "Design of Rectifier with Comparator using Unbalanced Body Biasing for Wireless Power Transfer", ISOCC, vol.2013, no.11 2013

[4] Truong Thi Kim Nga, Hyung-Gu Park, and Kang-Yoon Lee, "A High Efficiency Active Rectifier for 6.78MHz Wireless Power Transfer Receiver with Bootstrapping Technique and All Digital Delay-Locked Loop", IEIE Transactions on smart processing and computing, vol. 3, no. 6, December 2014

Leakage Control System Using Data Estimation of Resistive Memory

JunYoung Kweon[1], JunTae Choi[2], and Yun-Heup Song[2]*

[1]Division of Nanoscale Semiconductor Engineering
[2]Department of Electronics and Computer Engineering
Hanyang University, Seoul 04763, Korea
*E-mail : yhsong2008@hanyang.ac.kr

Tony Tae-Hyoung Kim
School of Electrical and Electronics Engineering
Nanyang Technological University
50 Nanyang Avenue, Singapore 639798
thkim@ntu.edu.sg

Abstract— **In this work, we present a data-aware read sneak current control technique for improving sensing margins in cross-point (X-point) resistive memories. We introduce a novel method of estimating resistive memory data distribution. The result from this estimation is utilized for determining adequate bias condition for reliable read operation. The adjusted bias condition will realize constant leakage current, which facilitates reliable sensing regardless of the data distribution in the resistive memory array.**

Keywords;Resistive memory, Leakage control, X-point

I. INTRODUCTION

X-point resistive memory [1] architecture has been evermore popular because it provides higher memory density compared to conventional architecture with one transistor in each memory cell. However, it is challenging to secure reliable read and write operation due to the variations in sneak current. This requires additional margins in read and write current, which is not desirable for low power consumption and can affect the reliability of resistive memory devices. One of the most critical issues is determining read current. Since only a part of read current flows through selected memory cells, the amount of sneak current affects sensing significantly. In this work, we presents a novel technique for simply estimating the data distribution of X-point resistive memory, whose result will be utilized for setting adequate bias condition to achieve reliable read operation.

II. PROPOSED DATA-AWARE LEAKAGE CONTROL TECHNIQUE

A. Array data distribution estimation technique

Fig. 1 illustrates a simplified X-point array structure. Four critical signals ($V_A \sim V_D$) are controlled properly for read and write operation. During read operation, read voltage ($V_A - V_B$ or $V_B - V_A$) is applied to a selected cell while V_C and V_D are floating [2][3][4]. Sneak current flows in other unselected cells. Fig. 1(a) and (b) depict two extreme cases generating maximum sneak current and minimum sneak current. The data-dependent sneak current is included in the current flowing between V_A and V_B, which makes sensing

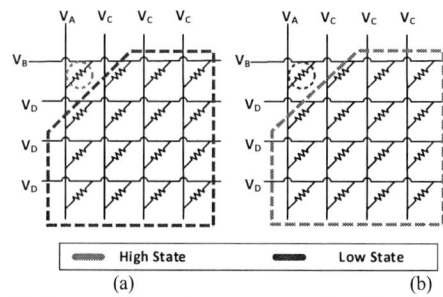

Figure 1. Sneak current dependency on array data: (a) maximum sneak current and (b) minimum sneak current.

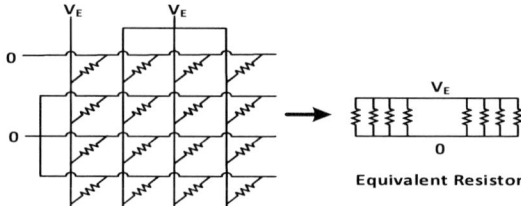

Figure 2. (a) Array Resistive model (b) Array select state diagram.

difficult. To remove the impact of the data-dependent sneak current on sensing, it is essential to estimate the amount of the sneak current. Fig. 2 illustrates the proposed method for array data distribution estimation, which will be used for sneak current estimation. This converts the array into an E.R(equivalent resistor) whose value can be calculated as given below.

$$1/R' = n/R_{Low} + m/R_{High} \quad (1)$$

Here, n is the number of devices with R_{Low} and m is the number devices with R_{High}. Measuring E.R, we can estimate how many states of R_{High}. Fig. 3 shows the relationship between the array data and the equivalent resistance. We simulated a 256 × 256 resistive memory array. It can be observed that the equivalent resistance tracks the array data properly. In actual implementation, the equivalent resistance can be measured by forcing voltage and measuring current or by forcing current and measuring voltage.

978-1-5386-7961-6/18 $31.00 © 2018 IEEE

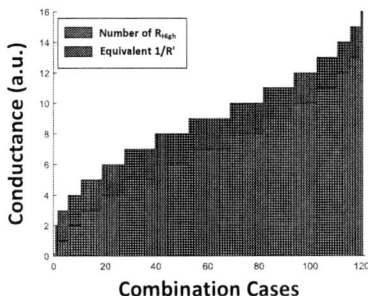

Figure 3 Relationship between the number devices with R_{High} and the equivalent resistance obtained through the proposed estimation method.

B. Data-aware sneak current control

Utilizing the array data information from the above section, we introduce a technique for mitigating the impact of data-dependent sneak current on sensing. Fig. 4 shows the bias condition for the proposed read operation. Unlike the conventional scheme where V_C and V_D are left floating, the proposed work adjusts V_X based upon the array data distribution obtained in Fig. 3. The properly adjusted V_X will make the sneak current almost constant, mitigating its impact on sensing. This concept is illustrated in Fig. 5, which shows current elements of cell. Fig. 6 presents simulated results showing the relationship between V_X and the array data for generating constant sneak current. Assuming the sneak current of 10 μA, the proposed biasing technique provides the sensing current margin of 3 μA. Note that when a larger number of memory devices are in the high resistance state, the required V_X becomes higher for the constant sneak current.

III. CONCLUTION

Data-dependent sneak current is a critical issue in sensing. Since the current flowing through an X-point resistive memory array includes both cell current and sneak current, it is highly important to know the amount of sneak current for reliable sensing. This work presents a novel technique for estimating array data and equalizing sneak current through optimal biasing. This improves sensing margin by mitigating the impact of data-dependent sneak current.

ACKNOWLEDGMENT

This research was supported by the MOTIE(Ministry of Trade, Industry & Energy (project number 10080625) and KSRC(Korea Semiconductor Research Consortium) support program for the development of the future semiconductor device.

REFERENCES

[1] DerChang Kau *et al.*, "A stackable cross point Phase Change Memory," *2009 IEEE International Electron Devices Meeting (IEDM)*, Baltimore, MD, 2009, pp. 1-4.

[2] A. Levisse, B. Giraud, J. P. Noël, M. Moreau and J. M. Portal, "SneakPath compensation circuit for programming and read operations in RRAM-based CrossPoint architectures," *2015 15th*

Non-Volatile Memory Technology Symposium (NVMTS), Beijing, 2015, pp. 1-4.

[3] Woorham Bae *et al.,* "A crossbar resistance switching memory readout scheme with sneak current cancellation based on a two-port current-mode sensing," *Nanotechnology,* 2016, vol 27 485201

[4] Mohammed Zackriya, Harish M. Kittur and Albert Chin, "A Novel Read Scheme for Large Size One-Resistor Resistive Random Access Memory Array," *Scientific Reports,* 2017, vol 7, 42375

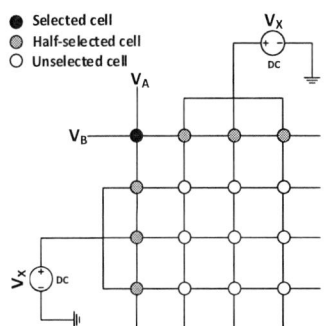

Figure 4 Proposed biasing technique for constant sneak current.

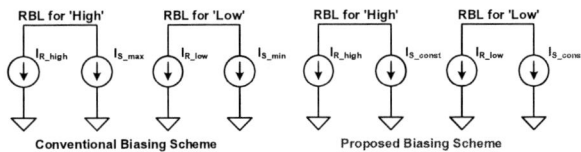

Figure 5 Concept of the proposed biasing technique for constant sneak current and sensing improvement.

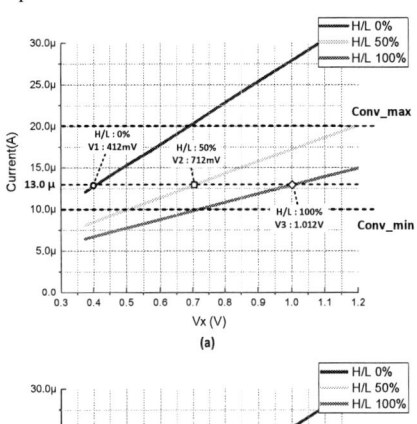

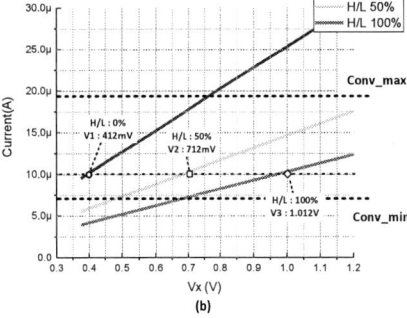

Figure 6 Simulated current result with different array data sweeping V_X: (a) reading low state and (b) reading high state. For the sneak current of 10 μA, V_X values are ~0.4 V, ~0.7 V, and ~1.0V the portions of memory devices with high resistance are 100%, 50%, and 0%, respectively.

Gap in pagination due to formatting issues.

Pages 198-203

Stereo vision-based Collision Avoidance for Unmanned Systems

Sungick Kong, Sang-Seol Lee and Sung-Joon Jang

Intelligent Image Processing Research Center, Korea Electronics Technology Institute, KETI

Seongnam-si, 13488, South Korea

kong@keti.re.kr, sslee81@keti.re.kr, sjjang0626@keti.re.kr

Abstract—**In the unmanned systems, the intelligent navigating technologies are very important. For this purpose, we introduce a navigation algorithm that includes collision avoidance using a stereo camera. To accurately avoid the obstacles, we propose the object detection, the navigable space searching and the efficient collision avoidance real-time algorithms. In this article, we focus on an unmanned aerial vehicle (UAV) system. The UAV basically flies along the global path towards the goal point entered by the user. If it is not possible, UAV return by performing the proposed local path planning algorithm. The generated local path is based on the new waypoint, and it safely flies to the final goal point using the calculated flight direction and distance information between waypoint and waypoint.**

Keywords; Unmanned Syatems; UAV(Unmanned Aerial Vehicle); Navigable Space; Collision Avoidance;

I. INTRODUCTION

A UAV is a vehicle that autonomously flies without human control. Nowadays, UAVs have been considered of various industrial fields, such as military, rescue, fire-detect and agricultural UAV. These UAVs are typically configured by the user to set the flight path and then fly. If a sudden object appears in the flight or there is a mistake made by the user, the UAV will not complete the flight mission. In order to solve this problem, we need a new intermediate waypoint considering the entire flight path while avoiding the obstacle, and an algorithm to update the path continuously. These intelligent autonomous flight algorithms protect UAVs and reduce financial losses.

In this paper, we propose a navigation algorithm based on stereo camera for UAVs. We constructed a system that uses a stereo camera, ZED, to fly more intelligently from the start point to the goal point via the navigation algorithm. We refer to the framework of local path planning using the obstacle avoidance of Rodrigeuz et al [1]. We were inspired by waypoint selection in the DeepFly [2] paper to find navigable space using depth information in depth images. To find the region, we have studied the depth image through CNN, but we used the connected component labeling and the threshold value according to disparity in the depth image very simply to define the distant area as the navigable space. We used the concept of bounding box for straightforward flightworthy area in the article [3] which attempted to avoid

obstacle avoidance in the indoor situation while constructing collision avoidance algorithm using ZED camera.

II. THE PROPOSED ALGORITHM

A. Navigation Algorithm

The proposed algorithm actively avoids locally occurring obstacles. This algorithm is shown in Fig. 1, RGB-D image is input, and Navigable space search and Object detection & Collision avoidance are operated to obtain waypoint information to be moved by the UAV. This waypoint information computes the distance (Z) value of the point corresponding to the center of the region and has rotated x and y angle information from the center position of the image. We obtained the disparity map through the calibration and rectification of the stereo camera. We consider the aerodynamic size of UAV as Fig. 2, the bounding box was constructed by using the principal point (Cu, Cv) which is the FOV center point of the image as the green box.

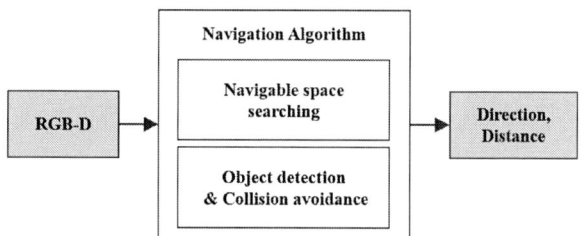

Figure 1. Navigation Algorithm

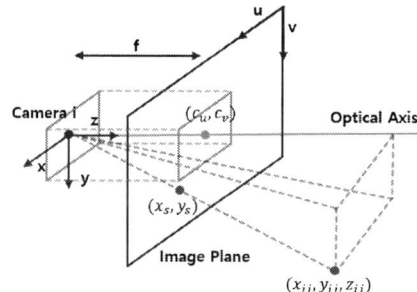

Figure 2. Projection of point j on image i

B. Navigable space

This algorithm works when an obstacle exists in a bubble that can fly straight and it is impossible to proceed. To find the navigable space, first segmentation the areas that are farthest from the depth map. Second, the segmentation region is connected component labeling, extracts large region candidates from the image, and adds the largest region as a waypoint.

| Segmented regions on input image | Depth map (threshold range) | Color depth map |

Figure 3. Segmented navigable space

C. Object detection & Collision Avoidance

This algorithm works when navigable space is not found or when dynamic obstacles that suddenly appear at close range should be avoided. To guide the safe route, we determine the avoidance direction using the detected obstacles information (size, distance). After avoid the obstacles, we register the new waypoint corresponding to that point.

The overall navigation algorithm operates according to the pseudo-code shown below.

Algorithm : Navigation algorithm

---

if goal reached **then**

    Update goal

else

    if Straight path is feasible **then**

        Waypoint ← Goal

    else

        Search Navigable space

        if Spaces are large enough **then**

            Choose the biggest one

        end if

        if Collision detected **then**

            Predict possible safe trajectory

        end if

        if Path found **then**

            Waypoint ← Next node in the path

        else

            Go around

        end if

    end if

end if

III. EXPERIMENTS

We implement the proposed algorithms based on WVGA (672x376) image of ZED camera. Fig. 4 shows a part of the experiment image taken by applying the algorithm. The green box represents the UAV's straight-forward bounding box. If there are obstacles within 3 meters, the red box is displayed. The blue box represents the navigable space, and the distances and the x-axis and y-axis information of the center point of the box are indicated.

Figure 4. Experiment images

IV. CONCLUSION

In this paper, we proposed an algorithm to find a navigable space to avoid a sudden dynamic object in order to plan a local path during a flight along a global path. It is expected that users will be able to use UAV more easily through intelligent autonomous flight algorithm. We will use this algorithm to experiment with UAV in a variety of environments. We will develop a system by linking with visual simultaneous localization and mapping (SLAM) algorithm using RGB-D in order to be able to return the UAV flying forward to the desired section again. We also want to reduce the weight of the UAV and develop it compactly using the light embedded platforms for more lightweight airframe hardware configuration during actual flight.

ACKNOWLEDGMENT

This work was supported by Institute for Information & communications Technology Promotion (IITP) grant funded by the Korea government (MSIP) (No.2017-0-00721, Development of intelligent semiconductor technology for vision recognition signal processing for vehicle based on multi-sensor fusion).

REFERENCES

[1] L. Rodriguez, JA. Cobano, and A. Ollero, "Efficient Local Path Planning for UAVs in unknown environments", XV Workshop of physical agents: book of proceedings, WAF 2014, June 12th and 13th, 2014 León, Spain, pp. 125-134.

[2] U. Shah, R. Khawad, and K M. Krishna, "DeepFly: towards complete autonomous navigation of MAVs with monocular camera", Proceedings of the Tenth Indian Conference on Computer Vision, Graphics and Image Processing, December 2016, p. 59.

[3] A U. Haque, A. Nejadpak, "Obstacle Avoidance Using Stereo Camera", arXiv preprint arXiv:1705.04114, May 2017.

Night-time Vehicle Detection Based on Brake/Tail Light Color

1st Thathupara Subramanyan Kavya
Electrical engineering
University of Ulsan
Ulsan, Republic of Korea
kavyats86@gmail.com

2nd Erdenetuya Tsogtbaatar
Electrical engineering
University of Ulsan
Ulsan, Republic of Korea
erdenetuya@nate.com

3rd Young-Min Jang
Electrical engineering
University of Ulsan
Ulsan, Republic of Korea
min-s2@nate.com

4th Sang-Bock Cho
Electrical engineering
University of Ulsan
Ulsan, Republic of Korea
sbcho@ulsan.ac.kr

Abstract – **In the statistical analysis, most of the road accidents happen during the night time. The major challenge for reducing night accidents is to detect vehicles at night time and keeping distance between the vehicles travelling in front. While driving in dark conditions, vehicles in front are generally visible by their tail and brake lights. The vehicle appearance is different at night time when compared to day light due to various factors like color of the vehicles, reflection of light from the vehicle body and natural light. In this paper, we are proposing a method for vehicle detection based on brake/ tail light color from the captured color image.**

Keywords—Vehicle detection,image processing,color detection,light detection

I. INTRODUCTION

In recent years numbers of vehicles in the roads are increasing exponentially. Due to this increased traffic, road crashes are reaching higher rates, day by day. Vehicle detection in daytime is very easier compared to night time. Visibility of the road ahead is very less due to appearance change and light and other environmental condition such as color of the vehicles, reflection of light etc. This is the main challenge during night-time driving, since the driver cannot see the whole body of the vehicle at the front [1], [2]. In most cases, the only visible region of the vehicles travelling at the front is the tail lights and brake light. To overcome this limitation, we are proposing a method for detecting vehicles based on brake/ tail light color.

The color emitting from brake/tail light is red. So, we are detecting red color from a colored image. Proposed method uses image processing system that can be easily implemented, and the proper detected results are obtained from the input images. The proposed system will help the drivers to avoid the forward system collision [3], [4], [5].

II. A ROPOSED METHOD

In this proposed method, the input we are considering is a color image. It contains all the three-color components. For detecting brake light or tail light we need to consider only the red component

A. RGB color Model

RGB model consists of red, green and blue components. Fig.1. shows the RGB color model. We have split the color model into three different components, as shown in.Fig.2.

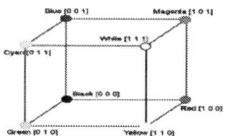

Fig.1. RGB color model

(a) Original image   (b) Red Band

(c) Green Band   (d) Blue Band

Fig.2. Different color components

B. Morphological Operations

Morphological filtering of a binary image is conducted by considering compound operations like opening and closing as filters. They may act as filters of shape. For example, opening with a disc structuring element smooths corners from the inside, and closing with a disc smooths corners from the outside. These operations can filter out the noise from an image. We are assuming that, any details that are smaller in size than the structuring element, is a noise element. Only those portions of the image that fit the structuring element are passed by the filter and smaller structures are blocked and excluded from the output image. The size of the structuring element is most important to eliminate noisy details but not to damage objects of interest. In Fig.3. shows the morphological closing operations

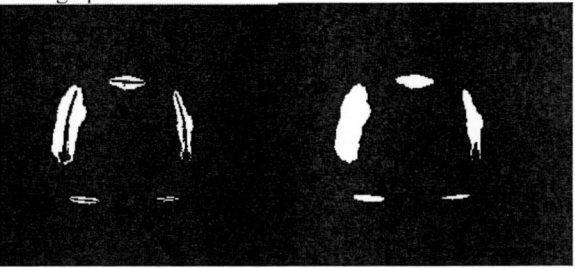

(a) before operation   (b) after closing operations

Fig.3. Morphological Closing Operation

The algorithm can be summarized as follows: In stage one, we are separating the input color image into blue, green, and red components. Once all the color bands are extracted, a lower and upper threshold value is applied to each component (select the threshold for each color band). For getting correct and clear detection, the small areas are filtered out and also for smoothing the border we are using a morphological operation. And finally, we can easily detect the vehicle from a colored image. Fig.4 shows the flowchart of the proposed method.

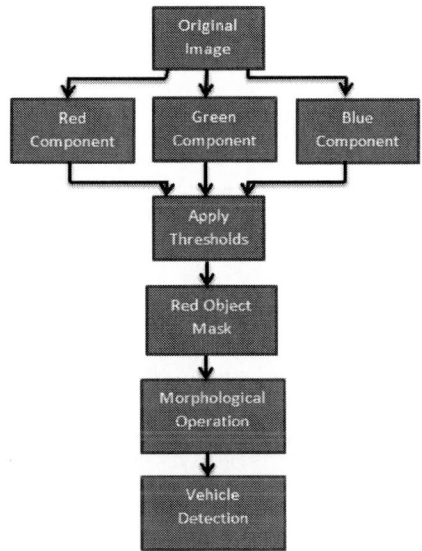

Fig.4. Flow chart of the detection procedure

III. EXPERIMENTAL RESULTS

The experimental results of the proposed method is as shown in Fig.5. For the implementation and verification of the algorithm we used Matlab 2018 software. It clearly detects the vehicles using color. The bounding box is detected and the driver is alerted that the vehicle is getting closer to the one travelling at the front.

Fig.5.(a) shows the original image Fig 5.(b) shows the results after the morphological operation and Fig.5 (c). shows the final result for vehicle detection.

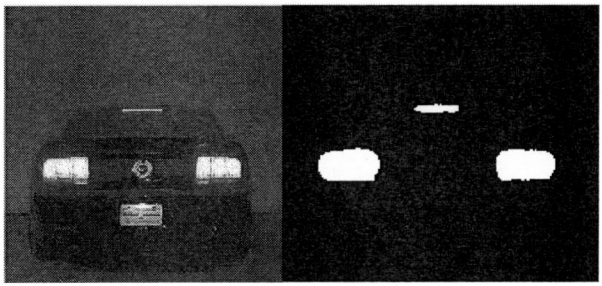

(a)   original Image   (b)   after morphological operation

(c)   vehicle detection

Fig. 5. The correct and error result in whole frame.

IV. CONCLUSION

With the increase in road traffic and road crashes, divers are thriving for accurate assistance system for warning about potential hazards. Brake/ tail lamps are considered as one of the vital elements for detection the vehicles at the front during night time. In this paper, we have proposed a color-based vehicle detection method. Tail light and brake light are red in color and using this method we can easily detect the vehicle from an image. The proposed method can easily avoid the accidents happened due to sudden brake down and also very helpful for detecting vehicles for a smooth night drive.

Future work in this direction may seek to improve the segmentation range and apply to the video sequences.

ACKNOWLEDGMENT

This work was supported by the 2016 fund of University of Ulsan, Ulsan, Republic of Korea.

REFERENCES

[1] Yen-Lin Chen,Hao-Yu Huang and Chung-Jui Fan, " A real-Time Vision System for Nighttime Vehicle Detection and Traffic Surveillance",IEEE Transactions on Industrial Electronics,Vol.58.No.5, May 2011
[2] Tuan-Anh Vu,Long Hoang Pham,Tu Kha Huynh and Synh Viet-Uyen Ha, " Nighttime Vehicle detection and Classification via Headlights Trajectories Matching", presented at the 2017 International Conference on System Science and Engineering (ICSSE)
[3] Swathy S Pillai and Radhakrishnan B, "Night vehicle Detection Using Tail Lights: A Survey", International Journal of Engineering Research and General Science Volume 4, Issue 2, March-April, 2016
[4] Chinmoy Jyoti Das, and Parismita Sarma , "Vehicle Recognition Based on Tail Light Detection", International Journal of Advanced Research in Computer and Communication Engineering Vol.5,Issue 3,March 2016
[5] Bhavinkumar M.Rohit and Mitul M.Patel . "Nighttime vehicle Tail light detection in low light video frames using Matlab", International Journal for Research in applied Science & Engineering Technology (IJRASET) Vol.3, May 2015.

978-1-5386-7961-6/18 $31.00 © 2018 IEEE   207   ISOCC 2018

Implementation of Multi-Channel FM Repeater using Digital Signal Processing Algorithm in FPGA

Mangi Han[1], Ji Min Song[2], Hoon Gee Yang[2], and Youngmin Kim[1]

[1]School of Computer and Information Engineering, Kwangwoon University,
[2]Department of Electronics Convergence Engineering, Kwangwoon University,
Nowon-gu, Kwangwoon-ro 20, Seoul 01897, Republic of Korea
youngmin@kw.ac.kr

Abstract— **In this study, we implemented a high performance and a multi-channel FM repeater using a Field Programmable Gate Array (FPGA). In a system for providing services using wireless communication, a radio-shaded area is generated due to various obstacles. Thus, an electronic device that receives weak or low-level signals and retransmits them at a higher level is crucial. In addition, a parallel implementation of digital filters and gain controllers is necessary for a multi-channel repeater. When power level is too low or too high, the repeater needs to compensate power and ensure a stable signal. The proposed circuit includes modulation, demodulation, Cascaded Integrator Comb filter (CIC filter), Automatic Gain Controller (AGC), and can control 40 channels in parallel.**

Keywords—FPGA, FM repeaters , Frequency Modulation (FM), Automatic Gain Controller (AGC), Filter.

I. INTRODUCTION

In a system for providing services using radio, such as mobile communication, a radio-shaded area is inevitably generated due to a special topography and a property between a base station and a base station constituting a wireless communication service network. To resolve this phenomenon, a repeater, an electronic device which can retransmits regenerated signals is required.

Traditionally, hams have wired together multiple standard integrated circuits to construct radios, TNCs and other devices. However, modern devices manufactured in high volume integrate the most functionality on a single chip that is customized for the purpose. Feature sizes in integrated circuits have now become so small that FPGA can be used to achieve what once required a custom chip. Since the logic in the FPGA can be used in parallel rather than sequentially, it is a much more powerful tool than a microprocessor. Recently, FPGAs also contain larger amounts of dedicated memory, making it possible to incorporate microprocessors within the FPGA. In this study, we design digital circuits for a multi-channel FM signal repeater by using digital signal processing algorithm and implement them in FPGA.

Digital signal processing algorithm includes frequency modulation, demodulation scheme that is widely used for supporting wireless communication and private mobile radio standards amongst others [1] [2]. In addition, a real time automatic gain controller designed to compensate the power. The digital system is designed to control 40 channels at once in parallel [3]. The designed digital circuit has been implemented

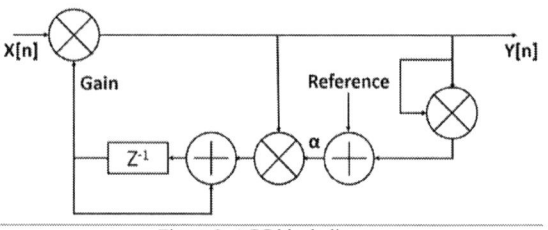

Figure 2. AGC block diagram

by using Verilog HDL and tested using Xilinx Kintex 7 - xc7k325tffg676 - 1 as a target device.

II. DIGITAL SIGNAL PROCESSING ALGOTIRHM

A block diagram of the digital signal processing is shown in Fig. 1. The channelized FM signal (analog) is converted into a digital signal through an analog to digital converter (ADC) having a sampling frequency of 150 MHz in this study, and a digital signal is modulated to convert a preselected channel into a baseband signal. The signal transferred to the baseband passes through multiple filters, which filter out adjacent channels. First, the baseband signal passes through the Cascaded Integrator Comb filter (CIC) filter followed by the Interpolated Second Order Polynomials (ISOP) filter. To filter out adjacent channels properly, the decimation ratio of CIC filter has to be determined. Matlab simulations are conducted to determine the decimation ratio of the CIC decimator filter and the ratio need to be 693 so that the frequency of the CIC filter to be -50 dB at the frequency of 0.1 MHz, which is the channel boundary frequency of the FM signal. Same filters are used both for the in-phase component and the quadrature component. An ISOP filter is used to compensate the droop due to the CIC filter. After that, the real time simple AGC, which is shown in Fig. 2, controls the power level of the digital signal. As shown, the power level of the input power is detected in the Power module first. The detecting window size in the Power module can be changed by users. After detecting the power level, the Controller module compare the detected power and the reference power. Then, a control parameter α, which is calculated by the ratio between the reference power and the input power, is multiplied to the input signal. After gain control, the CIC interpolator filter, which has the same interpolation ratio as the decimator filter, restores the original signal. Finally, the signal is recovered by

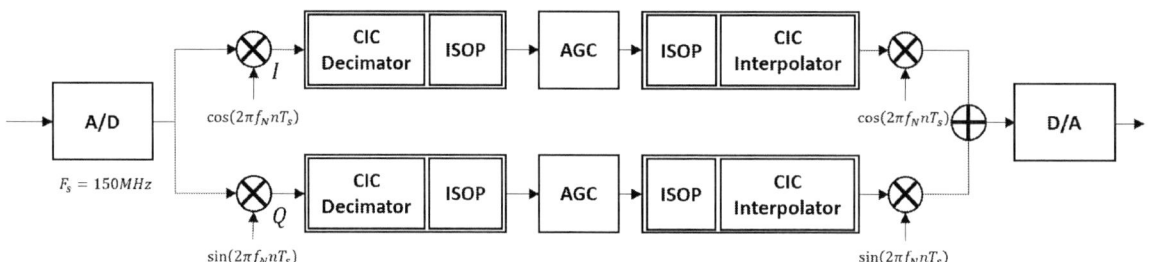

Figure 1. A block diagram of the FM digital signal processing

978-1-5386-7961-6/18 $31.00 © 2018 IEEE

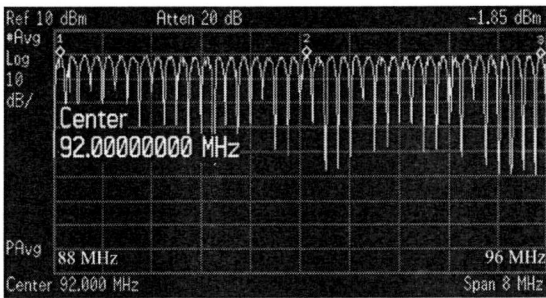

Figure 3. Filtered 40 FM signals from 88 MHz to 96 MHz

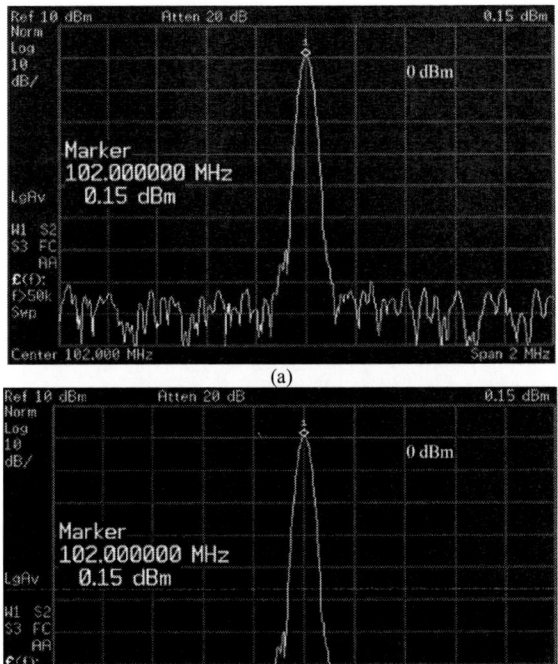

(a)

(b)

Figure 4. Output of the 102.1 MHz signal; (a) when input power is -60dBm and (b) when input power is -30dBm

combining the in-phase signal with the quadrature signal in the demodulation block and pass through the digital to analog converter (DAC) for RF transfer.

III. RESULT

We design and implement the digital circuit with Verilog HDL in Vivado 17.4 version. For verification, we focus on two realistic scenarios. First, we investigate whether multiple signals arrive at the same time through proper filters. Second, even if signals of different power levels come in, the AGC can control the power and a stable signal is always transmitted.

In experiment, we generate practical FM signals from 88 MHz to 96 MHz at the same time by using a signal generator and send them to the ADC of the FPGA board. As shown in Fig. 3, each channel is filtered properly at its center frequency, and the filter shapes also are constructed as intended.

And, the AGC is verified by using two different power values of the same frequency as shown in Fig. 4. A 102.1 MHz FM signal of -60 dBm and -30 dBm powers are inserted into the

TABLE I. UTILIZATION OF RESUORCE

	Utilization	Ratio (%)
LUT	35090	17.22
LUTRAM	1195	1.87
FF	26449	6.49
BRAM	19	4.27
DSP	150	17.85
BUFG	15	46.88

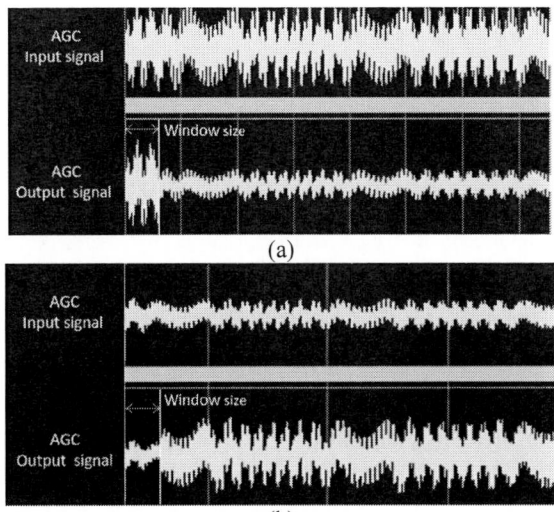

(a)

(b)

Figure 5. Output of AGC; (a) when α is smaller than 1 and (b) when α is bigger than 1.

repeater and the AGC correctly control the power level to be approximately 0 dBm in both cases. And Fig. 5 show that AGC output signal varied after detecting the power level in the window size. When α is smaller than 1, AGC output signal amplitude is reduced. While α is bigger than 1, AGC output signal amplitude is amplified. As shown, the AGC output signal amplitude is normalized to max power. Resource utilization of the design in FPGA is summarized in Table I.

IV. CONCLUSION

In this paper, we design a digital signal processing algorithm, and implement targeting FPGA-based repeater. FPGA-based repeaters are much smaller in size and can be processed in parallel than software or analog repeaters, resulting in better performance. Thus FPGA-based repeater is good candidate for analog repeaters and software repeaters.

ACKNOWLEDGMENT

This research was supported by the MIST (Ministry of Science and ICT), Korea, under the National Program for Excellence in SW supervised by the IITP (Institute for Information & communications Technology Promotion) (2017-0-00096). This work (Grants No. C0509906) was supported by Business for Cooperative R&D between Industry, Academy, and Research Institute funded Korea Small and Medium Business Administration in2017.

REFERENCES

[1] I. Hatai and I. Chakrabarti, "A new high-performance digital FM modulator and demodulator for softwrare-defined radio and its FPGA implementation," *IJRC,* vol 2011, no. 2, Jan 2011.

[2] I. Hatai and I. Chakrabarti, "FPGA Implementation of a Digital FM Modem," in *Proc. ICIMT,* pp. 475-479, 2009.

[3] C. Kim and S. Im, "Digital automatic gainconrol for software radio W – CDMA base stations," in *Proc. Electronics Letters,* vol. 39, no. 3, pp. 318-320, Feb 2003

Efficient Four-way Row-splitting Layered QC-LDPC Decoder Architecture

Tram Thi Bao Nguyen and Hanho Lee

Department of Information and Communication Engineering,
Inha University, Incheon, Korea
hhlee@inha.ac.kr

Abstract—**This paper presents a four-way row-splitting architecture for layered low-density parity-check (LDPC) decoder. Based on the proposed method, an efficient partially parallel pipelined QC-LDPC decoder is proposed. The synthesis results using TSMC 40-nm standard cell CMOS technology show that the proposed decoder achieves the maximum required throughput of 7.05 Gb/s and outperforms its predecessors in terms of area efficiency.**

Keywords; LDPC code; row-splitting; layered; pipelined; decoder

I. Introduction

Low-density parity-check (LDPC) codes, which were first proposed by Gallager in 1962, have been received widespread attention because of their superior error-correction performance, highly parallel implementation and high throughput potential. Therefore, LDPC codes are sufficient choices in many modern communications standards. The new generation of 5G standard, which focuses on improved error-correction performance, lower power consumption and higher throughput, is expected to begin rolling out worldwide in the early 2020s. LDPC codes have been accepted in 5G communication standards (3GPP) as channel coding schemes for data channels [1].

In this paper, in order to reduce the complexity while maintaining the required throughput, a four-way row-splitting pipelined layered QC-LDPC decoder architecture is proposed for multi-gigabit wireless communication.

II. Proposed Four-Way Row-Splitting Pipelined Layered LDPC Decoder Architecture

To increase parallelism and reduce decoder complexity, the proposed four-way row-splitting LDPC decoding algorithm partitions the parity-check matrix H into 4 sections which are denoted as splits A, B, C, and D. The four-way row-splitting architectures reduce the number of columns for each equally dimensioned section by a factor of 1/4. Each section in a layer is processed separately by the check node unit (CNU) and variable node unit (VNU). Let l be the layer number and let i be the iteration number. R denotes the check-to-variable message conveyed from check node c to variable node v. L represents the variable-to-check message from variable node v to check node c. In the ith iteration, the log-likelihood ratio (LLR) message from the $(l-2)^{th}$ layer to the next layer l for variable node v is represented by $L_v^i[l-2]$. The proposed algorithm is described as follows:

Algorithm 1: Proposed four-way row-splitting decoding algorithm

1. Initialization: LLR initialization at each VN, $v = 1, 2, ..., n$

$$L_v^0[0] = \frac{2R_v}{\sigma^2} \qquad (1)$$

for iteration $i = 1$ to I_{max} **do**
  for layer $l = 1$ to m **do**
    for section $u = 0$ to 3 **do**
      for VN $v = (u \times n/4)+1$ to $(u+1) \times n/4$ **do**

        2. Check node processing:

$$L_{vc}^i[l] = L_v^i[l-2] - R_{cv}^{i-1}[l] \qquad (2)$$

$$R_{cv}^i[l] = max\{ \min_{v' \in Row[c] \setminus v} \left| L_{v'c}^i[l] \right| - \beta, 0\} \times \prod_{v' \in Row[c] \setminus v} sign\left(L_{v'c}^i[l]\right) \qquad (3)$$

        3. Variable node processing:

$$L_v^i[l-1] = L_v^i[l-2] - R_{cv}^{i-1}[l-1] + R_{cv}^i[l-1] \qquad (4)$$

      end for
    end for

4. Hard decision: Hard decision at each VN

$$\hat{C}_v^i = \begin{cases} 0, & if\ L_v^i[l] \geq 0 \\ 1, & otherwise \end{cases} \qquad (5)$$

5. Termination check: Early termination check at each CN
  if $\hat{C} \oplus H^T = 0$ **then** terminate
  end for
end for

Fig. 1 depicts the architecture of the proposed four-way row-splitting pipelined layered QC-LDPC decoder. The proposed decoder consists of 42 parallel CNUs and VNUs since each sub-matrix size is $z = 42$ for the IEEE 802.11ad standard. Apart from the input and output pipeline stages, a total of three pipeline stages are inserted in both CNUs and VNUs. First, the input LLRs that are related to each section of the H-matrix are provided to barrel shifters, which rotate the LLRs for CNU and VNU processing. Note that two separate barrel-shifter blocks are utilized for CNUs and VNUs, because they operate on different layers in each clock cycle due to pipelining. In CNU, the first section *C1* processes the LLRs corresponding to the split *A* of the decoder. Then, the split *B* is processed by section *C1*, and LLRs corresponding to section *A* are passed to next pipeline stage of the second section *C2*. Similarly, in the VNU case, four sections of the H-matrix are processed by four pipeline stages. In the proposed decoder, a feedback is visible after the CNU pipeline to verify the correct calculation of $R_{cv}^i[l]$. The flip-flop at the end of the min-sorter in the CNU stores the intermediate min result of previous split for one clock cycle. The valid min result is produced in the next clock cycle after the processing of the latter split for each

---

This research was supported by the MSIT (Ministry of Science, ICT), Korea, under the ITRC support program (IITP-2018-2014-1-00729) supervised by the IITP, and in part by the WCSL research grant directed by Inha University.

TABLE I.   ANALYSIS IMPLEMENTATION AND COMPARISON RESULTS

Design	Proposed (2018)	Mao-Ruei [2] (JSSC 2017)	Sabooh [3] (IET 2013)	Hiroyuki [4] (GSIP 2015)
IEEE standard	802.11ad	802.15.3c	802.11ad	802.11ad
Decoding schedule	Layered	Layered	Layered	Flooding
Architecture	Split-row	Full row	Full row	Full row
Code length	672	672	672	672
Multi-mode	Yes	Yes	Yes	Yes
CMOS technology	40-nm	90-nm	65-nm	40-nm LP
Core area (mm^2)	0.195	2.25	1.1	0.8
Normalized core area (mm^2)	0.195	0.444	0.417	0.8
Clock freq. (MHz)	850	157	215	220
Quantization bits	5	4	8	5
Throughput (Gb/s)	7.05	5.28	6.0	6.16
Normalized throughput (Gb/s)	7.05	11.88	9.75	6.16
Area efficiency ($Gb/s/mm^2$)	36.15	2.35	5.45	7.70
Normalized area efficiency ($Gb/s/mm^2$)	36.15	26.76	23.38	7.70

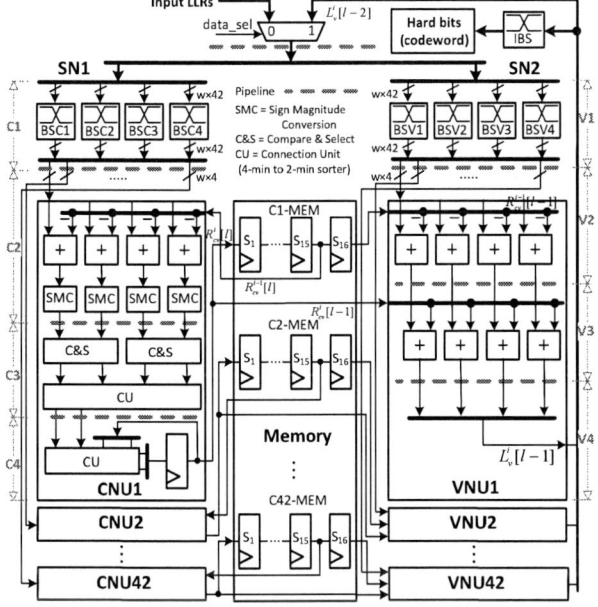

Figure 1.   Proposed four-way row-splitting based pipelined layered QC-LDPC decoder.

layer. Decoding is carried out through an iterative process of message exchange between the CNUs and VNUs. The CNU block implements Eqs. (2) and (3), while the VNU implements Eq. (4).

III. ANALYSIS AND COMPARISON RESULTS

Table I reports implementation and comparison results of the proposed decoder with various recently published multi-gigabit LDPC decoder implementations [2-4]. The decoder was modeled in Verilog hardware description language (HDL) and synthesized with TSMC 40-nm CMOS technology. For a fair comparison, the core area and throughput are normalized to 40-nm CMOS technology by using the scaling factors as suggested in [5]. The proposed multi-gigabit decoder consumes an area of

0.195 mm^2. Although the clock cycles are increased, it still achieves the maximum throughput of 7.05 Gb/s, which satisfies the throughput requirements of the IEEE 802.11ad standard. It can be observed that the proposed four-way row-splitting based pipelined layered QC-LDPC decoder is more hardware efficient compared to other LDPC decoders with a full-row architecture. Moreover, it outperforms all decoders in Table I in terms of area efficiency. Specifically, the normalized area efficiency of our work is 35.1% better than that of the next most efficient decoder [2], and 54.6% compared with the full-row LDPC architecture in [3].Therefore, the proposed decoder provides significant area saving over prior architectures.

IV. CONCLUSIONS

This paper presents an efficient four-way row-splitting pipelined layered QC-LDPC decoder architecture with much reduced hardware resource requirement. This decoder also provides better area efficiency compared to fully parallel designs. The proposed architecture is very suitable for high data rate communication systems employing LDPC codes for channel coding.

REFERENCES

[1] Session Chairman (Nokia), "Chairman's Notes of Agenda Item 7.1.5 Channel coding and modulation," 3GPP TSG RAN WG1 Meeting No. 87, R1-1613710, Reno, USA, Nov. 2016.

[2] M. R. Li, C. H. Yang and Y. L. Ueng, "A 5.28-Gb/s LDPC Decoder with Time-Domain Signal Processing for IEEE 802.15.3c Applications," IEEE Journal of Solid-State Circuits, vol. 52, no. 2, pp. 592-604, Feb. 2017.

[3] S. Ajaz and H. Lee, "Reduced-complexity local switch based multi-mode QC-LDPC decoder architecture for gigabit wireless communications," IET Electronics Letters, 49(2013) 1246-1248.

[4] H. Motozuka, N. Yosoku, T. Sakamoto, T. Tsukizawa, N. Shirakata and K. Takinami, "A 6.16Gb/s 4.7pJ/bit/iteration LDPC decoder for IEEE 802.11ad standard in 40nm LP-CMOS," in 2015 IEEE Global Conference on Signal and Information Processing (GlobalSIP), pp. 1289-1292, 2015.

[5] M. Weiner, et al., LDPC decoder architecture for high-data rate personal-area networks, in: 2011 IEEE International Symposium on Circuits and Systems (ISCAS 2011), pp. 1784-1787, 2011.

Reconfigurable Multi-Input Adder Design for Deep Neural Network Accelerators

Hossein Moradian, Sujeong Jo, Kiyoung Choi

Department of Electrical and Computer Engineering, Neural Processing Research Center (NPRC)
Seoul National University, Seoul, Republic of Korea
{hossein, sjjo, kchoi}@dal.snu.ac.kr

Abstract— **This paper proposes two efficient designs of reconfigurable multi-input adders for deep neural network accelerators. The reconfigurability allows us to use resources in different ways optimized to different applications. The designed adders enable bit-width adaptive computing in neural network layers, which improves computing throughput. The proposed designs are implemented with 45nm CMOS TSMC library and the results show that the proposed modules achieve throughput much higher than that of conventional designs even with the reconfigurability without significant hardware overhead.**

Keywords; reconfigurable computing; deep neural networks; high troughput

I. INTRODUCTION

The accuracy of DNN algorithms is achievable at the expense of computational cost. To reduce the cost, various methods have been suggested, one of which is reducing the bit-precision of weights and activations in a DNN. It also reduces the memory requirements and computational complexity with limited accuracy degradation [1][2], which in turn decreases power consumption and increases the performance.

Most DNNs have used 32-bit single precision floating-point operations, which require high cost of area and energy consumption. However, recent researches in DNNs using 16-bit fixed-point arithmetic operations demonstrate acceptable accuracy by using only 16-bit datapaths [3]. Dynamic fixed-point is one of the most effective attempts to reduce bit-precision [4][5]. More recent work shows that DNNs can be implemented with 4-bit precision and achieve accuracies similar to that with16-bit floating-point [6][7].

In general, different layers in a DNN need different precision. However, most of the conventional hardware accelerators support only fixed datapaths and precision; thus, for the layers that need less precision, the hardware module is unnecessarily large and inefficient.

Multiply-accumulation (MAC) operations for matrix-vector and matrix-matrix multiplications are the most time-consuming operations of DNNs and various algorithms have been devised to accelerate them. However, it is a challenging task to map such algorithms to general purpose or custom hardware architecture [8]. The bit-serial sum of the product is one of the methods to perform MAC operations. Some recent researches used bit-serial multiplication for MAC operations [9-14]. In these designs, multiplication is done by addition, requiring only adder modules. Bit-serial computation enables trade-offs between performance, energy, and area [9]. In bit-serial computation, multi-input

adders are the most critical modules and their throughput affects the overall performance of the network.

DNN accelerators require reconfigurability to maximize resource utilization and thus maximize performance. For them to be reconfigurable, they need to employ reconfigurable MAC modules to support various precisions that could be used in different layers.

II. PROPOSED METHODS

Fig. 1 illustrates the conventional multi-input adder. This adder receives m inputs with n bit-width each. The output of the adder needs $n + \lceil \log_2 m \rceil$ bits to keep the accuracy. We present two mechanisms for reconfigurable accumulators (we use the terms *multi-input adder* and *accumulator* interchangeably) supporting various precisions to substitute for conventional accumulators.

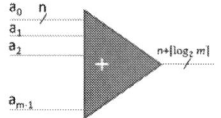

Figure 1.   Conventional multi-input adder.

A. Parallel Reconfigurable Multi-Input Adder (PR-Adder)

Suppose n and k are the highest and the lowest precision (bit-width) required in the network layers. In PR-Adder method, to accumulate m numbers, instead of using one multi-input n-bit adder, we use $\lceil \frac{n}{k} \rceil$ m-input k-bit adders, as presented in Fig. 2.

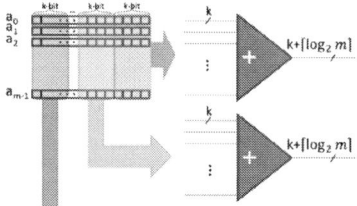

Figure 2.   PR-Adder design.

Using this design when n-bit precision is needed, $\lceil \frac{n}{k} \rceil$ adders will generate accumulation result for their own ranges, and thus for lower precisions (i.e., small n), fewer adders are involved to calculate the result. In addition, however, the output of the adders should be aggregated together to form the final result. For n-bit precision, $\lceil \frac{n}{k} \rceil$ outputs need to be added together with k-bit shift, as shown in Fig. 3.

978-1-5386-7961-6/18 $31.00 © 2018 IEEE

Assuming $k=4$, the PR-Adder approach first generates $4 + \lceil \log_2 m \rceil$ bits intermediate outputs and then depending on the input precision, the intermediate outputs are aggregated. For example, if we have four 4-bit adders in 16-bit precision, the four intermediate outputs are combined together to generate one output; in 8-bit precision, the four adders can be used to generate two outputs at the same time; in 4-bit precision, the outputs of four adders are four final outputs generated in parallel.

Figure 3. Calculating final accumulation result.

B. Serial Reconfigurable Multi-Input Adder (SR-Adder)

In this design instead of using $\lceil \frac{n}{k} \rceil$ modules in parallel to n-bit accumulation, we use one adder module, $\lceil \frac{n}{k} \rceil$ times (in $\lceil \frac{n}{k} \rceil$ cycles). Same as PR-Adder, we divide the input number into k partitions, but here each part is fed to the adder serially in each cycle. By decreasing the input number precision, fewer cycles are needed to get the final result. For example, to for 16-bit input precision with 4-bits grouping ($k=4$), four cycles are needed, and for 8- and 4-bit input precision with the same grouping ($k=4$), two and one cycles are needed, respectively.

In this design, after each cycle, except for the last cycle, $\lceil \log_2 m \rceil$ MSB bits of the adder output will be sent back to the input side of the adder to be added with the numbers of the next cycle and the rest k bits will be sent out as part of final accumulation result. On account of this, $k \geq \lceil \log_2 m \rceil$ inequation needs to be maintained. For the last cycle, all $k + \lceil \log_2 m \rceil$ bits are sent out to form the MSB part of the final accumulation result and the $\lceil \log_2 m \rceil$ bits returning line to the adder should be set to zero in order to have correct computation for the next incoming numbers. This can be done by using a two to one multiplexer and selecting zero input in the first cycle of calculation. Fig. 4 illustrates the described design.

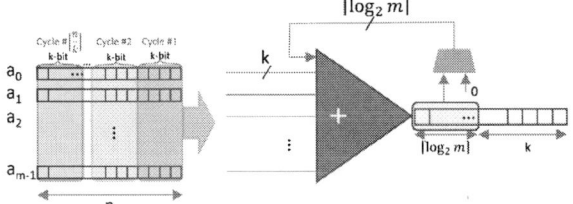

Figure 4. SR-Adder design.

III. EVALUATION

In this section, we compare the proposed designs with a conventional design. To attain area and power comparison, a 16-input 16-bit conventional adder, a PR-Adder with four 16-input 4-bit adders, and a SR-Adder with one 17-input 4-bit SR-Adder are designed in Verilog HDL and synthesized the corresponding hardware by Synopsys Design Compiler tools with TSMC 45nm technology process library. For all the three designs, clock period is fixed to 0.49ns (all designs are synthesized with the same delay constraint). The functional simulation to verify the functional correctness of the synthesized circuits is performed with ModelSim SE. Table I lists the synthesis results for the three designs. The results show that reconfigurable PR-Adder occupies 1.24% more area and consumes 0.9% more power compared to the conventional design but its throughput is $2\times$ and $4\times$ higher than that of the conventional adder when the input precision is eight and four, respectively. On the other hand, SR-Adder occupies 5.70% less area and consumes 0.1% more power compared to the conventional adder and has the same throughput as PR-Adder.

TABLE I.   AREA AND POWER CONSUMPTION OF THREE DESIGNS

	Bit-width	#input	Area	Power
Conventional	16	16	2500.47	3.0289
PR-Adder	4x4	16	2531.52	3.0561
SR-Adder	4	17	589.53 (×4: 2358.12)	0.7580 (×4: 3.0320)

IV. CONCLUSION

In this paper, we describe two designs for reconfigurable multi-input adder and compared them with the conventional design. Evaluation results state that applying the proposed methodologies do not add significant overhead to the system while adding flexibility to the architecture by making it reconfigurable. In addition, the proposed schemes have higher resource utilization and performance.

ACKNOWLEDGMENT

This work was supported by Samsung Advanced Institute of Technology (SAIT).

REFERENCES

[1] F. Li, and B. Liu, "Ternary weight networks," arXiv preprint arXiv:1605.04711, 2016.

[2] C. Zhu et al., "Trained ternary quantization," arXiv preprint arXiv:1612.01064, 2016.

[3] Y. Du et al., "A streaming accelerator for deep convolutional neural networks with image and feature decomposition for resource-limited system applications," arXiv preprint arXiv:1709.05116, 2017.

[4] M. Courbariaux, Y. Bengio, and J.P. David, "Training deep neural networks with low precision multiplications," arXiv preprint arXiv:1412.7024, 2014.

[5] D. Das et al., "Mixed precision training of convolutional neural networks using Integer operations," arXiv preprint arXiv:1802.00930, 2018.

[6] J. Choi et al., "PACT: Parameterized clipping activation for quantized neural networks," arXiv preprint arXiv:1805.06085, 2018.

[7] D. Shin et al., "14.2 DNPU: An 8.1 TOPS/W reconfigurable CNN-RNN processor for general-purpose deep neural networks," ISSCC 2017.

[8] T. SegFault, "Matrix-matrix multiplication using systolic array architecture," Bluespec 2015.

[9] P. Judd et al., "Stripes: Bit-serial deep neural network computing," MICRO 2016.

[10] A. Delmas et al., "Dynamic stripes: Exploiting the dynamic precision requirements of activation values in neural networks," arXiv preprint arXiv:1706.00504, 2017.

[11] J. Albericio et al., "Bit-pragmatic deep neural network computing," MICRO 2017.

[12] A. Delmas et al., "Tartan: Accelerating fully-connected and convolutional layers in deep learning networks by exploiting numerical precision variability," arXiv preprint arXiv:1707.09068, 2017.

[13] A. Delmas et al., "Dpred: Making typical activation values matter in deep learning computing," arXiv preprint arXiv:1804.06732, 2018.

[14] D. Lee, S. Kang, and K. Choi, "ComPEND: Computation pruning through early negative detection for ReLU in a deep neural network accelerator," ICS 2018.

A Method of Prevent Loss of Information in Ill-Posed Problem Based Application using Atmospheric Scattering Model

Geun-Jun Kim and Bongsoon Kang
Dept. of Electronic Engineering
Dong-A University
Busan, Korea
firstaccel@gmail.com and bongsoon@dau.ac.kr

Abstract— **Atmospheric scattering model is widely used in image signal processing. This is a model for the process by which an object is photographed. This model is also commonly used in modern advanced image processing technology. When using the atmospheric scattering model, you should consider loss of information. If you used the atmospheric scattering model and do not use the loss of information, you will lose information in a special case. In this paper, we consider about atmospheric scattering model based application and ill-posed problem.**

Keywords; Atmospheric Scattering Model, Ill-posed problem, Loss of inforamtion, Image signal processing.

I. INTRODUCTION

In these days, there are so many applications in the image signal processing area. Fortunately, we can divide it into large categories. If we divide into two categories with extreme clarity, there are well-posed problem and ill-posed problem. We called it well-posed problem when the problem has a unique solution. Ill-posed problem does not have solution or there are multiple solutions. In the image signal processing area, ill-posed problems are occurred often. Recently, the use of cameras has become more and more widespread. This required a technology to acquire images in various environments. This required a way to overcome the environment in which the image was taken. Atmospheric scattering model is widely used to solve this problem. Single image dehazing is one of the popular ill-posed problem with atmospheric scattering model. Fig. 1 presents example of single image dehazing [1].

(a)   (b)

Figure 1. Example of Single Image Dehazing [1]: (a) Hazy image, (b) Dehazed image.

In this paper, we propose methods to prevent loss of information caused by not considered by researchers using atmospheric scattering model.

II. BACKGROUDNS

A. Atmospheric Scattering Model

Atmospheric scattering model is also called attenuation model [2]. This model is mathematically defined as (1). μ, ν is a pixel location of image. I is captured image, J is scene radiance, T is transmission, and A is global air light.

$$I=JT+A(1-T) \tag{1}$$

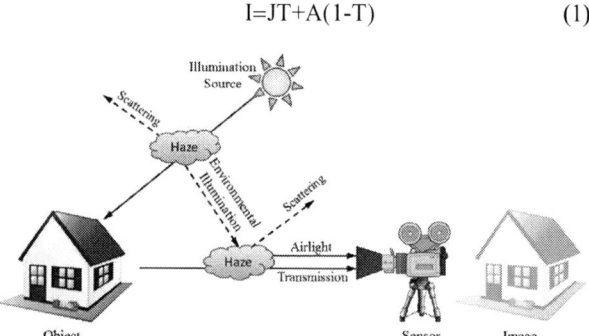

Figure 2. Graphical Atmospheric Scattering Model [2].

Fig. 2 also graphically describes atmospheric scattering model [2]. The camera takes the reflected light from the object. This reflected light is attenuated as it passes through the atmosphere. This is called direct attenuation. It is mathematically expressed as the scene radiance multiplied by the transmission. In addition, components by illumination are added. This is expressed global air light multiplied by one minus transmission. These two components make captured image I. We have only captured image I, but we have to find scene radiance, transmission, and global air light. This is why single image dehazing is classified as ill-posed problem.

978-1-5386-7961-6/18 $31.00 © 2018 IEEE   214   ISOCC 2018

B. Single Image Dehazing

Recently, the current main flow of dehazing is using a single image [3-6]. Various algorithms such as dark channel prior [3], guided filter [4], and machine learning [5] are used to remove haze using a single image. Dark channel prior is based on statistical experiment, guided filter is based on conventional image signal processing, and [5] is based on supervised learning. These three algorithms based on different fundamentals but has common side effect that is loss of information.

III. LOSS OF INFORMATION

Fig. 3(a) is an input hazy image. The center of this image has very high brightness. But bottom has very low brightness. When this image is solved by formula, information is lost. The degrees vary by algorithm, but information has been lost in areas where Fig. 3(b) to (d) all three methods had lower brightness in the input image. The loss of this information is caused by a formula for restoring scene radiance. When we calculate scene radiance, (1) is rewritten as (2). There is a process of removing global air light from molecules in fractions. Generally, single image dehazing algorithms use the brightest pixel in the image. In this process, pixels with low brightness in the input image will be reduced below zero. And then, when the value of this pixels become smaller in the calculation with the amount of transmission, the value of the pixels is not restored when the waiting light is added later. This is because only a single image dehazing is applied to the whole image by selecting light a large value as standby light considering only haze.

In this paper, we propose an adaptive based on image characteristics using the brightness characteristics of the image to prevent loss of information. It expressed mathematically as (3). A^a is an adaptive global air light, P_d is the lowest 0.1% dark pixels value and P_b is the brightest 0.1% brightest pixels value. Fig. 4 presents the result of proposed method on Fig. 3(d).

$$J = ((I-A)/T) + A \qquad (2)$$

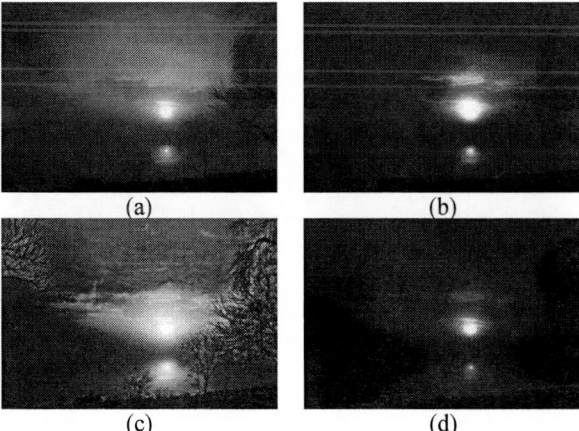

Figure 3. Hazy image and result of single image dehazing: (a) Hazy image, (b) Result of [3], (c) Result of [6], and (d) Result of [5].

Figure 4. Prevent loss of information on Fig. 3(d).

$$A^a = A(P_d / P_b) \qquad (3)$$

Comparing Fig. 4 and Fig. 3(d), Fig. 4 contains more information than Fig. 3(d). And comparing Fig. 4 and Fig. 3(a), Fig. 4 has the haze removed from Fig. 3(a) to have a richer sky, but the visibility of the image has been preserved.

IV. CONCLUSIONS

If an ill-posed problem, such as a single image dehazing, uses the atmospheric scattering model, a side effect will occur. This side effect is the loss of information on pixels with low brightness in the input image, which can have a significant impact on the quality of the image. In this paper, we propose an adaptive based on image characteristics using the brightness characteristics of the image to prevent loss of information. Through the simulation, we have confirmed that this method can protect information about pixels input at low brightness.

ACKNOWLEDGMENT

This research was supported by Basic Science Research Program through the National Research Foundation of Korea (NRF) funded by the Ministry of Education.

(NRF-2015R1D1A1A01060427)

REFERENCES

[1] J.P. Tarel et al. "Vision Enhancement in Homogeneous and Heterogeneous Fog," IEEE Intell. Trans. Sys. Magazine, vol. 4, no. 2, pp. 6-20, April, 2012.

[2] B. Cai et al, "Dehazenet: An End-to-End System for Single Image Haze Removal," IEEE Trans. Image Process., vol. 25, no. 11, pp.5187-5198, August, 2016.

[3] K. He, J. Sun, and X. Tang, "Single Image Haze Removal using Dark Channel Prior," IEEE Trans. Patt. Anal. Mach. Intell., vol.33, no.12, pp.2341-2353, December, 2011.

[4] K. He, J. Sun, and X. Tang, "Guided Image Filtering," IEEE Trans. Patt. Anal. Mach. Intell., vol.35, no.6, pp.1397-1409, 2013.

[5] Q. Zhu, J. Mai, and L. Shao, "A Fast Single Image Haze Removal Algorithm using Color Attenuation Prior," IEEE Trnas. Image Process., vol. 24, no. 11, pp.3522-3533, 2015.

[6] J.P. Tarel and N. Hautiere, "Fast Visibility Restoration from a Single Color or Gray Level image," IEEE Int. Conf. Comp. Vis., pp.2201-2208, Oct., 2009.

4-Bit Data Arrangement Algorithm for CAN Compression

Yeon-Jin Kim, Ho-Yun Lee, Jin-Gyun Chung
Division of Electronic Engineering
IT Convergence Research Center, Chonbuk National University, South Korea
E-mail: yang@jbnu.ac.kr, jgchung@jbnu.ac.kr

Abstract—**Controller Area Network (CAN) has become the most widely used protocol in automotive networks due to its high reliability and low cost. There can be more than 80 electronic control units (ECUs) distributed in a modern vehicle. As the number of ECUs connected to the CAN bus increases, so does the busload. Due to the real-time constraints and limited resources in the ECUs, reducing busload is becoming an important problem. In this paper, we propose a systematic data arrangement algorithm to efficiently reduce the CAN data frames. Experimental results show that higher compression ratios are obtained by the proposed algorithm compared with the existing algorithm.**

Keywords; Controller area network; data compression; in-vehicle communication; data arrangement

I. INTRODUCTION

CAN is a high-speed, serial communication protocol which has the capacity of real-time control [1]. Recently, various kinds of electrical and electronic systems have been installed in vehicles. As the number of ECUs increases, the data traffic on the CAN network will also increase. If the CAN bus is overloaded, it is not easy to transmit low-priority CAN messages due to the long waiting time. In addition, the probability of an error transmission in the CAN network is also increased.

The CAN busload can be effectively reduced by applying data compression technology to CAN data. In most CAN data compression algorithms, the compression efficiency depends on the accuracy of the maximum difference value predicted for a specific application. However, it is very difficult to select the maximum difference value optimally. In [2], it is shown that the improved CAN data reduction (ICANDR) algorithm eliminates the need to predict the maximum difference value and that the algorithm achieves the highest compression efficiency compared with other methods.

The compression efficiency of ICANDR algorithm depends on data arrangement. In [3], an 8-bit systematic data arrangement algorithm is presented.

In this paper, we propose a 4-bit data arrangement procedure for ICANDR algorithm.

II. EXISTING DATA ARRANGEMENT ALGORITHM FOR CAN COMPRESSION

To apply ICANDR algorithm, CAN data need to be arranged in the form shown in Fig. 1. Then, to increase the data compression efficiency, the rapidly changing data should be placed in the least significant part and the slowly changing data should be placed in the most significant part. The data arrangement algorithm in [3] determines and renames each byte of CAN data from *a* to *h* by a systematic algorithm.

III. PROPOSED DATA ARRANGEMENT PROCEDURE

In the proposed data arrangement method, data is arranged by the unit of 4-bits (half byte). Then, compared with the existing 8-bit data arrangement, more accurate data arrangement can be achieved.

The proposed data arrangement is shown in Fig. 2. Fig. 2 shows that it is required to determine the locations of 16 4-bit groups. In other words, input CAN data need to be divided into 16 4-bit groups and placed in different positions to increase the compression efficiency. The proposed algorithm is described using magnitude values and frequency values performed on each half byte as

$$S_{fm}(n) = \alpha \times S_{freq}(n) + \beta \times S_{mag}(n),$$
$$n = 0,1,\dots,16, \quad (1)$$

where $S_{mag}(n)$ is the sum of *n*-th half byte column of magnitude matrix and $S_{freq}(n)$ is the sum of *n*-th half byte column of frequency matrix. The weighting factors α and β are normally chosen as $\alpha=\beta=0.5$. Different weighting values can be chosen depending upon applications.

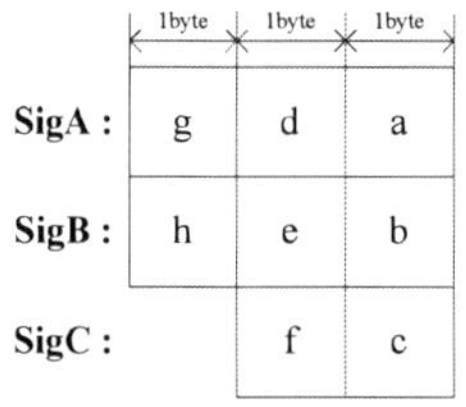

	1byte	1byte	1byte
SigA :	g	d	a
SigB :	h	e	b
SigC :		f	c

Fig. 1. Existing data arrangement

	4bit	4bit	4bit	4bit	4bit	4bit
Sig A:	o	m	j	g	d	a
Sig B:	p	n	k	h	e	b
Sig C:			l	i	f	c

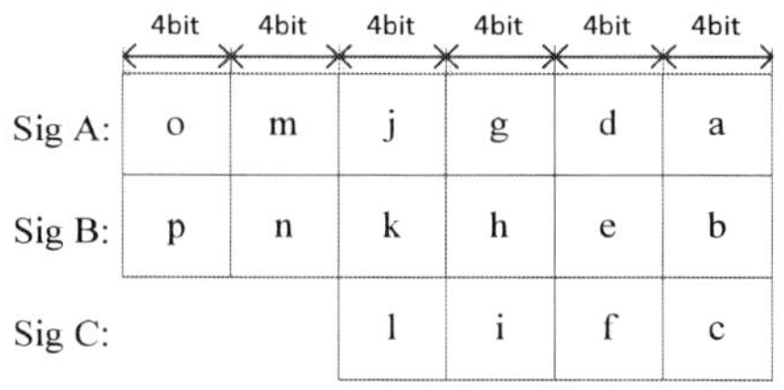

Fig. 2. Proposed data arrangement.

978-1-5386-7961-6/18 $31.00 © 2018 IEEE

The proposed algorithm is summarized as follows:

Data arrangement algorithm	
Step 1:	Perform XOR operations between each half byte of successive CAN data fields.
Step 2:	Transform XOR result of each half byte from hexadecimal number to decimal number, which is called magnitude value.
Step 3:	Determine the frequency value from a magnitude value. If a magnitude value is non-zero, the corresponding frequency value is 1. Otherwise, it is 0.
Step 4:	Add magnitude values of each half byte to get $S_{\mathrm{mag}}(n)$.
Step 5:	Add frequency values of each half byte to get $S_{\mathrm{freq}}(n)$.
Step 6:	Compute $S_{f_m}(n)$ using $S_{\mathrm{mag}}(n)$ and $S_{\mathrm{freq}}(n)$.
Step 7:	By comparing the $S_{f_m}(n)$ values for $n = 0, 1, \ldots 15$, determine the data rank $D_{\mathrm{rank}}(n)$.
Step 8:	Using $D_{\mathrm{rank}}(n)$, arrange each half byte in the form shown in Fig. 2.

TABELE I. CAN DATA EXAMPLE

Frame No.	B(7)	B(6)	B(5)	B(4)	B(3)	B(2)	B(1)	B(0)
F_i	0	F	4	1	D	A	E	5
F_{i+1}	1	0	0	0	F	0	D	0
F_{i+2}	0	F	3	1	D	9	E	6
Frame No.	B(15)	B(14)	B(13)	B(12)	B(11)	B(10)	B(9)	B(8)
F_i	1	0	0	9	1	6	7	1
F_{i+1}	0	0	C	7	1	D	5	1
F_{i+2}	0	0	0	9	1	0	0	1

TABELE II. XORED CAN DATA

	B(7)	B(6)	B(5)	B(4)	B(3)	B(2)	B(1)	B(0)
$F_i \oplus F_{i+1}$	1	15	4	1	2	10	3	5
$F_{i+1} \oplus F_{i+2}$	1	15	3	1	2	9	3	6
	B(15)	B(14)	B(13)	B(12)	B(11)	B(10)	B(9)	B(8)
$F_i \oplus F_{i+1}$	1	0	12	14	0	11	2	0
$F_{i+1} \oplus F_{i+2}$	0	0	0	14	0	13	5	0

TABELE III. $S_{f_m}(n)$ OF CAN DATA

	B(7)	B(6)	B(5)	B(4)	B(3)	B(2)	B(1)	B(0)
S_{fm}	2	16	4.5	2	3	10.5	4	6.5
	B(15)	B(14)	B(13)	B(12)	B(11)	B(10)	B(9)	B(8)
S_{fm}	1	0	6.5	15	0	13	4.5	0

TABELE IV. $D_{rank}(n)$ OF CAN DATA

	B(7)	B(6)	B(5)	B(4)	B(3)	B(2)	B(1)	B(0)
D_{rank}	l	a	g	k	j	d	i	f
	B(15)	B(14)	B(13)	B(12)	B(11)	B(10)	B(9)	B(8)
D_{rank}	m	n	e	b	p	c	h	o

TABLE V. COMPARISON OF THE COMPRESSION EFFICIENCIES BETWEEN EXISTING AND PROPOSED ALGORITHM

ID	Existing	Proposed	Diff
316	86.18%	87.21%	1.03%
545	79.56%	83.48%	3.92%
580	95.26%	96.17%	0.91%

$S_{f_m}(n)$ values denote the degree of change of each half byte. The slowly changing data (small $S_{f_m}(n)$ values) is placed in the most significant part while the frequently changing data (large $S_{f_m}(n)$ values) is placed in the least significant part. For example, while the half byte with the largest $S_{f_m}(n)$ value is placed at the location of a, the half byte with the smallest $S_{f_m}(n)$ value is placed at the location of p. The other half bytes are placed according to the magnitude of their calculated $S_{f_m}(n)$ values. Tables I ~ IV show an example for the calculation of $S_{f_m}(n)$ values, where sixteen half bytes are denoted by $B(n)$, $n = 0, 1, \cdots, 15$.

Table V compares the compression ratios obtained by existing algorithm and the proposed algorithm using the actual CAN signals from KIA Sorento vehicle. It is shown that compression efficiency can be improved up to 3.92% by the proposed method.

IV. CONCLUSIONS

In this paper, we proposed a 4-bit data arrangement algorithm for CAN to maximize the compression efficiency. Experimental results indicate that higher compression efficiency can be obtained by the proposed method.

REFERENCES

[1] Robert Bosch GmbH, "CAN specification version 2.0," Chuck Powers, *Motorola MCTG Multiplex Applications*, Apr. 1995.

[2] Y.-J. Wu and J.-G. Chung, "An improved controller area network data-reduction algorithm for in-vehicle networks," *IEICE Trans. Fundamentals*, vol. E100-A, No.2, pp. 346-352, Feb 2017.

[3] Y.-J. Kim, Y. Zou and J.-G. Chung, "Improved CAN compression algorithm by data reordering," *International Soc Design Conference (ISOCC)*, 2017, pp.264-265.

Generalized Adaptive Variable Bit Truncation Method for Approximate Stochastic Computing

Keerthana Pamidimukkala[1], Kyung Ki Kim[2], Yong-Bin Kim[3] and Minsu Choi[1]

[1]Dept of ECE, Missouri Univ of Science & Technology, Rolla, MO, USA, {kpt4b,choim}@mst.edu
[2]Dept of Electronic Eng., Daegu University, Gyeongsan, Korea, kkkim@daegu.ac.kr
[3]Dept of ECE, Northeastern University, Boston, MA, USA, ybk@ece.neu.edu

Abstract—**Stochastic computing as a computing paradigm is currently undergoing revival as the advancements in technology make it applicable especially in the wake of the need for efficient reduced precision computing for emerging applications. Recent research in stochastic computing exploits the benefits of approximate computing, called Approximate Stochastic Computing (ASC), which further reduces the operational overhead in implementing stochastic circuits. A new generalized adaptive method improving on ASC is proposed in this work. The proposed method has been discussed with two possible implementation variants - Area efficient and Time efficient. The proposed method has also been implemented in Matlab to compare against ASC and is shown to perform better than previous approaches for error-tolerant applications.**

I. INTRODUCTION

Stochastic computing refers to the method of using probabilistic representations of numbers to perform numerical operations. Stochastic computing was introduced as a concept of computing in 1956 [1]. This concept was further improved with implementations throughout the 1960s [2, 3]. To explain the concept of stochastic computing, it is necessary to understand stochastic representation of data.

To convert any number into a stochastic bit stream, a comparator and a Random Number Generator (RNG) can be used. It is to be noted that the input number and the RNG output need to be on the same scale for the conversion to be appropriate. The RNG can also be reused to represent multiple inputs in a stochastic system; this results in correlation between the stochastic bit streams which can be detrimental to the operation of certain stochastic circuits but it can also be exploited for benefits as detailed in [4].

Using a stochastic bit stream, it is possible to use the probabilistic nature of logic gates to perform arithmetic operations. Consider a simple AND logical gate. If the inputs are considered as streams of bits, It can be observed from this truth table that the inputs are of probability 0.5 each (denoting that the probability of finding bit '1'in the stream is 0.5). Similarly, the output stream has a probability of 0.25. Here the input probabilities seem to have been multiplied when enough number of input bits gets processed for convergence.

In stochastic computing, as a rule of thumb, at least 2^n bits are used in the stochastic domain to represent an n bit binary number for convergence. However, this exponential increase in the number of stochastic bit length has been known to be one of the major issues to address. Applying approximate

computing at the input level by discarding bits of lower significance form the input will further reduce the number of stochastic bits required since the number of bits required to represent the input is now lowered. If m bits are truncated from the input numbers, the required number of bits in the stochastic stream would approximately be 2^{n-m}. A case study of this concept is presented in [5]. Here, the author uses the values of $n = 4$ and $m = 4$ for implementation.

Additionally, to increase the efficiency especially for the Robert-Cross edge detection implementation for example, an adaptive mechanism has been discussed in [6]. This adaptive mechanism suggests adding an extra count of 1 to the output bit stream while converting back to the binary domain. This extra count is conditional and is only applied when the majority of the inputs have the bit 1 in their 5th most significant bit position. This work aims to propose an improvement over the concept of ASC.

II. THE PROPOSED GENERALIZED BIT TRUNCATION MODEL FOR ASC

Consider numbers stored in binary format (integers for simplicity). Information stored in such a format lies in between the most significant high bit and the least significant high bit in any number's representation. This can be represented as shown in Figure 1.

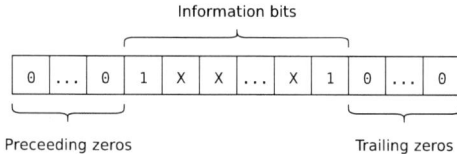

Fig. 1: Information in Binary format

It is imperative to perform mathematical operations on information stored in above format by using the bits shown in the information range. This also means that truncating certain bits based on position, in any number, may result in lost information. The proposed method aims to perform any operation in the information range of its operands instead of a fixed position range. Consider the following examples for $n = 8$ bit operands.

$$
\begin{aligned}
A &= 000000 \left[\begin{array}{c} 01 \\ 11 \\ 01 \end{array} \right.
\end{aligned}
\quad
\begin{aligned}
&= (1)_{base\ 10} \\
&= (3)_{base\ 10} \\
&= (1)_{base\ 10}
\end{aligned}
\quad (1)
$$

with $B = 000000$ and $\dfrac{A+B}{2} = 000000$.

$$A = 00 \begin{bmatrix} 0101 \end{bmatrix} 00 = (20)_{base\ 10}$$
$$B = 00 \begin{bmatrix} 1110 \end{bmatrix} 00 = (56)_{base\ 10} \quad (2)$$
$$\frac{A+B}{2} = 00 \begin{bmatrix} 1001 \end{bmatrix} 10 = (38)_{base\ 10}$$

The example shown in (1) shows that the operation is dominant in the range of two lowest significant bits (highlighted with brackets). Similarly the example in (2) shows that the operation is dominant in the range of 6th lowest significant bit to 3rd lowest significant bit. The dominant information among the operands takes precedence, which means the information bit range varies based on the operands for a given operation. The generic process of working with specific information bit ranges in any given set of operands can be done in the following three steps: 1) Extract information bits from all operands; 2) Perform the operation on extracted information bits; and 3) Modify the result to match original scale of the operands.

III. PERFORMANCE EVALUATION

To evaluate the proposed method in comparison with [5], a simple image edge detection method called Robert Cross filter was implemented Matlab to simulate this process in order to benchmark its performance. Robert Cross edge detection is an algorithm applied to gray scale images over a neighborhood of 2×2 pixels highlighting the changes in intensities which results in an image that highlights only the edges in the input image. The equation for this process is $Y(i,j) = 0.5 \times (|X(i+1,j+1) - X(i,j)| + |X(i,j+1) - X(i+1,j)|)$, where X is the input image, Y is the output image, (i,j) are pixel indices in vertical and horizontal direction respectively. The implementation uses the values of $n = 8$ (i.e., pixel grayscale depth) and $m = 4$ (i.e., truncated bits) resulting in $k = 4$ for fairness in comparison with approximate stochastic processing as defined in [5]. In order to better analyze the performance of the proposed design as compared with previous research, the same metrics as used in [5, 6] are used in the current document. They are explained as follows:1) Mean Squared Error (MSE) - This metric is chosen to show how different the output of proposed method (or any method being compared) is, as compared to the theoretical output of Robert-Cross algorithm; and 2) Peak Signal-to-Noise Ratio (PSNR) - This metric depicts the amount of noise present in the output as compared to the signal, due to MSE.

These metrics are calculated as follows: $MSE = \sum_{i=0}^{r} \sum_{j=0}^{c} |O(i,j) - O_t(i,j)|^2$ and $PSNR = 10 \times log_{10} \left(\frac{255^2}{MSE} \right)$, where r is the number of rows in the input image, c is the number of columns in the input image, O is the output being compared, and O_t is the theoretical Robert-Cross edge detection output. The number 255 in the PSNR equation refers to the peak possible signal value for an 8-bit grayscale image.

Figure 2 is an open source test image used in benchmarking image processing applications. It can be observed in Figure

2 that output in Fig. 2c contains visibly lesser noise in low intensity areas such as the coat and the sky, when compared to both Figs. 2d and 2e. This is also evident in the MSE and PSNR values: 147.8/26.4 for Fig. 2c; 183.8/25.4 for Fig. 2d; and 267.3/23.8 for Fig. 2e. The proposed method is able to produce acceptable approximate result via efficient reduced precision stochastic computing.

(a) Input image

(b) Conventional SC output

(c) Output using the proposed method

(d) ASC output with 16 cycles

(e) ASC output uing 17 cycles

Fig. 2: Comparing edge detection outputs for an open source image: Camera man

REFERENCES

[1] J. Von Neumann, "Probabilistic logics and the synthesis of reliable organisms from unreliable components," *Automata studies*, vol. 34, pp. 43–98, 1956.

[2] P. Mars and W. J. Poppelbaum, *Stochastic and deterministic averaging processors*. Peter Peregrinus Press, 1981, no. 1.

[3] B. R. Gaines, "Stochastic computing," in *Proceedings of the April 18-20, 1967, Spring Joint Computer Conference*, ser. AFIPS '67 (Spring). New York, NY, USA: ACM, 1967, pp. 149–156. [Online]. Available: http://doi.acm.org/10.1145/1465482.1465505

[4] A. Alaghi and J. P. Hayes, "Exploiting correlation in stochastic circuit design," in *Computer Design (ICCD), 2013 IEEE 31st International Conference on*. IEEE, 2013, pp. 39–46.

[5] R. Seva, P. Metku, K. K. Kim, Y.-B. Kim, and M. Choi, "Approximate stochastic computing (asc) for image processing applications," in *SoC Design Conference (ISOCC), 2016 International*. IEEE, 2016, pp. 31–32.

[6] R. Seva, P. Metku, K. K. Kim, Y. B. Kim, and M. Choi, "Approximate stochastic computing (asc) for image processing applications," in *2016 International SoC Design Conference (ISOCC)*, Oct 2016, pp. 31–32.

The Analysis of CNN Structure for Image Denoising

Jae Hyeon Park, Jeong Hyeon Kim, and Sung In Cho
Electronic Engineering
Daegu University
Daegu, Republic of Korea
qoreho0011@daegu.ac.kr, kjh1119@daegu.ac.kr, and csi2267@daegu.ac.kr

Abstract—**This paper proposes an optimal structure of a convolutional neural network (CNN) for image denoising by analyzing the conventional CNN denoisers. There are three main factors that can determine the denoising performance of the CNN denoiser: the number of feature dimensions of each convolution layer, the number of convolution layers, and the usage of dilated convolution. We analyze the denoising performance variations of the conventional CNN denoiser depending on the above three factors and propose the optimal structure of the CNN denoiser. Experimental results showed that the above three factors have a high correlation with the denoising performance. Based on the experimental results, we could provide the optimal structure of the CNN denoiser.**

Keywords; convolution neural network; image denoising; feature dimension; dilated convolution

I. INTRODUCTION

Image denoising is a process to restore an image by removing image noise from a given noisy image. The conventional denoising methods can be categorized into model-based [1-3] optimization methods and discriminative learning methods. The most popular methods of model-based optimization are non-local mean filter (NLM) [1], block-matching and 3D filtering (BM3D) [2]. The representative discriminative learning methods [4-6] are denoising convolutional neural network (DnCNN) [4], and image restoration CNN (IRCNN) [5]. These CNN denoisers train weights for convolution and bias values by using training and its ground-truth data pairs. Then, the trained weights and bias values are used for image noise elimination. These CNN-based methods significantly improved the denoising performance compared with the existing denoising methods. There are various factors that determine the denoising performance of the CNN denoiser. In this paper, we analyze the denoising performance variations of the CNN denoiser by using three main factors that can determine the denoising performance, and propose the optimal structure of the CNN denoiser.

II. CONVENTIONAL CNN DENOISERS

Fig. 1 shows the overall architecture of the conventional CNN denoiser [4-5]. As shown in this figure, convolution is applied iteratively on the RGB input image to extract the desired output. At this time, the depth of the spatial kernel is determined by the feature dimension of each convolution layer to be generated. DnCNN [4] and IRCNN [5] are the most popular CNN denoisers and have the same architecture with Fig. 1.

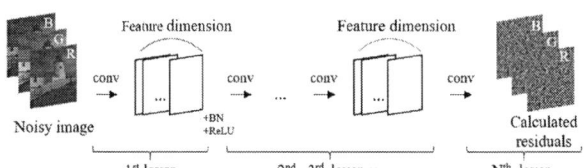

Figure 1. The conventional CNN denoiser, (conv: convolution).

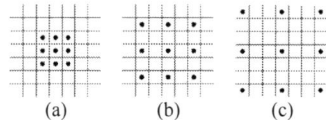

Figure 2. Weight configurations for (a) Normal 3 × 3 convolution, (b) Two-dilated convolution, (c) Three-dilated convolution.

A. DnCNN

DnCNN [4] consisting of 21 convolution layers, and batch normalization (BN) [7] is used between two convolutions to enhance the training speed and denoising performance. As an activation function, a Rectifier Linear Unit (ReLU) [8] is used. The CNN denoiser is trained by using a residual learning scheme [9] that trains weights and bias values to extract the difference between original and noisy images, which means image noise, in order to enhance the denoising performance compared to the general learning scheme that trains weights and bias values to extract the denoised image. For each layer, a convolution using 3 by 3 spatial kernel is applied, and 64 feature dimensions are used.

B. IRCNN

Compared with DnCNN, IRCNN [5] utilizes the dilated convolution [10] as shown in Fig. 2 to increase the receptive field of each convolution. By utilizing the dilated convolution, IRCNN effectively reduces the number of convolution layer (from 21 to 7) while maintaining or slightly increasing the denoising performance.

The factors that determine the denoising performance of the CNN denoiser can be classified into the following three categories: the number of convolution layer, the feature dimension, and dilated convolution. In the next section, we will analyze the denoising performance variation caused by the above factors.

III. EXPERIMENTAL RESULTS

We used a total of 3500 datasets for training images. It consists of 500 from the ImageNet database [11], and 3000 from

---

This work was supported by the National Research Foundation of Korea (NRF) grant funded by the Korea government (MSIP; Ministry of Science, ICT & Future Planning) (No. 2017R1C1B5075091).

978-1-5386-7961-6/18 $31.00 © 2018 IEEE

the Waterloo Exploration Database [12]. Additive white Gaussian noise (AWGN) was used as the image noise model, and Kodak testset was used. The peak signal to noise ratio (PSNR) values and the structural similarity index (SSIM) [13] were used for the objective evaluation of image quality.

A. The number of feature dimension

We analyzed the denoising performance depending on 32, 48, 64, 96, 128, and 160 feature dimensions with seven convolution layers as in IRCNN. The gradient of the PSNR value was decreased around 64 dimensions as shown in Fig. 3 (c). Therefore, it is recommended to set the feature dimension to 64 ~ 96 considering the computational complexity and the denoising quality.

B. The number of convolution layer

To analyze the denoising performance depending on the number of convolution layers, 64 feature dimensions were used as in DnCNN and IRCNN. As shown in Table II, as the number of convolution layers increased the PSNR values of the CNN denoiser increased. However, if the number of convolution layers exceeded seven, the PSNR value was rather reduced. This was caused by a side effect of the dilated convolution. Specifically, as the number of layers increases, the rate of the dilated convolution increases too much, and the relationship between adjacent pixels cannot be properly utilized for image denoising. When the dilated convolution was replaced with the normal convolution, PSNR value increased as the number of convolution layer increased as shown in Table II. Because the gradient of PSNR value was decreased around nine convolution layers, it is recommended to set the number of convolution layer to 7 ~ 9.

C. Dilated convolution

As shown in Fig. 3 (d), the CNN denoiser with the dilated convolution showed 0.82dB higher PSNR value than the CNN denoiser without the dilated convolution without increasing computational complexity. Therefore, it is highly recommended to use the dilated convolution for a CNN denoiser.

IV. CONCLUSION

In this paper, we analyzed the three factors affecting the denoising performance of the CNN denoiser and proposed the optimal CNN denoiser structure as follows. First, the use of 64-96 feature dimensions provides the most cost-effective denoising performance. Second, it is recommended to use around seven convolution layers when the dilated convolution is used. Third, the dilated convolution can effectively enhance denoising performance without increasing the computational complexity.

TABLE I.　DENOISER PERFORMANCE VARIATION BY DEPTH OF FEATURE DIMENSION

The number of feature dimension	Noise level (σ_n=25)					
	32	48	64	96	128	160
AVG. PSNR [dB]	31.775	31.971	32.108	32.205	32.285	32.302
AVG. SSIM	0.895	0.898	0.900	0.901	0.903	0.903

TABLE II.　DENOISER PERFORMANCE VARIATION BY THE NUMBER OF CONVOLUTION LAYER FOR KODAK IMAGE SET

The number of convolution layer		Noise level (σ_n=25)				
		3	5	7	9	11
With D-conv[a]	AVG. PSNR [dB]	31.141	31.887	32.108	32.054	32.879
	AVG. SSIM	0.881	0.896	0.900	0.899	0.898
Without D-conv	AVG. PSNR [dB]	30.839	31.619	31.924	32.084	32.180
	AVG. SSIM	0.875	0.891	0.896	0.899	0.901

a. Dilated convolution.

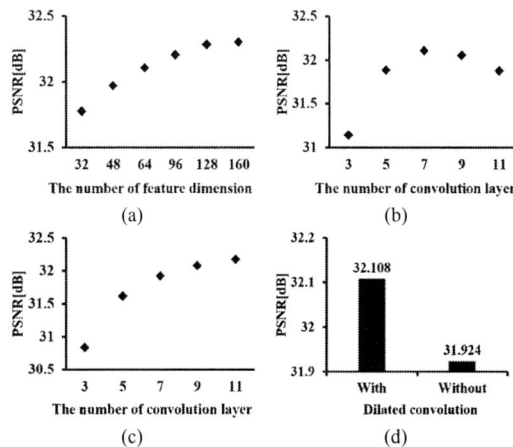

Figure 3.　Variations of PSNR values (a) PSNR vs. the number of feature dimension, (b) PSNR vs. the number of convolution layer (with dilated convolution), (c) PSNR vs. the number convolution layer (without dilated convolution), and (d) PSNR vs. with or without the dilated convolution.

REFERENCES

[1] A. Buades, B. Coll, and J. M. Morel, "A non-local algorithm for image denoising," in *Proc. IEEE Int. Conf. Comput. Vis.,* Jun. 2005, pp. 60–65.

[2] K. Dabov, A. Foi, V. Katkovnik, and K. Egiazarian, "Image denoising by sparse 3-D transform-domain collaborative filtering," *IEEE Trans. Image Process.*, vol. 16, no. 8, pp. 2080–2095, Aug. 2007.

[3] S. I. Cho and S.-J. Kang, "Geodesic path-based diffusion acceleration for image denoising," *IEEE Trans. on Multimedia.*, vol. 20, No. 7, pp. 1738-1750, Jul. 2018.

[4] K. Zhang, W. Zuo, Y. Chen, D. Meng, and L. Zhang, "Beyond a Gaussian denoiser: Residual learning of deep CNN for image denoising", *IEEE Trans. Image Process.*, vol. 26, no. 7, pp. 3142–3155, Jul. 2017.

[5] K. Zhang, W. Zuo, S. Gu, and L. Zhang, "Learning deep CNN denoiser prior for image restoration," in *Proc. IEEE Int. Conf. Comput. Vis.*, pp. 2808–2817, Jun. 2017.

[6] S. I. Cho and S.-J. Kang, "Gradient Prior-aided CNN Denoiser with Separable Convolution-based Optimization of Feature Dimension," *IEEE Trans. on Multimedia.*, 2018. [Online]. Available: https://ieeexplore.ieee.org/stamp/stamp.jsp?tp=&arnumber=8419273.

[7] S. Ioffe and C. Szegedy, "Batch normalization: Accelerating deep network training by reducing internal covariate shift," in *International Conference on Machine Learning.*, 2015, pp. 448–456.

[8] A. Krizhevsky, I. Sutskever, and G. E. Hinton, "ImageNet classification with deep convolutional neural networks," in *Proc. Adv. Neural Inf. Process. Syst.*, 2012, pp. 1097–1105.

[9] K. He, X. Zhang, S. Ren, and J. Sun, "Deep residual learning for image recognition," in *Proc. IEEE Conf. Comput. Vis. Pattern Recog.*, Jun. 2016, pp. 770–778.

[10] F. Yu, V. Koltun, "Multi-scale context aggregation by dilated convolutions," *in Proc. Int. Conf. Learn. Represent.*, 2016.

[11] J. Deng, W. Dong, R. Socher, L.-J. Li, K. Li, and L. Fei-Fei, "Imagenet: A large-scale hierarchical image database," In *Proc. IEEE Conf. Comput. Vis. Pattern Recog.*, 2009, pp. 248–255.

[12] K. Ma, Z. Duanmu, Q. Wu, Z. Wang, H. Yong, H. Li, and L. Zhang, "Waterloo exploration database: New challenges for image quality assessment models," *IEEE Trans. Image Process.*, vol. 26, no. 2, pp. 1004–1016, Feb. 2017.

[13] Z. Wang, A. C. Bovik, H. R. Sheikh, and E. P. Simoncelli, "Image quality assessment: from error visibility to structural similarity," *IEEE Trans. Image Process.*, vol. 13, no. 4, pp. 600–612, Apr. 2004.

Gap in pagination due to formatting issues.

Pages 222-223

Optimized Image Crop-based Video Retargeting

Jeong Hyeon Kim, Jae Hyeon Park, and Sung In Cho
Electronic Engineering
Daegu University
Daegu, Republic of Korea
kjh1119@daegu.ac.kr, qoreho0011@daegu.ac.kr, and csi2267@daegu.ac.kr

Abstract— **This paper proposes a new video retargeting method that utilizes the cropping and bilinear scaling techniques. Specifically, the cropping positions for a given image are determined by using results of the logo and saliency detection, and the strength of the bilinear scaling is determined depending on the size of cropping. As a result, the proposed method effectively reduces distortion of visually important contents and maintains a temporal consistency. Experimental results showed that the proposed method provided the highest preference compared with the existing methods in the paired-comparison-based user study evaluation.**

Keywords: video retargeting, seam carving, logo detection, saliency map

I. INTRODUCTION

Due to the developments of wireless communication and mobile devices, displays having various aspect ratios have been developed. Therefore, the aspect ratios of legacy video contents have to be resized to the aspect ratio of the target display device.

There are two approaches to the video retargeting: a simple video retargeting method that does not take into account the characteristics of a given image and a content-aware video retargeting method. The bilinear scaling-based method and letter box-based method are the most popular simple video retargeting methods. The bilinear scaling-based method can occur image distortion because it scales a selected axis regardless of content characteristics. The insertion of the letter box has the disadvantage that it cannot utilize all the area of the display. Seam carving [1] and Greisen's method [2] are the most famous content-aware video retargeting methods. The seam carving iteratively removes or duplicates a seam to adjust the aspect ratio of a given image. The seam is a continuous path with composed unimportant pixels. Greisen's method [2] adjusts the grid size of each pixel on the x- or y-axis based on the saliency of a given image. The content-aware video retargeting methods can be suitable for the image retargeting. However, they are very likely to create the jittery artifacts due to the lack of a temporal coherency when they are used for the video retargeting. Therefore, for the effective video retargeting, the distortion of the visually important area of a given image should be minimized while maintaining the temporal coherency of a given video.

In this paper, we propose a video retargeting method that minimizes image distortion of the visually important area and maintains the temporal coherency. The proposed method

Figure 1. The procedures of the proposed method (a) original image (aspect ratio = 16: 9), (b) the static area of the current frame, (c) logo detection result, (d) saliency map, (e) cropped image (aspect ratio = 19.1: 9), (f) bilinear scaling result (aspect ratio = 21: 9).

transforms the existing 16: 9 aspect ratio video to the 21: 9 aspect ratio video.

II. PROPOSED METHOD

As shown in Fig. 1, the proposed method increases the horizontal axis ratio by cropping the top and bottom of a given image (Fig. 1 (e)). Then, we transform the aspect ratio of the cropped image to the target aspect ratio (21:9) by using the bilinear scaling method (Fig. 1 (f)). The cropping positions of the top and bottom of a given image are determined by using the results of the logo and saliency detection (Figs. 1 (c) and (d)).

A. Logo detection

The logo in an image is the most important contents that should be preserved. Hence, to preserve this, we detect the logo location by using the local binary pattern (LBP) [4], frame difference, and edge detection. Firstly, the difference of *LBPs* of current and previous frames are defined as follows:

$$LBP_S = \begin{cases} 0, & LBP_C - LBP_P \geq TH_{LBP} \\ 1, & otherwise \end{cases}, \quad (1)$$

where $LBP_S=1$ denotes the static area. LBP_C and LBP_P are the LBP results for the current frame and the previous frame. The TH_{LBP} that is a threshold value for LBP difference was set to three empirically. Secondly, the frame difference is defined as follows:

$$F_S = \begin{cases} 0, & F_C - F_P \geq TH_{F_S} \\ 1, & otherwise \end{cases}, \quad (2)$$

This work was supported by the National Research Foundation of Korea (NRF) grant funded by the Korea government (MSIP; Ministry of Science, ICT & Future Planning) (No. 2017R1C1B5075091).

978-1-5386-7961-6/18 $31.00 © 2018 IEEE

Figure 2. Video retargeting results (a) original image, (b) by bilinear scaling method, (c) by seam caving method, (d) by Greisen's method, (e) by the proposed method.

where 1 denotes the static area. F_C and F_P denote the current frame and the previous frame, respectively. The threshold value in (2) was set to five empirically.

Thirdly, we detect the edge of the current frame by using the sobel edge detector [5]. The edge detection result is binarized by using a threshold value 10.

Lastly, the final static area in the current frame as follows:

$$FC_s(i) = LBP_s(i) \times F_s(i) \times E(i), \qquad (3)$$

$$S_{ACC} = \sum_{i=1}^{N} FC_s(i), \qquad (4)$$

where i denotes the frame index. E denotes the edge detection result. FC_S is the final detected static area in the current frame as shown in Fig. 1 (b). N denotes the total number of frames for each test set. S_{ACC} is the cumulative sum of the FC_S. S_{ACC} having the larger value than $0.93N$ is determined to the logo area as shown in Fig. 1 (c). The upper cropping position of a given image is determined above the line of the detected logo position.

B. Saliency map

We used the convolution of the image amplitude spectrum and the low-pass Gaussian kernel to find saliency region [6]. The extracted salience map is binarized using the threshold value that is the average value of the extracted saliency map of a given image. The binarized salience map is accumulated. If the pixel has the accumulation value larger than $0.3N$, it is classified to the final saliency pixel as shown in Fig. 1 (d). The lower cropping position is determined below the lowest line of the detected saliency pixel.

C. Final bilinear scaling

Finally, the cropped image is transformed to the target aspect ratio by using the bilinear scaling method as shown in Fig. 1 (f).

III. EXPERIMENTAL RESULTS

To evaluate our method, we compared the results of the proposed method with those of three benchmark methods which are the bilinear scaling, the seam carving [1], and Greisen's method [2] by performing the paired-comparison-based user study [7]. The paired-comparison-based user study evaluates a user's preference with respect to two results. A total of 12 participants attended in the experiment and six test sets were used. The test sets have an aspect ratio of 16: 9 and 60 frames per second.

As shown in Fig. 2 (b), the bilinear scaling method produced severe image distortion where the important content becomes fat because it scales a selected axis regardless of content

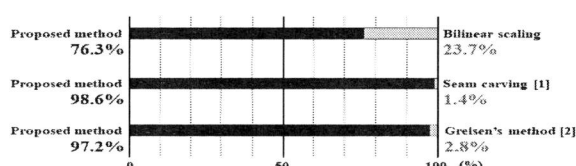

Figure 3. The paired-comparison-based user study results (users have a preference on the visual quality of our solution over the other three benchmarking methods the other three benchmarking methods)

characteristics. The seam carving and the Greisen's methods generated jittery artifacts because of the lack of temporal coherency. However, the proposed method reduced the artifact produced by the bilinear scaling method through cropping. In addition, the temporal coherency can be maintained perfectly. As shown in Fig.3, the proposed method was more preferred in 76.3% of the comparison with the bilinear scaling, in 98.6% of the comparison with the seam carving, and in 97.2% of the comparison with the Greisen's method.

IV. CONCLUSION

In this paper, we proposed a video retargeting method that crops the top and bottom of a given image and then transforms the cropped image to the target aspect ratio by using bilinear scaling. The cropping positions were determined by using results of the logo and saliency detection. Although the proposed method had lost some information in the upper and lower areas compared with the conventional methods, the distortion of the important contents was considerably reduced and the temporal coherency was maintained perfectly. In the paired-comparison-based user study, the proposed method was more preferred than the benchmark methods.

REFERENCES

[1] M. Rubinst ein, A. Shamir, S. Avidan, "Improved Seam Carving for Video Retargeting", *ACM Trans. Graphics*, vol. 27, no. 3, pp. 1-9, Aug. 2008.

[2] P. Greisen, M. Lang, S. Heinzle, A. Smolic, "Algorithm and VLSI architecture for real-time 1080p60 video retargeting", *Proc. High Perform. Graph. Conf.* Jun. 2012, pp. 57-66.

[3] G. J. Bae, S. J. Kang, Y. H. Kim, "Census Transform-based Static Caption Detection for Frame Rate Up-Conversion", *IEEE International Symposium on Circuits and Systems (ISCAS)*, May. 2017, pp. 1-4.

[4] T. Ahonen, A. Hadid, M. Pietikainen, "Face Description with Local Binary Patterns: Application to Face Recognition", *IEEE Trans. Pattern Analysis and Machine Intelligence*, vol. 28, no. 12, pp. 2037-2041, Dec. 2006.

[5] W. Gao, L. Yang, X. Zhang and H Liu,"Animproved Sobel edge detection," *Computer Science and Information Technology (ICCSIT)*, vol. 5, Jul. 2010, pp. 67-71.

[6] J. Li, M. Levine, X. An, X. Xu, and H. He, "Visual saliency based on scale-space analysis in the frequency domain," *IEEE Trans. Pattern Analysis and Machine Intelligence*, vol. 35, no. 4, pp. 996–1010, Apr. 2013.

[7] L. Zhang, M. Song, Q. Zhao, X. Liu, J. Bu, and C. Chen, "Probabilistic graphlet transfer for photo cropping," *IEEE Trans. Image Process.*, vol. 22, no. 2, pp. 802–815, Feb. 2013.

High-throughput HW-SW implementation for MV-HEVC decoder

WEI LIU
Dept. of Electronics
Engineering
Konkuk University
Seoul Korea
liuwei1108@konkuk.ac.kr

WEI LI
Dept. of Electronics
Engineering
Konkuk University
Seoul Korea
liwei@konkuk.ac.kr

Park Sang Un
Dept. of Electronics
Engineering
Konkuk University
Seoul Korea
kos8108@konkuk.ac.kr

Yong Beom Cho*
Dept. of Electronics
Engineering
Konkuk University
Seoul Korea
ybcho@konkuk.ac.kr

Abstract— The multi-view HEVC extension was completed in July 2014 by the Motion Picture Experts Group and the Video Coding Experts Group. Multi-view video based on stereo representation has become more and more popular. In addition, a variety of multimedia content can now be provided for mobile devices. Therefore, there is a need for a real-time multi-view video decoder. Obviously, because of the complexity of software-based methods, high-resolution MV-HEVC video cannot be decompressed in real time. Especially in mobile devices, there is no hardware that can support MV-HEVC decoding today. Hence, in this paper, we present a new parallel architecture for MV-HEVC and use FPGA to accelerate complex operations. The experimental results show that the proposed multi-view video coding can decompress 3 view video of 1920 x 1080 resolution in real time on the ZYNQ platform.

Keywords; HW-SW; MV-HEVC; high-throughput; real-time; parallel processing

I. INTRODUCTION

MV-HEVC which is based on the HEVC standard to further improve coding efficiency was established in July 2012 by the Joint Group of 3D Video Coding Extensions Development (JVT-3V) [1][2][3][4]. In the parallel method of MV-HEVC decoding, it can be divided into two levels, the task level and the data level. The task level can be divided into intra-frame parallel and inter-frame parallel solutions. In order to better realize the parallelism of intra-frame coding, HEVC introduces two new parallel approach, namely Tile [5], and Wavefront Parallel Processing (WPP) [6]. In order to solve the drawbacks of WPP parallelism, Chi et al [7] proposed a method called OWP. The core idea of the OWP approach is that when there has free thread, the free thread will start to decode the next frame. K. Chen et al [8] proposed an IFW approach. IFW implements a parallel method of WPP between multiple frames. IFW satisfies the efficient use of multi-thread while coding multiple frames in parallel with sufficient threads. With data level L. WEI LIU proposed a SIMD for MV-HEVC. Although All above method parallelism of HEVC can be well solved real-time decoding HEVC or low resolution MV-HEVC, the throughput cannot solve the real-time decompression of 1080p MV-HEVC. we present a new parallel architecture for MV-HEVC based on WPP and use FPGA to

accelerate inter prediction.

II. PREPARE PARALLEL INTER-VIEW DECODING METHOD

Whether it is OWP, or IFW or other parallel methods, such as the Frame-Based Parallel Framework proposed by Y.Z Duan et al [13], in order to improve parallelism, inter-frame parallelism is necessary. However, parallelism inter-view must face how to resolve data dependencies between frames. We have made novel optimizations for the parallel method between frames. The data between frames in the HEVC encoding and decoding process is derived from the motion vector d(MV) search and motion compensation (MC). That is, the data reconstruction of the decoded frame must depend on the reconstruction of the dependent frame and the filtered data. Therefore, the author of the IFW method proposes that a decoded frame is coded in parallel with respect to the manner in which several CTB rows are delayed by the previous frame. The number of delayed CTB rows is set according to the size of the compressed search range, which not only increases the thread synchronization time, but also increases the corresponding frame waiting time (FWT) which affecting the parallelism of codec. Not only this, relative to HEVC, MV-HEVC introduces motion prediction between views [4] in order to improve compression efficiency, which increases the difficulty of inter-view parallel implementation and inter-view waiting time. Hence, we propose a parallel approach by appropriately changing the order of decompression. Because MV-HEVC has multiple frames decoding at the same time and there is no date dependency between views on different time. Based on above feature, we implement an inter-frame parallel method that does not require consideration of data dependencies between frames. First introduce the two most common MV-HEVC GOP structure types, namely PIP and IPP structure. If the multi-frame decompression is performed in the original order, the data dependency of inter-view will produce the corresponding FWT. We set the FWT to F_{wt} and the frequency at which the FW appears is related to parallel the number of frames which set as F_n. If set the number of frames in AU to be v_n. When $F_n > v_n$, data dependencies between frames of the same Layer level are introduced. As shown Fig.1, when $F_n <= v_n$, there is only data dependency between views. However, the number of parallel views between frames will directly affect the degree of parallelism. Therefore, we set the number of parallel frames as $F_n = v_n$, which can reduce the complexity of inter-view date dependencies, which can guarantee multiple frames in parallel.

If the decoded video is 1080P, then RS = 17, the size of the DR is determined by the search range. If the search range is set to 64, the size of the DR can be calculated to be equal to 3, which is also available in paper [12]. The size of $p_{(wt)}$ is related to the video itself and the test environment. However,

after many tests we found that it was basically greater than 0.12. In this paper, we take $p_{(wt)} = 0.1$ for the convenience of calculation. As long as the PIP structure N > 170 and the IPP structure N > 85, the advantages of the our proposed in this paper are reflected. The size of N is much larger than 170 when most of the video is decompressed.

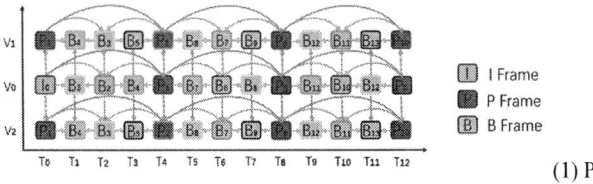

(1) PIP

(2) IPP

Figure 1. Proposed MV-HEVC inter-prediciton struct

III. HARDWARE ACCELERATION FOR INTER PREDICTION

In the hardware design we implemented the HW-SW design using Xilinx's SDSoC. We used HTM16.0 to implement the design presented in Chapter 2, and in the profile, we found a function that uses a lot of resources in MV-HEVC.From the profile results, it can be seen that **DecompressSlice** uses a lot of computing resources. We transform the multiplication and addition parts into HDL. We use SDSoC to automatically perform high-level hardware synthesis on the above-part C++ code to the RTL automatically generate a high-throughput AXI standard bus between ARM and FPGA. Fig. 2 shows the resources used in the Xilinx ZCU102 used by the generated RTL. In addition, the data size and type of bus used between ARM and FPGA is using AXIDMA_SG.

Utilization Estimates					
⊟ **Summary**					
Name	BRAM_18K	DSP48E	FF	LUT	URAM
DSP	-	759	-	-	-
Expression	-	-	0	25372	-
FIFO	-	-	-	-	-
Instance	-	31	0	1232	-
Memory	1	-	0	0	-
Multiplexer	-	-	-	5050	-
Register	-	-	42098	-	-
Total	1	790	42098	31654	0
Available	1824	2520	548160	274080	0
Utilization (%)	~0	31	7	11	100

Figure 2. Utilizaiton Estimates for MV-HEVC acceleration.

IV. EXPERIMENTAL RESULT

As mentioned in the previous section, we proposed a new parallel solution based on inter-view. We used the HTM 16.0 which is an open reference source for HEVC and compiling by SDSoC. In additional, we are using Xilinx ZCU 102 and set the number of layers as 3, CTU size is 64 x 64 which depth is 4, GOP size is 8, using TZ searching and the searching range is 64 and, in each frame have 1 slice. In additional, we enable RDOQ, DF, SAO and AMP. As shown in Table 1, we tested 8 kinds of video sequences. 3 of them have a resolution of 1024 x 768, and the other two have a resolution of 1920 x 1088. From Table 2, the FPS results show that the throughput of our proposed method is about 11 times higher than that of 3D-

HTM16, and the MV-HEVC of 3 views can be decompressed in real time at 1920x1088.

Table 1 Test video sequence

Sequence	Resolution	Frame Rate (FPS)	Total Frames
Balloons	1024 x 768	30	300
Shark	1920 x 1088	30	300
GhostTownFly	1920 x 1088	25	250
UndoDancer	1920 x 1088	25	250

Table 2 Frame per second in IPP and PIP struct

Sequence	QP	Bit-rate (kbps)	3D-HTM (s)PIP	HW+SW (s)PIP	3D-HTM (s)IPP	HW+SW (s)IPP
Balloons 1024 x 768	25	1722.4	68.814	10.008	66.078	6.426
	30	925.5	58.842	9.09	57.42	5.418
	35	482.8	51.93	4.932	51.156	5.076
	40	278.3	46.854	4.59	45.36	4.482
Shark 1920 x 1088	25	5589.8	134.55	9.306	130.572	9.306
	30	2753.4	120.276	8.514	116.748	8.388
	35	1192.3	108.486	8.19	106.434	7.74
	40	657.2	102.852	7.362	102.312	7.434
GhostTownFly 1920 x 1088	25	5081.4	111.888	7.794	109.17	8.19
	30	2156.7	101.088	7.398	99.072	7.398
	35	1015.6	93.834	7.02	92.934	6.912
	40	584.3	87.822	6.534	84.312	6.3
UndoDancer 1920 x 1088	25	6729.3	140.022	9.63	135.414	9.576
	30	2849	118.548	8.298	115.434	8.19
	35	1327.6	110.88	7.686	107.748	7.65
	40	683.5	105.282	7.38	102.456	7.29

CONCLUSION

This paper presents a high parallel and high throughput MV-HEVC implementation on the ZYNQ ZCU 102 processor. Based on the results, the proposed optimized result is about 11 times faster than the reference software HTM16. The proposed MV-HEVC optimization method can decompress 388p video of 3 views in real time in a low power processor. With the next generation of video coding standards, the promotion of MV-HEVC and the increasing number of mobile terminals, it is possible to watch full-HD multi-view video in real time.

ACKNOWLEDGMENT

This work was supported by the institute for information & communications Technology Promotion (IITP) grant funded by the Korea government (No. 2017-0-01736). We thank professor Cho. who provided insight and expertise that greatly assisted the research.

REFERENCES

[1] Y. Chen, Y.-K. Wang, K. Ugur, M. M. Hannuksela, J. Lainema, and M. Gabbouj, "The emerging MVC standard for 3D video services," EURASIP J. Appl. Signal Process., vol. 2009, pp. 8:1–8:13, Jan. 2008.

[2] G. Tech et al., "Overview of the multiview and 3D extensions of high efficiency video coding," IEEE Trans. Circuits Syst. Video Technol., vol. 26, no. 1, pp. 35–49, Jan. 2016

[3] L. WEI, L. JIAO, Cho Yong B., "A novel architecture for parallel multi-view HEVC decoder on mobile device," EURASIP Journal on Image and Video Processing, March. 2017.

[4] G. Tech et al., "Overview of the multiview and 3D extensions of high efficiency video coding," IEEE Trans. Circuits Syst. Video Technol., vol. 26, no. 1, pp. 35–49, Jan. 2016

[5] . Fuldseth, M. Horowitz, S. Xu, and M. Zhou, "Tiles," Joint Collaborative Team on Video Coding (JCT-VC), Document JCTVC-E408, Geneve, Switzerland, March 2011.

[6] F. Henry and S. Pateux, "Wavefront Parallel Processing," Joint Collaborative Team on Video Coding (JCT-VC), Document JCTVC-E196, Geneve, Switzerland, March 2011.

[7] C. C. Chi et al., "Parallel scalability and efficiency of HEVC parallelization approaches," IEEE Trans. Circuits Syst. Video Technol., vol. 22, no. 12, pp. 1827–1838, Dec. 2012.

[8] K. Chen, J. Sun, Y. Duan and Z. Guo, "A Novel Wavefront-Based High Parallel Solution for HEVC Encoding," in IEEE Transactions on Circuits and Systems for Video Technology, vol. 26, no. 1, pp. 181-194, Jan. 2016.

Accurate Stochastic Computing Using a Wire Exchanging Unipolar Multiplier

Hounghun Joe, Manhee Cho and Youngmin Kim

School of Computer and Information Engineering,
Kwangwoon University, Nowon-gu, Kwangwoon-ro 20, Seoul 01897, Republic of Korea
youngmin@kw.ac.kr

Abstract—**Stochastic computing is one of the approximate computing that provides low power, simple structure, and error tolerance. Stochastic circuits consist of random number generators, comparators to generate bitstream, basic logic gates for arithmetic operations, and counters. In this study, we propose a simple and powerful stochastic multiplier by exchanging wires in operation. Simulation shows that the proposed design reduces relative error, standard derivation compared with conventional designs, and has compact size compared to existing multipliers.**

Keywords—*Approximate Computing; Stochastic Computing; Stochastic Computing Correlation; Unipolar Multiplier*

I. INTRODUCTION

Recent IT technology such as image processing classifiers or deep neural networks is becoming more complex and requiring more energy. However, they do not require accurate results. From this aspect, a stochastic computing (SC), one of the approximate computing, which has an advantage in energy efficiency and high error tolerance attracts a lot of attention [1]-[3]. In SC, a simple logic can be used as a computation module to save energy. SC uses a sequence of bits, which is generally called a bitstream, and requires a proper operation circuit with a correct correlation depending on its own design. Even though the SC advantage includes small size, low power, and high error tolerance, accuracy and size in integrated circuit are major concerns in practical implementation. In this study, we propose a simple and powerful technique for higher accuracy in unipolar multiplier by exchanging wire method.

II. BACKGROUND

A. Stochastic Computing and Correlation

Stochastic computing (SC) is one of the approximate computing methodology and uses a bistream to represent a value. An integer X can be expressed by L bitstream, $Sx = \{Sx_1, Sx_2, \cdots, Sx_L\}$. For example, a stochastic number S has a probability of 1 at 25% and a probability of 0 at 75%, probability of observing 1 is $P = 0.25$.

In SC, unipolar or bipolar encoding can be used. In the unipolar encoded value, '1' has a weight of +1 and '0' has a weight of 0. Thus, the unipolar encoded value is limited to the range [0, 1]. For instance, the bitstream L = 01001100 has $P_L = 3/8$. By contrast, in the bipolar encoded value, '1' has a weight of +1 and '0' has weight of −1. Therefore, the bipolar encoded value is limited to the range [−1, 1]. For example, the bitstream L = 01001100 has $P_L = -1/4$ in bipolar.

When computing with bitstreams, basic gates can be used for arithmetic operations. The AND gate performs a multiplication role in the unipolar SC. Given input A, B, and output C, which have a probability of P_A, P_B, and P_C, the multiplication is defined

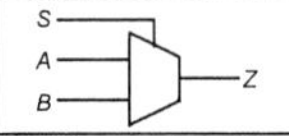

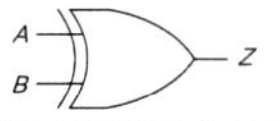

Fig. 1. Circuits and correlations of each operation in SC: (a) scaled add, (b) subtract, and (c) multiply.

as $P_C = P_A \times P_B$. For instance, when A = 01101010, B = 10111011, P_A is 0.5, and P_B is 0.75. According to the formula, C, which is (A & B), becomes 00101010 and P_C is 0.375.

In SC, the bitstream is generated through the stochastic number generator (SNG) including random number generators (RNG) and comparators; one bit in bitstream is '1' if input binary number is bigger than the generated random number from the RNG, otherwise it is '0'. The linear feedback shift register (LFSR) is widely used as a conventional RNG in SC. The generated bitstreams are used in the bitwise operation, and the counter counts '1' to restore the decimal value. When restoring decimals, we take log values for the multiplier and multiplicand, through a leading one detector (LOD) and a priority encoder. Then, the counted value is shifted by the larger value of the two log values through the shifter and the comparator.

The logic gates used in SC are sensitive to the stochastic computing correlation (SCC), which has a value between -1 and +1. The SCC value 0 means that the specific bitstreams X and Y are uncorrelated. The SCC value +1 and -1 means maximal positive and negative correlation, respectively. The SCC is defined as:

$$SCC(X, Y) = \begin{cases} \dfrac{ad-bc}{N \times \min(a+b,a+c)-(a+b)(a+c)} & ad > bc \\[2ex] \dfrac{ad-bc}{(a+b)(a+c)-N \times \max(a-d,0)} & ad \le bc \end{cases}$$

In this formula, *a* is the number of positions (NP) that X and Y are both '1', *d* is the NP that X and Y are both '0' in contrast. *b* is the NP that X is '0' and Y is '1', *c* is the NP that X is '1' and Y is '0' [5]. The process of creating the bitstream has an impact on the SCC value and a proper computation module with a correct SCC value should be used in SC. The circuits commonly used in SC and their correlations are shown in Fig. 1.

III. PROPOSED WIRE EXCHANGING UNIPOLAR MULTIPLIER

The main benefit of SC is that complex computations can be replaced by simple logic operations (AND or XOR). But, if many SNGs are required for higher input bits, the SNG occupies a large area in the SC. Thus, the overall area of the SC mainly depends on the SNGs. In addition, as mentioned previously, to obtain the correct result, it is necessary to use a proper operational circuit with a correct SCC.

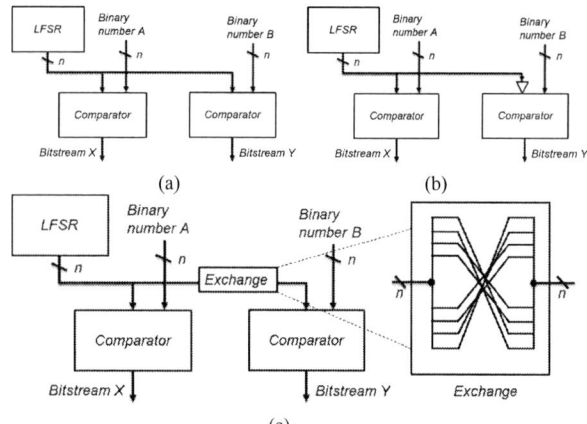

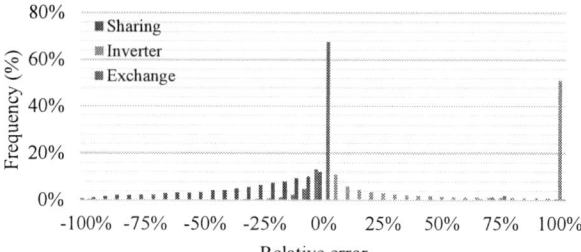

Fig. 2. Bitstream generation methods: (a) sharing, (b) sharing with an inverter, (c) the proposed sharing with a wire exchange

Fig. 3. Relative error distribution comparison.

Table I: Average stochastic computing correlation (SCC_AVG), average absolute relative error (ABS_RE_AVG), standard derivation (SD) and coefficient variability (CV) between the conventional and the proposed design.

	Sharing	Inverter	Proposed
SCC_AVG	1	-0.921	0.164
ABS_RE_AVG	0.342	0.648	0.104
SD	0.271	0.404	0.219
CV	0.794	0.623	0.475

Table II: Frequency, dynamic power, and logic utilization in FPGA of the accurate multipliers and the proposed SC based design.

	Accurate multiplier		Proposed stochastic multiplier
	Binary	Wallace	
Frequency (MHz)	102.51	150.33	417.54
Dynamic Power (mW)	1.1	2.15	0.56
Logic utilization	33/32070	70/32070	13/32070

We propose a wire exchanging technique, which is optimized for SCC in the unipolar multiplier. The proposed design provides more accurate results than the conventional design in smaller size. It uses a random number from one LFSR and pairs of even and odd wires are exchanged symmetrically. The bit exchanged in the proposed design is defined as follows:

$$\text{Exchange}(b_{n-1}b_{n-2}b_{n-3}b_{n-4} \dots b_3 b_2 b_1 b_0)$$
$$= b_1 b_0 b_3 b_2 \dots b_{n-3}b_{n-4}b_{n-1}b_{n-2}$$

b is each bit value of the random number (RN) from the LFSR. However, the SCC value is degraded, and the error will increase, when an original RN and the exchanged RN become identical. To compensate for this issue, we shift one more bit when restoring to decimal because of the non-zero n-2th bit of the input patterns. In conclusion, this design provides accurate computation in the unipolar multiplier, because the SCC is more uncorrelated than that of the conventional design.

IV. SIMULATION RESULT

To verify the proposed design, one million randomly generated 8-bits numbers are used. Three bitstream generation methods used in the simulations are shown in Fig. 2; a simple LFSR value sharing in Fig. 2(a), and inverting the shared value through an inverter in Fig. 2(b). They are compared with the proposed design shown in Fig. 2(c). SCC average is defined as:

$$\text{SCC average} = \frac{\sum_{i=0}^{n} SCC(X_i, Y_i)}{n}$$

A histogram of the relative error (RE) is drawn in Fig. 3 and the average values of the SCC and error statistics are summarized in Table I. The RE is defined as (ACC - APP) / ACC, where ACC is the accurate result and APP is the result by the stochastic computation. As shown in Table II and Fig. 3, SCC, RE, SD and CV are reduced significantly by the proposed design compared with the conventional design and the SCC is close to zero (i.e. uncorrelated), which is important for the unipolar multiplier. For example, up to 6x average SCC reduction and more than 4x reduction in average absolute RE can be achieved by the proposed design.

For hardware design metric comparison, the operating frequency and the dynamic power of the binary multiplier, Wallace tree multiplier, and the proposed design are analyzed by using a FPGA. The results in Cyclone V, 5CSEMA5F31C6 device, DE1-SoC are summarized in Table II. As shown, the proposed design is faster and consumes less power than other accurate multipliers. For example, only 1/4 of power in FPGA is required by the proposed design compared with the Wallace tree multiplier and the proposed stochastic multiplier is 4x faster than the accurate binary multiplier. In addition, 1/5 of logic utilization is needed compared to the Wallace tree multiplier.

V. CONCLUSION

In this paper, we propose a unipolar multiplier to improve accuracy compared with conventional stochastic structures by exploiting the exchanging wire method. Simulation results show that the proposed method provides a SC design with less power and smaller area over the other multipliers with higher accuracy than conventional methods.

ACKNOWLEDGMENT

This research was supported by the MIST (Ministry of Science and ICT), Korea, under the National Program for Excellence in SW supervised by the IITP (Institute for Information & communications Technology Promotion) (2017-0-00096).

REFERENCES

[1] Gaines, B. R., "Stochastic computing systems," *Advances in Information Systems Science*, vol. 2, pp. 37-172, 1969

[2] Alaghi, A., & Hayes, J. P., "Survey of stochastic computing," *ACM TECS*, 2012

[3] Han, J., & Orshansky, M., "Approximate computing: An emerging paradigm for energy-efficient design," *18th IEEE ETS*, pp.1-6, 2013

[4] Alaghi, A., "The Logic of Random Pulses: Stochastic Computing," PhD thesis, University of Michigan, 2015

[5] Lee, V. T., et al., "Correlation manipulating circuits for stochastic computing." *in Proc. DATE*, pp. 1417-1422, Mar 2018

[6] Ichihara, H., *et al.* "Compact and accurate stochastic circuits with shared random number sources," *in Proc. ICCD*, pp. 361–366, Oct 2014

True Random Number Generator Using Bio-related Signals in Wearable Devices

Hoyoung Yu and Youngmin Kim

School of Computer and Information Engineering,
Kwangwoon University, Nowon-gu, Kwangwoon-ro 20, Seoul 01897, Republic of Korea
youngmin@kw.ac.kr

Abstract— Random number generation (RNG) is the most important factor in a security system. Various RNGs are introduced for efficient and unpredictable random number generations, but a separate hardware is required to generate unpredictable random numbers, and efficient software-generated random numbers have low security. In this study, we propose an RNG which can generate unpredictable random numbers by utilizing bio-related signals embedded in wearable devices. The random numbers generated by the proposed structure are verified for excellent security through NIST test suite.

Keywords—TRNG, Bio signals, PPG, Wearable Devies, Security

I. INTRODUCTION

RNG is an important security block, which is widely used in most cryptographic applications such as AES, PBE, DES, etc. The RNG must operate independently and generate an unpredictable sequence of random numbers. RNG is largely divided into a pseudo random number generator (PRNG) and a true random number generator (TRNG).

PRNG uses mathematical algorithms to generate random numbers that seem to have randomness. Generally, a software algorithm is used when generating random numbers, and the sequence of random numbers is determined by the initial seed value. Thus, it is possible to predict the random number pattern by obtaining the seed value or collecting a lot of data. Therefore, even though PRNG has the advantage of easy to implementation, it is not suitable for use in finance or defense systems, which require high security because of the vulnerability of the PRNG random patterns. A typical type of PRNG is a linear feedback shift register (LFSR) [1]. The LFSR has a structure in which a value input to a register is calculated as a linear function of previous state values. The LFSR can generate a maximum length sequence of $2^n - 1$ if the linear function is well chosen, and it repeats the specific period continuously according to the PRNG characteristic. The LFSR is very simple to implement in circuitry and has the advantage of generating a maximum length sequence, but it is impossible to generate a complete random number due to the nature of the deterministic system.

TRNG generates random numbers based on unpredictable physical sources such as hardware temperature, jitter, noise, spectrum, etc. [2]. Thus, it is extremely difficult to predict random patterns. Due to the inability to predict the random number generation pattern, TRNG is well suited for use where higher security is required. However, TRNG requires separate hardware to generate random numbers, which is not easy to implement and is very unfit for use in small mobile devices because of its large chip area.

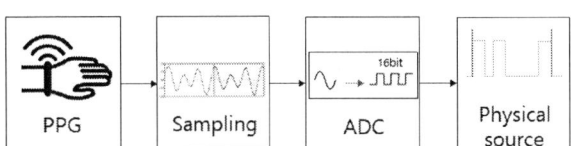

Fig. 1. Physical source acquisition diagram

In this paper, we propose a TRNG that can generate unpredictable random number patterns using the advantages of PRNG, which provides maximum length sequence generation and easy circuit implementation. The proposed TRNG can generate random number patterns by using of the Photoplethysmography (PPG) sensor value in the wearable device as a physical random source. The proposed TRNG generates 16-bit random numbers with a simple conventional LFSR and random bio signals. The TRNG is implemented in an FPGA and the random numbers generated in the FPGA are verified for excellent security randomness through the NIST test suite [3].

II. PHYSICAL SOURCE ACQUISITION

Nowadays, a PPG sensor is embedded in most wearable devices to measure the value of arterial blood flow as the heart moves and capture arterial blood flow changes. The PPG sensors have the disadvantage that the signals are very noisy depending on the physical and environmental factors when the fingers move or light changes [4]. This disadvantage can be advantageous when used as a physical source of TRNG. The value is keep changing every time according to the very fine movement of the person or the surrounding environment, and it is unpredictable.

In order to use the data value of the PPG sensor as a physical source of the TRNG, it is necessary to change the analog signal to a 16-bit digital data value. As shown in Fig. 1, the analog signal output from the PPG sensor is sampled, and the sampled value is converted to a 16-bit data value through the ADC module. In general, the ADC in the wearable device converts the PPG analog signal to digital for digital signal processing [5]. When a 16-bit data value is generated through the ADC, a timer interrupt is issued at a desired time to obtain a 16-bit data value, and the obtained 16-bit data is used as a physical source for generating a random number in LFSR.

III. TRUE RANDOM NUMBER GENERATOR USING PPG SENSOR

The proposed TRNG structure is illustrated in Fig. 2. As shown, the proposed TRNG uses the conventional LFSR module and the feedback polynomial of the LFSR is fixed to $x^{16} + x^{15} + x^{13} + x^{14} + 1$ to generate the maximum length sequence

978-1-5386-7961-6/18 $31.00 © 2018 IEEE

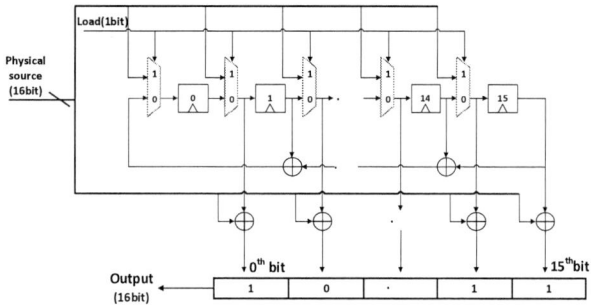

Fig. 2. TRNG using physical source of biosensor

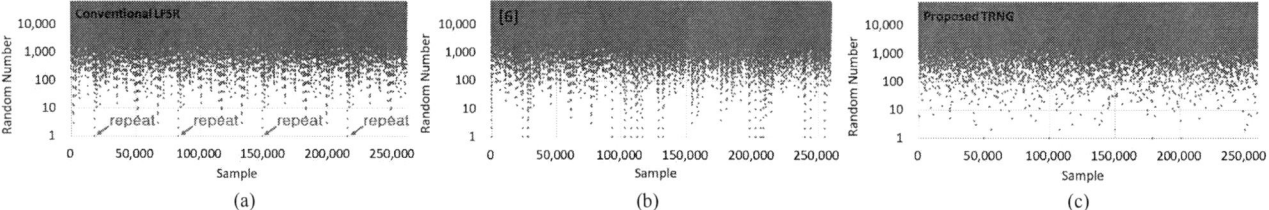

Parameters	Conventional	[6]	Proposed
Frequency	P	F	P
Block Frequency	P	F	P
Cumulative Sums	F	F	P
Runs	F	P	P
Longest Run	F	P	P
Non-Overlapping Template	F	P	P
Overlapping Template	F	P	P
Approximate Entropy	F	P	P
Random Excursions	F	F	P
Random Excursions Variant	F	P	P
Serial	F	F	P
Linear Complexity	P	P	P

TABLE I. NIST TEST SUITE RESULT

Fig. 3. Random number patterns; (a) conventional LFSR, (b) [6], and (c) proposed TRNG

of 65,535 random numbers in 16-bits. LFSR requires an initial seed value and the 16-bit digital values obtained from the PPG sensor is used in this study. In the LFSR with a fixed polynomial, the outputs are determined by the polynomial without duplication, and then the outputs repeat continuously after entire cycle of the LFSR operation. Thus, the next random number can be predicted in the conventional LFSR if the attacker obtains the maximum number of sequences more than once even if the seed value is periodically changed due to the fixed polynomials. To solve this problem, we insert XOR gates to mix the LFSR outputs with the 16-bit physical random source obtained from the PPG sensor as shown in Fig. 2. The LFSR outputs and the unpredictable physical source value are operated by XOR gate, resulting in a more complex random number pattern than the conventional LFSR. In this study, new physical source random value is obtained from the PPG sensor whenever the maximum length sequence of 16-bit LFSR is reached. Changing the physical source creates a perfect random number that is difficult to predict.

IV. SIMULATION RESULT

The Verilog and Intel Quartus2 design tools are used to implement the proposed TRNG. To verify the superiority of the proposed TRNG, we compare it with the conventional LFSR and the polynomial modifiable LFSR [6]. Simulations are conducted to generate four cycles of the maximum length sequence (i.e., over 250k numbers) in 16-bit LFSR and the random numbers are evaluated through NIST test suite. The test results are shown in Table I. The conventional LFSR generates exactly same 0 and 1 ratio patterns, which is the most basic random number condition in the test. However, it fails in most tests because the value is constantly repeated after the maximum length sequence is generated as shown in Fig. 3 (a).

Next random value is not predictable by continuously changing the polynomial in [6]. However, not all the polynomials can generate the maximum sequence of the LFSR. Because the length of the sequence is continuously changed, the 0 - 1 ratio in

the random number becomes different depending on the polynomials as shown in Fig. 3(b). Therefore, the frequency test (i.e., ratio of 0 to 1), which is the most basic random number characteristic in the NIST test suite does not meet. As a result, even though the random number patterns are more complicated than the conventional LFSR, it is still not enough to be applied to applications requiring true random numbers.

However, the proposed TRNG using the PPG sensor has passed all tests as shown in Table I. Since the proposed TRNG periodically changes the physical source and generates the maximum length sequence, the generated random numbers are truly unpredictable and have no periodicity as shown in Fig. 3(c).

V. CONCLUSION

In this paper, we propose a TRNG using a PPG sensor embedded in wearable devices. The TRNG use the physical source obtained by the PPG sensor for the seed value and mixing the outputs. The proposed TRNG is verified for excellent security through NIST test suite.

ACKNOWLEDGMENT

This research was supported by the MISP (Ministry of Science, ICT & Future Planning), Korea, under the National Program for Excellence in SW supervised by the IITP (Institute for Information & communications Technology Promotion) (2017-0-00096)

REFERENCES

[1] S. Hellebrand, et al., "Built-in test for circuits with scan based on reseeding of multiple-polynomial linear feedback shift registers," *IEEE Transactions on Computers*, vol. 44, no. 2, pp. 223-233, 1995.

[2] K. Lee, et al., "TRNG (True Random Number Generator) Method Using Visible Spectrum for Secure Communication on 5G Network," *IEEE Access*, vol. 6, pp. 12838-12847, 2018.

[3] A. Rukhin, et al., "A Statistical Test Suite for Random and Pseudorandom Number Generators for Cryptographic Applications," in *NIST*, 2010.

[4] P. T. Wood, et al., "Active motion artifact cancellation for wearable health monitoring sensors using collocated MEMS accelerometers," in *proc SPIE*, no. 5765, pp. 811-820, 2005.

[5] J. Jiang, et al., "The routing algorithm based on fuzzy logic applied to the individual physiological monitoring wearable wireless sensor network," *Journal of Elec. and Com. Eng*, vol. 2015, no. 63, pp. 1-7, 2015.

[6] M. G. Han and Y. Kim, "Unpredictable 16 bits LFSR-based true random number generator," *ISOCC*, pp. 284-285, 2017

Low-Power Null Convention Logic Multiplier Design Based On Gate Diffusion Input Technique

Prashanthi Metku[1], Kyung Ki Kim[2], Yong-Bin Kim[3] and Minsu Choi[1]

[1]Dept of ECE, Missouri Univ of Science & Technology, Rolla, MO, USA, {pmcmc,choim}@mst.edu
[2]Dept of Electronic Eng., Daegu University, Gyeongsan, Korea, kkkim@daegu.ac.kr
[3]Dept of ECE, Northeastern University, Boston, MA, USA, ybk@ece.neu.edu

Abstract—The increasing power consumption in the synchronous circuits is the major concern in the semiconductor industry. The major contributor to this power consumption is the clock generator and the clock distribution. This problem can be addressed by using the asynchronous circuits. Null Convention Logic (NCL) is one of the most commonly known delay insensitive approach for designing asynchronous designs. However, realizing the NCL circuits using the commonly used complementary metal oxide semiconductor (CMOS) technique is said to increase the area and the power consumption. The low power design technique known as Gate Diffusion Input (GDI) can be used for implementing the NCL circuits to reduce both the area and the power. Application of the external input to the sources of the pMOS and nMOS transistors, allows to reduces the area and the dynamic switching. Thus, decreasing the transistor count and the power. The proposed GDI NCL technique is used for designing the 4-bit un-pipelined NCL multiplier. The design was realized and simulated in gpdk045 Cadence Virtuoso. In comparison to the CMOS model, the GDI model shows 21.6 % in transistor count and the dynamic power is reduced by 13.7 %.

I. INTRODUCTION

Though the synchronous circuits are dominating the semiconductor industry, its continuous decreasing IC feature size has become the major limiting factor [1]. With the continuous decreasing size and increasing clock speed clock leads to the most challenging issues such as complex clock distribution, clock skew and power consumption in the synchronous designs [2]. The main contribution to the increasing power consumption is the dynamic switching power. The most predominant factors affecting the dynamic switching are the power supply and the threshold voltages [2]. The drastic decrease in this voltage leads to the significant leakage power which in turn causes large power consumption. Thus, leading to the renewed interest towards the asynchronous circuits.

Null Convention Logic

Null Convention Logic (NCL) is one of the widely accepted delay-insensitive (DI) paradigms for designing asynchronous circuit [3]. Irrespective of the input availability, the NCL circuits always produce the accurate results. Thus, deemed to be most efficient approach for designing clockless DI circuits [4]. NCL utilizes the symbolic completeness of expression to achieve the delay insensitive behavior. NCL circuits employs multi- rail logic i.e. dual-rail or quad-rail logic circuits to eliminate the time- reference problem [5]. The NCL gates, also known as threshold gates are designed with hysteresis loop to maintain the delay-insensitivity.

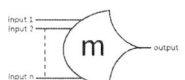

Fig. 1: $THmn$ Gate [2].

NCL gates consists of 27 fundamental gates [6]. Fig. 1 illustrate the primary type of threshold gate (THmn) where 1 le m le n [3]. Here, m represents total number of inputs and n constitutes the number of inputs to be asserted. In the NCL circuits, at least m out of n inputs are asserted for the output to be asserted [5]. The DI denominational logic is always enclosed between DI registers in the NCL designs. To ensure that no two DATA wavefront are overwriting, they are always separated by NULL wavefront [5].

MODIFIED GDI

Gate Diffusion Input (GDI) is a promising low power design technique for considerable area/power reduction. The modified GDI basic cell representation is depicted in the Fig. 2 [7].

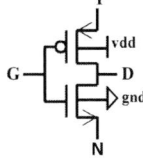

Fig. 2: Modified GDI Basic Cell [7].

The modified GDI and the conventional GDI have similar structures apart from the body connection. In the conventional GDI the body connection of the pMOS and nMOS transistors are connected to their respective sources which only can be implemented in silicon on insulator o twin-tub technology [8]. Whereas, in modified GDI body of the pMOS and nMOS are connected to ground and VDD respectively which makes it feasible to be implemented in the standard CMOS process [7].

As seen from the Fig. 2, the modified GDI consists of three terminals 1) P and N: pMOS and nMOS transistors outer diffusion node; 2) D: common drain for both the transistors and 3) G: common gate input for nMOS and pMOS transistors [9]. The three nodes N, P and D can be used as the input or the output ports, as per the circuit requirement. Since, the P and N nodes are not connected to the fixed VDD and ground, they tend to suffer the significant voltage drop at the output. This drawback of the modified GDI technique is addressed by using the regenerative inverters [9].

978-1-5386-7961-6/18 $31.00 © 2018 IEEE   233   ISOCC 2018

II. PROPOSED MODEL

The modified GDI approach used for designing 4-bit non-pipelined is presented. The basic NCL 4-bit multiplier design is presented in [5]. In this paper, two different models; the CMOS and GDI models of this multiplier are proposed and tested in this work. In the GDI model, NCL gates designed using low power modified GDI technique are utilized for implementing the designs. The proposed GDI model provides better performance in terms of power consumption and transistor count.

The 4-bit multiple consists of two 8-bit GDI registers, incomplete GDI AND, complete GDI AND gate, GDI half adders and GDI full adder. Fig. 3 and Fig. 4 illustrates the realization of the incomplete GDI AND and GDI half adder using GDI TH24COMP, GDI TH12 and GDI TH22 gates.

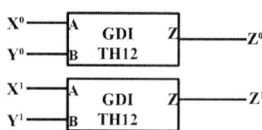

Fig. 3: GDI representation of Incomplete AND Gate.

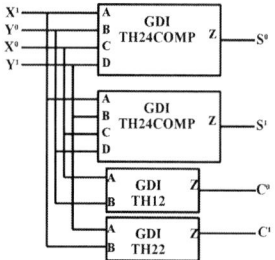

Fig. 4: GDI Implementation of Half Adder.

To overcome the limitation of voltage, drop at the output, restoration buffers are added at the output of the each NCL gate. Multi-threshold voltage logic is utilized to obtain the low power GDI NCL circuits. In this approach both the low-threshold voltage and the high threshold voltage transistors are utilized. Low-threshold voltage transistors are used in the path were voltage drop is excepted and the regenerative buffers are designed using the high threshold transistor. The implementation of the GDI NCL Th22 gate using the both the high and low threshold transistors is depicted in the Fig. 5.

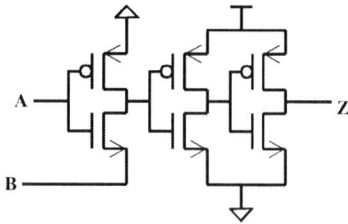

Fig. 5: GDI Implementation of Th22 Gate.

III. SIMULATION RESULTS

This section presents the simulation results of the 4-bit non-pipelined multiplier implemented using CMOS and GDI technology. Both the CMOS and GDI models are realized in $45nm$ technology using Cadence Virtuoso tool with $V_{DD} = 1V$ and temperature= $27°C$. The CMOS model is designed using the standard threshold transistors so that it has moderate delay and power. Performance metric is based on the static power, dynamic power and number of transistors used. Simulations were carried out on all the possible input combinations and their average values are presented.

TABLE I: Simulation Results of 4-bit NCL RCA implemented using CMOS HYBRID and GDI modles.

Design Technique	Static Power (nW)	Dynamic Power (μW)	Transistor Count
CMOS Model	1.66	24.27	2040
GDI Model	2.9	19.01	1760

As seen from the Table. I, the GDI model shows better performance in terms of dynamic power dissipation and transistor count. Whereas the CMOS model consumes low static power. In comparison with the CMOS model, the GDI model shows 21.6 % reduction in dynamic power and the transistor count is reduced by 13.7 %. However, the increase in the leakage power in the GDI model is due to the usage of regenerative buffers and designing the circuit using both high threshold and low threshold transistors.

REFERENCES

[1] S. C. Smith, "Design of an fpga logic element for implementing asynchronous null convention logic circuits," *IEEE Transactions on Very Large Scale Integration (VLSI) Systems*, vol. 15, no. 6, pp. 672–683, 2007.

[2] ——, "Gate and throughput optimizations for null convention self-timed digital circuits," Ph.D. dissertation, University of Central Florida Orlando, Florida, 2001.

[3] M. Choi, B.-H. Kang, Y.-B. Kim, and K. K. Kim, "Asynchronous circuit design using new high speed ncl gates," in *Proc. Int. SoC Design Conf. (ISOCC)*, Nov. 2014, pp. 13–14.

[4] S. K. Bandapati and S. C. Smith, "Design and characterization of null convention arithmetic logic units," *Microelectronic engineering*, vol. 84, no. 2, pp. 280–287, 2007.

[5] S. C. Smith, R. F. DeMara, J. S. Yuan, D. Ferguson, and D. Lamb, "Optimization of null convention self-timed circuits," *INTEGRATION, the VLSI journal*, vol. 37, no. 3, pp. 135–165, 2004.

[6] S. C. Smith, R. F. DeMara, J. S. Yuan, M. Hagedorn, and D. Ferguson, "Null convention multiply and accumulate unit with conditional rounding, scaling, and saturation," *Journal of Systems Architecture*, vol. 47, no. 12, pp. 977–998, 2002.

[7] A. Morgenshtein, I. Shwartz, and A. Fish, "Gate diffusion input (gdi) logic in standard CMOS nanoscale process," in *Proc. IEEE 26-th Convention of Electrical and Electronics Engineers in Israel*, Nov. 2010, pp. 000 776–000 780.

[8] A. Morgenshtein, V. Yuzhaninov, A. Kovshilovsky, and A. Fish, "Full-swing gate diffusion input logiccase-study of low-power cla adder design," *INTEGRATION, the VLSI journal*, vol. 47, no. 1, pp. 62–70, 2014.

[9] A. Morgenshtein, A. Fish, and A. Wagner, "Gate-diffusion input (gdi)-a novel power efficient method for digital circuits: a design methodology," in *Proc. 14th Annual IEEE Int. ASIC/SOC Conf*, 2001, pp. 39–43.

A Radiation Hardened SRAM with Self-refresh and Compact Error Correction

[1]Sultan M. Siddiqui, [1]Ruchi Sharma, [1]Van Loi Le, [1]Taegeun Yoo, [2]Ik-Joon Chang, and [1]Tony Tae-Hyoung Kim

[1]School of EEE, Nanyang Technological University, Singapore
[2]Kyunghee University, Suwon, Korea
thkim@ntu.edu.sg

Abstract— **This work presents a radiation resilient SRAM with a self-refresh scheme and error correction. The self-refresh scheme maintains the number of Single Event Upsets (SEUs) below a correctable value during the idle mode by checking and correcting stored data row by row. A 4KB SRAM test chip in 65nm CMOS technology demonstrates that the combination of the proposed self-refresh and the error correction improves the radiation tolerance significantly when the SRAM is under accelerated proton radiation. At 39.38 MeV of radiation energy and the operating frequency of 3.6MHz, the proposed schemes reduces the numbers of errors in the SRAM by 25× and 8× for the proton radiation durations of 10s and 50s, respectively.**

Index Terms — **Single Event Upset (SEU), single error correction double error detection (SECDED), SRAM**

I. INTRODUCTION

Single Event Upset (SEU) has become increasingly problematic for integrated circuits implemented in the deep submicron technology [1]. SEU usually causes bit flipping in memories (e.g. SRAM, Flash) when the charge accumulated at a storage node is greater than the critical charge. The scaling in the operating voltage and the process technology has decreased the critical charge, which makes memory cells more vulnerable to SEU [2]. SRAMs have been widely used in space applications. Various radiation hardened SRAM cells such as Dual Interlocked Storage Cell (DICE) [3] and Quatro-10T [4] can effectively reduce SEUs at the cost of additional transistors but are not fully radiation resilient. Other widely used solutions for SEUs are Error Correction Codes (ECCs) such as Parity and Hamming codes [5], [6]. A larger number of SEUs can be handled if complex ECC algorithms such as low-density parity check (LDPC) code [7] are used. However, these complex ECC algorithms degrade the SRAM performance significantly and require substantial silicon area overheads. Therefore, it is imperative to maintain the number of SEUs in a row within a detectable or correctable level without significant overheads in silicon area and performance.

In this paper, we propose a radiation resilient SRAM with self-refresh and Single Error Correction and Double Error Detection (SECDED), which maintains the number of SEUs small enough to be detected or corrected by SECDED during normal SRAM operation. The proposed self-refresh is similar to the refresh operation in DRAMs except that error correction is incorporated. To further improve the radiation resilience at the circuit level, an 8T SRAM cell with decoupled read port and extended diffusion area.

II. PROPOSED RADIATION HARDENED SRAM WITH SELF-REFRESH AND SECDED

Fig. 1(a) shows the 8T SRAM cell with a decoupled read port. To increase the critical charge for better SEU resilience, the diffusion areas of Q and QB are extended as depicted in Fig 1(b). This work employs simple SECDED to handle SEUs without significant overheads in silicon area and performance. However, this requires the number SEUs in each row below two so that they can be corrected or detected by SECDED. To realize this, this work proposes a self-refresh scheme that limits the number of SEUs in each row by periodically reading data and correcting

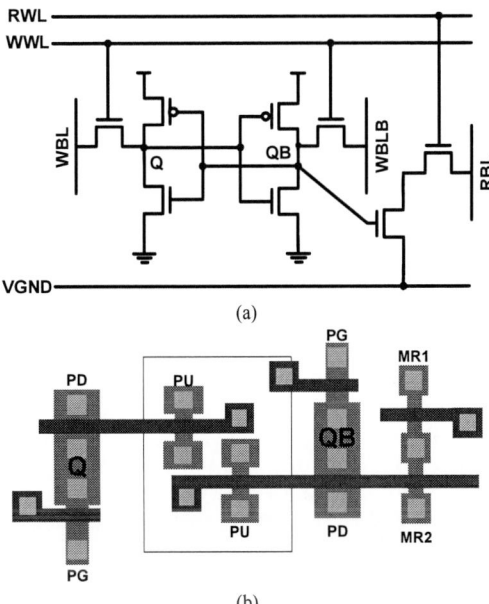

(a)

(b)
Figure 1. (a) 8T SRAM cell schematic and (b) layout.

them if necessary. Fig. 2 illustrates the principle of the proposed self-refresh scheme. The SEU probability of an SRAM cell increases linearly with time. If the SEU probability is too high, the number of SEUs in each row can be more than the target number (e.g. two in this work), which cannot be corrected nor detected. The SEU probability can be maintained low by checking data in each cell periodically. The period between subsequent checking operations for a cell determines the upper limit of the SEU probability. In this work, this self-refresh operation is executed when the SRAM is in the idle mode. The self-refresh operation is described in Fig. 4. When the memory

This work is supported by OSTIn-SIAG under the grant number of NRF2014SAS-SRP001-057.

978-1-5386-7961-6/18 $31.00 © 2018 IEEE   235   ISOCC 2018

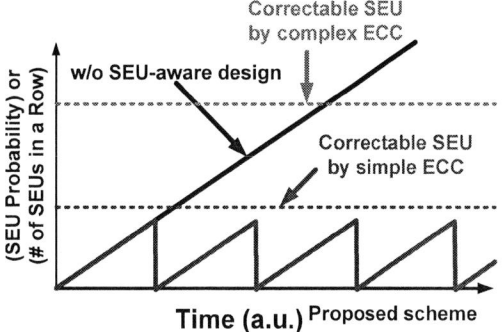

Figure 2. Concept of the proposed self-refresh.

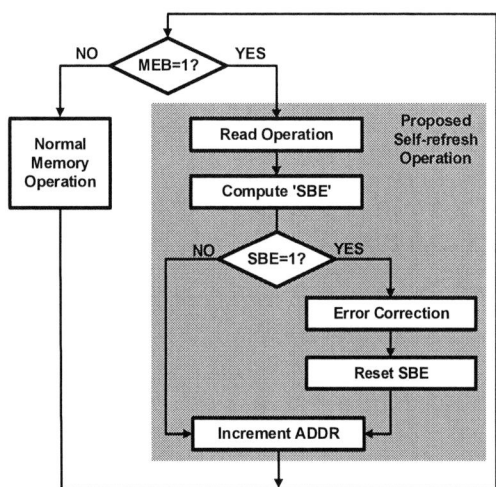

Figure 3. Proposed SRAM self-refresh concept flow diagram.

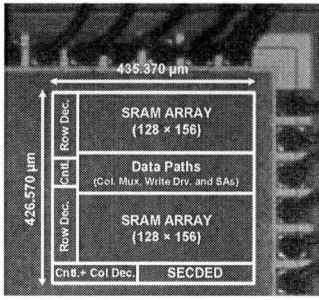

Feature	Value
Technology	65nm UMC
Address Bus & Data Bus	10-bit & 32-bit
Power supply	1.2V core, 3.3V I/O
Access time	0.8ns
Leakage	14.4uw
Write Power	178.08uw
Read Power	138.6uw
Self-refresh Power	157.2uw

Figure 4. Test chip micrograph and summary table.

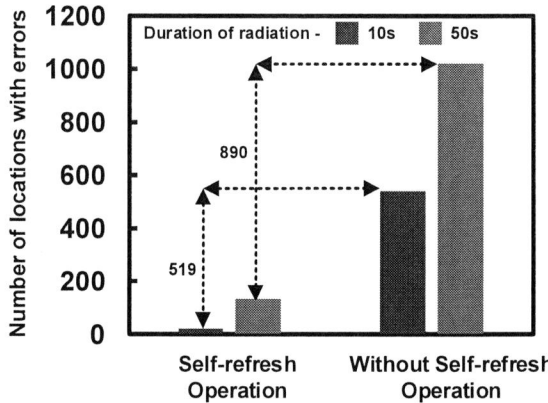

Figure 5. Measured result of the number of rows with errors with and without the proposed self-refresh (Energy - 39.38 MeV; Frequency - 3.6 MHz).

enable signal (MEB) is '0', the memory can be accessed for normal read and write operation (normal mode). During the normal mode, the SECDED ECC is active and the read-out data can be corrected for any single error occurrence. Single Bit Error (SBE) signal shows whether the read-out data has single error or not. When MEB is high, the SRAM is in the self-refresh mode. During the self-refresh mode, SRAM row data are read and checked for SECDED. If SBE='0', no error occurs and row data using the next address is read out and checked. If SBE='1', SECDED ECC corrects the error and overwrites the erratic data with this corrected data. This self-refresh operation continues until the self-refresh mode is disabled.

III. RADIATION MEASUREMENT RESULTS

A 4KB radiation resilient SRAM test chip was fabricated in 65nm CMOS technology. Fig. 4 shows the test chip die photo. Fig. 7 depicts the number of rows with errors with and without the proposed self-refresh at the energy of 39.38 MeV for two different radiation durations (10s and 50s). The number of rows with errors decreases significantly after employing the proposed self-refresh scheme. This result clearly shows that the SRAM with the proposed self-refresh and SECDED is a promising memory solution for space applications without complex ECC and significant area and performance overhead.

REFERENCES

[1] R. C. Baumann, "Radiation-Induced Soft Errors in Advanced Semiconductor Technologies," in *IEEE Trans. on Device and Material Reliability*, vol. 5, no. 3, pp. 305–316, 2005.

[2] P. E. Dodd and L. W. Massengill, "Basic mechanisms and modeling of single-event upset in digital microelectronics," *IEEE Trans. Nucl. Sci.*, vol. 50, no. 3, pp. 583–602, 2003.

[3] T. Calin, M. Nicolaidis, and R. Velazco, "Upset hardened memory design for submicron CMOS technology," *IEEE Transactions on Nuclear Science*, vol. 43, no. 6, pp. 2874–2878, 1996.

[4] S. M. Jahinuzzaman, D. J. Rennie, and M. Sachdev, "A soft error tolerant 10T SRAM bit-cell with differential read capability," *IEEE Transactions on Nuclear Science*, vol. 56, no. 6, pp. 3768–3773, 2009.

[5] R. W. HAMMING, "Error detecting and error correcting codes," *the Bell System Technical Journal,* April, 1950.

[6] C. L. Chen and M. Y. Hsiao, "Error-correcting codes for semiconductor memory applications: A state-of-the-art review," *IBM J. Res. Dev.*, vol. 28, no. 2, pp. 124–134, 1984.

[7] S. Lin and D. J. Costello, "Error Control Coding," *2nd ed. Englewood Cliffs, NJ, USA: Prentice-Hall*, 2004.

Gap in pagination due to formatting issues.

Pages 237-238

Experimental Verification of a Simple, Intuitive, and Accurate Closed-Form Transfer Function Model for Diverse High-Speed Interconnects

Kyunghyun Lim[1], Minsoo Choi[1], Myat-Thu-Linn Aung[2], Kyunghwan Kim[1], Ji-Seong Kim[1], Rock-Hyun Baek[1], Ho-Jin Song[1], Tony Tae-Hyoung Kim[2] and Byungsub Kim[1]

[1]Dept. of EE, Pohang University of Science and Technology (POSTECH), Pohang, Korea, byungsub@postech.ac.kr
[2]School of EEE, Nanyang Technological University, Singapore

Abstract—**This paper experimentally verifies a simple, intuitive, and accurate closed-form transfer function model for diverse high-speed interconnects for the first time. Recently, an approximate closed-form transfer function model, which is simple, intuitive, and accurate, as well as applicable to diverse high-speed interconnects, was proposed. However, because the model was only verified by SPICE simulation, the experimental verification is required to guarantee the reliability of the model in practical high-speed interconnects. To this end, an example interconnect was designed and fabricated, and its characteristic impedance and propagation constant were extracted. Then, the transfer function calculated from the extracted parameters by using the model was compared with measurement. The result shows that the model can describe the behavior of the fabricated example interconnect and thus is reliable in practical high-speed interconnects.**

Keywords—*Approximate transfer function, channel model, experimental verification, interconnects*

I. INTRODUCTION

As data rate in wireline communication increases recently, interconnects have become a major bottleneck limiting the link system performance. To overcome such a limitation, various link circuit techniques such as feed-forward equalization (FFE) [1] and decision feedback equalization (DFE) [2] have been developed. Utilizing the link circuit techniques requires circuit designers to clearly understand and to accurately analyze the interconnect behaviors. To aid circuit designers, an approximate closed-form transfer function model, which is simple, intuitive, and accurate, as well as applicable to diverse high-speed interconnects, was recently proposed in [3].

In [3], however, the model was only verified by SPICE simulation. SPICE simulation does not include frequency-dependent losses in practical high-speed interconnects such as surface roughness and fiber weave effect [4]. In addition, the result of SPICE simulation is inherently the same numerical solution of the same equation used in the model [3]. Therefore, verification by SPICE simulation cannot guarantee the reliability of the model in practical high-speed interconnects.

This paper experimentally verifies the simple, intuitive, and accurate closed-form transfer function model in [3] for the first time. Through the experimental verification, we show that the model is reliable in the practical high-speed interconnects too.

II. A SIMPLE, INTUITIVE, AND ACCURATE CLOSED-FORM TRANSFER FUNCTION MODEL

Manuscript received, 2018. This work was supported in part by National Research Foundation (NRF) of Korea (No. 2018R1A2A2A16022248), in part by Ministry of Science and ICT (MSIT) of Korea under the "ICT Consilience Creative Program" (IITP-2018-2011-1-00783) supervised by the Institute for Information & Communications Technology Promotion (IITP), and in part by Nanyang Technological University and the Agency for Science, Technology and Research, Singapore through Silicon Technologies Centre of Excellence under Grant 1123515003.

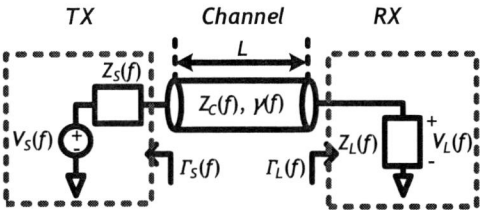

Figure 1. A simplified interconnect with a *Tx* and an *Rx*.

In this section, we review the closed-form transfer function model that was proposed in [3]. Fig. 1 shows a simplified interconnect model used to derive of the transfer function [3]. The transmitter (*Tx*) is modeled as a voltage source $V_S(f)$ and an impedance $Z_S(f)$. The receiver (*Rx*) is modeled as an impedance $Z_L(f)$. $V_L(f)$ is the received voltage at the *Rx*. The wire with length L can be characterized by a characteristic impedance $Z_C(f)$ and a propagation constant $\gamma(f)$. The transfer function $V_L(f)/V_S(f)$ of the interconnect (Fig. 1) can be derived as

$$\frac{V_L(f)}{V_S(f)} = \frac{Z_C(f)}{Z_S(f)+Z_C(f)} 2e^{-L\gamma(f)} \frac{Z_L(f)}{Z_C(f)+Z_L(f)} \left\{ \frac{1}{1-\eta(f)} \right\},$$
(1)

where $\eta(f)=\Gamma_S(f)\Gamma_L(f)e^{-2L\gamma(f)}$, $\Gamma_S(f)=(Z_S(f)-Z_C(f))/(Z_S(f)-Z_C(f))$, and $\Gamma_L(f)=(Z_L(f)-Z_C(f))/(Z_L(f)-Z_C(f))$. If $|\eta(f)|\ll 1$, equation (1) can be approximated into

$$\frac{V_L(f)}{V_S(f)} \approx \frac{Z_C(f)}{Z_S(f)+Z_C(f)} 2e^{-L\gamma(f)} \frac{Z_L(f)}{Z_C(f)+Z_L(f)}.$$
(2)

Equation (2) accurately describes the transfer function of an interconnect as long as the interconnect satisfies $|\eta(f)|\ll 1$ [3]. The relative error of (2) with respect to (1) is equal to $\eta(f)$ [3].

III. EXPERIMENTAL VERIFICATION

In this section, we experimentally verify the simple, intuitive, and accurate closed-form transfer function model in [3].

A. Extraction of Line characteristics

For verification of the model, we firstly extracted the line characteristics ($Z_C(f)$ and $\gamma(f)$) of an example interconnect by using a de-embedding method in [5]. Fig. 2(a) shows fabricated test-wire structures used for the extraction. We measured S parameters of the fabricated structures, and then applied a de-embedding method proposed in [5] to cancel the pad parasitic

978-1-5386-7961-6/18 $31.00 © 2018 IEEE

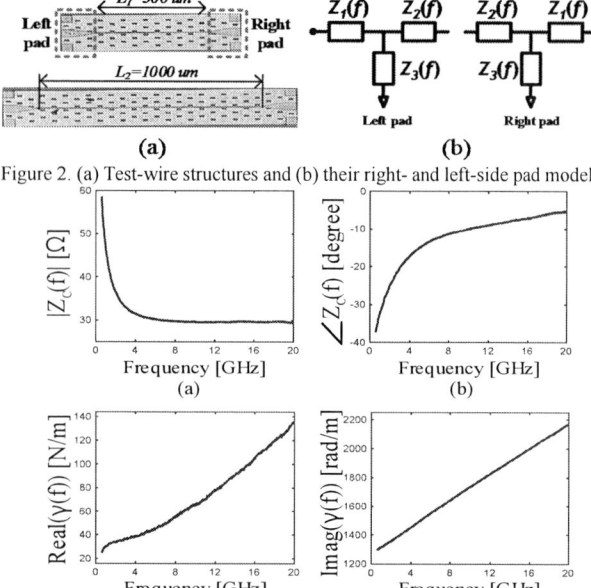

Figure 2. (a) Test-wire structures and (b) their right- and left-side pad models.

Figure 3. Extracted line characteristics: (a) $|Z_C(f)|$, (b) $\angle Z_C(f)$, (c) the real part of $\gamma(f)$, and (d) the imaginary part of $\gamma(f)$.

elements out. In this de-embedding, we modified the models of the pad parasitic elements used in [5] to the ones shown in Fig.

2(b) considering the contact impedance $Z_2(f)$ since it affects the measured data in our measurement. Fig. 3 shows the extracted line characteristics of the example interconnect. To verify that the extracted line characteristics are reliable, the transfer function of the 500-um test-wire structure (Fig. 2(a)) calculated from the extracted $Z_C(f)$ and $\gamma(f)$ (Fig. 3) using the rigorous closed-from transfer function (1) is compared with measurement (Fig. 4(a)). As shown in Fig. 4(a), the calculated transfer function accurately estimates the measurement result. The maximum relative error between them is only 1.4% (Fig. 4(b)). This shows that the line characteristics (Fig. 3) extracted by the method [5] are reliable. The extracted line characteristics will be used to experimentally verify the model in [3] in the next subsection.

B. Experimental Verification

Our experiment verifies that the model in [3] can describe the behavior of the fabricated example interconnect well, and thus is reliable in practical high-speed interconnects. For verification, we calculated the transfer function of the 500-um test structure (Fig. 2(a)) from the extracted line characteristics (Fig. 3) using the approximate transfer function model (2), and then compared the calculated value with the measurement result. As shown in Fig. 5(a), the transfer function calculated using (2) matches with the measurement result for $f<8$ GHz because $|\eta(f)|$ is significantly small (≈0.05) (Fig. 5(b)). However, for $f>8$ GHz, because $|\eta(f)|>0.05$, the transfer function calculated using (2) deviates from the measurement result (Fig. 5(a)) as discussed in [3]. Also, Fig. 5(b) shows that the relative error of the transfer function calculated using (2) with respect to the measurement result is equal to $\eta(f)$ calculated using the extracted line characteristics plotted in Fig. 3. This shows that the error of (2) is bounded by the small value of $\eta(f)$ and thus is predictable by

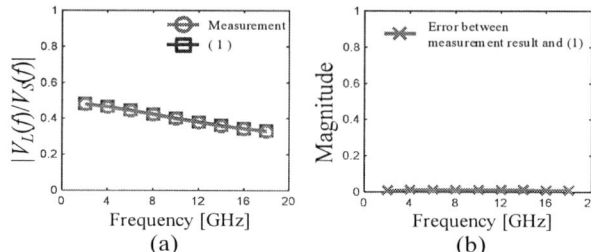

Figure 4. (a) Comparison of the transfer functions acquired by measurement and the rigorous closed-form transfer function (1), and (b) the relative error between them. A 500-um test structure was used.

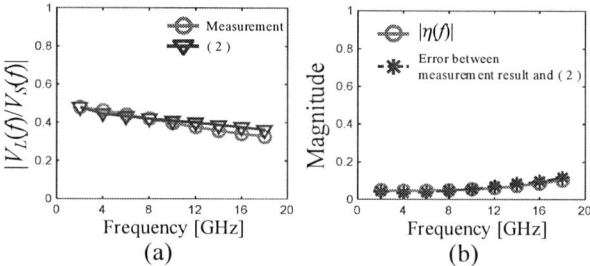

Figure 5. (a) Comparison of the transfer functions acquired by measurement and the simple, intuitive, and accurate transfer function model (2), and (b) the relative error between them. The calculated $|\eta(f)|$ is also plotted in Fig. 5(b). A 500-um test structure was used.

calculating $|\eta(f)|$ as discussed in [3]. This result experimentally verifies that the model in [3] can describe the behavior of an interconnect well, and thus is reliable in practical high-speed interconnects.

IV. CONCLUSION

This paper experimentally verifies the simple, intuitive, and accurate closed-form transfer function model in [3] for the first time. For verification, an example interconnect was designed and fabricated, and its line characteristics were extracted. Compared with the measurement result, the transfer function calculated using the model estimates the frequency response of the fabricated example interconnect well. This result verifies that the model in [3] is reliable in practical high-speed interconnects too.

ACKNOWLEDGMENT

The authors thank IDEC for CAD tool support.

REFERENCES

[1] S. Han, S. Lee, M. Choi, J.-Y. Sim, H.-J. Park, and B. Kim, "A coefficient-error-robust feed-forward equalizing transmitter for eye-variation and power improve-ment," *IEEE J. Solid-State Circuits*, vol. 51, no. 8, pp. 1902–1914, Aug. 2016.

[2] B. Kim, Y.Liu, T. O. Dickson, J. F. Bulzacchelli, and D. J. Friedman, "A 10-Gb/s compact low-power serial I/O with DFE-IIR equalization in 65 nm CMOS," *IEEE J. Solid-State Circuits*, vol.44, no. 12, pp. 3526–3538, Dec. 2009.

[3] M. Choi, J.-Y. Sim, H.-J. Park, and B. Kim, "An approximate closed form channel model for diverse interconnect applications," *IEEE Trans. Circuits Syst. I, Reg. Papers*, vol. 61, no. 10, pp. 3034–3043, Oct. 2014.

[4] *Star-HSPICE User's Manual*, Avant Corp., Fremont, CA, USA, Dec. 1999.

[5] H.-Y. Cho, J.-K. Huang, C.-W. Kuo, S. Liu and C.-Y. Wu, "A novel transmission-line deembedding technique for RF device characterization," *IEEE Trans. Electron Devices*, vol. 56, no. 12, pp. 3160-3167, Nov. 2009.

A Wearable Electrocardiogram Monitoring System Robust to Motion Artifacts

Taeryoung Seol[1] , Sehwan Lee[2] , Junghyup Lee[3]
Integrated Nano-Systems Laboratory
Daegu, Republic of South Korea
E-mail:[1]str96@dgist.ac.kr, [2]lsh7815@dgist.ac.kr, [3]jhlee1@dgist.ac.kr

(Abstract) This study proposes a wearable system that can measure electrocardiogram (ECG) signals reliably in an environment with high motion induced noise. This system employs a motion artifact extraction method based on a triple-axis accelerometer attached to each electrode independently to remove motion artifact from ECG signals with high performance. Recursive Least Square (RLS) and Least Mean Square (LMS) algorithms remove extracted noise from the source signals, thereby obtaining a mean square error (MSE) of 0.0166 when using RLS and 0.0160 when using LMS. This means that the performance improved respectively by approximately 5.1% and 8.6% compared to that of the recently developed ECG monitoring system.

Keywords- Electrocardiogram(ECG); Motion artifact; Adaptive filtering; Mean Squared Error(MSE)

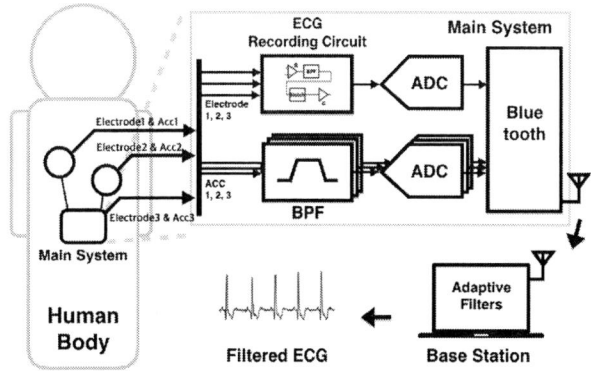

Fig. 1. The proposed wearable ECG monitoring system that is robust to motion noise.

I. INTRODUCTION

Electrocardiogram (ECG) monitoring is one of the most important and simplest methods that can determine and prevent cardiac illness. But aside from the simplicity of the method, the ECG must be monitored frequently and may require up to 4 weeks of monitoring for careful chekups. That is, patients have to attach such measurement devices to their body for a long time in their daily lives. Thus, a small wearable ECG monitoring system needs to be developed to resolve the inconvenience of such measurement.

Such wearable devices suffer from continual exposure to human movements, which can significantly disturb the accuracy of ECG measurements. For example, ECG results are invalid when the subject is monitored while performing excessive movements such as running or jogging. One fundamental reason that motion noise occurs is the changes in the electrode–tissue impedance (ETI) between the electrode and the human body caused by movement. Thus, a wearable ECG monitoring system that is robust to motion noise requires the precise observation of electrode movement and a technique to remove motion noise from ECG signals based on observations.

Studies on the removal of motion noise for wearable ECG monitoring devices are at primitive stages. Among these, studies using accelerometers have been published in [2] [3]. A system for the removal of motion noise from ECG signals

using single and dual-axis accelerometers has been conducted in [2]. A similar system that implements a triple-axis accelerometer has been conducted in [3]. However, since these did not consider each electrode's movements, which can cause motion noise due to the use of only a single accelerometer, they have limited ability to measure ECG precisely. The proposed wearable ECG monitoring system has an independent triple-axis accelerometer attached to each electrode, thereby measuring each electrode's movements accurately to extract the motion noise. Removing the extracted noise data from the source signals using Recursive Least Square (RLS) and Least Mean Square (LMS) algorithms, which employ an adaptive filtering technique, means that ECG signals can be measured successfully even in environments where motion noise is high such as during exercise.

II. ECG MONITORING SYSTEM

Fig. 1 shows the wearable ECG monitoring system proposed in this study. As shown in Fig. 1, this system consists of two units: the main system that reads the ECG signals from the human body using three electrodes and transfers the read signals wirelessly after their conversion to digital data, and a base station that displays the recovered ECG signals after adaptively filtering the transferred data wirelessly.

978-1-5386-7961-6/18 $31.00 © 2018 IEEE

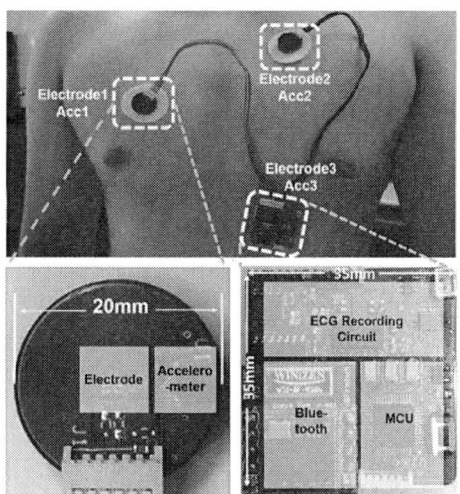

Fig. 2. Wearing example of the proposed system and the fabricated circuit board.

TABLE I.   MSE COMPARISON

	[2]	[3]	This work	
			RLS	LMS
MSE	0.0749	0.0175	0.0166	0.0160

In order to have robust characteristics to motion noise, an independent triple-axis accelerometer is attached to each of the three electrodes, thereby measuring the electrode movement directly to have a close correlation with both the three ECG signals and with the motion noise that is included in the signals at the same time. The measured ECG signals and movement signals of each electrode are transferred to the main system, which is implemented in Electrode 3 in Fig. 1.

The ECG recording circuit, which has a bandwidth of 0.5–100 Hz, amplifies the cardiac signal in the main system and transfers to the base station wirelessly through the module. Finally, the base station removes motion noise from the measured ECG signals and the results are as shown in the display. The adaptive filtering technique is applied using RLS and LMS algorithms to eliminate motion noise with high performance. The used parameter values are FIR filter order = 10 and forgetting factor (λ) = 0.98 for RLS, and FIR filter order = 10 and adaptive stem size parameter (μ) = 0.05 for LMS.

Fig. 2 shows the proposed wearable ECG monitoring system and a worn example system. The size of Electrode 3 in the main system is 35 × 35 mm and the sizes of accelerometer-attached Electrodes 1 and 2 are each 20 mm

The ECG took measurements while doing sit ups using the implemented system and this paper compared the results with those using the existing method in the same conditions as shown in Fig3; the figure shows the recovery of the distorted ECG signals to some extent in all used methods but the accuracy was not determinable.

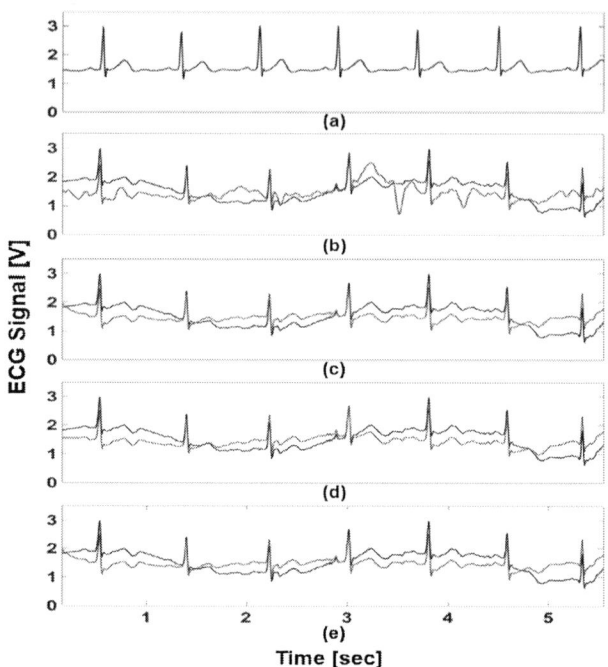

Fig. 3. (a) ECG signal graph in a static state, (b) ECG signal graph mixed with the motion artifact during sit ups and using the method proposed in [2], (c) using the method proposed in [3], (d) ECG graph when applying the RLS algorithm proposed in this study, and (e) when applying the LMS algorithm.

Thus, this paper uses MSE to compare the performance objectively and Table 1 summarizes the results [2–3]. Compared to the method proposed in [2], the proposed system has an improved performance by approximately 77.8% compared to using the RLS algorithm and 78.6% compared to using the LMS algorithm.

Compared to the result using the method proposed in [3], which is the most recently published, the performance improved by approximately 5.1% and 8.6% compared to when using RLS and LMS, respectively.

III. CONCLUSIONS

This study proposes a wearable ECG monitoring system that recovers ECG signals distorted by motion noise through an accelerometer attached to the electrodes and adaptive filtering. Using this system, ECG signals were successfully measured in a high-motion-noise environment, and the motion noise removal performance was improved by >8% compared to using a recently developed ECG monitoring system. In addition, this paper expects that the proposed technique can easily be applied to other wearable bio-medical systems.

REFERENCES

[1] Dilpreet Buxi. et al, "Correlation Between Electrode-Tissue Impedance and Motion Artifact in Biopotential Recordings," IEEE SENSOR JOURNAL, VOL.12, NO 12, 2012
[2] Mary Anne D, Raya & Luis G, Sison, "Adaptive Noise Cancelling of Motion Artifact in Stress ECG Signals Using Accelerometer," IEEE EMBS/BMES, 1756-1757, 2002
[3] Shing-Hong Liu, "Motion Artifact Reduction in Electrocardiogram Using Adaptive Filter," JMBE 31(1), 67-72 ,2010

Infrared and Visible Image Fusion using Multi-Scale Decomposition and Visual Saliency Map

Yunfan Chen, Han Xie, Donghoon Yeo, and Hyunchul Shin

Division of Electrical Engineering, Hanyang University, Ansan, Korea
shin@hanyang.ac.kr

Abstract— **In this paper, a novel infrared (IR) and visible (VIS) image fusion algorithm is proposed by using a joint-histogram weighted median filter (JH-WMF) and a mean filter (MF). The proposed method is based on a multi-scale decomposition (MSD) of an image into a base layer, a texture layer, and an edge layer, which is helpful to suppress the artifacts. Furthermore, a novel saliency detection algorithm is proposed for base layers fusion by combining JH-WMF and MF, which is more effective to reduce the loss of contrast. Experiments show that the proposed approach achieves better performance than other methods, in terms of subjective visual effect and objective assessment.**

Keywords — Image fusion; multi-scale decomposition; joint-histogram weighted median filter; mean filter

I. INTRODUCTION

Infrared and visible image fusion technology has wide applications in image processing and computer vision areas. Recently, many methods have been presented for IR and VIS image fusion. Among these, the multiscale image fusion [1-3] and optimization-based image fusion [4] are very successful methods. However, due to lack of consideration of spatial consistency during the process of fusion, these methods may produce distortions in brightness and color. Moreover, optimization-based methods may result in a blurred image. In this study, a novel image fusion algorithm is introduced that apply multi-scale decomposition (MSD) and joint-histogram weighted median filter (JH-WMF) [5] to overcome the aforementioned problems.

II. PROPOSED METHOD

The flowchart of the proposed approach is shown in Fig. 1. First, the input IR and VIS images are decomposed into base, texture, and edge layers. Second, the weight maps for the three layers fusion are caculated. Finally, multi-scale image reconstruction is implemented to generate the fused image.

A. Multi-scale image decomposition

As shown in Fig. 2, the input images are first decomposed into three scale representations by using JH-WMF and MF. JH-WMF is an edge-aware image filter that can well maintain edge features while smoothing small-scale details on each source image. Additionally, JH-WMF is very efficient since the computation complexity is $O(r)$ (r is the kernel size). The base layer of each input image is calculated by the following equation:

$$B_n = I_n * M \quad (1)$$

where I_n is the nth input image, M is the MF with the size of 31x31. Then, the detail layer can be obtained:

$$D_n = I_n - B_n \quad (2)$$

The edge layers are calculated by:

$$E_n = D_n * J \quad (3)$$

where J is the joint-histogram weighted median filter which is set with a default parameter. Then the texture layer can be easily obtained.

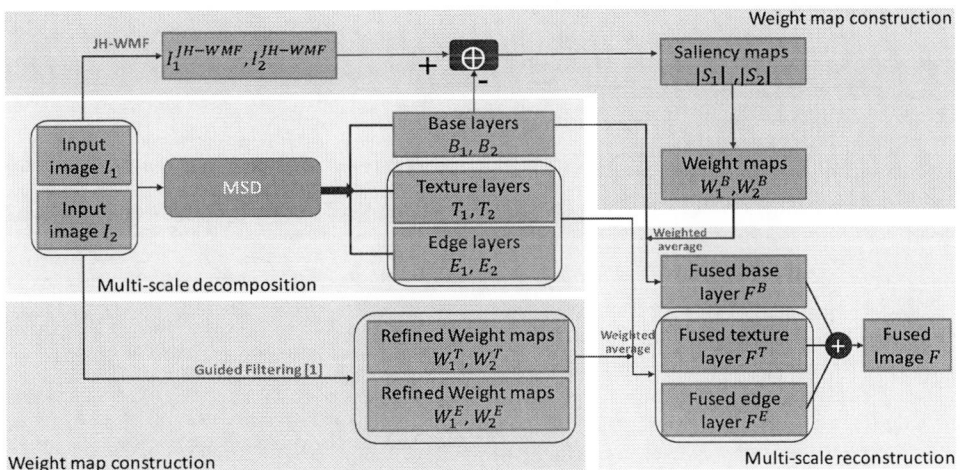

Fig. 1. The framework of our proposed method.

This material is based upon work supported by the Ministry of Trade, Industry & Energy (MOTIE, Korea) under Industrial Technology Innovation Program (10080619).

978-1-5386-7961-6/18 $31.00 © 2018 IEEE

As shown in Fig. 1. First, a JH-WMF is applied on each input image to remove noise or artifacts:

$$I_n^{JH-WMF} = I_n * J \tag{4}$$

where J is a JH-WMF. Then, the saliency map of each input image is calculated as follows:

$$S_n = |I_n^{JH-WMF} - B_n| \tag{5}$$

Next, the weight maps are determined by comparing each saliency map:

$$P = \begin{cases} |S_1|, & if\ |S_1| - |S_2| > 0 \\ 0, & else \end{cases} \tag{6}$$

$$W^B = g_{\sigma_c} * S_\lambda(\bar{P}) \tag{7}$$

where the Gaussian g_{σ_c} is applied for convolution to remove noise and locally smooth the weighting coefficients. Here, σ_c is set to 2. The S_λ is defined as

$$S_\lambda(\alpha) = \arctan(\lambda\alpha)\,/\arctan(\lambda) \tag{8}$$

The weighting maps W_1^T, W_2^T for texture layers fusion, and W_1^E, W_2^E for edge layers fusion are calculated by [6].

B. Multi-scale image reconstruction

First, the base, texture, and edge layers fused by:

$$F^B = W^B B_1 + (1 - W^B)B_2 \tag{9}$$

$$F^T = W_1^T T_1 + W_2^T T_2 \tag{10}$$

$$F^E = W_1^E E_1 + W_2^E E_2 \tag{11}$$

Then, the fused image F is obtained:

$$F = F^B + F^T + F^E \tag{12}$$

III. EXPERIMENTAL RESULTS

Our experiments are implemented on three IR-VIS image pairs (Camp, kayak, and road) from multi-modal image database [7] and processed on a computer with 4.0 GHz CPU and 32 GB RAM. The proposed method is compared with two recent published IR-VIS fusion works GTF [4] and MSD-SR [1]. The metrics Q_G and Q_{CB} are used to assess the fusion performance. The metric Q_G is based on the gradient, which assesses how much the edge information of the source image successfully transferred to the fused image. Human perception inspired metric Q_{CB} is based on human visual system (HVS) models. For all the metrics, larger values express better performance. A good survey and comparative study of these quality metrics can be found in Z. Liu *et al.*'s work [8].

As shown in Table 1, we can see that our approach is superior or comparable to the other methods. In Table 2, we report a comparison between our method and the other methods in terms of their computation efficiency. Average computational time values are calculated over three pairs of the input images. Table 2 reveals that our approach has a better balance between the runtime and the fusion performance.

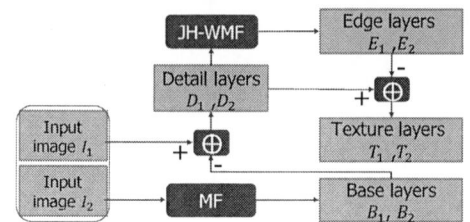

Fig. 2. Multi-scale decomposition (MSD)

TABLE I. THE QUANTITATIVE PERFORMANCE ASSESSMENTS OF DIFFERENT FUSION METHODS ON THE THREE PAIRS OF SOURCE IMAGES

Images	Index	Proposed	GTF	MST-SR
Camp	Q_G	**0.496**	0.397	0.370
	Q_{CB}	**0.541**	0.429	0.528
Kayak	Q_G	**0.604**	0.248	0.501
	Q_{CB}	**0.585**	0.330	0.517
Road	Q_G	**0.650**	0.344	0.479
	Q_{CB}	0.519	0.364	**0.519**

TABLE II. RUNTIME COMPARISION (UNIT: SECOND)

Images	Proposed	GTF	MST-SR
Camp	0.87	3.07	0.20
Kayak	1.08	4.76	0.24
Road	0.70	2.86	0.15
Average	**0.88**	3.56	0.20

IV. CONCLUSION

A novel infrared and visible image fusion method based on multi-scale image decomposition and visual saliency detection is introduced. Experimental results show that our approach achieves comparable or better performance when compared to those of the existing methods.

REFERENCES

[1] Liu, Yu, Shuping Liu, and Zengfu Wang. "A general framework for image fusion based on multi-scale transform and sparse representation." Information Fusion 24 (2015): 147-164.
[2] Bavirisetti, Durga Prasad, and Ravindra Dhuli. "Two-scale image fusion of visible and infrared images using saliency detection." Infrared Physics & Technology 76 (2016): 52-64.
[3] Gan, Wei, et al. "Infrared and visible image fusion with the use of multi-scale edge-preserving decomposition and guided image filter." Infrared Physics & Technology 72 (2015): 37-51.
[4] Ma, Jiayi, et al. "Infrared and visible image fusion via gradient transfer and total variation minimization." Information Fusion 31 (2016): 100-109.
[5] Zhang, Qi, Li Xu, and Jiaya Jia. "100+ times faster weighted median filter (WMF)." In Proceedings of the IEEE Conference on Computer Vision and Pattern Recognition, pp. 2830-2837. 2014.
[6] Li, Shutao, Xudong Kang, and Jianwen Hu. "Image fusion with guided filtering." IEEE Transactions on Image Processing 22.7 (2013): 2864-2875.
[7] http://www.escience.cn/people/liuyu1/index.html
[8] Liu, Zheng, et al. "Objective assessment of multiresolution image fusion algorithms for context enhancement in night vision: a comparative study." IEEE transactions on pattern analysis and machine intelligence 34.1 (2012): 94-109.

Weather Classification using Convolutional Neural Networks

Jehong An, Yunfan Chen, and Hyunchul Shin

Division of Electrical Engineering, Hanyang University, Ansan, Korea

shin@hanyang.ac.kr

Abstract— **Deep learning is one of the most popular fields in recent years and is applicable to many fields, such as computer vision and image processing. In this paper, we describe a new weather image classification technique using Alexnet and Resnet convolutional neural networks (CNN) combined with a multi-class Support Vector Machine (SVM). Experimental results in weather database, desnownet, and d-hazy datasets demonstrate that the proposed method achieves good performance in weather classification.**

Keywords—vision, convolutional neural networks, deep learning, image classification

I. INTRODUCTION

Image classification is used in many computer vision fields, such as ADAS, drone applications, image retrieval, and object recognition. In this paper, we describe a new multi-class classification scheme developed by using CNN combined with multi-class SVM [3]. For training and testing, we used the weather database [5], desnownet [6], and d-hazy dataset [4].

II. CNN ARCHITECTURE AND MULTI-CLASS SVM

A. CNN

In this paper, we adopt two CNN architectures, Alexnet [1] and Resnet [2] for feature extraction, respectively. Then, we use multi-class SVM [3] instead of existing SVM to classify images of various classes. Alexnet [1] uses ReLU (Rectified Linear Unit) activation function after convolution in the input image with 25 layers in total. Resnet [2] uses a total of 347 layers and 379 connections. The following sections describe details of Alexnet [1], Resnet [2], and multi-class SVM [3] used in this paper.

B. Alexnet

Alexnet [1] consists of a total of 5 convolutional layers and

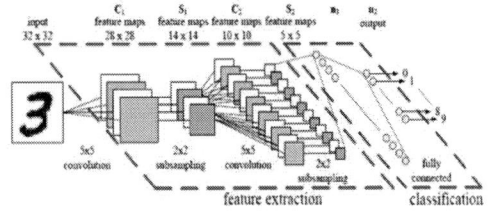

Fig. 1. CNN architecture

This work was supported by Basic Research Project in Science and Engineering through the Ministry of Education of the Republic of Korea and National Research Foundation of Korea (National Research Foundation of Korea 2017-R1D1A1B04-031040).

The equipments for experiments are supported by IDEC at Hanyang University.

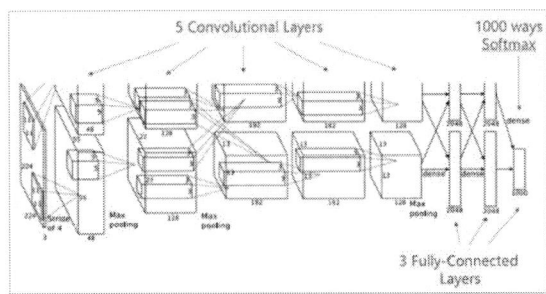

Fig. 2. Alexnet architecture

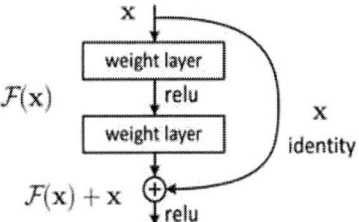

Fig. 3. Resnet architecture

3 fully connected layers and uses softmax for classification of 1000 categories. When designing CNN structure, the calculation problem is solved by using GPU. Alexnet [1] is the cornerstone of utilizing Deep Learning architecture. Fig. 2 shows Alexnet[1] architecture.

C. Resnet

The core idea of Resnet [2] is the residual block, which is shown in Fig. 3. It is a technique to combine input and output. The dimensions of input and output must be equal. By learning the difference between input and output, the middle layer can easily be optimized for deep neural networks, and the accuracy is improved due to the increased depth.

D. Multi-class SVM

A multi-class SVM [3] was used to classify images combined with CNN structures. SVM is one of the linear classifiers such as logistic regression, neural network, and Bayes classifier. The SVM algorithm is based on statistical learning theory and shows excellent generalization performance. It can be applied to various pattern recognition problems. It is also one of the popularized techniques because it can be used simultaneously with classification problems and prediction problems and has a high accuracy of prediction. However, the binary classifier has a disadvantage that it cannot be directly applied to the general multi-class problem. Therefore, in this paper, multi-class SVM [3] is used instead of existing SVM to enable classification in various classes. As the complexity of multi-class SVM increases, the SVM also increases proportionally.

III. DATASETS

The experimental datasets are weather database [5] (sunny, cloudy), desnownet: Context-Aware Deep Network for Snow Removal [6] (snowy), and d-hazy dataset [4]. The weather database is used for two-class weather classification and the number of images of two classes is sunny 5000, cloudy 5000, and total 10000. Desnownet dataset [6] is created by desnownet [6]. This dataset consists of training, test, and realistic 3 subsets, and only 1329 realistic images are used. D-hazy dataset [4] is created in d-hazy and only 1500 NYU_Hazy out of 4 classes of middlebury_GT, muddlebury_Hazy, NYU_GT, and NYU_Hazy are used.

IV. EXPERIMENTAL RESULTS

In Weather database [5], 5,000 images out of two classes were applied, and 70 % of the 5,000 images were used for training and the other 30 % were used for testing. As shown in Table 1, Image classification accuracy using Alexnet combined with multi-class SVM (Alexnet-MCSVM) obtained 86 % sunny and 75 % cloudy. The accuracy of Resnet combined with multi-class SVM (Resnet-MCSVM) is sunny 92%, cloudy 88%.

DesnowNet [6] and D-Hazy dataset [5] are combined with weather database [5] for 4 classes (snowy, sunny, cloudy, and hazy) classification. Each class has 1,000 images, of which 700 images were used for training and 300 images were used for testing. As shown in Table 2, the image classification accuracy is 97 % sunny, 100 % cloudy, 96 % hazy, and 95 % snowy.

Fig. 4 and Fig. 5 show the correct and incorrect classification results, respectively.

Experimental results show that our method works well on weather classification. We believe that our work will make a valuable contribution to the area of weather classification.

REFERENCES

[1] Krizhevsky, Alex, Ilya Sutskever, and Geoffrey E. Hinton. "Imagenet classification with deep convolutional neural networks." Advances in neural information processing systems. 2012.

[2] Kaiming He, Xiangyu Zhang, Shaoqing Ren, Jian Sun "Deep Residual Learning for Image Recognition" Computer Vision and Pattern Recognition (cs.CV). 2015.

[3] Chih-Wei Hsu, Chih-Jen Lin 'A Comparison of Methods for Multiclass Support Vector Machines "IEEE Transaction on Neural Networks.2002

[4] Cosmin Ancuti, Codruta O. Ancuti, Christophe De Vleeschouwer "DHAZY: A Dataset to Evaluate Quantitatively Dehazing Algorithms" IEEE International Conference on Image Processing 2016

[5] C. Lu, D. Lin, J. Jia, and C.-K. Tang, "Two-class weather classification," IEEE Transactions on Pattern Analysis and Machine Intelligence. 2017.

TABLE I.   CLASSIFICATION ACCURACY COMPARISION OF ALEXNET [1] AND RESNET [2] IN WEATHER DATABASE [5]

Methods	Alexnet-MCSVM	Resnet-MCSVM
Sunny	86	92
Cloudy	75	88

TABLE II.   4 CLASS WEATHER CLASSFIATION ACCURACY

Methods	Sunny	Cloudy	Hazy	Snowy
CNN-MCSVM	97	100	96	95

Fig. 4. Example of correct classification results

Fig. 5. Example of incorrect classification results

[6] Yun-Fu Liu, Da-Wei Jaw, Shih-Chia Huang, Jenq-Neng Hwang "DesnowNet: Context-Aware Deep Network for Snow Removal" Computer Vision and Pattern Recognition 2017

Korean Traffic Sign Detection Using Deep Learning

Prateek Manocha, Ayush Kumar
Dept. of Electronics and Communication Engineering,
Dept. of Biotechnology
Indian Institute of Technology, Guwahati, India
prateek.manocha4@gmail.com, ayushkum10@gmail.com

Jameel Ahmed Khan, Hyunchul Shin
Division of Electrical Engineering, Hanyang University,
Ansan, South Korea
jameelkhan@hanyang.ac.kr, shin@hanyang.ac.kr

Abstract— **In this paper, we present a new optimized architecture modified from YOLOv3 to detect three different classes of challenging Korean Traffic Sign Detection (KTSD) dataset. We optimized the new neural network called TS detector with denser grid size, and optimized anchor box size to detect prohibitory, mandatory, and danger classes of KTSD dataset. We trained this architecture on our Korean traffic sign dataset to achieve the mAP value of 86.61%. Our results are significantly better than original YOLOv3 and D-Patches algorithm in terms of mAP value and CPU time.**

Keywords; Korean Traffic Signs; D-Patches; YOLOv3.

I. INTRODUCTION

In advanced driver assistance system (ADAS), traffic sign detection plays an important role to receive the road and speed limit information in advance. Traffic sign detection is a challenging problem due to occlusion with trees, poles and huge vehicles. In addition, there is a large variation in shapes, colors, and size of traffic signs, due to changing scenery and illumination conditions of the road.

Figure 1: Three classes of Korean Traffic Sign Dataset

Figure 2: Detection example of our optimized algorithm

This work has been done while IIT students study at Digital Systems Lab., Hanyang University ERICA, Korea, as summer interns.

This material is based upon work supported by the Ministry of Trade, Industry & Energy (MOTIE, Korea) under Industrial Technology Innovation Program (10080619)

A. Related work

D-Patches approach by Yawar *et al* used classification from discriminative patches (or regions) in traffic sign images that are highly discriminative from their surroundings [1].

Training of D-Patches algorithm is on German Traffic Sign Detection Benchmark [2]. Test dataset (KTSD) used by Yawar *et al* is a challenging dataset of road images taken in Korea, containing three classes of traffic signs. We used the same dataset for testing our algorithm.

II. DETECTION WITH GRID SIZE OPTIMIZATION

YOLOv3 by Joseph *et al* [3] uses up-sampling and down-sampling in each layer by using filters to form an image grid. This architecture includes three detection stages. YOLOv3 resizes the input image into S×S grid in each detection stage.

For detecting small traffic signs, we optimized YOLOv3 to down-sample it by factor of 32, 16, and 4, to obtain a denser grid at three detection stages. Table 1. shows comparison of our approach (TS Detector) with original YOLOv3.

Table 1: Dividing the image into denser grid for small object detection

	Input Image Resolution	Grid Size First Stage	Grid Size Second Stage	Grid Size Third Stage
Orignal YOLOv3	1120×1120	35×35	70×70	140×140
TS Detector	1120×1120	35×35	70×70	280×280

At each detection stage, YOLOv3 uses three anchor boxes for detecting objects. Total nine anchor boxes are used with pre-calculated height and width from COCO dataset. For training the algorithm on traffic signs, we used K-mean clustering to calculate the anchor box size suitable for our training dataset. We also changed the number of boxes at each detection stage. In 1st detection stage, we used 5 boxes, and in 2nd and 3rd detection stage, we used 2, 2 anchor boxes respectively. By optimizing grid size, anchor box size, number of anchor boxes in each detection stage, and by training it on our self-made dataset, we achieved significant improvement in mAP value for three classes.

978-1-5386-7961-6/18 $31.00 © 2018 IEEE   247   ISOCC 2018

III. DATASET DETAILS

For getting good detection results, we made our own dataset for training the neural network. We collected videos from roads in Korea and extracted the frames. We manually selected frames with clear traffic signs of three classes. We manually annotated the selected frames to make training dataset. Our training dataset contain 3300 images of Korean roads. Each image is at different scene of road containing prohibitory, mandatory and danger classes of traffic signs. We ensured to include small, medium, and large size of traffic signs.

IV. EXPERIMENT DETAILS

We used Core i7 computer with nvidia Titan X GPU for our experiment. We continued training by varying learning rates, until the average loss reduced. After completing training, we did mAP analysis on different iteration values on validation data to avoid overfitting of training. We tested the algorithm on KTSD dataset containing 498 images. We also trained original YOLOv3 to compare the results.

V. RESULT COMPARISON

We tested our algorithm on KTSD dataset and compared the results with ground truth of given dataset. We calculated Intersection Over Union (IOU) of detected bounding box and ground truth bounding box to define true positives. We have set the threshold value of IOU as 40. Figure 4 shows that IOU value is above 40% and confidence score is 33.90% for danger class traffic sign. We also calculated the mAP values of original YOLOv3 and D-Patches algorithm.

Figure 5: Failure example of TS detector in occlusion situation

Table 2: Mean Average Precision (mAP) comparison

	D-Patches	Original YOLOv3	TS detector
Danger Class	88%	89%	91%
Prohibitory Class	78%	67%	90%
Mandatory Class	78%	66%	79%
mAP	81.19%	73.94%	86.61%

Table 3: CPU time comparison

	D-Patches on CPU	Original YOLOv3 on GPU	TS detector on GPU
Detection time on a frame of 1120×1120	2.5 Seconds	101 m-sec	119 m-sec

VI. CONCLUSION

By using TS detector, we can achieve better mAP with less computation time on challenging KTSD dataset. The mAP value of original YOLOv3 is less than that of the D-patches. However, after optimization of grid sizes for small traffic signs, it is significantly improved, so that the new TS detector outperforms both of YOLOv3 and D-Patches.

REFERENCES

[1] Y. Rehman, I. Riaz, X. Fan and H. Shin. Sneddon, "D-Patches: Effective Traffic Sign Detecion with Occlusion Handling," in *IET Computer Vision*, vol. 11, no. 5, pp. 368-377, 8 2017.

[2] J. Stallkamp, M. Schlipsing, J. Salmen and C. Igel, "The German Traffic Sign Recognition Benchmark: A multi-class classification competition," *The 2011 International Joint Conference on Neural Networks*, San Jose, CA, 2011, pp. 1453-1460.

[3] https://pjreddie.com/darknet/yolo/

Figure 4. Example of danger class prediction with IOU 74.51%

Design of Road Surface Lighting System for Rear Lamp using Automotive Ultrasonic Sensor

Donghee Han
Department of Automotive Convergence
Korea University
Seoul, Korea
dhhan430@korea.ac.kr

Hyo Bin Choi and Yong Sin Kim
School of Electrical Engineering
Korea University
Seoul, Korea
{chb2011, shonkim}@korea.ac.kr

Abstract— **There are quite a lot of accidents with pedestrians when backing a car. To prevent such accidents, recently, road surface lighting has appeared in the automotive industry. This technology is used to inform pedestrians that the vehicle is approaching by projecting light onto the road surface. In this paper, we propose the technology that changes a light pattern according to pedestrian's position to improve his visibility by linking existing automotive ultrasonic sensors and road surface lighting module.**

Keywords: Road surface lighting; Automotive Ultrasonic Sensor; Pedestrian Safety; Rear lamp;

I. INTRODUCTION

Conventional technologies such as rear camera or rear detection sensor for preventing rear-end collisions are designed to allow the driver to know and avoid the position of the pedestrians. However, in the case of road surface lighting, it is technique which informs the pedestrian that the vehicle is approaching. An important aspect of road surface lighting is how much it can improve pedestrian visibility. This technology is mainly used in front and rear lamps. In the case of a front lamp, many LED matrix technologies have recently appeared so that various road surface lighting patterns can be projected depending on various situations and improve the visibility of pedestrian. In the case of rear lamps, such a LED matrix technology is not applicable because the LED matrix technology is designed to ensure that the driver is able to efficiently have better visibility according to various driving situations [1]. But, the rear lamp only needs to perform a few functions such as notifying the following driver that you are going to stop or change direction. Therefore, the rear lamp should efficiently inform the pedestrians of vehicle access with a small number of light sources. In this paper, we propose the technique to change road surface light pattern according to pedestrians position by using a small number of light sources and existing automotive ultrasonic sensor.

II. CONVENTIONAL TECHNOLOGY

A. Road Surface Lighting

The road surface lighting in the rear of the car uses a fixed device for direct light [2]. These optical devices are classified into Fresnel Lens and Parabolic Reflector. As shown in Fig. 1,

Both methods mount the desired pattern film on the optics and project the pattern on the road surface. In this paper, we propose to use three direct light devices A, B, C with a pattern film as in Fig. 2. B is the light source responsible for the central area. A and C are the side area. The reason for using the three light sources is to improve the visibility of the pedestrian by changing the pattern area irradiated on the road according to his position.

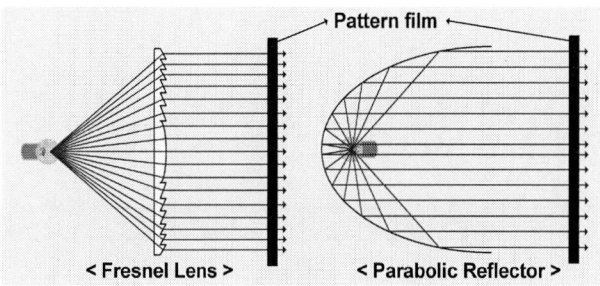

Fig. 1. A fixed device for direct light.

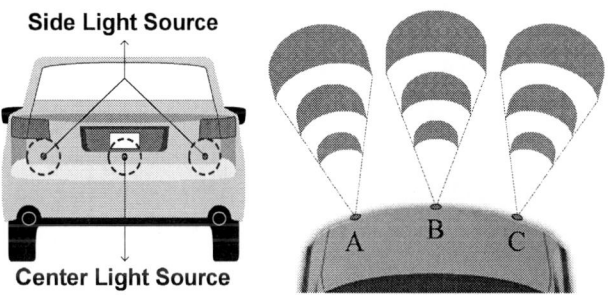

Fig. 2. Proposed road surface lighting module.

B. Automotive Ultrasonic Sensor

Generally, there are four ultrasonic sensors built into the rear bumper of a vehicle. Each sensor covers different areas and can detect objects within 5 meters as in Fig. 3 [3]. Also, These sensors can detect rod-shaped obstacles of 10 centimeters in diameter located within 2 meters and are robust to external noises such as rain [4]. We propose a technique to change the area irradiated on the road by the lighting module proposed in Fig. 2 depending on the position of the pedestrian using these existing sensors.

978-1-5386-7961-6/18 $31.00 © 2018 IEEE

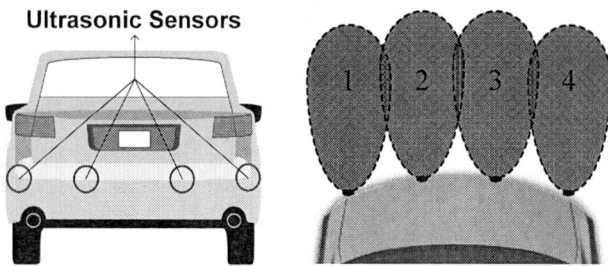

Fig. 3. Detecting area of automotive ultrasonic sensors.

III. PROPOSED SYSTEM

The system block diagram proposed in this paper is shown in Fig. 4. This system receives signals from the four ultrasonic sensors installed in the existing automobile and controls the light sources A, B, C through the MCU. If either sensor 2 or 3 detects the object, the MCU turns on the light source B to project the light pattern only in the central area as in Fig. 5(a). In addition, as shown in Fig. 5(b) and Fig. 5(c), if sensor 1 or 4 detects the objects, the MCU projects the light pattern on side areas by turning on source A or C. Of course, it's also possible that all the light sources A, B, C are turned on if there are objects in both central and side areas as in Fig. 5(d).

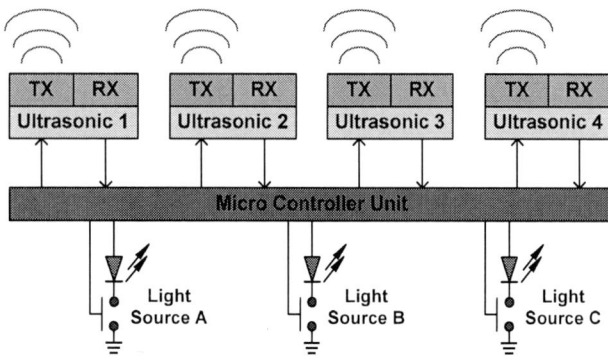

Fig. 4. Block diagram of the proposed system.

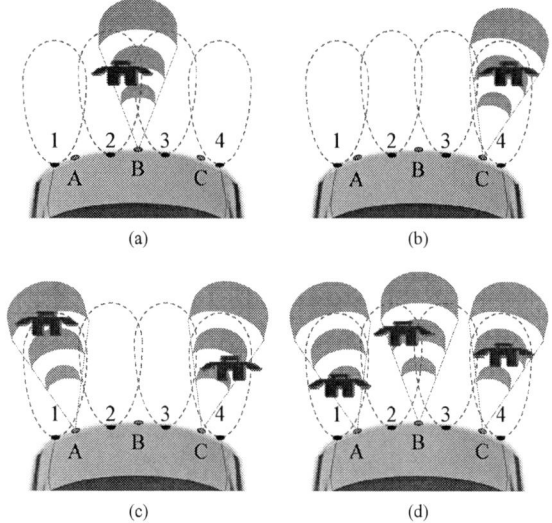

Fig. 5. Lighting pattern depending on pedestrian's position.

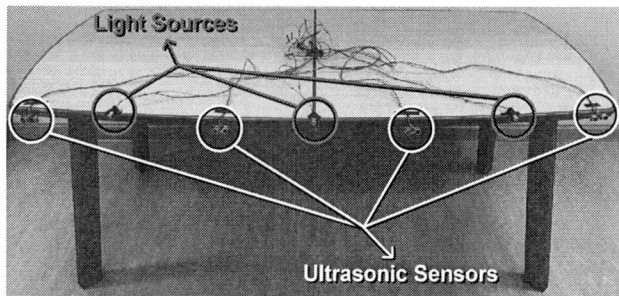

Fig. 6. Arrangement of light sources and ultrasonic sensors for experiment.

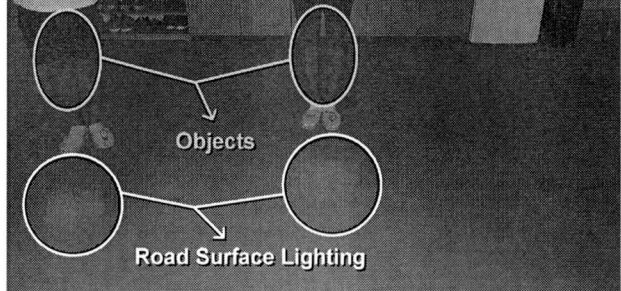

Fig. 7. Road surface lighting pattern by sensor detection.

IV. DISCUSSION AND SUMMARY

Almost all cars have built-in ultrasonic sensors on the rear, so the technology proposed in this paper can be easily applied. Also, controlling the light pattern according to the position of the pedestrian would be more helpful in improving his visibility than simply projecting the static light pattern.

ACKNOWLEDGMENT

This work is partly supported by the Technology development Program(S2522887) funded by the Ministry of SMEs and Startups(MSS, Korea) and also in part by the Basic Science Research Program through the National Research Foundation of Korea within the Ministry of Science, ICT and Future Planning under Grant NRF-2014R1A1A1003771.

REFERENCES

[1] Gordon Elger, Benno Spinger, Nico Bienen and Nils Benter, "LED Matrix Light Source for Adaptive Driving Beam Applications", Proceedings of the 63rd IEEE Electronic Components & Technology Conference, pp. 535-540, 2013.

[2] Shawn Knight, (October 11, 2017), "Mitsubishi's Safe and Secure Lighting system illuminates driver actions", Retrieved from https://www.techspot.com/news/71335-mitsubishi-safe-secure-lighting-system-illuminates-driver-actions.html.

[3] Jiaying Yu, Shengbo Eben Li, Chang Liu and Bo Cheng, "Dynamical Tracking of Surrounding Objects for Road Vehicles using Linearly-Arrayed Ultrasonic Sensors", Proceedings of IEEE Intelligent Vehicles Symposium, pp. 72-77, 2016.

[4] Manabu Ishihara, Makoto Shiina and Shin-nosuke Suzuki, "Evaluation of Method of Measuring Distance Between Object and Walls Using Ultrasonic Sensors", Journal of Asian Electric Vehicles, vol. 7, No. 1, pp. 1207-1211, June 2009.

Design of 3D Inductors for IoT Security

Bruce Kim
Department of Electrical Engineering
City University of New York
New York, New York USA
Bruce.kim@ieee.org

Sang-Bock Cho
Electrical Engineering Department
Ulsan University
Ulsan, South Korea

Abstract— This paper describes the design of through-silicon via (TSV)-based inductors for security of Internet-of-Things (IoT). We designed 3D inductors with security hardware using physically unclonable function (PUF) circuit. The secure 3D inductor could be tuned to desirable frequency by using MEMS switches.

Keywords; MEMS, PUF, 3D TSV, 3D Inductors.

I. INTRODUCTION

Internet-of-Things are becoming part of our everyday lives in consumer electronics as well as medical and avionics. IoT devices make up an information infrastructure; however, this infrastructure could be in danger due to cyber hardware uncertainties related to encryption, verification, Trojan hardware, and remote hacking. These uncertainties can lead to an unsafe regulatory environment, an unstable economy, increased cyber-security threats, and other major issues. Therefore, we need to protect the hardware from the circuit design phase to the manufacturing. [1].

II. APPROACH

A. Inductor Tunability

Tunable 2D on-chip inductors using various configurations of MOSFET switching have been reported. The conventional designs [2,3] used for switch inductors are illustrated in Figure 1. In Pham's model [2], 2n numbers of control units are augmented for additional n numbers of inductor segments. Off-state non-zero leakage current is one of the major problems with MOSFET-based switches. Due to the leakage current from drain to source, there is always unwanted power loss, which can be recovered using MEMS-based switches. This is due to MEMS high off-state impedance. Therefore, MEMS switches are highly desirable for making tunable inductors using configuration shown in Figure 1.

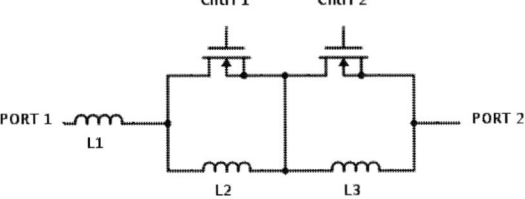

Figure 1. Conventional model 1 for variable inductance [2].

We have used MEMS switches for tunable inductors. The model presented in this paper optimizes the number of switches within the inductor segments and the area required to implement MEMS cantilevers. The model also completely cuts off the target inductor from the residual circuit at the off state.

B. TSV-Based Spiral Inductor

To model a TSV-based inductor, the key control factors are the diameter and length of the TSV; the distance between two unit inductors; and the length, width, and thickness of the metal interconnects. In most cases, the diameter and length of the TSV and the thickness of the metal interconnects are limited by the foundry process. However, a thickness up to 8 µm for the metal layer can be considered in specific cases [3]. Due to mechanical stress, spacing D has a certain minimum threshold limit [4].

C. MEMS Switch Design

A magnetically actuated switch was chosen for its high speed and insulation capability in the presence of high voltages. The design consists of a cantilever bridge structure, a permanent magnet, and an electrostatically actuated coil magnet. The working principle of the switch can be described in terms of the magnetic force acting on it when dc voltage is applied. The total force exerted on each switch is largely dependent upon the intensity of the magnetic field produced by the external permanent magnet and electromagnetic coil, the volume of the ferromagnetic material embedded on the bridge, and the spring constant of the flexible bridge. Figures 2 shows a top view of fabricated magnetically actuated MEMS switch [5].

Figure 2. 4X4 MEMS switch array in closed position [5].

The tunable inductor is tested on a low-noise amplifier (LNA) circuit to compare the performance of the MEMS and MOSFET switches. The LNA is designed for applications from 3 to 4 GHz. Our results show that the MEMS switch's performance is better than that of the high-leakage off-state MOSFET switch in terms of gain and reflection coefficient. Figure 3 illustrates the inductance of 3D TSV inductors in three different switch positions. We achieved up to 6nH in 3D.

978-1-5386-7961-6/18 $31.00 © 2018 IEEE 251 ISOCC 2018

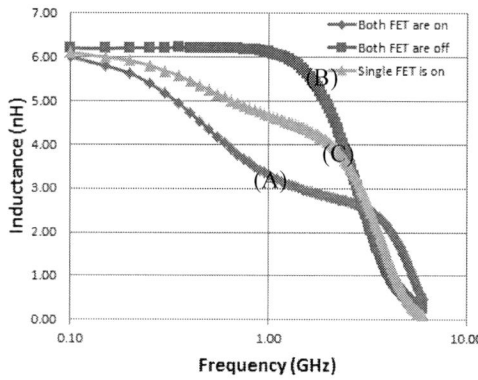

Figure 3. Inductance with N=12 for different switching operations (A) Both FETs are on; (B) Both FETs are off; (C) A single FET is on.

III. TUNABLE IOT SECURITY USING PUF CIRCUITS

In communications hardware, frequency tunability is highly desirable for multi-band communications. We were able to achieve tunable LNA using MEMS switches with TSV inductors. However, tunable inductance through MEMS switches involves security vulnerability. In order to mitigate this problem, we designed a physically unclonable function (PUF) circuit to protect the hardware security of the 3D inductors used in low-noise amplifiers. PUF circuits provide unique challenge-response digitized bits that depend on temperature, supply voltage, and EMI.

We used unique PUF response bits to tune the inductance circuit shown in Figure 1. The unique response bits worked together with the circuit combination to make up the desired inductance. The ring oscillator (RO) PUF is designed and prototyped. The prototype was assembled with discrete parts on a 3M breadboard, and the concept was tested for feasibility of the circuit. We measured unique responses based on the challenge bits. These PUF circuits could be used to control the MEMS switches to change the tunability of our LNA by correcting the inductance values in the desired communication band. This technique will ensure the security of IoT devices. We did CMOS layout of the entire PUF circuit and Fig. 4 is part of the entire PUF layout.

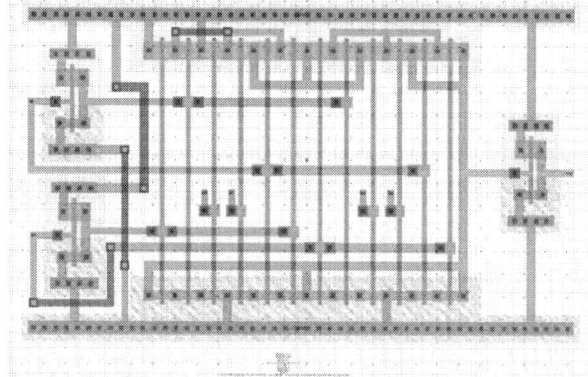

Figure 4. CMOS layout of a 4-input MUX.

IV. MATERIAL DESIGN

Ferrite materials such as MnZn-ferrite, and NiZn-ferrite became an alternative choice due to their high resistivity in 3D tunable inductors. The advantage of NiZn-ferrite over other ferrite materials is that it offers low magnetic loss tangent at high frequency. Permeability up to 450 can be achieved at 10 MHz using ferrite materials. Certain compositions of Ni and Zn in NiZn-ferrite can offer magnetic loss tangent as low as 0.027 at 1-1.5 GHz. If we use nickel, to decrease the eddy current loss, the core needs to be patterned to prevent large eddy current loops. On the other hand, with NiZn-ferrite materials, there is no need to pattern as the resistivity is very high (10^6 Ω-cm). There are already a few fabrication techniques to deposit NiZn-ferrite materials: spin spray coating [6], magnetron sputtering and Chemical Vapor Deposition. We chose NiZn-Ferrite over nickel due to its compatibility for fabrication and advantageous material properties. For the TSVs and metal interconnects, copper was considered.

IV. Conclusions

In this paper, we proposed a secure tunable inductor design for Internet-of-Things. This was accomplished with 3D TSV inductor with MEMS switch configuration by applying response bits from the PUF circuit. We observed varying inductances with MEMS-based switches. We believe that many IoT devices could be protected by using PUF circuits to tune inductors in the RF front-end circuits.

Acknowledgement

This work is supported by the 2014 Fund of the University of Ulsan. We would like to thank former students Saikat Mondal, Daniel Hirsh, and Jonathan Gamboa for their work on the 3D models and data.

References

[1] U. Tida et al., "Through-silicon-via inductor: Is it real or just a fantasy?" 19th Asia and South Pacific Design Automation Conference (ASP-DAC), pp. 20-23 Jan. 2014.

[2] P. Piljae, C. Kim, P. Y. Mun, S. D. Kim and H. K. Yu, "Variable inductance multilayer inductor with MOSFET switch control," Electron Device Letters, IEEE, vol.25, no.3, pp.144,146, 2004.

[3] D. P. Khoa, K. Okada and K. Masu, "On-chip variable inductor using MOSFET switches," European Microwave Conference, vol.2, pp. 4-6, 2005.

[4] C. Bousey, "Design and realizations of integrated filters on silicon for TV on mobile," in Proc. Of National Micro-Wave Days, (JNM 07), pp.1-4, 2007.

[5] R. Kasim, B. C. Kim and J. Drobnik, "Advanced MEMS for high power integrated distribution systems," International Conference on MEMS, NANO and Smart Systems, pp. 247-254, 2005.

[6] O. Obi, M. Liu, J. Lou et. al., "Spin-spray deposited NiZn-ferritefilms exhibiting μ_r'>50 at GHz range", Journal of Applied Physics, vol. 109, pp.07E527-3, 2011.

Low Power Near-sensor Coarse to Fine XOR based Memristive Edge Detection

Kamilya Smagulova, Aidana Irmanova, Alex Pappachen James
School of Electrical and Computer Engineering
Nazarbayev University, Astana, Kazakhstan
Email: apj@ieee.org

Abstract—**In this paper, we propose XOR based memristive edge detector circuit that is integrated into a near sensor log-linear CMOS pixel. Memristor threshold logic was used to design NAND gates, which serve as a building block for XOR gates. For validation of proposed circuit functionality hardware simulation of logic gates with a pixel pair was conducted using TSMC 0.18um technology and system-level simulation of the proposed circuit using SPICE models. The proposed method operates in low power and takes a small area on chip. The power consumption of one pixel is 1.16uW and total area 36.72 um^2 without photosensing component. The power consumption of NAND circuit is 1.11pW and total area 32.4um^2.**

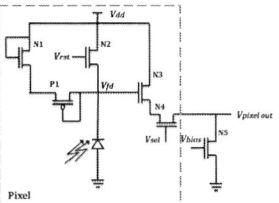

Fig. 1. Schematic of the log-linear pixel circuit

I. INTRODUCTION

In this work, we propose analog circuit implementation of near-sensor edge detection which provides faster computation without the need for analog-to-digital conversion. The core of the edge detection engine is based on XOR operation that was described in [1]. The proposed neuromorphic analog design presents novel use of memristive threshold logic gates for image processing. In the following section, the proposed architecture of the circuit that consists of image sensing circuit and memristive XOR logic gates are presented. Further, the simulation results of these components are discussed, comparing to the competing approaches of edge detection operation.

II. PROPOSED ARCHITECTURE

A. Active pixel sensor circuit

Depending on configuration, CMOS active pixel sensors (APS) is capable of yielding either linear or logarithmic output characteristics. Standard 3T linear APS has good sensitivity at low light illumination with DR about 80dB. While logarithmic 3T APS show better performance at high illumination with wider DR about 100dB. One of the approaches to extend pixel dynamic range is to combine log-linear response.

Fig.1 shows a wide dynamic range 5T active pixel with log-linear output characteristic proposed by [2]. n-type MOSFET transistors N_1 and N_2 in the circuit are used to convert generated photodiode current into voltage producing logarithmic or linear responses respectively. PMOS transistor P_1 serves as a switch to control the operation mode depending on illumination intensity. At low illumination conditions pixel operates as linear APS, whereas with high light intensity it produces logarithmic output. The resulting voltage at floating diffusion node V_{fd} is amplified by source follower N_3 and N_4 and read by transistor N_5.

B. Memristive XOR gate

XOR logic can be implemented using NAND gates as in Fig 2a. In [3], authors proposed a configuration of resistive memory threshold logic gate cell circuit as in Fig. 2b, that can be adjusted to operate as NAND or NOR gates. The gate was set up in the voltage divider configuration followed by a CMOS inverter. In this circuit memristors are used instead of resistors due to its small size and low leakage current. The mode of logic gate depends on the input of the third M_3 memristor: If the input is grounded then it is NAND, while driving it to High Voltage sets the mode to NOR. But the same NAND mode of the gate can also be used as NOR gate if the threshold of inverter will be adjusted to specific level. In Table I the analog output V_o of the resistive divider component of the gate are shown. To set the gate to NAND mode it is required to set the threshold of the inverter $V_{low} + (V_{high} - V_{low})/2$, while for NOR it is sufficient to set it $V_{low} + (V_{high} - V_{low})/4$.

TABLE I
ANALOG OUTPUT V_o OF MEMRISTIVE LOGIC GATE

V_{in1}	V_{in2}	V_{out}
V_{low}	V_{low}	$2V_{low}/3$
V_{low}	V_{high}	$(V_{low}+V_{high})/3$
V_{high}	V_{low}	$(V_{high}+V_{low})/3$
V_{high}	V_{high}	$2V_{high}/3$

Fig. 2b shows the NAND logic gate. To operate in NAND mode memristors are set to the same values $R_1=R_2=R_3=R$ and the inverter is utiilized to convert V_o to binary state V_{out} as in Fig 3.

978-1-5386-7961-6/18 $31.00 © 2018 IEEE

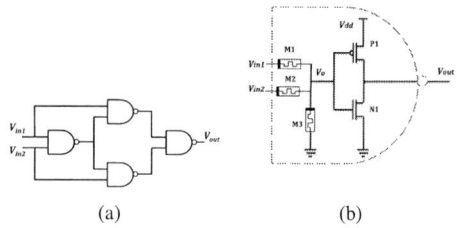

(a)   (b)

Fig. 2. Building (a) XOR Logic gate using (b) memristive NAND gates

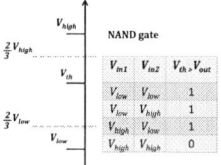

Fig. 3. Setting the threshold value for NAND operation

III. RESULTS AND DISCUSSION

A. Hardware simulation

Fig. 4 shows output voltage variation of the CMOS pixel for photocurrent between 2.5fA to 250nA. Corresponding generated voltage range varies from 50mV to 650mV. DR of the given pixel is of the order of 140dB and utilization of five transistors allows to keep high fill factor.

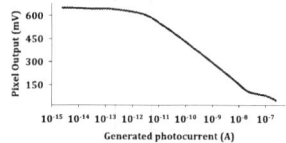

Fig. 4. 5T Pixel sensor output

The output response of memristive XOR circuit for the two pixels outputs is shown in the Fig.5 which demonstrates functionality of XOR gate for analog input values.

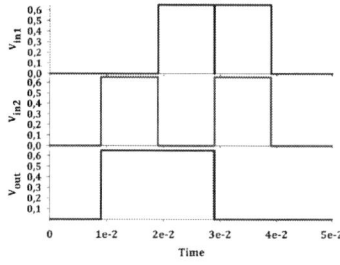

Fig. 5. Output response of XOR circuit

Simulations were performed in SPICE using 0.18 μm TSMC CMOS technology model and HP memristor model [4]. The power consumption of one pixel is $1.16\mu W$ and total area 36.72 μm^2 without photosensing component. The power consumption of NAND circuit is 1.11pW and total area 32.4 μm^2.

TABLE II
COMPARISON OF QUALITATIVE MEASURES OF EDGE DETECTION
OPERATORS AT DIFFERENT NOISE CONDITIONS

Measure	XOR		Canny		Sobel	
	Noise type					
	Speckle	Salt and Pepper	Speckle	Salt and Pepper	Speckle	Salt and Pepper
PSNR	17.2840	25.4213	18.2223	24.1921	15.9374	15.8522
MSE	0.0187	0.0029	0.0151	0.0038	0.0255	0.0260

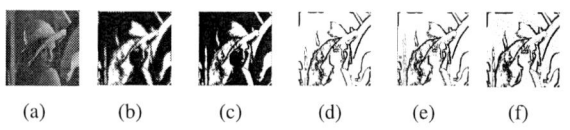

(a)   (b)   (c)   (d)   (e)   (f)

Fig. 6. Original image (a); binarized images with a shift of 15 pixels (b) and 3 pixels (c) detected edges of original b and c with proposed XOR circuit (d) detected edges of b and c with added *Speckle* noise (e) detected edges of b and c with added *Salt and Pepper* noise (f)

To demonstrate the functionality of memristive XOR circuit for edge detection operation the system simulation was performed in MATLAB. In Table II the comparison of qualitative measures for memristive XOR circuit and conventional edge detection operators are given. As it can be seen, near-sensor hardware edge detection circuit based on XOR operation demonstrates competitive results compared to Canny and Sobel operators, outperforming both of the approaches in edge detection of images with *Salt and Pepper* type of noise. Figs.6 (a-g) illustrate performance of XOR circuit for edge detection task with different noise types.

IV. CONCLUSION

In this paper, XOR-based edge detecting circuit design is proposed. The edge detection was performed with CMOS active pixel sensor array with log-linear response followed by XOR memristive circuits that were constructed from memristive NAND threshold logic gates. Presented approach provides near-sensor decision making which enables fast computation at low power cost and small on-chip area due to use of hybrid CMOS-memristor devices. Comparison of experimental results show that presented circuit design provides competitive performance compared to conventional operators. Although XOR operation stands behind some of the conventional edge detecting filters, its straightforward implementation on hardware and usability for data processing in analog domain make it more attractive compared to digital filters.

REFERENCES

[1] Diaconu A. V., Sima I. Simple, XOR based, image edge detection //Electronics, Computers and Artificial Intelligence (ECAI), 2013 International Conference on. IEEE, 2013. . 1-6.

[2] Abedin M. I., Islam A., Hossain Q. D. A self-adjusting Lin-Log active pixel for wide dynamic range CMOS image sensor //Telecommunications and Photonics (ICTP), 2015 IEEE International Conference on. IEEE, 2015. . 1-4.

[3] Maan A. K., James A. P. Voltage controlled memristor threshold logic gates //Circuits and Systems (APCCAS), 2016 IEEE Asia Pacific Conference on. IEEE, 2016. . 376-379.

[4] Abdalla, Hisham, and Matthew D. Pickett. "SPICE modeling of memristors." Circuits and Systems (ISCAS), 2011 IEEE International Symposium on. IEEE, 2011.

A Ka-band low noise amplifier in 0.15μm GaAs E-mode pHEMT technology

Jeongsoo Park, Jinhyun Kim and Jeong-Geun Kim
Department of Electronic Engineering
Kwangwoon University
Seoul, Korea
junggun@kw.ac.kr

Abstract— This paper presents Ka-band low noise amplifier in 0.15μm GaAs E-mode pHEMT Technology. The low noise amplifier has three-stage which used the microstrip line for low loss. The low noise amplifier input stage is a common source configuration for noise optimal matching. The proposed Ka-band Low Noise Amplifier is controlled only with single positive supply voltage. The measured result of the low noise amplifier shows the gain of 25.8 dB at 30 GHz and the simulated noise figure of 2 dB at 30 GHz. The proposed Ka-band low noise amplifier chip size is 1.7 x 0.8 mm².

Keywords : Ka-band; low noise amplifier; microstrip line; 0.15μm GaAs E-mode pHEMT Technology

I. INTRODUCTION

Currently, there is an increasing demand for Ka-band low noise amplifier for satellite and 5G communication. Most of silicon based beamforming IC shows low output power and high noise performance. The GaAs front-end IC is the solution to improve linearity and low noise performance [1-3]. Fig. 1 shows the system level diagram of the beamforming system with GaAs front-end IC. In this paper presents a Ka-band low noise amplifier in 0.15μm GaAs E-mode pHEMT Technology. The low noise amplifier has three-stage which used the microstrip line for low loss. The low noise amplifier measured gain of 25.8 dB at 30 GHz and the simulated noise figure of 2 dB at 30 GHz.

II. DESIGN OF THE KA-BAND LOW NOISE AMPLIFIER

We have designed a three-stage Ka-band low noise amplifier and it was considering the important conditions such as noise figure, in/output match, gain and stability in desired frequency band. The three-stage design is consisting of 1st low noise input stage and 2nd, 3rd gain stages. Fig. 2 shows the Topology of the Ka-band low noise amplifier which used the microstrip line for low loss. The low noise amplifier input stage is a common source configuration for noise optimal matching. For the matching networks, we preferred transmission lines because the simulation of an inductor requires a detailed knowledge of the substrate as well as structures in close proximity to the inductor that contribute to return current paths. The transmission lines are better suitable to realize small inductances in this frequency band. Also, the lengths are not prohibitively long and bending them further saves more area. The transmission lines have the same reactance at multiple frequencies. At very low frequencies, the transmission lines behave like a small inductor. So, it care should be taken to use a matching network of the transmission line type to avoid operation at undesired frequency bands. It can provide unexpected matching in undesired frequency bands under these conditions and the amplifier will have unexpected gain. Because of the gain in the undesired band around 5GHz in the simulation, a 5k Ohm resistors was added to the gate bias line to de-Q the matching network and completely defeat the un-wanted gain at 5GHz. The LNA satisfies the unconditionally stable condition in simulation results in various bias conditions for the maximum oscillation frequency [4], [5].

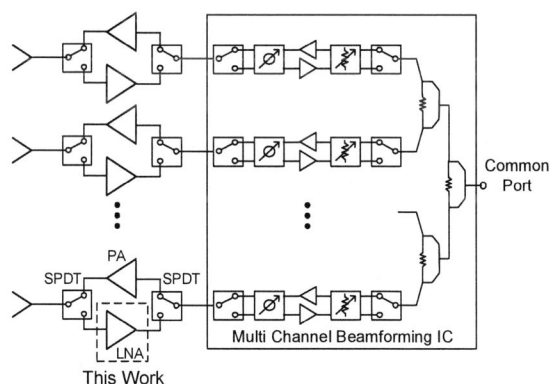

Fig. 3. Block diagram of the beamforming system with GaAs front-end IC

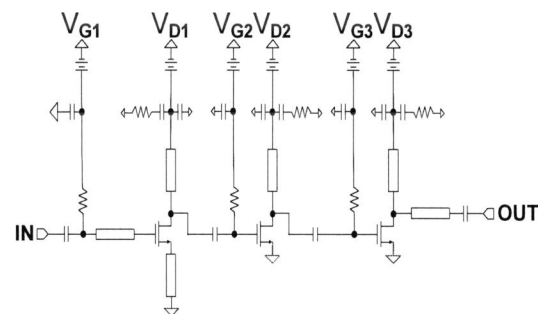

Fig. 4. Topology of the Ka-band low noise amplifier

This work was supported by Institute for Information and Communications Technology (IITP) grant funded by the Korea government (MSIT) (17-110-76-584, Development of Communication-Sensing Converged B5G Millimeter Wave System)

978-1-5386-7961-6/18 $31.00 © 2018 IEEE

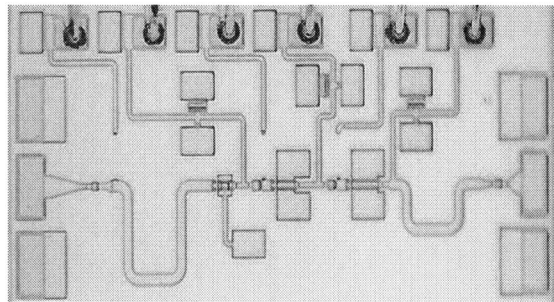

Fig. 3. Photograph of the low noise amplifier

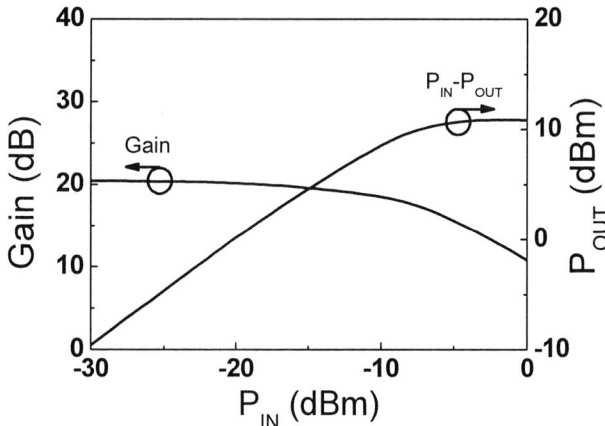

Fig. 5. Measured power characteristics of the low noise amplifier at 28 GHz

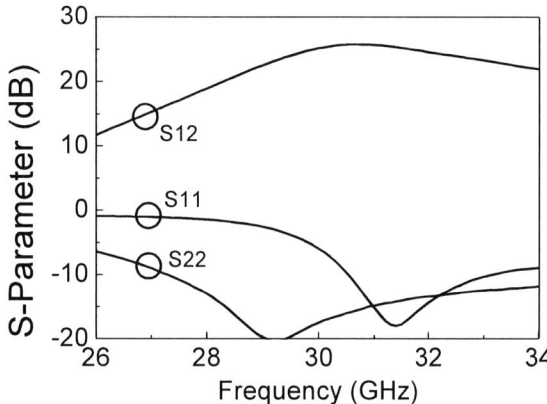

Fig. 4. Measured S-parameter of the low noise amplifier

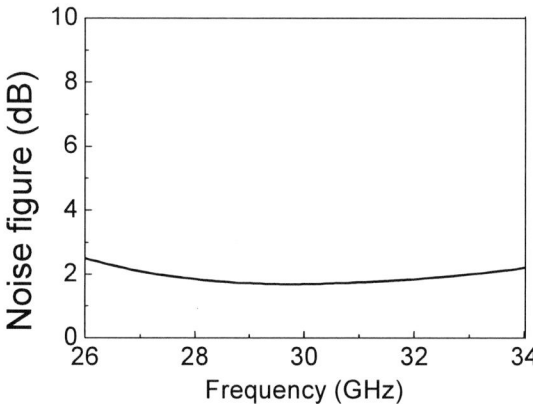

Fig. 6. Simulated noise figure of the low noise amplifier

III. MEASURED RESULTS OF THR 28 GHZ FRONT-END IC

The Ka-band low noise amplifier is fabricated in 0.15 μm GaAs E-mode pHEMT technology. Fig. 3 shows the microphotograph of the fabricated Ka-band low noise amplifier. The chip size is 1.7 x 0.8 mm². The Ka-band low noise amplifier is tested with spectrum analyzer and network analyzer with an SOLT calibration technique. Fig. 4 shows the measured small signal S-parameter of Ka-band low noise amplifier. The Ka-band low noise amplifier measured gain of 25.8 dB at 30 GHz. Fig. 5 shows the measured large signal power characteristics of Ka-band low noise amplifier at 28 GHz. The Ka-band low noise amplifier can be with an input P1dB of -14 dBm. Fig. 6 shows the simulated noise figure of Ka-band low noise amplifier. The Ka-band low noise amplifier simulated noise figure of 2 dB at 30 GHz. The amplifier consumes 25 mA of current from a 2 V supply voltage.

IV. CONCLUSION

This paper presented Ka-band low noise amplifier in 0.15μm GaAs E-mode pHEMT Technology. The low noise amplifier has three-stage which used the microstrip line for low loss. The Ka-band low noise amplifier measured gain of 25.8 dB at 30 GHz and the input P1dB of -14 dBm. The Ka-band low noise amplifier simulated noise figure of 2 dB at 30 GHz. Since the proposed Ka-band low noise amplifier with low loss and low noise performance, it can be applied satellite and 5G wireless communication system.

REFERENCES

[1] D.-W. Kang, J.-G. Kim, B.-W. Min, and G. M. Rebeiz, "Single and four-element Ka-band transmit/receive phased-array silicon RFICs with 5-bit amplitude and phase control", IEEE Trans. Microw. Theory Tech., vol. 57, no. 12, pp. 3534–3543, Dec. 2009.

[2] Y. Yuan, Y. Fan, Z. Chen, L. Li and Z. Yang. "Ku band 2 watt TR chip for phased array based on GaAs technology", IEICE Electronics Express, vol. 13, no. 7, pp 1-6, March 2016.

[3] J. Curtis, H Zhou and F Aryanfar. "A Fully Integrated Ka-Band Front End for 5G Transceiver", IMS Dig. Tech. Papers, May 2016.

[4] E. Adabi, B. Heydari, M. Bohsali and A. M. Niknejad. "30 GHz CMOS low noise amplifier", RFIC Symp. Dig., June 2007.

[5] M.A.T.Sanduleanu, G. Zhang, and J. R. Long, "31-34GHz Low Noise Amplifier with on-chip microstrip Lines and Inter-stage Matching in 90nm Baseline CMOS", RFIC Symp. Dig., July 2006.

978-1-5386-7961-6/18 $31.00 © 2018 IEEE

A 28-GHz 28.5-dBm power amplifier using 0.15-μm InGaAs E-mode pHEMT technology

Hui Dong Lee, Sunwoo Kong, Bonghyuk Park, Kwang Chun Lee
Future Mobile Communication Research Division
Electronics and Telecommunications Research Institute
Daejeon, Republic of Korea

Jeong-Soo Park, Jeong-Geun Kim
Department of Electronic Engineering
Kwangwoon University
Seoul, Republic of Korea

Abstract— **This paper describes the design of a 28-GHz 28.5-dBm power amplifier using a 0.15-μm InGaAs E-mode pHEMT technology. We have sought a method to obtain the required output power through the characteristics of the unit transistor provided by the manufacturer. For this purpose, the PA circuit is configured to effectively combine the output signals of eight unit transistors. As a result of the verification, 28.5-dBm output was obtained at 28-GHz and the maximum efficiency was more than 24.5%. The power amplifier draws 335 mA under a 6.4 V supply at 28.5-dBm output.**

Keywords; 28 GHz; 5G; millimeter wave; power amplifier; pHEMT;

I. INTRODUCTION

The RF technology for the 5G mobile communication is being actively implemented at millimeter wave band in the world due to the limitation of the bandwidth in the conventional below 6 GHz mobile communication system [1-4]. The frequency around 28-GHz has advantages of light weight of antennas and transistors because of the short wavelength, and it is advantageous in that gigabit data transmission can be achieved due to wide bandwidth. A 5G technology with a phased array antenna is being studied as a technology to improve the communication efficiency with the system [5-6]. In response to this demand, the 28-GHz band MMIC capable of high-performance and low-cost mass production has been developed. Particularly, it is urgent to develop an efficient power amplifier (PA) that consumes the largest DC power among the RF building blocks.

II. PA CIRCUIT DESIGN

To develop the MMIC of the 28-GHz band, it is important to select the transistor above all among GaAs MESFETs, HEMTs, pHEMTs, HBTs, etc. Considering the development of MMICs operating up to the 28-GHz band, the GaAs E-mode pHEMTs, which provide excellent noise, gain and frequency characteristics, would be the most suitable transistors without using a negative supply voltage. We have designed to achieve a power amplifier with a 28-dBm output and a gain over 23 dB. Figure 1 shows the structure of the proposed power amplifier. In this design, a binary tree combiner structure with a symmetrical output stage is used, and the chip size is minimized by applying a bias to the combine portion.

In order to achieve more than 28-dBm output from the transistors provided by WIN semiconductors, a 4-finger 75-μm transistor was determined as a unit transistor. The load-pull

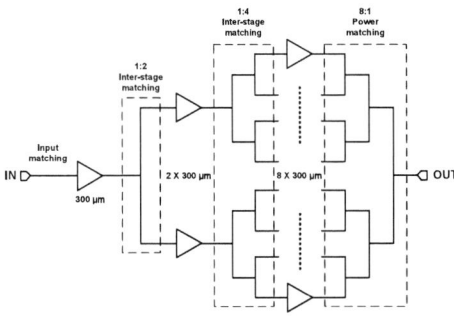

Figure 1. Structure of the proposed power amplifier

simulation result of the unit transistor presents that the optimal output impedance is 17.3 + j6.1, the output P1dB is around 21 dBm, and the saturated power is reached to 23 dBm at 28-GHz. The gain at the 1 dB compression point and the PAE at the point are 9 dB and 33.8%, respectively. From the simulation results of the unit transistor, the 3-stage PA topology was adopted and the 8 unit transistors at the final stage were combined in parallel. This topology is because a gain of more than 23 dB and an output of more than 28 dBm can be obtained even if a 1 dB loss in the matching circuit is considered.

The matching circuit in the output stage is critical because it is important that the power amplifier draw the maximum output of the transistor. Therefore, the matching of the power amplifiers is performed in the order of the output stage, the intermediate stage, and the input stage. The matching circuit used a low-pass filter type due to nonlinearity of transistor can be eliminated. In the case of a power transistor, the input and output impedances are small because several fingers are connected in parallel. Therefore, it is difficult to match the small input impedance and the output impedance to 50 ohms with a wide bandwidth. To solve this problem, matching is performed using a multi-stage low-pass filter. In order to avoid electro-migration of the metal used in the drain, since a large amount of current flows into the drain because eight transistors are combined in the output terminal, a microstrip line having a width slightly larger than the maximum current density. For the sable PA operation, the oscillation was suppressed by using a resistance of several ohms in the gate and drain bias lines.

Final simulation results present that input and output stages are matched to less than -10 dB at 27.5 GHz to 28.5 GHz. The average output power is equal to 24 dBm. The efficiency at the maximum output power was 30%, but the efficiency at the

Figure 2. Chip photograph and measurement module

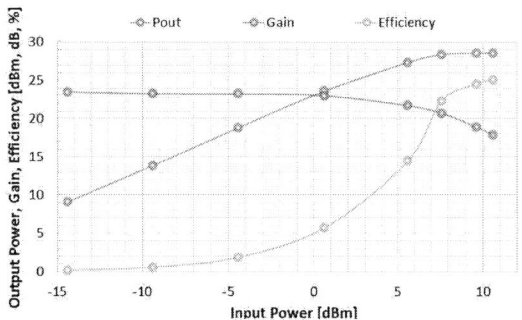

Figure 3. Measurement results of the implemented PA

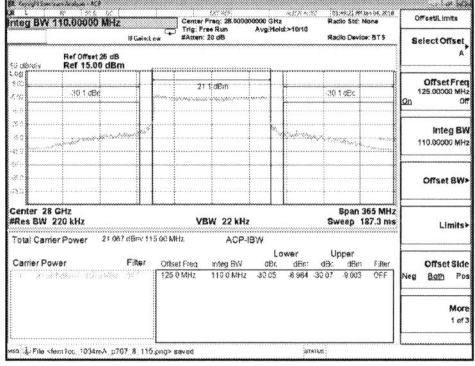

Figure 4. ACLR performance of the implemented PA

average power was simulated at 10.5%. The IMD3 characteristic was simulated at -29.2 dBc at the average power.

III. CHIP FABRICATION AND MEASUREMENT RESULTS

This 28-GHz power amplifier has been implemented in an E-mode pHEMT process. The overall MMIC size of the 28-GHz power amplifier is 2.3 mm x 2.2 mm. The decoupling capacitance was added to the gate and drain bias pads to remove low frequency oscillations from the DC bias and the bias was applied through EMI(electromagnetic interference) filter. Figure 2 shows the fabricated 28-GHz power amplifier chip and measurement module. The size of the manufactured 28-GHz PA module is 3 cm x 4 cm.

On-wafer measurements for the S-parameter analysis were performed using GSG probes of 150 μm pitch. S11 and S22 were less than -10 dB and S21 was around 24 dB at 28 GHz. Because the output power measurement cannot be completed with the GSG probe, a 28-GHz power amplifier module was manufactured. The input and output is connected to the 50 ohm line of a PCB board using the wire-bonding, and the input and output ports are configured using a K-connector. In order to measure Pin-Pout characteristics of the manufactured module, the power was applied by using Keysight E8267D. Since the output power is high, a 20 dB attenuator is connected to the output stage. The output power was measured using a spectrum analyzer (Agilent N9030A). Figure 3 shows the output power, gain, and PAE measurement results at 28-GHz. The Pin-Pout characteristics at 28-GHz were measured as the output P1dB of 25.5 dBm and the saturated power of 28.5 dBm.

The implemented PA was tested using an LTE signal with a 125 MHz bandwidth and 7 dB PAPR (peak to average power ratio) at 28 GHz. Figure 4 shows the results of ACLR measurements. The output powers of both adjacent channels are less than -30 dBc at the output power of 21 dBm.

IV. CONCLUSION

In this study, we developed a 28-GHz power amplifier MMIC. To increasing output power, the amplifier connects several transistors in parallel. For the size reduction, a symmetrical binary-tree power combiner was used and the PA bias line was shared at the combine portion. The size of the PA is 2.3 x 2.2 mm^2 and the size of the PA module is 3.0 x 4.0 cm^2. The small signal gain is 23.5 dB and the input and output matching is around -10 dB. The output P1dB is 25.5 dBm and the saturated power is 28.5 dBm. The maximum PAE is reached to 24.5% and the PA linearly operated to 21 dBm output power with -30 dBc ACLR performances.

ACKNOWLEDGMENT

This work was supported by Institute for Information & communications Technology Promotion(IITP) grant funded by the Korea government(MSIT). (No.2017-0-00409, Development on millimeter-wave beamforming IC for 5G mobile communication)

REFERENCES

[1] M. E. Leinonen, et al., "28 GHz Wireless Backhaul Transceiver Characterization and Radio Link Budget," *ETRI Journal*, vol.40, no.1, pp. 89-100, Feb. 2018.

[2] S. Kong, et al., "A 21.9-dB Gain 18.9–35.9-GHz low noise amplifier using InGaAs E-mode 0.15-um pHEMT technology," *IEEE Asia Pacific Microwave Conference*, pp.1203-1206, Nov. 2017.

[3] H. D. Lee, et al., "A low-power 50-GHz LC-VCO in a 65-nm CMOS technology," *IEEE Asia-Pacific Microwave Conference*, vol. 3, pp. 1–3, Nov. 2015.

[4] J.-N. Lee, et al., "Dual-Polarized Small Base Station Antenna Integrated RF Module Applicable to Various Cell Environments for Next-Generation Mobile Communication Service," *ETRI Journal*, vol.39, no.3, pp. 383-389, June 2017.

[5] J. D. Dunworth, et al., "A 28GHz Bulk-CMOS dual-polarization phased-array transceiver with 24 channels for 5G user and basestation equipment," *IEEE International Solid-State Circuits Conference*, pp. 70-72, Feb. 2018.

[6] J. Han, et al., "A Ka-band 4-ch bi-directional CMOS T/R chipset for 5G beamforming system," *IEEE Radio Frequency Integrated Circuits Symposium*, pp.41-44, June 2017.

A CMOS Rectifier with 72.3% RF-to-DC Conversion Efficiency Employing Tunable Impedance Matching Network for Ambient RF Energy Harvesting

Donggu Lee[1], Taejong Kim[1], Sinyoung Kim[2], Kanghyeon Byun[1], and Kuduck Kwon[1]

[1] Department of Electronics Engineering, Kangwon National University
Chuncheon-si, South Korea
kdkwon@kangwon.ac.kr

[2] Samsung Electronics Co. Ltd., Hwaseong,
Gyeonggi-Do 443-742, South Korea

Abstract — This paper presents a CMOS RF-to-DC converter with a tunable impedance matching network to widely harvest RF energy of 3G/4G cellular low-band frequency range. The proposed converter consists of the differential-drive cross-coupled rectifier and the matching network with a 4-bit capacitor array. The proposed converter is designed using 130nm CMOS process. It has a peak RF-to-DC conversion efficiency of 72.25%, 64.97 %, and 66.28 % at 700 MHz, 800 MHz, and 900 MHz with a load resistance of 10kΩ, respectively.

Keywords— *CMOS, RF-to-DC converter, cross-coupled rectifier, RF harvest, impedance matching, power conversion efficiency*

I. INTRODUCTION

Recently, self-sustainable low-power electronic devices require to extend battery life or to avoid the needs of a battery. RF energy harvesting technology is the main key to make self-sustainable low-power electronic devices feasible. Ambient RF signals such as TV, GSM, WCDMA, LTE, FM, and Wi-Fi are promising sources due to its availability and the amount of power present. RF energy harvesting over a wide frequency band improves conversion efficiency. Therefore, the wideband power matching is required. In this paper, a CMOS RF-to-DC converter with a tunable impedance matching network is presented to widely harvest RF energy of 3G/4G cellular low-band frequency range from 700 MHz to 900 MHz.

II. PROPOSED RF-TO-DC CONVERTER

The block diagram of the proposed RF-to-DC converter, which consists of a tunable impedance matching network and a rectifier is shown in Fig. 1. The proposed impedance matching network can perform power matching for the frequency range from 700 MHz to 900 MHz with a 4-bit capacitor array, which is switched by a digitally controlled signal. The rectifier employs a differential-drive cross-coupled topology as shown in Fig. 2 [1]. The cross-coupled topology can operate at low input power and achieve a high peak PCE. The resistive and imaginary components of the rectifier vary with the input power. To optimize the RF-to-DC conversion efficiency, the

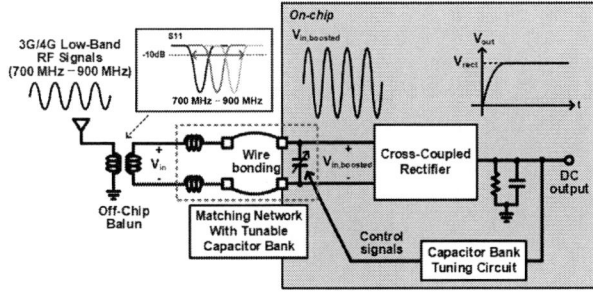

Fig. 1. Block diagram of the proposed RF to DC converter

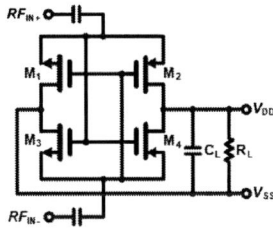

Fig. 2. Schematic of differential-drive cross-coupled rectifier

power matching of the RF-to-DC converter is performed near the input power when the rectifier has the peak power-conversion efficiency (PCE). Considering the sensitivity, RF-to-DC conversion efficiency and power matching of the RF-to-DC converter, a load resistor of 10kΩ was chosen.

III. SIMULATION RESULTS

Fig. 3 shows simulated S_{11} of the proposed converter. The S_{11} is less than -10 dB from 700 MHz to 900 MHz. Fig. 4 presents the simulated output voltage of the proposed RF-to-DC converter. The proposed RF-to-DC converter can achieve almost 800 mV output voltage at 0 dBm input power on matched frequency bands. Fig. 5 presents RF-to-DC conversion efficiency. The peak RF-to-DC conversion efficiency at -15 dBm input power is 72.25%, 64.97% and

978-1-5386-7961-6/18 $31.00 © 2018 IEEE

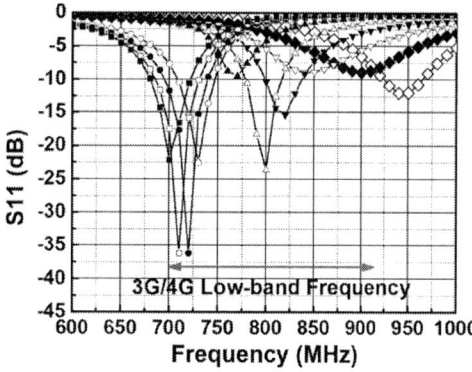

Fig. 3. Simulated S_{11}

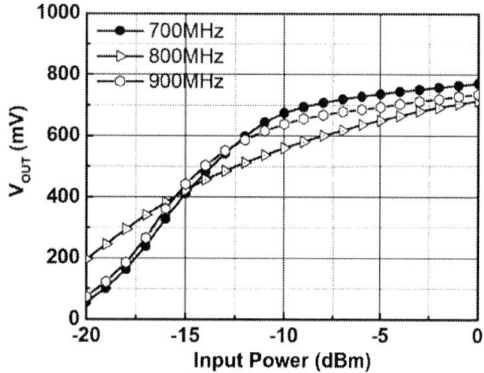

Fig. 4. Simulated V_{OUT}

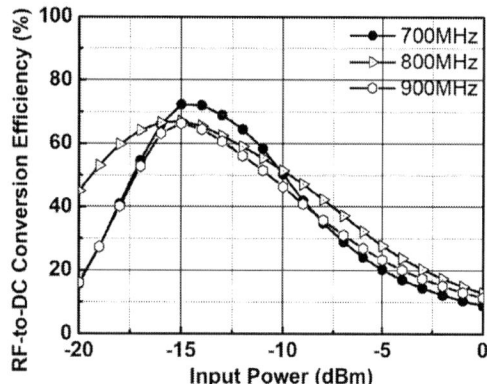

Fig. 5. Simulated RF-to-DC conversion efficiency

66.28% at 700 MHz, 800 MHz and 900 MHz, respectively. Table I shows performance summary and comparison of the proposed RF-to-DC rectifier with previous works. This work has high peak RF-to-DC conversion efficiency at low power and the tunable impedance matching network that covers 700 MHz to 900 MHz.

TABLE I. PERFORMANCE SUMMARY AND COMPARISON

Parameters	[1]	[2]	[3]	[4]	This work
Technology	180 nm	Off-Chip SMD	Off-Chip SMD	130 nm	130 nm
RF Frequency (GHz)	0.1, 0.5, 0.953, 2	0.55, 0.75, 0.9, 1.85, 2.15, 2.45	1.8, 2.2	0.953, 2	0.7 – 0.9
Matching network (Tunability)	No (No)	Yes (No)	Yes (No)	No (No)	Yes (Yes)
Rectifier topology	Cross-coupled rectifier	Voltage doubler	Diode (rectenna)	DDCC	Cross-coupled rectifier
RF-to-DC conversion efficiency (%)	82.6@ -25 dBm P_{in} and 953 MHz (Rectifier only)	67@ -5 dBm P_{in} and 0.9 GHz	55@ -5 dBm P_{in} and 2.14 GHz	73.9@ 4.5 dBm P_{in} and 953 MHz	72.25@ -15 dBm P_{in} and 700 MHz
R_L (KΩ)	5-100	10-75*	5	2	10

IV. CONCLUSION

In this paper, the proposed RF-to-DC converter performs RF energy harvesting of 3G/4G low-band frequency range from 700 MHz to 900 MHz. The proposed tunable impedance matching network with 4-bit digitally controlled on-chip capacitor array achieves higher integration and high peak RF-to-DC conversion efficiency from 700 MHz to 900 MHz RF signals.

ACKNOWLEDGMENT

This research was supported by Basic Science Research Program through the National Research Foundation of Korea (NRF) funded by the Ministry of Education (NRF-2018R1D1A1B07042804). This research was also supported by the MSIT(Ministry of Science and ICT), Korea, under the ITRC(Information Technology Research Center) support program(IITP-2018-0-01433) supervised by the IITP(Institute for Information & communications Technology Promotion) This work was also supported by IDEC (EDA tool and MPW).

REFERENCES

[1] K. Kotani, Atsushi, and T. Ito, "High-efficiency-drive CMOS rectifier for UHF RFIDs", *IEEE J. Solid-State Circuits*, vol. 44, no. 11, pp. 3011-3018, Nov. 2009.

[2] C. Song, Y. Huang, P. Carter, J. Zhou, S. Yuan, Q. Xu, and M. Kod, "A novel six-band dual CP rectenna using improved impedance matching technique for ambient RF energy harvesting", *IEEE Trans. on antennas and propagation*, vol. 64, no. 7, pp. 3160-3171, July. 2016.

[3] H. Sun, Y. Guo, M. He, and Z. Zhong, "A dual-band rectenna using broadband Yagi antenna array for ambient RF power harvesting", *IEEE antennas and wireless propagation*, vol. 12, pp. 918-921, Dec. 2013.

[4] A. Moghaddam, J. Chuah, H. Ramiah, J. Ahnadian, P.-I. Mak, and R. Martins, "A 73.9%-efficiency CMOS rectifier using a lower DC feeding (LDCF) self-body-biasing technique for far-field RF energy-harvesting systems", *IEEE Trans. Circuits Syst. I, Reg. Papers*, vol. 64. No4 pp. 992-1002, April. 2017.

Design and Analysis of Digital PID Controller in MCU and FPGA

Kyungnam Lee and Youngmin Kim

School of Computer and Information Engineering,

Kwangwoon University, Nowon-gu, Kwangwoon-ro 20, Seoul 01897, Republic of Korea

youngmin@kw.ac.kr

Abstract - **The main purpose of this study is to design an effective realization of digital proportional-integral-derivative (PID) control algorithms using field programmable gate array (FPGA) technology. Recently, FPGAs have become an alternative solution for the realization of digital control algorithm systems, which were previously dominated by general-purpose microprocessor systems. In comparison with conventional approaches, FPGA-based controllers offer various advantages such as high speed, complex functionality, and low power consumption. In addition, an FPGA-based platform has a capability to execute concurrent operations, allowing parallel architectural design of different digital controller systems. In this study, we analyze one application of hardware and software module development for realization of digital PID control algorithms in dynamic systems. We have successfully implemented and verified the FPGA-based PID controller for high speed DC motors which delivers an optimal balance of low risk, low cost, and low power. Also, we compare the performance with an MCU-based design.**

Keywords—PID, FPGA, MCU, dynamic systems, response time;

I. INTRODUCTION

The most of the automatic control loops in the process industries (over 90%) still rely on various forms of the ubiquitous PID controller which has been commercially available for over 70 years [1]. For many batch processing operations, the process control is achieved via infrequent manual adjustments by plant operators.

The PID control is the most commonly used control algorithm in the industry today. PID controller popularity can be attributed to the effectiveness of the controller in a wide range of operation conditions, its functional simplicity, and the ease with which engineers can implement it using current computer technology. The PID controllers are the most adopted controllers in industrial settings because of the advantageous tradeoff between cost and benefit they can provide. Despite their long history and the know-how gained from years of experience, the availability of microprocessors and software tools and the increasing demand for higher product quality at reduced cost have stimulated researchers to devise new methodologies to improve their performance and make them easier to use.

There are two approaches for implementing of the control systems using digital technology. The first approach is based on software which implies a memory-processor interaction. The memory holds the application program while the processor fetches, decodes, and executes the program instructions. Programmable Logic Controllers (PLCs), microcontrollers, microprocessors, Digital Signal Processors (DSPs), and general-purpose computers are tools for software implementation. On the other hand, the second approach is

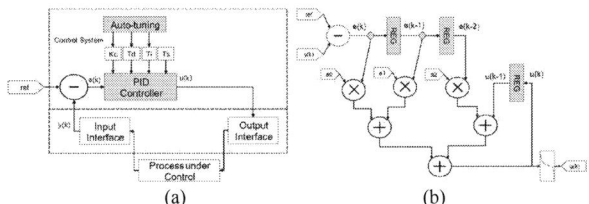

(a)   (b)

Fig. 1. (a) Typical control loop with a PID controller; (b) PID architecture (simplified version without including the circuitry to calculate a0, a1, and a2).

based on hardware. Early hardware implementation is achieved by magnetic relays extensively used in old industry automation systems. Then, hardware implementation by means of digital logic gates has become popular.

Recently, the specifications for control systems has grown to include a certain degree of intelligence. Thus, these systems must also be capable of intelligent sensor selection, remote monitoring and operation, and must be able to implement sophisticated control algorithms that require adaptation. Therefore, to meet these specifications, a new approach in terms of either hardware software co-design or reconfigurable hardware like FPGA that allow such a type of hardware/software co-design needs to take place.

In this study, we have implemented the PID algorithms both in software and hardware and the performance in a dynamic system is evaluated in MCU and FPGA based system.

II. DESIGN OF THE PID CONTROLLER

A typical closed loop system using a PID controller is shown in Fig. 1(a). The control system usually requires units to interface it to the environment. For instance, a converter to Pulse-Width Modulation (PWM) is needed when controlling DC motors.

The conventional PID controller equation is given as follows [1]:

$$u(t) = K_p \left\{ e(t) + \frac{1}{T_i} \int_0^t e(t)dt + T_d \frac{de(t)}{dt} \right\}$$

where K_p is the proportional gain, T_i is integral time constant, T_d is derivate time constant, e is the error signal and

978-1-5386-7961-6/18 $31.00 © 2018 IEEE   261   ISOCC 2018

u is the control output. The digital PID controller can be modified by the following difference equation:

$$u(k) = u(k-1) + a_0 e(k) + a_1 e(k-1) + a_2 e(k-2)$$

This type of difference equation is called the auto-tuning PID algorithm [4], in which the coefficients a_0, a_1, and a_2 are evaluated by the following expressions:

$$a_0 = K_c \left(1 + \frac{T_d}{T_s}\right) \quad a_1 = -K_c \left(1 + 2\frac{T_d}{T_s} - \frac{T_s}{T_i}\right) \quad a_2 = K_c \frac{T_d}{T_s}$$

The K_C, T_i, and T_d are PID parameters for tuning, and T_s is the sampling period in seconds. There are several methods for evaluating the PID parameters, generally called PID tuning methods [1]. When controlling time-invariant processes, the PID parameters can be constants and evaluated off-line, so, the PID architecture may use fixed values for the a_0, a_1, and a_2 coefficients. Otherwise, for time-variant processes there is a need to update those parameters; in this case the PID architecture has K_C, T_i, and T_d as parameters that can be automatically updated during runtime by auto-tuning algorithms. Fig. 1(b) shows a simple PID architecture with the a_0, a_1, and a_2 coefficients. This architecture uses three adders, three, and three registers. The arithmetic operations may multipliers have saturation behavior so that whenever the magnitude of the result of an operation is not represented by the output representation (overflow), the output result is the largest or the smallest representable value.

III. SIMULATION RESULT

In this study, we propose a methodology for design and implementation of PID controllers in FPGA (e.g., Spartan6) and MCU (e.g., STM32F103) with exploitation of the number of bits for fixed-point representations.

Waveforms of The PID controls measured by the oscilloscope both in MCU and FPGA-based systems are shown in Fig. 2. The operating frequency for MCU and FPGA are 72 MHz and 50 MHz, respectively. The supply voltage is fixed to 3.3 V. As shown in the figures, the MCU running at 72 MHz is able to execute each PID control iteration in 0.3 ms. While the FPGA running at 50 MHz can execute each PID control iteration in only 680 ns. Thus, the latency, which is the most important parameter for dynamic systems, is 400x faster by FPGA than MCU. For example, it takes 300 us to calculate new response based on the current value with feedbacks by the MCU, but, this time can be reduced up to 0.68 us in FPGA. Also, the output signal which remain high for 1 us to generate an idle time in both MCU and FGPA for separation between each calculation.

For power consumption, the efficient FPGA-based system implementation is done which use only 0.0264 W of the power compared with the 0.0333 W in MCU-based system. Design parameters and measurement results are summarized in Table I. The result clearly indicate that the FPGA-based system is an efficient solution for high-speed and low-power applications. In addition, the advantage of the FPGA-based system is that it can operate parallel controls, which several channels need PID controls at the same time and real-time.

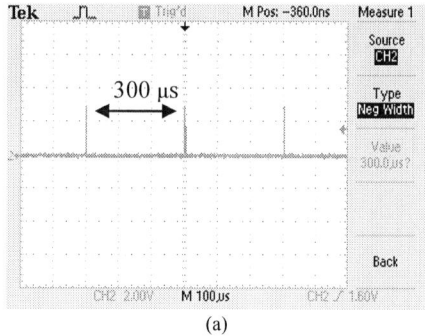

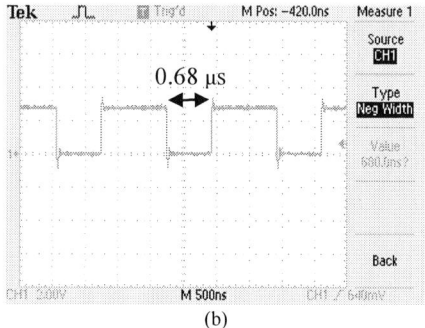

Fig. 3. Measured waveforms of the PID controls in (a) MCU and (b) FPGA

TABLE I. DESIGN PARAMETERS AND MEASURMENT RESULTS OF MCU AND FPGA-BASED PID CONTROLLERS

	Using MCU	Using FPGA
Operating frequency[MHz]	72	50
Latency [μs]	300	0.680
Power consumption [W]	0.0333	0.0264

IV. CONCLUSION

This paper presents a methodology to design PID controllers when targeting FPGA-based systems. The performance of the FPGA-based design is compared with MCU-based design for the same PID algorithms. The FPGA-based design provides a fast response time with less power consumption compared with MCU-based design. Thus, the controller implementation using an FPGA chip is advantageous and applicable to the real time dynamic systems.

ACKNOWLEDGMENT

This research was supported by the MISP (Ministry of Science, ICT & Future Planning), Korea, under the National Program for Excellence in SW supervised by the IITP (Institute for Information & communications Technology Promotion) (2017-0-00096)

REFERENCES

[1] K. Åström, T. Hägglund, "PID Controllers: Theory, Design, and Tuning," in *Instrument Society of America*, Research Triangle Park, NC, USA, 2nd, 1995.
[2] Michael A Johnson, Mohammad H Moradi, and J Crowe, PID control : *new identification and design methods*, Springer, New York, 2005.
[3] F. Michaud, *et al.*, "Mobile robot that can read symbols," in *Proc. IEEE ISCI Robotics & Automation*, 29 July-1 Aug. 2001, pp. 338-343.
[4] G. Reynoso-Meza, *et al.*, "Evolutionary auto-tuning algorithm for PID controllers", *IFAC Conf. on Advances in PID control PID*. Vol. 12. 2012.

Optimal Model Analysis
for Denoising Monte Calro Rendering Noise

Min-Cheol Kim
Dept. of Computer Engineering
Seokyeong University
Seoul, Korea
cioneraria219@skuniv.ac.kr

Kwang-Yeob Lee
Dept. of Computer Engineering
Seokyeong University
Seoul, Korea
kylee@skuniv.ac.kr

Abstract— **Recently monte carlo rendering has been used a lot for rendering realistic images. However, in order to obtain the film-quality result, We need a large amount of sample per pixel (spp), so many methods to postprocess a rendered image at a low spp. A denoising method with auxiliary buffers (normal, depth, albedo etc.) is often used. There are many filters-based methods that remove noises, the performance is good, but it is not easy to find the optimal filter parameter because it is too slow. In order to solve this problem, we propose a deep learning method which can find optimal filter parameters. In this paper, we find an optimal model for removing noise. We experiment with VGGNet, DnCNN, ResNet, DenseNet, and Recurrent autoencoder as a model to compare, and consequently Recurrent autoencoder can be seen to be the best model with this loss value 0.0649.**

Keywords; Monte Carlo Rendering, Denoising, Convolutional Neural Network

I. INTRODUCTION

Ray tracing or path tracing has been widely used for a rendering animation and photorealistic images. For rendering scenes without noises, it requires a lot of time. So a lot research have been done to achieve high quality images using the rendering method with less sample per pixel(spp), many renderers has used postprocess. A method filtering noise is often used with scene's auxiliary feature buffers (albedo, specular, diffuse, etc.) like Stein's unbiased risk estimate (SURE) used to minimize denoising error, additional features. [1][2] The result depends on the filter parameter and the processing to be performed on each feature. But there is a limit to manually determining the parameter. To address this problem, a method using machine learning was proposed, the performance was good but since It learned the parameter of non-local means filter or cross-bilateral filter to be a fixed filter, it has still a limit. [3] Similarly, a method to make a filter by convolutional neural networks (CNNs) using auxiliary buffer was suggested and learned in 24 hours and showed good results. [4] And There is another method using a recurrent autoencoder, which is a more complicated model, to remove interactively noise in real time. [5] In this paper, we will use an auxiliary buffer based on such a method, experiment to find an optimal model so that interactive filtering can be performed.

II. RELATED WORK

Since methods of removing monte carlo noise have too large to be covered, this paper focuses only using deep learning. This paper is aimed at finding an optimum deep learning model.

A. Learning Based Filtering

It uses multi-layer perceptron to obtain optimal parameters with 20 scenes in training set, and 8 with test set. In order to feed the features of scene in MLP input, it extracts secondary feature, and it has 36-10-6 layers. It determines parameters of non-local mean filter or cross cross-bilateral filter

B. Kernel Predicting Convolutional Networks

Fully convolutional network consists of 9 layers, models for specular and diffuse are separated respectively. It uses 600 scenes for training set and 25 test set. It is too large to use the whole input image, so images are cropped into 65 x 65 patches. As preprocessing, diffuse is divided into albedo and specular logarithm transformed. Then they are combined at the last layer and output the final filter to remove image noise.

C. Recurrent Denoising Autoencoder

It uses autoencoder which consists of encoder and decoder in fully convolutional layer for interactive reconstruction method at low sampling rate. Then it puts put recurrent at only encoder, and it has skip connection every two layers.

III. DATASET

The dataset consists of about 1500 scenes, provided by Disney. [6] Each scene has albedo, depth, diffuse, normal, color, specular, visibility, and variances of each channel. In this paper, divide diffuse into albedo and logarithmically transform specular. Then, 66 inputs were used up to variance, the difference between the x and y axes of each channel. Input, output images are rendered by 128 spp, 8192 spp, respectively. And we use about 800 scenes for training, 300 scenes for validation, and 300 scenes for test.

Identify applicable sponsor/s here. If no sponsors, delete this text box. (sponsors)

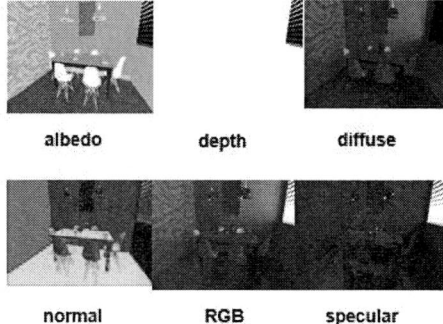

Figure 1. Auxiliary feature buffers of a scene in dataset

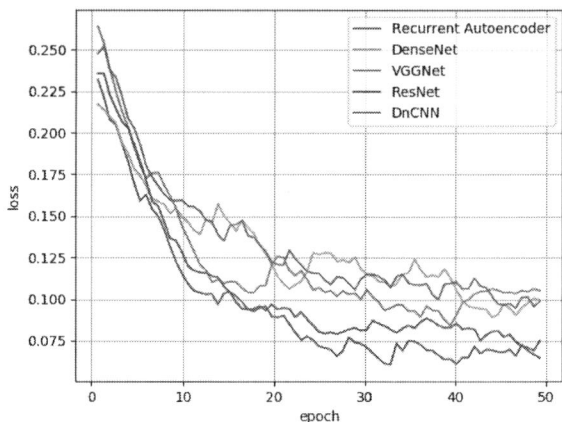

Figure 2. The loss comparison of each model

IV. EXPERIMENT

We use five models for denoising. Fully connected layer and pooling layer has been removed to output filtered images.

A. VGGNet

Using 1 layer with 3 x 3 size kernels can use fewer parameters that using multiple layer with 5 x 5 or 7 x 7, while get the same receptive field effect. All layers have the same resolution, and all convolutions use 3 x 3 kernels

B. ResNet

When the model has deep layer, there are vanishing, exploding gradient problem. These problems are solved by using identity shortcut connection [7] In addition, it adds batch normalization so that learning can be done faster, more stable. It can be created more than 100 layers, but we just add shortcut connection and batch normalization to VGGNet for compare with the model

C. DenseNet

Since ResNet has a summation between layers, information becomes smaller and interferes with flow. [8] However, DenseNet can maintain features while being stacked. To compare with other models, we used three dense blocks so that each layer has 6 layers.

D. DnCNN

It is aimed at extracting the noise by finding the difference from the final layer output instead of short connection at each layer for removing gaussian noise which is not intended for Monte carlo noise. [9] Similar to ResNet but focused on detecting noise.

E. Recurrent Autoencoder

For Denoising purpose there is autoencoder. Among them, recurrent is added to the encoder part of the model, and there is a skip connection between encoder and decoder. In order to keep exceptionally resolution, max pooling layer added for two convolution layers.

TABLE I.   THE RESULT OF COMPARISON EXPERIMENT

Model	Loss	Model	Loss
VGGNet	0.1051	DnCNN	0.0985
ResNet	0.0750	Recurrent Autoencoder	0.0649
DenseNet	0.0996		

ACKNOWLEDGMENT

This research was supported by the MOTIE(Ministry of Trade, Industry & Energy) (10080568) and KSRC(Korea Semiconductor Research Consortium) support program for the development of the future semiconductor device.

REFERENCES

[1] LI, Tzu-Mao; WU, Yu-Ting; CHUANG, Yung-Yu. SURE-based optimization for adaptive sampling and reconstruction. ACM Transactions on Graphics (TOG), 2012, 31.6: 194.

[2] Rousselle, Fabrice; MANZI, Marco; ZWICKER, Matthias. Robust denoising using feature and color information. In: Computer Graphics Forum. 2013. p. 121-130.

[3] Kalantari, Nima Khademi; BAKO, Steve; SEN, Pradeep. A machine learning approach for filtering Monte Carlo noise. ACM Trans. Graph., 2015, 34.4: 122:1-122:12.

[4] Bako, Steve, et al. Kernel-predicting convolutional networks for denoising Monte Carlo renderings. ACM Trans. Graph, 2017, 36.4: 97.

[5] CHAITANYA, Chakravarty R. Alla, et al. Interactive reconstruction of Monte Carlo image sequences using a recurrent denoising autoencoder. ACM Transactions on Graphics (TOG), 2017, 36.4: 98.

[6] https://www.disneyresearch.com/research/dataset-terms/

[7] HE, Kaiming, et al. Deep residual learning for image recognition. In: Proceedings of the IEEE conference on computer vision and pattern recognition. 2016. p. 770-778

[8] HUANG, Gao, et al. Densely Connected Convolutional Networks. In: CVPR. 2017. p. 3.

[9] ZHANG, Kai, et al. Beyond a gaussian denoiser: Residual learning of deep cnn for image denoising. IEEE Transactions on Image Processing, 2017, 26.7: 3142-3155.

A Software-based Scan Chain Diagnosis for Double Faults in A Scan Chain

[1]Hyeonchan Lim, Seokjun Jang, and [2]Sungho Kang

Departments of Electrical & Electronic Engineering

Yonsei University

Seoul, Korea

[1]{lhcy92, onics1492}@soc.yonsei.ac.kr and [2]shkang@yonsei.ac.kr

Abstract—**This paper presents a simulation-based scan chain diagnosis algorithm for double faults in a scan chain. With an ordinary flush test pattern, only the last fault can be observed while the others are masked by the last fault. Considering the multiple faults, the forward fault and the rearward fault are decided by analyzing signatures of a faulty chain and good chains, separately. The results of analysis are upper bound (rearward fault) and lower bound (forward fault). If only one fault exists, there is only a candidate in the boundary. The proposed technique can be applied for both stuck-at and transition faults. ISCAS'89 benchmark circuits verify the proposed method and the experimental results show that the diagnosis resolution is increased compared to the conventional method.**

Keywords; Diagnosis; scan chain; fault simulation; double faults;

I. INTRODUCTION

For improving testability of very-large-scale-integration circuits, scan architecture has been adopted widely. In a general scan architecture, the hardware area related to scan chains accounts for about 30 percent of the total circuit area [1]. Hence, about 30 percent of defects that affect logic lead scan chain to failure. A scan chain is crucial for the testing of other parts of the circuits, such as memories. Therefore, it is important to find the location of the fault in a faulty scan chain to determine the cause of the failure.

Scan chain diagnosis is the procedure where the location of the defective scan cell in a scan chain is identified. For the exact identification, several scan chain diagnosis methods have been proposed. However, the previously proposed methods need additional hardware overhead, have still low diagnosis performance, target only some kinds of fault types, or don't consider multiple faults [2-5]. In order to overcome these limitation, a simulation-based scan chain diagnosis method for double faults in a scan chain is proposed. It can handle not only stuck-at faults but also transition faults. In the double faults condition, the effect of the fault which is located in the forward scan chain can be masked by the fault which is located in the rearward scan chain, which is unobservable at the scan out port. Thus, the first fault and the last fault in the scan chain are considered and detected separately. The proposed method analyzes the netlist of the circuits under test (CUT) before the testing and the fail log of the testing. Using the results of the

simple analyzation, the locations of the double faults can be decided exactly and rapidly.

II. PROPOSED METHOD

The flow of the proposed method is divided into 3 steps; 1) finding faulty scan chain, 2) estimating forward/rearward faults, and 3) deciding upper/lower boundary. Since the fault type isn't decided in the first step, the proposed method can be applied to not only stuck-at fault but also transition fault.

A. Finding Faulty Scan Chain

The decision process of the faulty scan chain is equal to the conventional method. A flush test pattern (or chain check pattern) which is repetition of 0's and 1's, such as '00110011' is shifted into the scan chains and is shifted out without capture response. In a good scan chain, the scan-out response is identical to the scan-in stimuli. However, in a faulty scan chain, the response will be different with the stimuli, as some bits are affected by the faults in the scan chain. For example, with stuck-at-0 fault, the response will be '00000000'.

B. Estimating rearward fault

The rearward fault can be estimated by analyzing the fail signatures of the faulty chain. Some capture responses which are in front of the rearward fault will be affected by the fault during shift-out operation. Thus, the fail signatures of the according scan flip-flops will frequently exist in the fail log. However, the others which are behind the rearward fault will not be affected and normally observable at the scan-out port. Hence, the fail signatures of the according scan flip-flops rarely exist in the fail log. By analyzing how frequently the fail signatures exist in the fail log, the location of the rearward fault can be estimated as shown in Fig. 1.

C. Estimating forward fault

As mentioned above, the effect of the forward fault cannot be observed at the scan-out port of the faulty scan chain, as it is masked by the rearward fault. Therefore, in order to find the masked fault, the fail signatures of good scan chains should be analyzed. During the scan-in operation, the scan-in stimuli which are behind the forward fault will be affected by the fault. The affected values propagate to the scan flip-flops of good scan chain through the combinational logic during the capture operation. Hence, the fail signatures of the scan flip-flops of

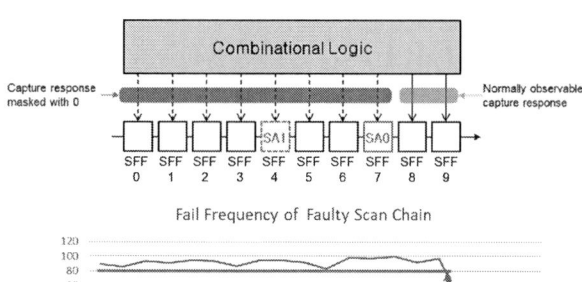

Fig. 1. An example of estimating rearward fault

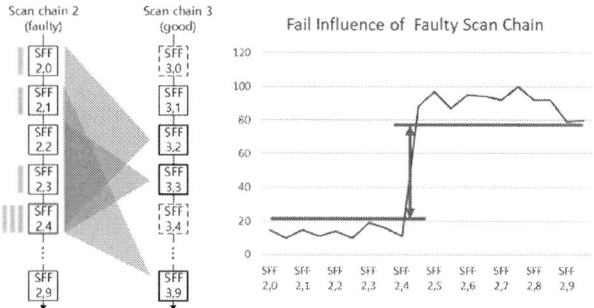

Fig. 2. An example of estimating forward fault

the good scan chains which are connected to the affected scan flip-flops will frequently exist in the fail log. By back-tracing from the scan flip-flops of the good scan chains, the scan flip-flops which are behind the forward fault and cause failures at the good scan chains can be estimated. The forward fault is estimated by two matrices; scan flip-flops relation matrix, SR, and fail signature frequency matrix, FF. SR is computed by determining whether the scan flip-flops of the faulty scan chain are connected to the scan flip-flops of the good scan chains. For n scan chains with each m scan flip-flops, the size of SR is $(m \times (n - 1)m)$. FF whose size is $((n - 1)m \times 1)$ shows how frequently the signatures of the scan flip-flops in the good scan chains exist in the fail log. By multiplying SR and FF, $(m \times 1)$-size failure influence matrix, FI, is computed, which shows how frequently the scan flip-flops in the faulty scan chain affect the failures in the good scan chain. The scan-in stimuli of the scan flip-flops behind the forward fault will be affected by the fault, and the elements of FI which corresponds the scan flip-flops will have higher values, as shown in Fig. 2.

III. EXPERIMENTAL RESULTS

In the experiments, ISCAS'89 benchmark circuits were used, and commercial automatic test pattern generator (ATPG) tool-TETRAMAX of Synopsys was used for fault simulation and for fail log generation. Table I shows the experimental results on the benchmark circuits. Column 1 shows name of the circuits. Column 2 shows composition of the scan chains; the length of the scan chains and the number of the scan chains. Column 3 shows the number of ATPG patterns used for the fault simulation and the fail log generation. Column 4 and 5 show diagnosis resolution and accuracy, respectively. For the

Table I Experimental Results on ISCAS'89 Benchmark Circuits

Circuits	Scan Chain	# Patterns	Average Resolution	Average Accuracy
s298	7 x 2	22	2.57	100%
s1423	19 x 4	43	2.90	94%
s38417	82 x 20	82	2.51	96%
s38584	73 x 20	115	2.58	98%
average	-	-	2.64	97%

convenience of the experiments, two random faults are inserted into two adjacent scan flip-flops in a scan chain. The results are from the average values of every adjacent scan flip-flop pair. Since two adjacent scan flip-flops are faulty, the resolution must be bigger than 2; one for rearward fault and the other one for forward fault. It shows 2.64 average diagnosis resolution which is closed to the minimum resolution, 2, and 97% average diagnosis accuracy.

IV. CONCLUSION

In this paper, a software-based scan chain diagnosis method for double faults in a scan chain is proposed. It doesn't require additional hardware which can be fatal to the yield of semiconductor at the early part of technology. By analyzing the netlist of CUT and the fail log, the double faults are diagnosed exactly. Also, it does not require complex computation or several fault simulations, but simple matrices multiplication, leading to short computation time.

REFERENCES

[1] Wei-Hen Lo, Ang-Chih Hsieh, Chien-Ming Lan, Min-Hsien Lin, and TingTing Hwang, "Utilizing circuit structure for scan chain diagnosis," IEEE Transactions on Very Large Scale Integration (VLSI) Systems, Jan. 2014, vol. 22, no. 12, pp. 2766-2778.

[2] Ruifeng Guo and Srikanth Venkataraman, "An Algorithmic Technique for Diagnosis of Faulty Scan Chains," IEEE Transactions on Computer-Aided Design of Integrated Circuits and Systems, Sep. 2006, vol. 25, no. 9 pp. 1861-1868.

[3] Helen-Maria Dounavi and Yiorgos Tsiatouhas, "Stuck-at Fault Diagnosis in Scan Chains," in Proc. 9th International Conference on Design & Technology of Integrated Systems in Nanoscale Era (DTIS), 2014, pp. 1-6

[4] HeRuifeng Guo, Yu Huang, and Wu-Tung Cheng, "Fault Dictionary Based Scan Chain Failure Diagnosis," in Proc. 16th IEEE Asian Test Symposium, 2007, pp. 1-6

[5] Mingjing Chen and Alex Orailoglu, "On Diagnosis of Timing Failures in Scan Architecture," IEEE Transactions on Computer-Aided Design of Integrated Circuits and Systems, Jun. 2012, vol. 31, no. 7 pp. 1102-1115.

Low Power Scan Chain Architecture Based on Circuit Topology

Heetae Kim, Hyunggoy Oh, Sangjun Lee, and Sungho Kang
Electrical & Electronic Engineering
Yonsei University
Seoul, Korea
{kht2161, kyob508, lsj920807}@soc.yonsei.ac.kr, and shkang@yonsei.ac.kr

Abstract— **Scan-based test is widely used method to test the digital circuits, increasing the controllability of the circuit under test. However high controllability can cause fatal problems by excessive test power consumption. To resolve these problem, this paper proposes a scan chain architecture which reduces the test power consumption. The proposed method decreases the test data volume by partitioning a scan chain into many scan segments. The scan chain partitioning is performed based on circuit topology, and it increases the number of scan segments that can be bypassed. Simulation results show that the proposed method reduces the test power consumption up to 43.33% compared to the previous work.**

Keywords—Scan partitioning, test power reduction, scan chain bypass, scan-based test.

I. INTRODUCTION

Since the evolution of integrated circuits has increased the size and complexity of circuits, the amount of power consumption during the circuit test also has grown by lots of test patterns. In addition, power consumption during the circuit testing is greater than the power consumed in normal operation due to the higher switching activity. Because excessive test power consumption can lead some to serious problems such as test overkill and IR drops, it is necessary to reduce the test power consumption [1].

Scan chain partitioning methods have been studied to resolve the problems induced by excessive power consumption. [2] proposed a method that divided a scan chain into several segments and bypassed the segments whose all test patterns are X bits. The method reduced the power consumption by decreasing the test data volume. However [2] didn't mention how to determine the scan segments. [3] proposed reconfigurable scan architecture which can change the length of the scan segments flexibly. But [3] still didn't present the criteria to determine the length of the segments.

This paper proposes a scan chain partitioning method based on circuit topology. The proposed method forms scan segments depending on circuit topology which can effectively reduce the test data volume by increasing the number of segments that can be bypassed.

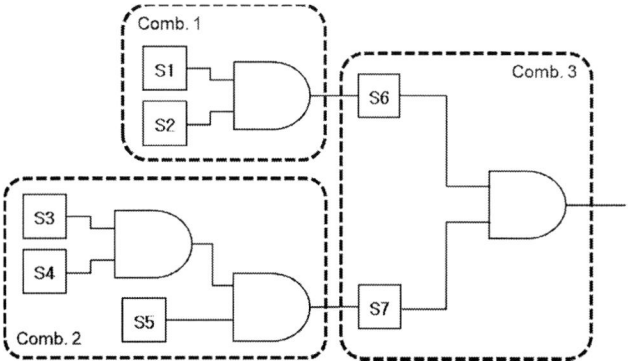

Figure 1. Example circuit for scan partitioning

II. SCAN PARTITIONING METHOD BY CIRCUIT TOPOLOGY

This paper proposes a new scan partitioning method to reduce the scan shift power. The proposed scan partitioning method reduces the shift power by increasing the number of segments which can be bypassed. In order to increase the number of segments to bypass, the scan cells which have don't care bit are assigned to one segment and circuit topology helps enables these scan cells to be grouped into one segment.

Fig. 1 is an example of digital circuit to illustrate the proposed method and Table 1 shows the test patterns of the example circuit in Fig. 1. Since the test patterns P1, P2 and P3 in Table 3 have only 0 or 1 bits, scan bypass technique cannot be performed for test pattern P1, P2 and P3. However, the test pattern P4 has the potential for scan segment bypass. Fig. 2 shows two methods of scan partitioning. The method in Fig. 2(a) cannot bypass any scan segment, but the method in Fig. 2(b) can bypass the first scan segment which has only don't care bits in the test pattern. Therefore, it is possible to reduce test data volume by forming the scan segments can be bypassed.

Table 1 Test patterns of the example circuit

	S1	S2	S3	S4	S5	S6	S7
P1	0	1	0	1	1	0	1
P2	1	0	1	0	1	1	0
P3	1	1	1	1	0	1	1
P4	X	X	1	1	1	X	X

978-1-5386-7961-6/18 $31.00 © 2018 IEEE   267   ISOCC 2018

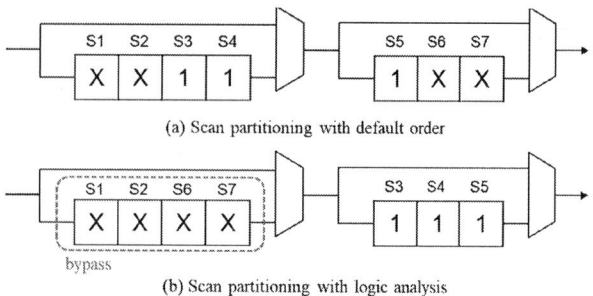

(a) Scan partitioning with default order

(b) Scan partitioning with logic analysis

Figure 2. Illustration of scan partitioning

The example circuit in Fig. 1 is comprised of 3 combinational logics, which are marked as comb. 1, comb. 2 and comb. 3. The value of S1 and S2, which are test stimuli of comb. 1, can only test the AND gate in comb. 1 and the value of S1 and S2 not affect other combinational logics. In the same way, S6 and S7 are stimuli of comb. 3 and can test only AND gate in comb. 3. Since the purpose of test pattern P4 is testing the AND gates in comb. 2, the values of S1, S2, S6 and S7 in test pattern P4 are don't care bits.

In order to increase the number of segments can be bypassed, scan cells connected to the same combinational logic are classified into the same scan segment. Additionally, all segment's length should be similar for effective scan bypass technique. Therefore, the proposed method classifies scan cells according to the combinational logics which include the scan cells. Then, place the classified cells so that the length of the segment is as similar as possible. The proposed method has disadvantage that the minimum length of segment is larger than or equal to the maximum number of scan cells which are included one combinational logic.

III. SIMULATION RESULTS

Simulations are performed to compare the test power consumption of the previous method [3] and the proposed method. In this paper, we use a weighted transition metric (WTM) to evaluate the test power consumption. WTM represents sum of switching activities during scan shift operation and WTM can be calculated as Equation (1) where L is length of scan segment and v_i is the value of the i-th scan cell. Commercial tools, Design Compiler and TetraMax are used and 4x4 Multiplier and Accumulator (MAC) design with 132 scan cells per chain is used as the circuit under test.

$$WTM = \sum_{i=1}^{L-1}(v_i \oplus v_{i+1}) \times i \qquad (1)$$

Table 2 shows a comparison of WTM of the full scan method, the previous method and the proposed method and notation L in Table 2 represents the maximum length of segments. It shows that WTMs of the conventional method and the proposed method are decreased as increasing the number of segments, whereas WTM of method using full scan is not changed. It is because the more number of segments makes L

in equation (1) lower. Therefore, the effect of L should be also considered to reduce WTM.

Table 2. Comparison with the conventional method

# of Segments	Full Scan	Conventional [3]		Proposed		
	WTM (L = 132)	WTM	L	WTM	L	WTM reduction
5	145,798	25,807	27	32,955	33	-27.70%
6	145,798	17,270	22	17,206	25	0.37%
7	145,798	12,474	19	9,997	25	19.86%
8	145,798	4,920	17	2,788	17	43.33%

The proposed method reduces WTM compared to the conventional method except when the number of segments is 5. Because L of the conventional method is smaller than L of the proposed method, WTM of the proposed method increases 27.70% compared to that of the conventional method. On the other hand, although L of the proposed method is larger than that of the conventional method, 0.37% of WTM reduction is performed when the numbers of segments is 8. Moreover, WTM decrease to 19.86% when the number of segments is 7. It shows that balance between the size effect of L and the effect of how the segments are organized is important when L of the proposed method is larger than that of the conventional method. When the number of segments is 8, both methods have the same size of L and the proposed method reduces 43.33% of WTM compared to the conventional method.

IV. CONCLUSION

This paper proposed a low power scan architecture to reduce the test power consumption. The proposed method forms scan segments to decrease the test data volume. The segments made by circuit topology are effective to use scan bypass technique. Simulation results show WTM comparison between the previous method and the proposed method. Although the proposed method might increase the maximum length of scan segments, the proposed method reduces the test power consumption up to 43.33%.

ACKNOWLEDGMENT

This work was supported by the IT R&D program of MOTIE/KEIT. [10052716, Design technology development of ultra-low voltage operating circuit and IP for smart sensor SoC].

REFERENCES

[1] A. S. Abu-Issa, "Energy-efficient scheme for multiple scan-chains BIST using weight-based segmentation," IEEE Trans. Circuits and Systems II : Express Briefs, vol. 65, no. 3, pp. 361–365, March 2018.

[2] E. Arvaniti, and Y. Tsiatouhas, "Low-power scan testing: a scan chain partitioning and scan hold based technique," Journal Electron. Testing, vol. 30, no. 3, pp. 329-341, 2014

[3] H. Oh, H. Kim, J. Lim, and S. Kang, "Reconfigurable scan architecture for test power and data volume reduction," IEICE Electronics Express, vol. 14, no. 13, 2017.

Lifetime Improvement Method using Threshold-based Partial Data Compression in NoC

Ju Sung Kim[1], Jeong Beom Hong[2], Ju Yeon Kang[1], Tae Hee Han[1]

[1]Department of Electrical and Computer Engineering, Sungkyunkwan University

[2]Department of Semiconductor and Display Engineering, Sungkyunkwan University

Suwon, South Korea

{karlema, adfffsa, kjuyun2000, than}@skku.edu

Abstract— Network-on-chip (NoC) is widely used in the multiprocessor system-on-chip (MP-SoC) because it has advantages in parallel processing and scalability. However, as the transistor density and the volume of communication increase, maintaining the reliability is becoming a major challenge. Especially, heat generation due to heavy traffic is one of the main causes of reliability degradation, that accelerates aging of chips and results in permanent faults. In previous studies, mapping and routing methods had been used to distribute unbalanced traffic for extending the lifetime. However, these methods have a limitation because the overall amount of traffic is not changed. In this paper, we propose a method to extend the lifetime by threshold-based partially applied Huffman coding in order to reduce the total amount of traffic considering implementation overhead. Experimental results show that the traffic volume is reduced by 17.24 % on average, mean time to failure (MTTF) is increased by 13.55 % compared to that of the aging aware fully adaptive routing without compression.

Keywords; Compression, Network-on-chip(NoC), Lifetime improvement, Huffman coding

I. INTRODUCTION

NoC is an interconnection structure that has advantages for parallel processing and scalability over existing bus architecture [1]. However, as the feature size continues shrinking and the number of cores is increasing, the power density in a chip is becoming a big challenge [2]. As the power density increases gradually with shrinking feature size, increased temperature exacerbates the reliability of a chip. It is reported that the failure rate of electronic components increases by 316 % as the feature size decreases by 64 % [2]. Electromigration(EM), negative bias temperature instability (NBTI), hot carrier injection (HCI) and time-dependent dielectric breakdown (TDDB) accelerate an aging effect which results in higher failure rate. To mitigate this problem, several studies have been conducted in system-level by distributing heavy traffics by detouring packets or remapping cores. However, reliability improvement using these methods have a severe restriction because it does not change the overall amount of traffic. As a result, it can be improved by traffic compression to effectively mitigate the aging. Because, compression incurs additional overhead, and thus the compression ratio should be adjusted considering additional compression/decompression logic. In this paper, we propose a method to increase the chip lifetime by reducing the overall traffic volume using a partial Huffman compression scheme based on the threshold thus to minimize the overhead.

II. LIFETIME ESTIMATION FAILURE MODELS

An estimation of the lifetime reliability of micro architecture systems can be carried out by simulation or analysis methods. Reliability aware microprocessor (RAMP) is one of the representative lifetime analytic methods. The mean time to failure (MTTF) models are used in RAMP. For estimating the failure model, we combine TDDB and NBTI effects (Eq. (1), (2)) with the sum-of-failure-rate (SOFR) model. TDDB has been become critical as the gate oxide thickness decreases due to ongoing technology shrinkage. NBTI continues to become more severe due to increased field and operating voltage reduction applied to gate oxide. The following equations are the MTTF of TDDB and NBTI.

$$\text{MTTF}_{TDDB} \propto \left(\frac{1}{V}\right)^{(a-bT)} e^{\frac{(X+\frac{Y}{T}+ZT)}{kT}} \quad (1)$$

where generally used values as follows : $a = 78$, $b = -0.081$, $X = 0.759\text{eV}$, $Y = -66.8\text{eVK}$, $Z = -8.37e^{-4}\text{eV/K}$, k is Boltzmann's constant, and T is the operating temperature of the circuit based on the data from [3]

$$\text{MTTF}_{NBTI} \propto \left[\ln\left(\frac{A}{1+2e^{\frac{B}{kT}}}\right) - \ln\left(\frac{A}{1+2e^{\frac{B}{kT}}-C}\right) \times \frac{T}{e^{\frac{D}{kT}}}\right]^{\frac{1}{\beta}} \quad (2)$$

where A, B, C, D, and β are model fitting parameters. Typical values of the parameters are as follows: $A = 1.6328$, $B = 0.07377$, $C = 0.01$, $D = -0.06852$ and $\beta = 0.3$ based on the data from [3][4].

III. PROPOSED ARCHITECTURE

There are two major methods to compress data on a NoC: i) Cache compression (CC) scheme for inserting compression encoding/decoding circuits between the core and the network interface, ii) NIC compression scheme (NC) for adding compression encoding/decoding circuits in the network interface.

We adopt CC method because it has advantages to design by including additional logics independently without revising existing architecture. Furthermore, According to Ping Zhou, data traffic corresponding to frequently repeated values (FV) in the NoC takes up a large portion [5]. For this reason, Huffman coding was used an effective compression method for FV. However, there is an additional implementation overhead between core and NI. To reduce this overhead and improve the lifetime effectively, we apply Huffman coding to only some cores which have more traffic than threshold.

978-1-5386-7961-6/18 $31.00 © 2018 IEEE

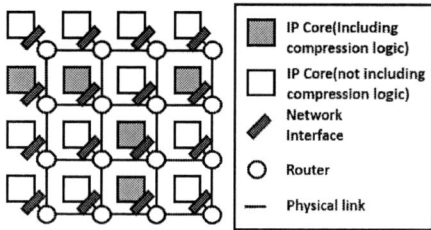

Figure 1. Partially applied Huffman coding in NoC.

The MTTF, a criterion for evaluating reliability, is highly dependent on the temperature inside the chip, which is related to traffic. Lifetime is reduced due to hotspot incurred by traffic concentration. Therefore, we set the threshold based on the traffic volume, and the compression efficiency should be increased as much as possible considering the overhead. The equations for determining the threshold value are Eq. (3) and (4). Eq. (3) is a formula for setting the weight of the average value. This method determines the ranking of links from the top to the bottom by the amount of traffic volume. The maximum weight is obtained by dividing the total number of links by the number of links that are higher than the average of the traffic, and when the ranking is low, the weight is gradually decreased. Eq. (4) sets the final threshold by multiplying the previously calculated weights by the average. In this case, since bidirectional links require more compression logic, the threshold is finally determined considering this.

$$w_{i,j} = \left(1 - \left(\frac{1}{C} \times (r_{i,j} - 1)\right)\right) \times \frac{C}{A} \qquad (3)$$

$$Threshold = \sum_{\forall i,j} \frac{e_{i,j} \times w_{i,j}}{C \times d_{i,j}} \qquad (4)$$

Core communication graph (CCG) represents the amount of communication between cores is showed by *CCG (V, E, C, D)*. Each vertex $v_i \in V$ indicates a core, $e_{i,j} \in E$ is the communication volume from v_i to v_j, $c_{i,j} \in C$ means the number of connected edges. C refers to the total number of edges connected between cores. $d_{i,j} \in D$ indicates whether the connected edge is bidirectional or unidirectional. $r_{i,j} \in R$ is the rank of the amount of communication from v_i to v_j, and A is the number of edges whose traffic is above average.

IV. EXPERIMENTAL RESULTS

The whole traffic benchmark was configured, and the traffic volume calculated by using NoXIM simulator. The temperature related to the traffic volume was verified using Hotspot 5.0 simulator. The Huffman coding was implemented in Cycle accuracy C++ simulator. Based on Ping Zhou research, an average of 42.88% was supposed for ratio of top 8 FV [5].

TABLE I. COMPARISON OF THRESHOLD EFFICIENCY BETWEEN AVERAGE AND PROPOSED METHOD

	DVOPD	MPEG4	MWD	VOPD	MP3 ENC DEC
Average	0.7767	0.7410	0.4692	0.8078	0.9164
Proposed	0.8361	0.9231	0.4692	0.8228	0.9164

TABLE II. COMPRESSISON RATIO AND OVERHEAD REDUCTION RATIO

	DVOPD	MPEG4	MWD	VOPD	MP3 ENC DEC
Compression ratio (%)	13.61	17.52	23.34	16.14	15.63
Overhead reduction ratio (%)	67.44	61.53	27.27	60	69.23

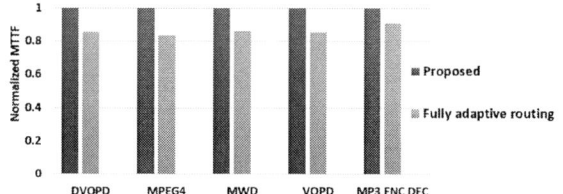

Figure 2. Normalized MTTF.

Table 1 shows the energy efficiency comparison between Eq. (3) based threshold and the simple threshold using the average value. As a result, it is exhibited that the proposed method decreases the energy consumption by 5.13 %. As shown in Table 2, the average compression ratio of the proposed scheme is 17.24%, and the overhead reduction ratio is 57.09% as compared with the non-partial compression scheme. Figure 2 shows the normalized MTTF of proposed scheme. Results reveal that our method increases MTTF by 13.55% than fully adaptive routing.

V. CONCLUSION

Traffic distribution methods through mapping and routing have limitations to improve lifetime. In this paper, we proposed an architecture that reduces the amount of traffic using data compression to solve these problems. To minimize the overhead due to additional compression logic, Huffman coding is partially applied based on threshold. Our experimental results show that the proposed architecture not only improves efficiency than applying compression logic to all nodes, but also improves chip lifetime compared to existing mapping and routing methods.

ACKNOWLEDGMENT

This work is supported in part by the IT Research and Development Program of MSIP/IITP under Grant 2016-0-00088.

REFERENCES

[1] P. Pande, C. Grecu, M. Jones, A. Ivanov, and R. Saleh, "Performance Evaluation and Design Trade-Offs for Network-on-Chip Interconnect Architectures," IEEE Trans on Computers, Vol. 54, Issue 8, pp. 1025-1040, August. 2005.

[2] L. Wang, X. Wang, and T. Mak, "Adaptive Routing Algorithms for Lifetime Reliability Optimization in Network-on-Chip," IEEE Transactions on Computers, Vol. 65, Issue 9, pp. 2896-2902, September. 2016.

[3] J. Srinivasan, S. V. Adve, P. Bose, and J. A. Rivers, "The Case for Lifetime Reliability-Aware Microprocessors," International Symposium on Computer Architecture, Vol. 32, Issue 2, pp. 276-287, March. 2004.

[4] S. Zafar, B. Lee, J. Stathis, A. Callegar, and T. Ning, "A model for negative bias temperature instability (NBTI) in oxide and high k pFETs," Int. Symposium on VLSI Technology, pp. 208-209, 2004.

[5] P. Zhou, B. Zhao, Y. Du, Y. Xu, Y. Zhang, J. Yang, and L. Zhao, "Frequent Value Compression in Packet-based NoC Architectures," Proceedings of the 2009 Asia and South Pacific Design Automation Conference, pp. 13-18, January, 2009.

Energy Efficient Analog Synapse/Neuron Circuit for Binarized Neural Networks

Jaehyun Kim, Chaeun Lee, and Kiyoung Choi
Department of Electrical and Computer Engineering
Seoul National University, Seoul, Korea
{jayc0de, mki11, kchoi}@snu.ac.kr

Abstract— **Energy efficiency is one of the most important factors to make deep neural networks viable in embedded systems. In this paper, we propose an analog synapse circuit using resistive random access memory (ReRAM) which operates with a switched capacitor neuron for binarized neural networks (BNNs). Thanks to the compact and energy efficient ReRAM synapses, the circuit simulation results of an MLP implemented with the proposed synapse and neuron circuits show 2.5ns classification latency and very high energy efficiency of 1536TOPS/W on 32nm technology.**

Keywords; deep neural network; binarized neural network; reram; analog neuron; energy efficient

I. INTRODUCTION

Deep neural networks (DNNs) have achieved great successes in image classification [1] and many other areas [2-3]. However, they accompany tremendous number of computations and excessive power consumption, which limits the broad use of DNNs especially in embedded systems.

To alleviate the problem, binarized neural network (BNN) has been proposed [4], which uses binary numbers (-1 and 1) for weights and activations instead of the conventional 32-bit floating point numbers, while incurring only marginal accuracy loss. Due to the use of binary numbers, BNNs can reduce the size of network parameters and lower computational overhead compared to the conventional neural networks.

Thanks to those advantages, BNN is suitable to implement deep neural networks in embedded systems. The work in [5] proposed a switched capacitor based analog neuron to implement a BNN for embedded systems with high energy efficiency. However, the use of SRAM and latches for weight storages and the standard cells used to implement the XOR operations between weights and activations incur large overhead in area and energy consumption.

In this paper, we propose an energy efficient analog synapse and neuron circuits for BNN implementation. The contributions of this paper are as follows:

- We propose a synapse circuit exploiting resistive random access memory (ReRAM) [6] for compact area and low energy consumption.

- We achieve classification accuracy of 98.60% and 1536 TOPS/W energy efficiency with MLP on MNIST.

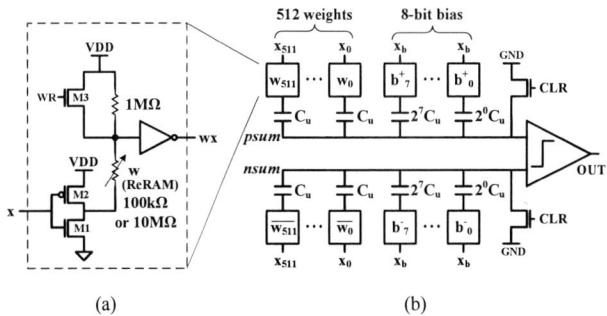

Figure 1. The proposed circuits of (a) a synapse exploiting ReRAM for weight storage and (b) a neuron using switched capacitors accompanying 512 weights and an 8-bit bias. C_u is set to 1fF.

II. HARDWARE ARCHITECTURE

A. Use of 0 and 1 Activations

We train the BNN with 0 and 1 activations while the original BNN uses -1 and 1 activations. Then the XOR operations between weights and activations can be replaced by AND operations, which enables low cost hardware implementation. Table I compares the classification accuracy of the original BNN with -1 and 1 activations and the modified BNN with 0 and 1 activations after training with MLP (784-512-512-10) on MNIST. The use of 0 and 1 activations does not incur any accuracy loss.

TABLE I. CLASSIFICATION ACCURACY AFTER TRAINING

Accuracy	Original BNN (-1 and 1 activations)	Modified BNN (0 and 1 activations)
Avg.	98.51%	98.60%

Network topology: MLP 784-512-512-10, dataset: MNIST

B. Analog Synapse and Neuron Circuits

Fig. 1 shows the proposed analog circuits for a synapse (a) and a neuron with many synapses (b). The synapse circuit consists of three transistors, a resistor, and a ReRAM cell. In a read mode (WR=0), it performs AND operation between a weight (w) and an input activation (x). If the weight is 1, the ReRAM cell is programmed to low resistance state (LRS); otherwise, the cell is programmed to high resistance state (HRS). We use 100kΩ for LRS and 10 MΩ for HRS. To represent a

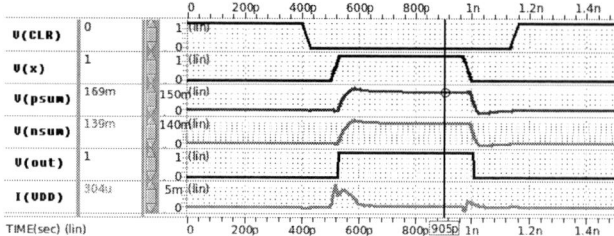

Figure 2.   Operation of the neuron circuit. 130 positive weights and 100 negative weights are activated. Bias is set to 0. PVT condition is TT/1V/25℃.

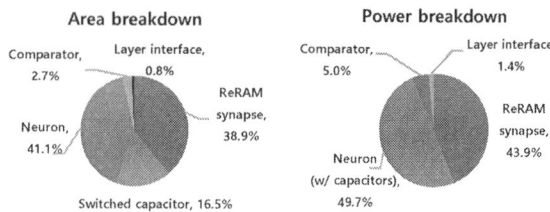

Figure 3.   Area and power breakdown results of MLP based on ReRAM synapses.

negative weight, a pair of synapse circuits is used with opposite ReRAM configurations (HRS for the synapse circuit connected to *psum* and LRS for the synapse circuit connected to *nsum*). In a write mode (WR=1), M3 transistor is activated and the source node of M2 transistor is set to high impedance in order to program the ReRAM cell by using M1 transistor as a selector.

The neuron circuit computes the weighted sum of 512 input activations including an 8-bit bias. Fig. 2 shows the operation of a neuron simulated by a circuit simulator with 32nm technology library. Initially, all inputs are deasserted and CLR signal is asserted to discharge all the capacitors to GND, which makes the neuron output 0. If input activations become active, their corresponding capacitors are pushed up and charge redistribution occurs through *psum* and *nsum* nodes, respectively. In this example, voltage level of *psum* is higher than that of *nsum* because the number of activated positive weights are larger than that of activated negative weights. Thus, the comparator produces high output.

Compared to the switched capacitor neuron with digital synapses implemented with latches and XOR gates [5], the proposed circuit performs more efficiently in terms of energy and area due to the use of ReRAM synapses which have much less area and energy consumption.

III.   EXPERIMENTAL RESULTS

A.   Experimental Setup

Experiments were performed with MLP (784-512-512-512-10) on MNIST dataset. We used the PyTorch framework and TITAN-X GPUs to train the network with 0 and 1 activations. To evaluate the power, performance, and area of the network implemented with the proposed neuron circuit, FineSim circuit simulator is used with Synopsys 32nm generic technology library. The first layer of the network is omitted in the evaluation since the proposed architecture cannot accept multi-bit inputs.

B.   Results

Table II shows the power, performance, and area of the MLP networks based on XOR synapses in [5], AND synapses (uses 0/1 activations instead of -1/1 activations), and the proposed ReRAM synapses. The network based on the proposed neuron with ReRAM synapses takes 9x smaller area while providing 13x higher energy efficiency compared to the one in [5]. The average power consumption of the network is somewhat large (0.277W) because the neurons in a layer operate concurrently. However, the short classification latency of 2.5ns leads to the high energy efficiency of 1536 TOPS/W.

Fig. 3 shows the area and power breakdown results of the network implemented with the proposed scheme. We consider the first inverter in Fig. 1(a) as a part of ReRAM synapse and the second as a part of neuron since the former drives current through the ReRAM for AND operation but the latter charges the capacitor to perform integration. The layer interface consists of a latch and an AND gate to capture the neuron output and generate the input signal for next layer. The switched capacitors and neuron circuits consume the most area and power.

TABLE II.   COMPARISON OF DIFFERENT BNN IMPLEMENTATIONS

Metric	BNN Implementations (32nm)		
	XOR synapse [5]	AND synapse	ReRAM synapse
Latency	2.5 ns	2.5 ns	2.5 ns
Area	10.2 mm²	7.3 mm²	1.1 mm²
Power	3.617 W	1.743 W	0.276 W
Energy	9.04 nJ	4.36 nJ	0.69 nJ
Energy. eff.	117 TOPS/W	243 TOPS/W	1536 TOPS/W

TT corner, 1V VDD, 25℃

IV.   CONCLUSION

We have introduced a new switched capacitor neuron circuit with ReRAM synapses for binarized neural networks. Thanks to the compact and energy efficient ReRAM synapse circuit, the BNN implementation of MLP with the proposed synapse and neuron circuits achieves short classification latency and very high energy efficiency, which makes it promising option for embedded systems.

ACKNOWLEDGMENT

This work was supported by the KIST Institutional Program (Project No. 2E27810-18-P034).

REFERENCES

[1]   A. Krizhevsky, I. Sutskever, and G. Hinton, "Imagenet classification with deep convolutional neural networks," in *Proc. NIPS*, 2012.

[2]   G. Hinton *et al.*, "Deep neural networks for acoustic modeling in speech recognition: The shared views of four research groups." IEEE Signal Processing Magazine, vol. 29, no. 6, pp. 82-97, 2012.

[3]   R. Collobert and J. Weston, "A unified architecture for natural language processing: Deep neural networks with multitask learning," in *Proc. ICML*, 2008.

[4]   M. Courbariaux *et al.*, "Binarized neural networks: Training deep neural networks with weights and activations constrained to +1 or -1," arXiv, 2016.

[5]   D. Bankman *et al.*, "An always-on 3.8 μJ/86% CIFAR-10 mixed-signal binary CNN processor with all memory on chip in 28nm CMOS," in *Proc. ISSCC*, 2018.

[6]   S. Mittal, "A survey of ReRAM-based architectures for processing-in-memory and neural networks," *MAKE*, vol. 1, no. 1, 2018.

Method of RTL Debugging When Using HLS for HW Design

Different Simulation Result of Verilog & VHDL

Sang Un Park	Tae Pyeong Kim	Mee Zee Lee	Yong Beom Cho*
Dept. of Electronics Engineering	Dept. of Electronics Engineering	Dept. of Electronics Engineering	Dept. of Electronics Engineering
Konkuk University	Konkuk University	Konkuk University	Konkuk University
Seoul Korea	Seoul Korea	Seoul Korea	Seoul Korea
kos8108@konkuk.ac.kr	Pilgrim82@gmail.com	miziyam@konkuk.ac.kr	ybcho@konkuk.ac.kr

Abstract— **HLS (High-Level Synthesis) is useful hardware design technologies using High level language (C, C++ and System C) higher than RTL (Register Transfer Level). But there is error case in which the different simulation result of Verilog & VHDL when verifying in the RTL simulation step using HLS. In this paper, We propose the causes of these problems and how to resolve them. Furthermore, we suggest how to solve similar problems with ASIC design using HLS**

Keywords; HLS (High-Level Synthesis), High level language (C, C++ and System C), RTL (Register Transfer Level), Hardware Design, ASIC

I. INTRODUCTION

HLS is a process that automatically synthesizes RTL with same specific functions written high level language such as C, C++ [1]. High level language is easier to design and modify than hardware technology language such as Verilog or VHDL. The latest HLS synthesis tools has led to fewer considerations in hardware optimization than using hardware technology languages. In addition, when verifying a converted RTL, use C/C++ test bench simulation to automatically generate it to Verilog or VHDL test bench simulation.

A key point of High-level synthesis is how efficiently code written in High level language is implemented in hardware design [2]. But inefficient syntax can occur because a code is automatically synthesized into a hardware rather than a person directly implementing the hardware. Due to inefficient syntax of synthesized RTL, there are cases in which verification often results in a different simulation result from Verilog and VHDL.

Therefore, Section 2 identifies why the results of the RTL converted from the same High-level language are different. Next Section 3, we present two ways to solve this problem. Then Section 4, Suggests how to solve ASIC design similar problems using HLS. Finally, With the conclusion, we complete this paper.

II. PROBLEM

Fig. 1(a), (c) are the results of a simulation after converting the same high-level language (C) into Verilog and VHDL using HLS. At this time, the test bench uses an automatically converted same RTL testbench using C test bench. However, the two simulation show different results. Fig. 1(a) can see the

stopped state as a result of Verilog simulation. Fig. 1(c) can be seen to operate normally as a result of the VHDL simulation. The above error is caused by the following reasons:

A. Uninitialized Variables

A problem occurs when the converted Verilog code uses the signal for condition of If loop of other sub modules while the signal is not initialized. If the variable use condition with logic equivalent operator (==) in Verilog, the possible logical values are 0, 1, and X. In some cases, the unknown signal fails to meet the condition by X and does not move to the next state. How to resolve this issue is described in Section 3.

B. Reason of VHDL simulation passed

Reason why only VHDL simulation in the same testbench passed is also variable type. The converted VHDL code is also used for condition of If loop of other sub modules without initializing on the same signal. However, when this signal is processed through an equivalent operator in condition of If loop, the signal appears to have only a value of true, false. Because of this, the VHDL simulation passed on to the next state even if the uninitialized variable is used for the condition of if loop.

Verilog is very simple in data type compared to VHDL and is very easy in hardware modeling, but VHDL is available in a variety of languages and user defined data types, showing different characteristics [3].

III. DEBUGGING METHOD

There are two ways to debug these problems. The First is to modify the High-level language before converting to RTL using HLS, and secondly, there is a way to modify it after the RTL is convert. The following are descriptions of each method.

A. High-Level Language

The problem shown earlier is that the uninitialized variable is used as a variable in an instantiated module and the variable is used in the condition of if loop and does not perform the desired action.

Therefore, before RTL synthesis, variables used in the High-level language must be initialized before being used as variables in the function called. If this is the case, Initialization

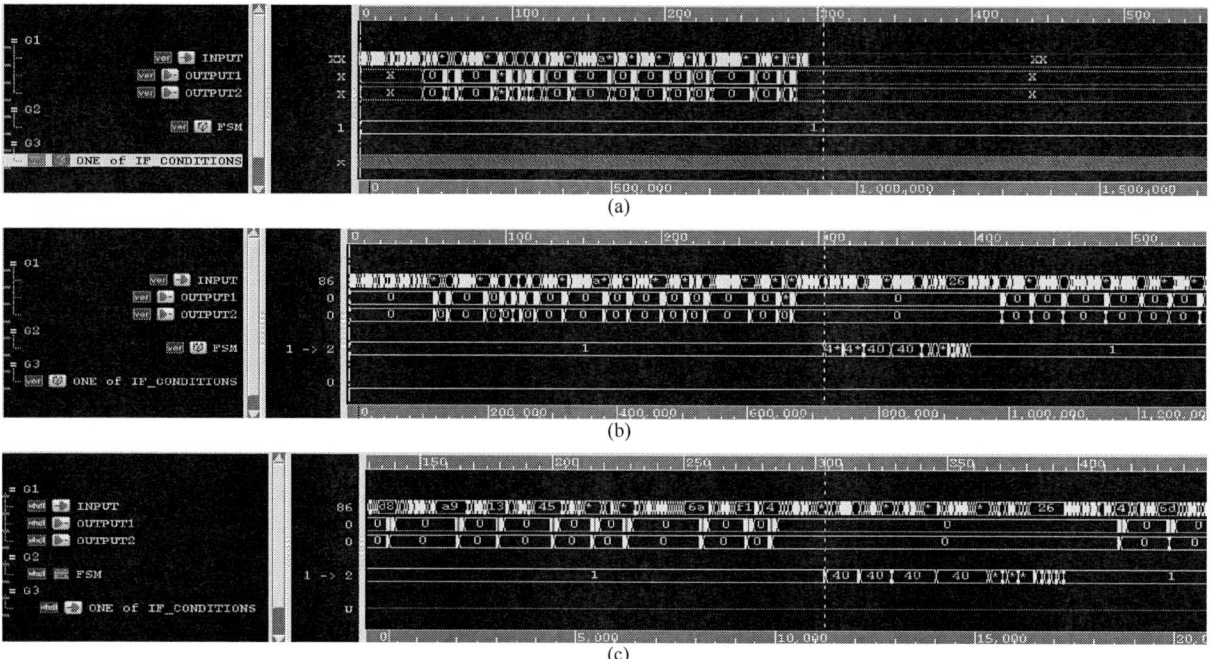

Fig. 1. RTL Simulation Results; (a) Problem Verilog (b) Resolved Verilog (c) VHDL

of the variable will be performed when the RTL is synthesized, thus solving the above problem.

B. Register Transfer Level

If resolve the error after it is converted already to RTL using HLS, Check the status of the point at which the simulation stopped and the point at which it does not move to the next state, and identify the variables. Then, trace the variable to find the part of the initialization did not occur, and proceed with the reset to the desired value and check again.

The simplest way to resolve this is to use case equivalent operators (===) instead of logical equivalent operators. Case equivalent operators can resolve these errors because logical values are only true and false. This method is simple, but it is suitable for simulation purposes because it is an incorrect syntax for the synthesis of the gate level.

In the above methods, trace and initialize the problematic variables then perform a simulation, we can see that the code that did not work as shown in Fig. 1(a) is operating correctly as shown in Fig. 1 (b).

IV. USING HLS FOR ASIC DESGIN

To synthesize converted RTL using HLS into a gate level for ASIC design. There is a similar problem as described above. Variables that require initialization in converted RTL using HLS are often not initialized. It is not compatible with synthesizable syntax because it is declared directly in the variables even though it is initialized. This must be initialized according to the initial reset signal. Also, since HLS tool converts System Verilog syntax or system task, it is necessary to convert it to a synthesizable syntax.

CONCLUSION

In this paper, we proposed that when verifying an RTL converted through HLS, how to resolve the problem different simulation result of Verilog & VHDL. But There are many improvements that need to be made to hardware design entirely using HLS techniques. Also, HLS verification for the ASIC design of the gate level is not mature, meaning it will be a barrier to using HLS in the ASIC design [4]. If these points are improved, it is expected that the design of ASIC will be possible only with HLS in the future.

ACKNOWLEDGMENT

This research was supported by the MISP(Ministry of Science, ICT & Future Planning), Korea, under the National Program for Excellence in SW)(2018-0-00213) supervised by the IITP(Institute for Information & communications Technology Promotion)(2018-0-00213). We thank professor Cho. Who provided insight and expertise that greatly assisted the research.

REFERENCES

[1] S. Hellebrand, et al., "The Research of Hardware Design Methodology Using High Level Synthesis Technique," IEIE, pp. 539-540, 2008.

[2] P. Coussy, D. D. Gajski, M. Meredith, and A. Takach, "An Introduction to High-Level Synthesis," IEEE Des. Test Comput., vol. 26, no. 4, pp. 8–17, 2009.

[3] Douglas J. Smith, "VHDL & Verilog Compared & Contrasted - Plus Modeled Example Written in VHDL, Verilog and C," DAC '96 Proceedings of the 33rd annual Design Automation Conference, vol. 6, pp. 771-776, 1996.

[4] A. Rukhin, et al., "High-Level Synthesis for FPGAs: From Prototyping to Deployment," IEEE Transactions on Computer-Aided Design of Integrated Circuits and System, vol. 30, no.4, pp. 473 - 491, 2011.

978-1-5386-7961-6/18 $31.00 © 2018 IEEE   274   ISOCC 2018

AUTHOR INDEX

Abbasizadeh, Hamed 60
Adiono, Trio 36, 115, 125
Ahn, Namhyun 127
An, Jehong 245
Asahara, Hiroyuki 158
Attili, Imtinan B. 95
Aung, Myat-Thu-Linn 239
Avitabile, G. 34
Bae, Gyujin 74
Baek, Rock-Hyun 239
Bi, Ziqiang 176
Byun, Kanghyeon 259
Byun, Younghoon 129
Cai, Yujie 123
Cha, Ji-Hyoung 68
Chae, Kwanyeob 140
Chang, Chun-Hao 15
Chang, Ik-Joon 235
Chang, Yeong-Jar 15
Che, Wei 27
Chen, Jian-Kai 50
Chen, Kuan-Hung 82
Chen, Yunfan 243, 245
Cheng, Wei-Kai 13, 50
Cheong, Minho 11
Chiou, Lih-Yih 15
Cho, Keewon 21
Cho, Manhee 229
Cho, Sang-Bock 206, 251
Cho, Sung I. 220, 224
Cho, Yong B. 226, 273
Choi, Byeong-Ho 151
Choi, Byungjun 46
Choi, Hyo B. 249
Choi, Jaegyeong 109
Choi, Jun R. 99
Choi, Juntae 196
Choi, Ken 44, 151
Choi, Kiyoung 72, 212, 271
Choi, Minsoo 239
Choi, Minsu 218, 233
Choi, Soyeon 121
Chung, Jin-Gyun 216
Coviello, G. 34
Diab, Maha S. 103
Edahiro, Masato 117
Fan, Yibo 123
Fang, Dan 27
Fuada, Syifaul 115
Fujii, Kyohei 164, 168
Gonuguntla, Venkateswarlu 99
Guo, Gan 27
Hamedi-Hagh, Sotoudeh 3
Han, Donghee 249
Han, Jun 123

Han, Mangi 208
Han, Seok-Kyun 105, 107
Han, Su-Hyun 68
Han, Tae H. 25, 269
Hanyu, Masami 48
Harimurti, Suksmandhira 125
Hashimoto, Shuhei 164
Hayashikoshi, Masanori 48
Hejazi, Arash 60
Ho, Yuan-Chen 131
Hong, Jeong B. 269
Hong, Zhiliang 27
Hosokawa, Yasuteru 162
Hsu, Chun-Lung 23
Huang, Jiun-Lang 9
Huang, Shih-Hsu 13, 50
Huang, Shi-Yu 90
Huang, Yujie 123
Hwang, Dongil 40
Hwang, Seokha 129, 133
Ido, Masamichi 48
Im, Hyejin 101
Irmanova, Aidana 253
Ishida, Ryuta 38
Ito, Daisuke 135
James, Alex P. 253
Jang, Seokjun 265
Jang, Sung-Joon 151, 204
Jang, Young-Chan 1
Jang, Young-Min 206
Je, Minkyu 68
Jiang, Zhenzhen 178
Jin, Minhyun 101
Jing, Minge 123
Jo, Junseo 133
Jo, So Y 127
Jo, Sujeong 212
Joe, Hounghun 229
Juang, Tso-Bing 170
Jung, Hyunki 105, 107
Kametaka, Yuki 158
Kang, Bongsoon 214
Kang, Jin-Ku 143
Kang, Ju Y. 269
Kang, Seokhyeong 149
Kang, Suk-Ju 127
Kang, Sungbum 72
Kang, Sungho 7, 11, 21, 265, 267
Kavya, Thathupara S. 206
Khan, Jameel A. 247
Kim, Bohun 46
Kim, Bruce 251
Kim, Byungsub 17, 239
Kim, Geun-Jun 214
Kim, Heetae 7, 267

AUTHOR INDEX

Kim, Jaehyun 271
Kim, Jeong H. 220, 224
Kim, Jeong-Geun 255, 257
Kim, Jinhyun 255
Kim, Ji-Seong 239
Kim, Jongsik 109
Kim, Ju S. 269
Kim, Jungah 109
Kim, Jusung 105, 107
Kim, Kwangmin 17
Kim, Kyung K. 218, 233
Kim, Kyunghwan 239
Kim, Min 143
Kim, Min-Cheol 263
Kim, Namhoon 143
Kim, Seong-Jin 68
Kim, Seungsoo 109
Kim, Sinyoung 259
Kim, Soo Y. 101
Kim, Sunwoo 42
Kim, Tae P. 273
Kim, Taehwan 19
Kim, Taejong 259
Kim, Tony T.-H. 196, 235, 239
Kim, Yeon-Jin 216
Kim, Yong S. 249
Kim, Yong-Bin 218, 233
Kim, Young Hwan 74, 76, 149
Kim, Youngbae 44, 151
Kim, Youngmin 208, 229, 231, 261
Kishine, Keiji 135
Kiyota, Yukiyoshi 48
Kong, Byeong Y. 119
Kong, Sungick 204
Kong, Sunwoo 257
Koo, Billy 140
Kousaka, Takuji 158
Kumar, Ayush 247
Kumawat, Renu 145
Kweon, Junyoung 196
Kwon, Hyunjeong 149
Kwon, Kuduck 259
Lai, Tzu-Yi 82
Le, Van L. 235
Leach, Mark 178
Lee, Chaeeun 42
Lee, Chaeun 271
Lee, Donggu 259
Lee, Eunchong 151
Lee, Hanho 210
Lee, Ho S. 76
Lee, Ho-Yun 216
Lee, Hui D. 257
Lee, Hyunhoon 129
Lee, Ingeol 11

Lee, Jhonson 125
Lee, Jinyong 52
Lee, Junghyup 241
Lee, Kang Y. 60
Lee, Kwang C. 257
Lee, Kwangjin 25
Lee, Kwang-Yeob 263
Lee, Kyungnam 261
Lee, Mee Z. 273
Lee, Sang-Gug 105, 107
Lee, Sanghun 42
Lee, Sangjae 80
Lee, Sangjun 7, 267
Lee, Sang-Seol 204
Lee, Sehwan 241
Lee, Sunggu 129, 133
Lee, Taeju 68
Lee, Yongho 109
Lee, Youngjoo 129, 133
Lee, Young-Woo 21
Lee, Yunsoo 127
Li, Bo-Yi 9
Li, Jin-Fu 23
Li, Wei 226
Li, Yu-Ting 23
Lie, Donald Y. C. 111
Lim, Eng G. 172, 174, 178
Lim, Hyeonchan 265
Lim, Kyunghyun 239
Lin, Chen-Hsien 13
Lin, Cong-Yi 170
Lin, Guan-Zhong 170
Liu, Wei 226
Lopez, Jerry 111
Lu, Juin-Ming 15
Lu, Liang-Ying 15
Ma, Jieming 176
Maddisetti, Lavanya 84
Mahmoud, Soliman 95, 103
Man, K. 172, 174, 176, 178
Manocha, Prateek 247
Mayeda, Jill C. 111
Metku, Prashanthi 233
Miyata, Yuichi 166
Moon, Byungin 80
Moon, Yong 194
Moradian, Hossein 212
Nabavi, Morteza 92
Naito, Fumiya 135
Nakamura, Makoto 135
Nakashima, Yasuhiro 48
Nguyen, Tram T. B. 210
Nishio, Yoshifumi 162, 164, 166, 168, 180
Oh, Hyunggoy 7, 267
Oh, Hyunyoung 40

AUTHOR INDEX

Oh, Jihun 140
Oh, Myungwoo 42
Ohsato, Tatsuki 160
Ohtagaki, Hirokazu 158
Oura, Akari 168
Paek, Yunheung 40
Pamidimukkala, Keerthana 218
Park, Bonghyuk 257
Park, In-Cheol 119
Park, Jae 220, 224
Park, Jaehong 140
Park, Jeongsoo 255
Park, Jeong-Soo 257
Park, Jongsun 19, 46
Park, Joonho 194
Park, Junmo 40
Park, Sang U 273
Park, Sanghune 140
Patel, Shreyash 44
Piccinni, G. 34
Punekar, Gauri 99
Rad, Reza E. 60
Ravindra, J. 84
Rho, Chang H. 143
Rikan, Behnam S. 60
Saini, Jitendra K. 145
Saito, Tatsuya 48
Sato, Toshinori 38, 86
Sawan, Mohammad 92
Sekiya, Hiroo 160
Seo, Sungyoul 21
Seo, Youngho 42
Seol, Taeryoung 241
Setiawan, Erwin 36, 115
Shams, Maitham 92
Shao, Haijun 27
Sharma, Ruchi 235
Shen, Zong J. 172
Shi, C.-J. R. 54
Shin, Hyunchol 109
Shin, Hyunchul 243, 245, 247
Shin, Jongshin 140
Shin, Saebyeok 105, 107
Siddiqui, Sultan M. 235
Smagulova, Kamilya 253
Smith, Jeremy S. 176
Son, Kyung-Sub 143
Song, Ho-Jin 239
Song, Ji M. 208
Song, Minkyu 101
Song, Yun-Heup 196
Srinivasulu, A. 145
Sumimoto, Shu 166
Sun, Chi-Tien 23
Sungchung, Chester 42

Suzuki, Junichi 48
Talarico, C. 34
Tamura, Kojiro 158
Tanaka, Tomotaka 135
Teoh, Melody 3
Tong, Qiang 151
Tsai, Tsung-Han 78, 131
Tsai, Yih-Ru 131
Tsogtbaatar, Erdenetuya 206
Ukezono, Tomoaki 38, 86
Un, Park S. 226
Utomo, Dzuhri R. 105, 107
Uwate, Yoko 162, 164, 166, 168, 180
Vaddi, Ramesh 99
Wang, Jingchen 178
Wang, Jooho 42
Wang, Yu 54
Wang, Zhao 178
Wei, Xiuqin 160
Wei, Yu-Chi 90
Xiang, Yingfei 54
Xie, Han 243
Xue, Pan 27
Yamada, Yuta 160
Yamashita, Junichi 48
Yang, Hoon G. 208
Yang, Li 172, 174
Yang, Myonghoon 40
Yang, Wei-Hsuan 15
Yao, Chia-Hsiang 78
Yellappa, Palagani 99
Yeo, Donghoon 243
Yogatama, Bobbi W. 125
Yoo, Hoyoung 121
Yoo, Taegeun 235
Yoshimura, Ryuta 166
Youn, Eunji 1
Youn, Yelim 17
Yu, Hoyoung 231
Yu, Joonsang 72
Yue, Yong 176
Zeng, Xiaoyang 123
Zhang, Heng 178
Zhao, Ce Z. 172, 174
Zhao, Chun 172, 174
Zhong, Zhaoqian 117
Zhou, Guang Y. 172, 174

IEEE
445 Hoes Lane
Piscataway, NJ 08854-4141

ISBN 978-1-5386-7961-6